Microbial Genetics

The Jones and Bartlett Series in Biology

Microbial Genetics

David Freifelder
University of California, San Diego

Jones and Bartlett Publishers, Inc.
BOSTON   PORTOLA VALLEY

Editorial office:
Jones and Bartlett Publishers, Inc.
30 Granada Court
Portola Valley, CA 94025

Sales and customer service office:
Jones and Bartlett Publishers, Inc.
20 Park Plaza
Boston, MA 02116

Library of Congress Cataloging-in-Publication Data

Freifelder, David Michael, 1935–
    Microbial genetics.

    Includes index.
    1. Microbial genetics.   I. Title.
QH434.F74   1987   576′.139   87-3783

ISBN: 0-86720-076-6

Cover art   Molecular model for the complex of the bacteriophage 434 repressor with its operator, as determined by x-ray crystallography. Courtesy J. E. Anderson, M. Ptashne and S. C. Harrison; computer graphics illustration by H. Holley. Department of Biochemistry and Molecular Biology, Harvard University.

Cover design   Rafael Millán

Illustrators   Donna Salmon, Cyndie Clark-Huegel, Evanelle Towne, John and Judy Waller, Dorothy Beebe, Kelly Solis-Navarro, Brenda Booth, and Jack Vitkus

Production   Bookman Productions

Composition   Graphic Typesetting Service

Printed in the United States of America
10 9 8 7 6 5 4

To my parents,
MORRIS and FLORENCE,
and to my children,
RACHEL and JOSHUA

Contents

PART 2
MOLECULAR ASPECTS OF GENE EXPRESSION   113

Chapter 6
Gene Expression   **115**

Chapter 9
DNA Damage and Repair   191

Chapter 10
Mutagenesis, Mutations and Mutants   213

Chapter 12
Transposable Elements
281

Chapter 17
Phage Genetics III: Lysogeny   **417**

PART 5
THE NEW MICROBIAL GENETICS 485

Preface

Experiments in microbial genetics carried out in the 1940s and 1950s gave birth to what ultimately became known as molecular biology. The elucidation of the structure of DNA by James Watson and Francis Crick, neither of whom were geneticists, had enormous impact on all of genetics and biology. Many physicists, fascinated by the new discoveries and attracted by the quantitative aspects of microbial genetics joined classical microbiologists, geneticists, and biochemists, and what has been called the Golden Age of Biology began. During this period genetic analysis took on new importance in that it provided the insights for developing detailed theories of gene action. Genetic results obtained with bacteria and bacteriophages led to hypotheses about gene organization, chromosome structure, regulation of gene action, genetic recombination, and many other phenomena. Although genetic results could not rigorously prove a molecular mechanism, they placed limits on hypotheses and were enormously suggestive about mechanisms. In fact, hypotheses based solely on genetic results so often proved to be correct that it became commonplace among molecular biologists to assume the validity of a mechanism before biochemical proof was obtained; certainly, mechanisms that were inconsistent with genetic results were not thought worthy of consideration. A great deal of the early work in microbial genetics was compiled in an excellent and influential text—*The Genetics of Bacteria and Viruses* by William Hayes—the most recent edition of which was published in 1968. Since that time, few texts on microbial genetics have appeared, the most recent one being more than five years old. *Microbial Genetics* represents my attempt to bring this still-useful subject up to date.

The application of microbial genetics led to the accumulation of a huge body of knowledge and a continually greater understanding of the nature of the gene. In the early 1970s microbial genetics itself underwent a revolution, with the development of the recombinant DNA technology. This collection of remarkable but straightforward techniques, usually called genetic engineering, allows manipulation and exchange of fragments of DNA outside of the cell and reintroduction of recombinant DNA into a new cell. In this way, novel organisms can be created, with characteristics drawn from distant species and genera. Even human genes can be transferred to a bacterium, and genes of microorganisms can be placed in animal cells. In fact, the glitter of this new technology has led to the conversion of hundreds of research laboratories to gene-cloning factories, and to the development of a new industry, known as bioengineering. This new technology is the topic of the later chapters in this book.

Microbial Genetics is the result of a good deal of prodding by many colleagues who urged me to write such a book. Although I am a "modern" molecular biologist (though I have never cloned a gene), I did acquire my research experience during the first Golden Age and worked intensively in various aspects of microbial genetics. Furthermore, since I have had the experience of writing several books, I was told that the task was clearly mine. Thus, several months ago I started to re-read ancient books and papers. This brought back many pleasant memories of lectures and meetings that were thoroughly exciting, because in retrospect we seem to have known, comparatively speaking, very little about the details of genetic processes. These recollections, as well as thoughts about many former colleagues, made the process of book-writing more of a pleasure than a task.

Microbial Genetics begins with five review chapters. These include basic material about classical genetics, DNA and protein structure, and the basic biology of bacteria and bacteriophages. Students who have had a course in biochemistry can surely omit Chapters 2 and 3, and those having studied classical genetics can undoubtedly skip Chapters 1, 4, and 5. Since these chapters are meant as a review, the material is presented fairly succinctly. Furthermore, only topics needed in later discussions have been included.

Every chapter in this book ends with a substantial set of questions and problems. Most of the beginning questions simply test the memory of the student; they are designed to ensure that the reader has learned the definitions and understood the most elementary concepts. Later questions are more difficult. Complete answers to all questions and problems, with explanations, are given in the Answers section at the back of the book. A rather extensive bibliography is also given at the end of the book.

I was fortunate to have the help of many reviewers, of whom the following deserve special mention: Mary Jane Voll (University of Maryland, College Park), who also reviewed the first edition of my book *Molecular Biology*; Douglas Smith (University of California, San Diego); and Stephen Pruett (Mississippi State University). To each I owe my thanks. I also would like to thank Arthur Bartlett (Jones and Bartlett, Publishers), who has published most of my books, for making the life of an author a pleasure; and Hal and Robin Lockwood (Bookman Productions), with whom I have worked several times and who always do a great job.

David Freifelder
February 1, 1987

PART 1

ESSENTIALS OF GENETICS AND MICROBIOLOGY

Chapter 1

Essentials of the Genetics of Haploid Organisms

An important concept of modern biology is that living systems persist over long periods of time through heredity but also change by evolution. **Heredity** is the process by which all living things produce offspring like themselves. This capacity for self-reproduction involves the transmission from parent to offspring of information that specifies a particular pattern of growth and organization. **Genetics** is the science of heredity; it is concerned with the physical and chemical properties of the hereditary material, how this material is transmitted from one generation to the next, and how the information it contains is expressed during the development of an individual.

Organisms resemble one another because they have received hereditary information from a common ancestor. Since species continue to exist through many generations, the hereditary material must be inherently stable. However, within most species, immense hereditary diversity is present; for example, although most humans have the same overall appearance, even persons as closely related as sisters or brothers usually look quite different. Hereditary variability is an essential factor in evolution, which is the cumulative change in the genetic characteristics of populations of organisms through time. Evolution is the most important generalization in biology; it is the process that enables us to understand the origin of the enormous number of species of living things and the relation between them. Evolution provides a logical basis for understanding how the complex organisms existing today were derived through continuous lines of descent, with modification, from the first primitive organisms.

Genetic analysis is an abstract and logical method by which the genetic factors determining certain properties of living cells can be located with respect to one another and by which changes in these properties can be expressed in quantitative terms. Furthermore, it provides information about functional relations between genes and their products. Genetic analysis has been a driving force in the development of molecular genetics. Alone it cannot prove anything about molecular mechanisms, because the conclusions it leads to are independent of the molecular basis of biological phenomena. However, genetic analyses have been the major source of intuition about molecular mechanisms and have forced one to consider phenomena that might otherwise be ignored. Genetics-based arguments can suggest a particular mechanism so emphatically, and the mechanism suggested has so often turned out to be the correct one, that molecular biologists often view alternative and conflicting ideas as not even worthy of consideration.

The following sections are an introduction to genetic analysis of organisms as well as to the genetic phenomena most commonly used in such analyses. Much of this material will be familiar to the reader who has previously studied genetics or college-level biology, though the point of view may be different from that found elsewhere, being specifically that of the genetics of prokaryotic microorganisms.

GENETIC NOTATION, CONVENTIONS, AND TERMINOLOGY

In order to discuss genetics one must first distinguish the phenotype of a cell from its genotype and develop a consistent notation for describing the genetic properties of an organism. The notation we present is used for bacteria, but not for all organisms.

The **phenotype** is an observable property of an organism. Thus, a cell that can synthesize the amino acid leucine is denoted Leu$^+$; if it cannot do so, it is denoted Leu$^-$. Note that the symbol has three letters, is capitalized, and is not italicized. **Genotype** refers to the genetic composition of an organism. A Leu$^-$ cell certainly must have some defective gene that keeps it from synthesizing leucine; this would be denoted *leu$^-$* (or in some books, *leu*, without the minus sign). Note that the genotype is written with three lowercase letters and in italics. Several genes may be required to synthesize leucine. These are usually denoted *leuA*, *leuB*, . . . , with all letters italicized and a fourth capitalized letter. A Leu$^+$ (leucine-synthesizing) cell must have one functional copy of every requisite gene; thus, its genotype must be *leuA$^+$leuB$^+$*, . . . , which is normally summarized by writing *leu$^+$* unless it is important for some reason to state the genotype of each gene.

A typical bacterium has only a single set of genes; it is **haploid.** Eukaryotic organisms (whose cells contain a nucleus bounded by an enclosing membrane) usually have two sets of genes—they are **diploid.** A diploid cell might have one normal copy of a particular gene and one defective copy. In this case, the genotype includes both copies, and the gene symbols would be separated by a diagonal line, for example, *leu*⁺/*leuA*⁻. A cell could also have two normal copies or two defective copies.

Some genes are responsible for resistance and sensitivity to certain extracellular products such as antibiotics. The genotypes of a penicillin-resistant cell and a penicillin-sensitive cell are written *pen-r* and *pen-s*, respectively; the phenotypes are correspondingly Pen-r and Pen-s.

In summary, for bacteria the following conventions are used:

1. Abbreviations of phenotypes contain three roman letters (the first capitalized) with a superscript + or − to denote presence or absence of the designated character and a nonsuperscript "s" and "r" for sensitivity and resistance, respectively.
2. The genotypic designation is always lowercase and, except for superscripts, all components of the symbol are italicized.

This convention has not been adopted for bacteriophages, for which one- and two-letter symbols are widely used.

Geneticists also have occasion to designate particular mutations of a gene. Mutations are usually numbered in the order in which they have been isolated. Thus, the leucine mutations numbered 58 and 79 are written *leu58* and *leu79* (or occasionally *leu-58* and *leu-79*). If one wanted to denote mutations in a particular gene, one would write *leuA58* and *leuB79*, if the mutations were in the *leuA* and *leuB* genes, respectively.

It is often necessary to denote the protein products of a particular gene. No notation has ever become standard, but in this book the following conventions are used. A gene product, for example, substance Q, for which the genetic symbol might be *kyu*, will be written out as substance Q, or it will be abbreviated and termed the *kyu* product or the Kyu protein, or it will simply be designated Kyu. All three designations are convenient and will be used in keeping with the three common modes seen in scientific literature. Unfortunately, mutations are frequently isolated before gene function is known; in these cases, common usage has given rise to annoying notational differences, such as *polA* for DNA polymerase I (Pol I) and *nal* for a DNA gyrase subunit. Sometimes the genotypic designations are changed, but usually we are forced to live with the differences.

The term **genome** is useful, though it has evolved to have various meanings. It is correctly defined as the genetic complement (the set of all genes and genetic signals) of a cell or virus. In laboratory jargon, when discussing bacteria or phages, the term has come to be used to refer to their single DNA molecule, a classically incorrect but now accepted usage.

Numerous abbreviations are commonly used for various genes and substances. These are listed in Table 1-1.

We have indicated that a gene can be either functional or nonfunctional and that these states are denoted by a superscript + or − after the gene

Table 1-1 Standard substances used in microbial genetics which are encountered in this book, and their abbreviations

Substance	Genotype symbol[1]	Substance	Genotype symbol
Amino acids		*DNA and RNA bases*[2]	
Alanine	*ala*	Purine	*pur*
Arginine	*arg*	Pyrimidine	*pyr*
Asparagine	*asn*	Adenine	*ade*
Aspartic acid	*asp*	Cytosine	*cyt*
Cysteine	*cys*	Guanine	*gua*
Glutamic acid	*glu*	Thymine	*thy*
Glutamine	*gln*	Uracil	*ura*
Glycine	*gly*	*Vitamin*	
Histidine	*his*	Biotin	*bio*
Isoleucine	*ile*	*Sugars*	
Leucine	*leu*	Arabinose	*ara*
Lysine	*lys*	Galactose	*gal*
Methionine	*met*	Glucose	*glu*
Phenylalanine	*phe*	Lactose	*lac*
Proline	*pro*	*Antibiotics*	
Serine	*ser*	Ampicillin	*amp*
Threonine	*thr*	Chloramphenicol	*cam*
Tryptophan	*trp*	Kanamycin	*kan*
Tyrosine	*tyr*	Penicillin	*pen*
Valine	*val*	Streptomycin	*str*
		Tetracycline	*tet*

[1] When stating a substance or a phenotype rather than a genotype, the same three-letter symbol is used, but it is capitalized and not italicized—for example, His for histidine.

[2] When stating base sequences, a purine or a pyrimidine is usually abbreviated Pu or Py respectively, if the identity of the base is unimportant. Particular bases will be denoted by A (adenine), G (guanine), T (thymine), U (uracil), and C (cytosine).

abbreviation. The functional form of a gene is sometimes called **wild type** because presumably this is the form found in nature. However, this quite useful term is ambiguous, because with some organisms the − form of the gene is the one that is prevalent; for example, many bacterial species isolated from nature carry the genes for metabolism of lactose, yet in many strains the gene is defective—that is, *lac*⁻. In this book we shall minimize the use of the term.

The precise genetic term **allele** is used to indicate that there are alternative forms of a gene; sometimes the + and − forms are called the + allele and the − allele, respectively. An organism in which the members of a pair of alleles are different is said to be **heterozygous,** and one in which the two alleles are alike (either both wild type or both mutant) is **homozygous.** In a heterozygote the expression of one of the genes usually predominates; that allele is said to be **dominant,** and the other allele is **recessive.** For example, a *leu*⁺/*leu*⁻ heterozygote has a Leu⁺ phenotype, so the *leu*⁺ allele is dominant. It is frequently, but not always, the case that the wild-type allele is dominant in bacteria.

It is very common to use the term mutant for an allele that is not the wild-type form. Strictly speaking, a **mutant** is an *organism* whose genotype (or, more precisely, its DNA base sequence) differs from that found in nature. The difference is the mutation. However, it is more convenient (and definitely common jargon) to use the term mutant gene only when defective gene is meant, and we shall often use the word mutant in that sense in this book.

MUTANTS AND MUTATIONS

Mutations are the raw material of evolution, providing the changes on which selection can act. For the microbial geneticist mutations are tools, for example, **genetic markers** for recognizing a chromosome or a region of a chromosome. Furthermore, study of the changes induced by a mutation gives a great deal of information about genetic processes, as will be seen shortly.

The process of formation of a mutant organism is called **mutagenesis.** In nature and in the laboratory, mutations sometimes arise spontaneously, without any help from the experimenter. This is called **spontaneous mutagenesis.** Mutagenesis can also be induced by the addition of chemicals called **mutagens** or by exposure to radiation, both of which result in chemical alterations of the genetic material. Mutagenesis and mutations will be discussed in detail in Chapter 10.

Absolute and Conditional Mutants

Mutants can be classified in several ways. One classification is based on the conditions in which the mutant character is expressed. In a haploid organism such as a bacterium an **absolute defective mutant** displays the mutant phenotype under all conditions; that is, if a mutant bacterium requires leucine for growth in all culture media and at all temperatures, it is an absolute defective. A **conditional mutant** does not always show the mutant phenotype; its behavior depends either on environmental conditions or on the presence of other mutations. An important example of a conditional mutant is a **temperature-sensitive (Ts) mutant,** which behaves normally at one temperature, e.g., below about 34°C, and as a mutant at another temperature, for example, above 39°C; an intermediate state is usually observed between these temperatures. Note that the gene itself is not altered above 39°C; rather, the *product* of the gene is inactive above 39°C. Temperature-sensitive mutants have been of great use in the laboratory because the activity of a gene product can be turned off simply by raising the temperature. In some cases the temperature-sensitive defect is reversible, so normal activity can be restored by lowering the temperature.

The most widely encountered conditional mutants of bacteria and bacteriophage are the **suppressor-sensitive mutants;** these exhibit the mutant phenotype in some bacterial strains but not in others. The difference is a consequence of the presence of particular gene products, called **suppressors.** These suppressor-gene products either compensate for the defect in the mutant or, in a variety of ways, enable an altered gene to produce a functional gene product. In the jargon of molecular genetics, one says that the phenotype of a suppressor-sensitive mutant "depends on the genetic background." A mutation is sometimes designated as a suppressor-sensitive one by adding the roman letters Am (refers to amber), Oc (for ochre), or Op (opal) in parentheses. (Amber and ochre are terms that arose from a private laboratory joke.) The symbols are roman and capitalized because they represent a phenotype; if the mutation has a number, this is also added. Thus, if mutation 35 in the *leu* gene is a suppressor-sensitive Am type, it would be written *leu*35(Am); if the mutation were temperature-sensitive, it would be written *leu*35(Ts). Two outdated symbols are *leu am*35 and *leu ts*35. Suppressors are discussed in Chapter 10.

The Number of Mutations in a Mutant

An important classification of mutants is based on the number of changes that have occurred in the genetic material (that is, the number of DNA base pairs that have changed). If only one change has occurred, the organism is said to possess a **point mutation.** If two changes are present, the organism is a **double mutant.** Sometimes the mutation occurs by means of removal of all or part of a gene; in that case, the mutation is a **deletion.** Occasionally, other material replaces the deleted gene or gene segment. Such a mutation is called a **deletion-substitution.** If genetic material is added without removal of any other material, it is an **insertion.** Some of these types of mutations are shown schematically in Figure 1-1.

Revertants and Reversion

A mutant organism sometimes regains its wild-type characteristics. This occurs by means of chemical changes in the mutant genetic material, which restore function. The process of regaining the original phenotype is called **reversion,** and an organism that has reverted is a **revertant.**

Reversion can be either spontaneous or induced by mutagens. The **reversion frequency**—that is, the fraction of cells in a population of mutants that after many generations of growth have regained the functional phenotype—is a useful criterion for identifying a point mutation, because one of a small number of alterations in the nucleic acid is sufficient to revert a point

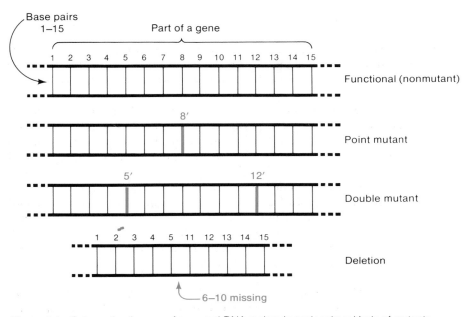

Figure 1-1 Schematic diagram of a normal DNA molecule and various kinds of mutants. In the point mutant, base pair 8′ replaces the normal base pair 8. The double mutant has two base pairs (5′ and 1′) not present in the nonmutant organism. In the deletion, five base pairs (6–10), present in the nonmutant, are absent.

mutant, and such a change can occur at some measurable, albeit low, frequency. With a deletion, however, the probability that the missing DNA will be replaced with material having an equivalent base sequence, thereby restoring a functional gene, is virtually zero. Thus, *reversions are not observed with deletions*.

With bacteria many reversion events can be detected by measuring the ability of a population of bacteria to form colonies on solid growth medium (colony formation is discussed in Chapter 4). For example, if 100 cells of a leu^- bacterial strain are placed on a solid medium lacking leucine, no colonies will form. However, if 10^7 leu^- bacteria are plated, about 50 colonies arise; these colonies must consist of Leu^+ bacteria (cells able to grow without an external supply of leucine) and are spontaneous revertants. The reversion frequency in this case is $50/10^7 = 5 \times 10^{-6}$. Such a frequency value is characteristic of the reversion of point mutants. The production of a spontaneous revertant is a random process, so the reversion of a double mutant would require two independent events. If each were to occur at a frequency of 5×10^{-6}, then the frequency of reversion of a double mutant would be $(5 \times 10^{-6})^2 = 2.5 \times 10^{-11}$. Reversion will be discussed in detail in Chapter 10.

Uses of Mutations: Some Examples

Some of the most significant advances in microbial genetics have come about by the use of mutations, as will be seen repeatedly in this book. In the following, the kinds of approaches that have been taken are described.

1. *A mutation defines a function*. For example, the sugar lactose might be taken up by a bacterium by passive diffusion through the cell membrane, or a particular system might be responsible for the process. Wild-type *E. coli* can take up lactose from a $10^{-5} M$ solution, but mutants have been found that cannot do so even in much higher concentrations. This finding indicates that a genetically determined system for lactose intake exists, though the observation does not tell what this system is. Temperature-sensitive mutant organisms are especially useful in defining functions. For example, Ts mutants of *E. coli* have been isolated that fail to synthesize DNA. The mutations fall into at least ten distinct classes, suggesting that there may be at least ten different proteins required for DNA synthesis.

2. *Mutations can introduce biochemical blocks that aid in the elucidation of metabolic pathways*. The metabolism of the sugar galactose, for example, requires the activity of three distinct genes called *galK*, *galT*, and *galE*, which has been determined by the isolation of three classes of Gal$^-$ mutants. If radioactive galactose ([^{14}C]Gal) is added to a culture of Gal$^+$ cells, many different radioactive compounds can be found as the galactose is metabolized. At very early times after addition of [^{14}C]Gal, three related compounds are detectable: [^{14}C]galactose-1-phosphate (Gal-1-P), [^{14}C]uridine diphosphogalactose (UDP-Gal), and [^{14}C]uridine diphosphoglucose (UDP-Glu). Mutations in different genes will block distinct steps of the metabolic pathway. If the mutant cell possesses a *galK*$^-$ mutation, the [^{14}C]Gal label is found only in galactose. Thus, the *galK*-gene product is known to be responsible for the first metabolic step. A *galT*$^-$ mutant accumulates Gal-1-P. Thus, the first step in the reaction sequence is found to be the conversion of galactose to Gal-1-P by the *galK* gene product (namely, the enzyme galactokinase). If a *galE*$^-$ mutant is used, some Gal-1-P is found but the principal radiochemical is UDP-Gal. Thus, the biochemical pathway must be

$$\text{Gal} \xrightarrow[\substack{galK \\ \text{product}}]{} \text{Gal-1-P} \xrightarrow[\substack{galT \\ \text{product}}]{} \text{UDP-Gal} \xrightarrow[\substack{galE \\ \text{product}}]{} \text{X}$$

The identity of X cannot be determined from these genetic experiments.

3. *Mutants enable one to learn about metabolic regulation*. Many mutants have been isolated in which there is an alteration in either the amount of a particular protein that is synthesized or the way the amount synthesized responds to external signals. The mutations define regulatory systems. For

example, the enzymes corresponding to the *galK*, *galT*, and *galE* genes are normally not present in bacteria but appear only after galactose is added to the growth medium. However, mutants have been isolated in which these enzymes are always present, whether or not galactose is also present. This indicates that some gene is responsible for turning the system of enzyme production on and off, and this regulatory gene must be responsive in some way to the presence and absence of galactose.

    4. Mutations enable a biochemical entity to be matched with a biological function or an intracellular protein. For many years an *E. coli* enzyme called DNA polymerase I was studied in great detail. Purified polymerase I is capable of synthesizing DNA, so it was believed that this enzyme was also responsible for bacterial DNA synthesis. However, an *E. coli* mutant (*polA*⁻) was isolated in which the activity of polymerase I was reduced 50-fold, yet without any detectable effect on either the growth rate of the cells or the ability to synthesize DNA. This observation suggested strongly that DNA polymerase I could not be the only enzyme that synthesizes intracellular DNA. Indeed, biochemical analysis of cell extracts of the *polA*⁻ mutant showed the existence of two other enzymes, DNA polymerase II and DNA polymerase III, which could, when purified, also synthesize DNA. In further study, a temperature-sensitive mutation in a gene called *dnaE* was found to be unable to synthesize DNA at 42°C, though synthesis was normal at 30°C. The three enzymes—DNA polymerases I, II, III—were isolated from cultures of the *dnaE*⁻(Ts) mutant and each enzyme was assayed. Although polymerases I and II were active at both 30°C and 42°C, polymerase III was active at 30°C but not at 42°C; therefore, polymerase III was determined to be both the product of the *dnaE* gene and the enzyme responsible for intracellular DNA synthesis.

    5. Mutants locate the site of action of external agents. The antibiotic rifampicin prevents synthesis of RNA. When first discovered, it was not known whether rifampicin might act by preventing synthesis of precursor molecules, by binding to DNA and thereby preventing the DNA from being copied to make RNA, or by binding to RNA polymerase, the enzyme responsible for synthesizing RNA. Mutants were isolated that were resistant to rifampicin. These mutants were of two types—those in which the bacterial cell wall was altered such that rifampicin could not enter the cell (an uninformative type of mutant) and those in which the RNA polymerase was slightly altered. The finding of the latter mutants proved that the antibiotic acts by binding to RNA polymerase.

    6. Mutants can indicate relations between apparently unrelated systems. Bacteriophage λ, which normally adsorbs to and grows in *E. coli*, fails to adsorb to a bacterial mutant unable to metabolize the sugar maltose. Such failure is not associated with mutants incapable of metabolizing other sugars or with any other phages, so this knowledge implicated some product or agent of maltose metabolism in the adsorption of λ. Similarly, *E. coli* mutants unable to adsorb the phage φ80 require exceedingly high concentrations of the Fe^{2+} ion in the growth medium if the bacterium is to grow. This is because the adsorption site of φ80 and the protein responsible for transport of the Fe^{2+} ion are structurally related.

7. *Mutations can indicate that two proteins interact*. How this occurs is best shown by a hypothetical example. Suppose that mutants carrying mutations in either of the two genes *a* and *b*, which are responsible for synthesizing the proteins A and B, fail to carry out a particular process. Clearly, the products of both genes are necessary for the process to occur. A and B may act consecutively in the process or interact to form a single functional unit consisting of both proteins. Interaction as a single unit is often indicated by reversion studies. When revertants of an a^- mutant are sought, it is sometimes found (by additional genetic analysis) that reversion is a result of production of a mutation in gene *b*. When reversion of this type occurs, it is usually observed that reversion of some b^- mutants (those not formed by reversion of an a^- mutant) is often the result of (different) mutations in gene *a*. The interpretation of these results is the following. Proteins A and B interact to form a protein aggregate (Figure 1-2), which is the active struc-

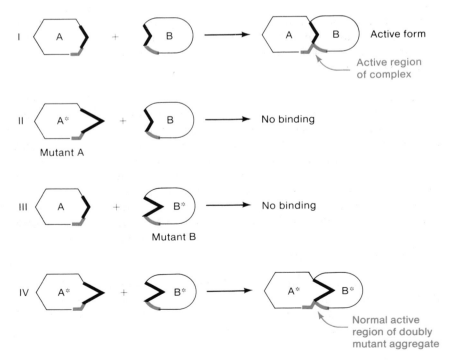

Figure 1-2 Schematic diagram showing how two separately inactive mutants can combine to make a functioning protein complex. Sites of interaction of proteins A and B are denoted by heavy lines. Components of the active site of the A-B complex are shown in red. Only complexes I and IV have active binding sites.

ture. Some alterations in either A or B prevent aggregation inasmuch as the protein will have been changed at a binding site. A compensating alteration in the other protein can then enable the interaction to occur again. Such an interpretation has frequently been found to be correct. For example, some mutations in a phage-λ DNA-replication gene (P) are compensated for by a nonlethal mutation in an E. coli DNA-replication gene (dnaB), and the corresponding gene products, P protein and DnaB protein, bind together to form a protein active in the replication of λ DNA.

GENETIC ANALYSIS OF MUTANTS

A great deal of information about molecular mechanisms can be obtained by making various combinations of separately isolated mutations. This is done by **genetic recombination.** Another important use of genetic recombination is the construction of an array that indicates the positions of genes on a chromosome with respect to one another. When this is done by genetic techniques, the array is called a **genetic map. Physical maps** have also been constructed by using various physical techniques, and some elegant experiments have shown that in some organisms the gene positions in the two maps are identical, or at least nearly so. Another genetic technique is **complementation.** By this procedure it is possible to determine the number of genes responsible for a particular phenotype and to distinguish genes from regulatory sites. In the following sections both genetic recombination and complementation will be described. Numerous examples of the use of these techniques will be found throughout the book.

Genetic Recombination

Genetic recombination is the process of combining two genetic loci, initially on two different chromosomes, onto a single chromosome. The molecular mechanisms, which are very complex, are not well understood and vary in different organisms and with different types of recombination; for the present purposes it is sufficient to refer to the "scissors-and-tape" mechanism. In this simple model, one assumes that two chromosomes align with one another, that a cut is made in both chromosomes at random but matching points, and that the four fragments are then joined together to form two new combinations of genes. This is a naive and incorrect model; it enables one to account for only the simpler features of genetic exchange, but these features are in fact the only ones that are of concern at this time. The process is depicted in Figure 1-3. Two parental chromosomes having the genotypes a^+b^- and a^-b^+ pair, are cut and are then joined to form two recombinant chromosomes whose genotypes are a^+b^+ (wild type) and a^-b^- (double mutant).

Genetic recombination frequently occurs when one bacterium is simultaneously infected with two mutant phage. That is, if the parental phage have

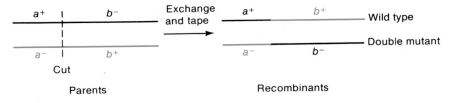

Figure 1-3   A schematic diagram showing genetic exchange.

the genotypes a^+b^- and a^-b^+, as in Figure 1-3, among the hundred-or-so progeny phage released when the infected bacterium lyses, there will be a few a^+b^+ and a^-b^- recombinant phage. The ratio

Number of recombinant phage/Number of total phage

is called the **recombination frequency.** More generally, for all organisms in which recombination occurs the recombination frequency is (the number of recombinant types)/(number of parental plus recombinant types). A more precise expression that is valid for distant markers will be given shortly.

Genetic Mapping

In genetic mapping, the distance along the chromosome between two recombining genetic loci (or mutations) determines the recombination frequency. As long as the two loci are not too near one another, *the recombination frequency is proportional to distance*, because chromosomal cuts are made at random. Thus, in the following crosses between chromosomes, the genotypes of which are $a^+b^-c^-$ and $a^-b^+c^+$, and the genes of which are in alphabetical order and equally spaced,

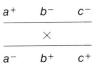

there will be twice as many a^+c^+ recombinants as a^+b^+ recombinants, because loci a and c are twice as far apart as loci a and b.

Because the recombination frequency is proportional to distance, recombination frequency can be used to determine the arrangement of genes on the chromosome. This can be seen in a simple example (Figure 1-4). Consider three genes a, b, and c, whose arrangement is unknown. Using the notation $p \times q = m$ percent to denote a recombination frequency of m percent between genes p and q, we assume it has been observed that $a \times b = 1$ percent and that $b \times c = 2$ percent. The two arrangements shown in the figure are consistent with these values and can be distinguished

Figure 1-4 Two arrangements of three genes for which $a \times b = 1$ percent and $b \times c$ = 2 percent. See text for discussion.

by determining the recombination frequency between a and c. Let us assume that this is 1 percent. If that is the case, only arrangement II is possible. The order $c\ a\ b$ for these genes and the relative separation constitute a genetic map.

Any number of genes can be mapped in this way. For instance, consider a fourth gene d, in the preceding example. If $d \times b = 0.5$ percent, d must be located 0.5 unit either to the left or to the right of b. If $a \times d = 1.5$ percent, d is clearly to the right of b, and the gene order is $c\ a\ b\ d$. If $a \times d = 0.5$ percent, the gene order would be $c\ a\ d\ b$.

The analysis just given has been oversimplified because the occurrence of multiple exchanges has been ignored. This will be discussed shortly.

The unit of distance in a genetic map is called a **map unit**; it is defined as 1 percent recombination. Thus, two genes that recombine with a frequency of 2.8 percent are said to be located 2.8 map units apart. One map unit amounts to a physical length of the chromosome in which an exchange will occur once in 100 pairings. In eukaryotic genetics a map unit is sometimes called a centimorgan (cM), but this term is rare in microbial genetics.

Linkage and Multifactor Crosses

Mapping can be carried out with a fewer number of crosses if the cross is done with three loci simultaneously (a **three-factor cross**). The procedure utilizes an effect known as **linkage,** a term used in several contexts. Strictly speaking, linkage means that two genes rarely segregate from one another. This is perhaps the most common usage in microbial genetics, in which two genes or markers are considered linked if they are fairly near one another in a map. In classical Mendelian genetics linkage refers to the state of being on a single chromosome, so genes are considered linked if they segregate from one another *only* by recombination. In this context one refers to linkage groups. In prokaryotic microbial genetics, for all practical purposes, an organism has a single linkage group, because, except for the existence of plasmids (Chapters 5 and 11), all prokaryotes contain a single chromosome.

Consider the cross shown in Figure 1-5 between two parents whose genotypes are $a^+b^-c^-$ and $a^-b^+c^+$. Instead of measuring recombination frequencies, let us just select all recombinants that are a^+c^+ and then in a

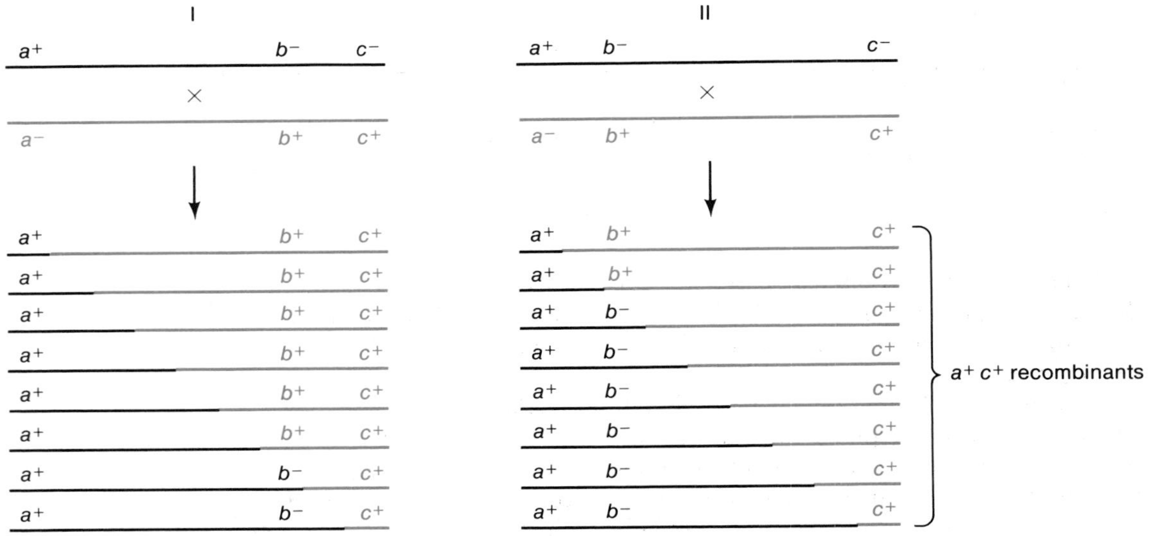

Figure 1-5 A three-factor cross that shows linkage. Eight possible a^+c^+ progeny arising from equally spaced exchange points are shown for each arrangement. In I, 6 of 8 a^+c^+ are $a^+b^+c^+$; in II, 6 of 8 a^+c^+ are $a^+b^-c^+$.

second test ask whether they are also b^+ or b^-. If the genes were arranged as in panel I, the recombinants would mostly be b^+, because the distance from a to b is greater than that from b to c. On the contrary, if the genes were located as in panel II, most of the a^+c^+ recombinants would be b^-. This simple analysis locates b with respect to a and c. For arrangements I and II, one says that b and c or a and b, respectively, are linked.

The procedure just described does not give the gene order. That is, if b and c are linked, the gene order could be $a\ b\ c$ or $a\ c\ b$ and these two orders would not be distinguished by the analysis just given. Gene order can in theory always be determined by the mapping procedures that we have used to distinguish the two arrays shown in Figure 1-4; in the case of two closely linked loci, however, determination of gene order is not feasible from data derived from two-factor crosses, because the data are usually not sufficiently accurate. For example, suppose $a \times b = 7$ percent and $b \times c = 0.2$ percent. If $a \times c = 6.8$ percent, the order is $a\ c\ b$, and if it is 7.2 percent, the order is $a\ b\ c$. However, experimentally observed values might be $a \times b = 7.0 \pm 0.3$ percent and $a \times c = 7.1 \pm 0.3$ percent, so that the order would not be established with certainty in this way. Figure 1-6 shows how a three-factor cross clearly gives the order. A cross is performed between parents having genotypes $a^+b^-c^+$ and $a^-b^+c^-$. Two types of data can be obtained. In the first, the linkage method is used; that is, b^+c^+ recombinants are selected and these are tested to determine whether they are a^+ or a^-. They

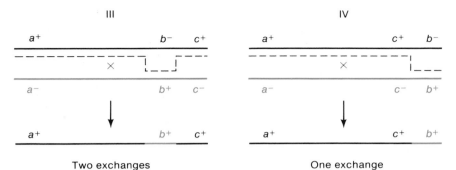

Two exchanges   One exchange

Figure 1-6 Determination of gene order by a three-factor cross. The frequency of appearance of $a^+b^+c^+$ is much higher for order IV because only one exchange is required.

are usually a^- for order III and a^+ for order IV. In the second and more rapid method, the frequency of $a^+b^+c^+$ recombinants is measured. These recombinants arise by two exchanges in case III but only one exchange in case IV. The frequency for a double exchange is the product of the frequencies of each simple exchange. Thus, for case III, the recombination frequency is $(0.07)(0.002) = 0.00014 = 0.014$ percent, but for case IV, in which only a single exchange is needed, the frequency is 0.2 percent (the same as that for forming b^+c^+ recombinants). Hence one three-factor cross yields the gene order unambiguously; in the next section it will be seen that the cross also generates a quantitative genetic map for the three markers.

Multiple Exchanges and the Recombination Frequency for Distant Markers

When two genes are quite distant in a chromosome, more than one exchange may often occur in a single pairing event. This phenomenon complicates the interpretation of recombination data, primarily because if two exchanges occur between two markers, recombination will not be observed. If three exchanges occur, recombinants form. In general, recombination is not observed in a two-factor cross if the number of exchanges is an even number. Because of the failure to detect all exchanges an observed recombination value will always be an underestimate of the true exchange frequency and hence of the map distance between two genes.

There are two ways around this problem. First, one can avoid the use of distant markers and instead make use of the additivity of recombination frequencies, described earlier. This procedure minimizes the number of multiple exchanges. Roughly speaking, if observed recombination values are a few percent or less, multiple exchanges will be unimportant; however, if the observed recombination value is greater than 10 percent, the additivity method will be preferable. For example, consider the markers a, b, c, d, and e, which map in alphabetical order and are separated by 2, 7, 9, and 4 map units, respectively, determined from the two-factor crosses $a \times b$, $b \times c$, $c \times d$, and $d \times e$. Additivity shows that the most distant markers, a and e, are separated by $2 + 7 + 9 + 4 = 22$ map units. However, the observed recombination value in the cross $a \times e$ might be 20 percent, a 2-percent underestimate of the correct value.

The second method avoids the problem by including a third marker in the cross as a measure of double exchanges. Consider the cross $a^+ b^- c^+ \times a^- b^+ c^-$, in which the gene order is $a\ b\ c$. Data for the cross are given in Table 1-2. The frequency of exchanges between a and b calculated from the data is $(264 + 275 + 8 + 7)/7680$, or 7.21 percent, and between b and c it is $(223 + 220 + 8 + 7)/7680$, or 5.96 percent. That is, the map distances a–b and b–c are 7.21 and 5.96 map units, respectively. Note that in determining these frequencies we have counted the 15 double exchanges, even though a and c are not in a recombinant array, because the recombination frequency for a particular interval is based on the number of progeny in which an exchange has occurred anywhere in the interval.

If the recombination frequency between a and c were calculated just on the basis of the observed ac recombinants (that is, if the marker b had not been present), the observed recombination value would have been 264 $+ 275 + 223 + 220)/7680$, or 12.79 percent, which is slightly smaller than the more correct value of $7.21 + 5.96 = 13.17$ percent obtained by adding the frequencies a–b and b–c. The difference is just twice the $(8 + 7)/7680 = 0.20$ percent frequency of the double exchanges (the values are not precise because of rounding-off errors), since this value was used twice—once in determining the a–b distance and once for the b–c distance.

Table 1-2   Data for calculation of recombination frequencies

Genotype	Number	Class
$a^+ b^- c^+$	3408	Parental
$a^- b^+ c^-$	3275	Parental
$a^+ b^+ c^-$	264	Single exchange between
$a^- b^- c^+$	275	a and b
$a^+ b^- c^-$	223	Single exchange between
$a^- b^+ c^+$	220	b and c
$a^+ b^+ c^+$	8	Double exchange
$a^- b^- c^-$	7	Double exchange
Total	7680	

A recombination frequency represents the probability of an exchange in a particular interval. If exchanges occur at random, the probability of two exchanges in the interval a–c should be the product of the recombination frequencies for the two subintervals a–b and b–c. That is, if R is used for recombination frequency, $R_{\text{doubles}} = R_{ab}R_{bc}$. Examination of the data of Table 1-2 shows that this is not the case, for $R_{ab}R_{bc}$ is $(0.0721)(0.0596) = 0.43$ percent, not the observed value of 0.20 percent. The value of the observed frequency divided by the calculated frequency of double recombinants is called the **coefficient of coincidence, S,** and it is frequently less than 1. In fact, usually the value decreases as nearer markers are used in the cross. Note that $S < 1$ means that the presence of one exchange reduces the probability of finding another exchange nearby, a fact that must be incorporated into all models for the molecular basis of recombination.

Deletion Mapping

Figure 1-1 described a deletion mutation in which a portion of the genetic material is missing. It is usually desirable to know the extent of a deletion, and this can be determined by recombination experiments. The principle is straightforward: in a cross between a deletion mutant and a second mutant carrying a point mutation, wild-type progeny cannot be formed if the deletion spans the region of the map that includes the point mutation. This principle is illustrated in Figure 1-7. A variety of crosses are shown between (1) a mutant carrying a deletion that includes genes c and d and (2) mutants that have a defective allele of one of the genes a–e. Wild-type progeny are formed when the defect is in genes a, b, or e, because the deletion does not span these genes. However, in the crosses with the c^- or d^- mutation, wild-type progeny are not produced because the deletion does not have any genetic material corresponding to the mutant alleles.

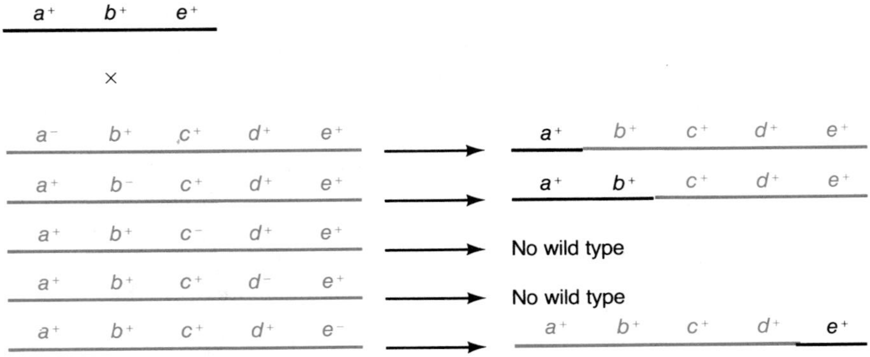

Figure 1-7 Mapping of a deletion.

Collections of deletions in which various amounts of genetic material are missing are exceedingly valuable for genetic mapping, for they vastly reduce the number of crosses required to map a new mutation. For example, an uncharacterized mutation can be mapped by crossing it with a set of deletions having different and known boundaries. The lack of wild-type recombinants in a cross with a particular deletion (e.g., I) indicates that the mutation is in the region spanned by the deletion. Furthermore, if wild-type are produced in crosses with a slightly smaller deletion (II), the mutation must be located between the boundaries of deletions I and II. We will see an elegant example of deletion mapping in Chapter 15, where phage genetics is discussed.

Complementation

As we have seen earlier, a particular phenotype is frequently the result of the activity of many genes. In the study of any genetic system it is always important to know the number of genes and regulatory elements that constitute the system. The genetic test used to determine this number is called **complementation.**

Complementation is best explained by example. The test requires that two copies of the genetic unit to be tested be present in the same cell. In bacteria this can be done by constructing a **partial diploid**—that is, a cell containing one complete set of genes and a duplicated copy of part of the genome. How cells of this type are constructed is described in Chapter 14. A partial diploid is described by writing the genotype of each set of genes on either side of a diagonal line. As an example of this, $b^+ c^+ d^+ / a^+$ $b^- c^- d^- e^+ \ldots z^+$ indicates that a chromosomal segment containing genes b, c, and d is present in a cell whose single chromosome contains all of the genes a, b, c, $\ldots$, z. Usually, only the duplicated genes are indicated, so this partial diploid would be designated $b^+ c^+ d^+ / b^- c^- d^-$. We now consider a hypothetical bacterium that synthesizes a blue pigment from the combined action of genes a, b, and c. The genes encode the enzymes A, B, C, which we assume to act sequentially to form the pigment. If there is a mutation in any of these genes, no pigment is made. However, pigment is made by the partial diploid $a^- b^+ c^+ / a^+ b^- c^+$, because the cell contains a set of genes that produce functional proteins A, B, and C; B will be made from the $a^- b^+ c^+$ chromosome and A from the $a^+ b^- c^+$ chromosome. In a partial diploid $a_1^- b^+ c^+ / a_2^- b^+ c^+$, in which a_1^- and a_2^- are two different mutations in gene a, no pigment can be made because the bacterium will not contain a functional A protein. The two mutations a^- and b^- of the diploid are said to complement one another because the phenotype of the partial diploid containing them is $A^+ B^+$; the mutations a_1^- and a_2^- do not complement one another because the phenotype of the partial diploid containing these mutations is A^-.

Suppose that now a mutation x^- has been isolated but the gene in which the mutation has occurred has not been ascertained. By constructing a set of partial diploids, this gene can be identified by a complementation test. As a start, we might test the genes a, b, and c with the partial diploids a^-/x^- (I), b^-/x^- (II), and c^-/x^- (III). These diploids are shown schematically in Figure 1-8. If diploids I and II make pigment, the mutation cannot be in genes a or b; if no pigment is made by diploid III, the mutation must be in gene c. If pigment were made in all three diploids, then the important conclusion that mutation x^- is in none of the genes a, b, or c could be drawn. Furthermore, since we have assumed that a, b, and c are each pigment genes, the fact that x is not in any of these genes but prevents pigment formation would be evidence that pigment formation requires at least four genes ("at least" because more genes might still be discovered).

A common approach to the initial characterization of a genetic system is to isolate about 50 mutations and perform complementation tests between them, as will be shown in the example of the next section. The analysis, which may be tedious, is nonetheless a straightforward one. The basic rule is the following:

Rule I. If x^- and y^- complement, they are in different genes. The converse statement—that is, if two mutations do not complement, they are in the same gene—does not always follow.

There are three explanations for the lack of complementation of two mutations; these are stated in the following rule:

Rule II. If mutations x^- and y^- fail to complement, then one of the following is true:

(a) They are in the same gene, or
(b) At least one of the mutations is in a regulatory site for the other gene, or
(c) At least one of the mutations yields an inhibitory gene product.

There is one interesting exception to rule I. Some proteins consist of identical polypeptide chains. In some cases, mutant chains of two different types can

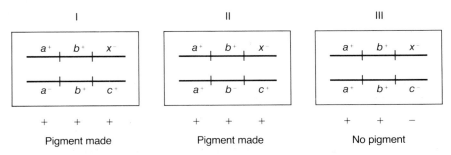

Figure 1-8 Complementation test showing three different diploids, I, II, and III. For illustrative purposes, the x^- mutation is in each case in gene c and is in the upper chromosome. The lower chromosome has a known mutation in either gene a (I), b (II), or c (III). To make pigment, a functional product must be made ($+$) from one of each of the three genes in a homologous pair.

aggregate to form a functional or partly functional protein. This phenomenon, which is fairly uncommon, is called **intragenic complementation.** It will not be encountered again in this book.

Complementation Analysis: An Example

Use of rules I and II allows every mutation to be placed in a mutually exclusive complementation group. Let us first examine these rules for a set of mutations, 1 through 8, in which rules I and II(a) account for all of the data. After studying this simple example, we will add first one mutation that requires that rule II(b) be invoked and then a second mutation that can be explained by II(c). The data are presented in Table 1-3. To analyze the first example, use rows 1 through 8 of the table. Each + or − entry designates a single pairing of the mutation numbers shown at the top and side of the table. Any − entry denotes that the pair does not complement; any + entry denotes complementation. In particular, notice any − entries that can be aligned in a single direction (i.e., that align down one column or across one row). These are noncomplementing mutations within a single gene. For example, in column 1, mutations 1 and 5 fail to complement; other noncomplementing (aligning) mutations can be found in columns 2, 3, 4, and 6, and in rows 4, 5, 6, 7, and 8. Three genes can be identified in this way: the first contains mutations 1 and 5; the second contains mutations 2, 4, and 8; and the third

Table 1-3   An example of complementation data

	Mutation number									
	1	2	3	4	5	6	7	8	9	10
Mutation number										
1	−									
2	+	−								
3	+	+	−							
4	+	−	+	−						
5	−	+	+	+	−					
6	+	+	−	+	+	−				
7	+	+	−	+	+	−	−			
8	+	−	+	−	+	+	+	−		
9	−	−	−	−	−	−	−	−	−	
10	±	−	±	−	±	±	±	−	−	−

Note: + = complementation; − = no complementation; ± = weak complementation.
An entry at the intersection of a horizontal row and a vertical column represents the result of one complementation test between two mutations. For example, the + entry at the intersection of row 3 and column 2 indicates that mutations 2 and 3 complement; the corresponding entry for row 2 and column 3 is not given since it is the same as the one just stated—that is, the table is symmetric about the diagonal. The analysis first discussed in the text uses only mutations 1–8, which is the reason for the dashed line between rows 8 and 9.

contains mutations 3, 6, and 7. (Inasmuch as the table is symmetrical, a row or column need be counted just once.) Single mutations (a row or column) that intersect at a + sign are usually complementing. The three genes (or groups of mutations) found in the table are therefore complementing. We can give each complementing gene an identifying letter for convenience:

Group	Mutation
A	1, 5
B	2, 4, 8
C	3, 6, 7

The simplest explanation for the data is that the phenotype being studied consists of at least three genes.

We now examine the effect of a mutation in a regulatory site rather than in a gene (rule II(b)). Suppose a ninth mutation had been isolated having the complementing property shown below:

	1	2	3	4	5	6	7	8	9
9	–	–	–	–	–	–	–	–	–

Alone these data might suggest that all of the mutations are in the same gene, but the data just analyzed show that this is not the case, for clearly, when mutation 9 is present in a particular chromosome, that chromosome will not yield the products of genes A, B, and C. Mutation 9 is called a **cis-dominant mutation,** because such a mutation prevents expression of related genes residing on the same chromosome in which the mutation is located. Such mutations invariably prove to be regulatory site mutations; that is, they are in some *site* on the chromosome that must be active if gene products in the *same* chromosome and in the particular gene system are to be made. For instance, the mutation might be in a site that signals start of synthesis of the gene products.

This interpretation of mutation 9 has not been done realistically, because mutations 1 through 8 were classified first and then 9 was introduced. Let us consider how all nine mutations would have been classified if they were examined simultaneously. Once again, – entries would be observed and clustered. However, mutation 9 would immediately be anomalous because it would appear in more than one complementation group—in fact, in all groups. Whenever a mutation appears in more than one group, the need for rule II(b) should be suspected and that mutation should temporarily be ignored in making the basic classification. It is, of course, not always the case that a mutation in a regulatory site will fail to complement all other mutations being examined, because the site might be required for activity of only one, or a few, genes in the set. That is, the regulatory mutation would only fail to complement mutations in the genes regulated. Many examples of mutations of this type will be seen in Chapter 7.

Let us now consider the rare case in which rule II(c) would be applied.

Inasmuch as mutation 10 fails to complement mutations 2, 4, and 8, it might appear to be a separate, additional mutation in the group B gene. The + response is weak with mutations 1, 3, 5, 6, and 7, though, and mutation 10 cannot be considered complementary to groups A and C because, tested against the mutations in A and C, 10 would also yield a − . The correct interpretation is that 10 is not only a B-group mutation but that the mutant gene product also inhibits the activity of a good copy of B. Thus, in each cell in which one would expect B to be fully active, the + response is weak because the total activity of the inhibited B is not adequate for a normal + response. In this case, mutation 10 is said to be anticomplementing.

THE NEED FOR ISOGENICITY IN STRAIN CONSTRUCTION

In an earlier section we have seen that mutations have great value in elucidating both genetic and biochemical properties of living systems. However, in order to use the reasoning in the seven examples given in that section, it is essential that the mutant organism differ from the reference (normal) organism by only one mutation. Clearly, if several mutations were present, one would not know which one was responsible for an observed phenotypic change. Two organisms that differ by only a single genetic locus are said to be **isogenic** or to have the same genetic background.

A variety of procedures are used to ensure that strains are isogenic. When mutations have been introduced by mutagenesis, the reversion frequency can be measured in order to ascertain that the mutation is a point mutation. This procedure is fairly reliable but will not indicate the presence of other mutations that do not directly affect the phenotype; however, such mutations are usually not significant. Mutations obtained in one strain are also frequently introduced into a desired strain by recombination. This would be done, for example, if the effect of two different mutations in a single organism is to be tested. In this case, a single exchange would normally result in replacement of a large portion of the genome of one organism by that of another. This can be avoided if several markers are used in the cross and double recombinants are selected so that only a small region of the genome of the "donor" organism replaces the same region of the recipient. Strain construction will be discussed further in Chapter 19. We will see in Chapters 18 and 19 that transduction is a fairly reliable procedure for achieving isogenicity, failing only if an unobserved mutation is very closely linked to a mutation of interest.

PROBLEMS

1. Does the symbol *lac*⁺ refer to the genotype or the phenotype of a bacterium?

2. What notation would indicate that a bacterium is resistant to the antibiotic penicillin?

3. If a cell has three genes involved in synthesis of proline, namely, *proA*, *proB*, and *proC*, would there be any difference in the genotype of a cell designated *proA$^+$ proB$^+$ proC$^+$* and one designated *pro$^+$*?

4. Referring to Problem 3, how would you write the phenotype of a cell whose genotype is *proA$^+$ proB$^-$ proC$^+$*?

5. How does an absolute defective mutant differ from a conditional mutant?

6. Insert the terms *pen-r* and Pen-r into the following sentence. A . . . . mutant contains a . . . . mutation.

7. What is the phenotype of a *leu$^-$*(Ts)/*leu$^+$* heterozygote, and does the phenotype depend on temperature?

8. The frequency of producing mutant *arb$^-$* is 2×10^{-6} per generation and mutant *snd$^-$* is 8×10^{-5}. What is the frequency of producing an *arb$^-$ snd$^-$* mutant in a single event?

9. Rank the following mutations in order of reversion frequency: deletion, point mutation, double mutation.

10. A cell carrying *v$^-$*(Ts), a temperature-sensitive mutation in gene *v*, has lost the ability to synthesize substance V. An enzyme Vase, which is involved in synthesizing V, is isolated from a mutant containing *v$^-$*(Ts) and shows normal activity over a wide range of temperatures. Is Vase the product of gene *v*?

11. A mutation prevents the synthesis of substance Z and results in the accumulation of large amounts of substance R, which is normally present in only very small amounts. What can probably be said about the relation between R and Z and about the gene product inactivated by the mutation?

12. Two phages with genotypes *x$^+$y$^-$* and *x$^-$y$^+$* are crossed. What are the genotypes of the possible recombinant types? Are any of them wild-type?

13. The recombination frequencies between three genes are: *a–b*, 2.6%, *b–d*, 1.4%, and *a–d*, 1.2%. What is the gene order?

14. The recombination frequency between *q* and *r* is 3.7%. Does it matter whether the map is represented *q*-3.7-*r* or *r*-3.7-*q*?

15. In a phage cross *ABD* × *abd*, what are the genotypes of the double-exchange classes, if the map order is *A D B*?

16. Mutants that fail to synthesize a substance X have been found in four complementation groups, none of which are *cis*-acting. How many proteins are required to synthesize X?

17. In a haploid organism mutation *x$^-$* eliminates production of a pigment; another mutation *y$^-$* does the same. A partial diploid *x$^-$/y$^-$* cell makes pigment. Do *x* and *y* complement? How many genes are required for pigment formation?

18. Study of a temperature-sensitive mutant in the gene *memA* indicates that there is no active MemA enzyme at 42°C, although at 34°C active enzyme is present. The enzyme is purified from cells grown at 34°C and its activity is tested at 42°C. It remains active at this temperature. What conclusion can be drawn about the relation between the *memA* gene and the MemA enzyme?

19. In phage BX4 some mutations in the phage gene G are compensated for by an additional mutation in gene H and some mutations in gene H are compensated for by mutations in gene G. What do these facts indicate?

20. Given the recombination frequencies between genes a and b, a and c, and c and d, how many more values are needed to construct a map?

21. The following recombination frequencies occur between the indicated markers: $a \times c$, 2%; $b \times c$, 13%; $b \times d$, 4%, $a \times b$, 15%, $c \times d$, 17%; $a \times d$, 19%.

    (a) What is the gene order?
    (b) In the cross $aBd \times AbD$, what is the frequency of getting ABD progeny?

22. In the cross $abcd \times ABCD$, the following recombinants were found at the indicated frequencies: $ABCd$, 3%; $abcD$, 3%; $AbcD$, 0.03%; $AbCd$, 0.0006%; $ABcD$, 0.06%. What is the gene order?

23. Four genes have the order $a\ b\ d\ e$. The recombination frequencies between various pairs are: $a \times b$, 1%; $b \times d$, 2%; $d \times e$, 3%.

    (a) What is the frequency of production of AE recombinants in the cross $AbDe \times aBdE$?
    (b) What fraction of the AE recombinants will be $ABDE$?

24. In a cross between two phage having genotypes EFG and efg, 1000 progeny were analyzed. The number of phage having each of the eight possible genotypes were as follows: efg, 396; EFG, 406; eFg, 23; efG, 1; EfG, 25; Efg, 75; eFG, 73; EFg, 1. Construct a map showing the positions of the genetic markers.

25. A three-factor phage cross $ABD \times abd$ is carried out which yields the fact that the recombinant AbD is formed at a much higher frequency than Abd. What is the gene order, assuming equal spacing of the genes?

26. A mutation j^- (Ts) eliminates the ability of a cell to synthesize substance J above 40°C. The enzyme Jase, which is needed to synthesize J, is isolated from the mutant and found to be functional above 40°C. What is the minimum number of proteins needed to synthesize J?

27. Four genes $kyuA$, $kyuB$, $kyuC$, and $kyuQ$ are known to be required to synthesize substance Q, and each biochemical reaction can be detected. The reaction sequence is $P \rightarrow B \rightarrow A \rightarrow Q$ in which the product of a gene $kyuX$ is needed to synthesize substance X. Addition of ^{14}C-P yields ^{14}C-Q.

    (a) A mutant is found for which addition of ^{14}C-P yields ^{14}C-A but no ^{14}C-Q. In what gene is this mutation?
    (b) Another mutant is found for which there is no conversion of ^{14}C-P to any other substance. Furthermore, addition of ^{14}C-A fails to yield ^{14}C-Q. What kind of mutant is this one?

28. The complementation data shown in Table P1-28 are observed. The numbers refer to particular mutations. The symbols + and − indicate that the two mutations do and do not complement, respectively. How many genes are represented? Assign the mutations to the genes.

Table P1-28

	Mutants						
	1	2	3	4	5	6	7
1	−	+	+	+	+	+	−
2		−	+	−	+	+	+
3			−	+	+	−	+
4				−	+	+	+
5					−	+	+
6						−	+
7							−

Nucleic Acids

All hereditary information resides in nucleic acids. In most organisms, genes are segments of DNA molecules, but in a few phages and many animal and plant viruses, RNA is the genetic material. A great deal of what we know about the nature of genes and gene expression has been obtained either from knowledge of the structure of DNA or RNA, from experiments in which DNA is used to carry genetic information from one organism to another, or from studies in which DNA and RNA are manipulated and altered. Hence, a review of the properties of these molecules is provided in this chapter.

COMPONENTS OF NUCLEIC ACIDS

A nucleic acid is a polynucleotide—that is, a polymer consisting of nucleotides. Each **nucleotide** has the three following components (Figure 2-1):

1. *A cyclic five-carbon sugar.* This is ribose, in the case of ribonucleic acid (RNA), and deoxyribose, in deoxyribonucleic acid (DNA). The difference in the structures of ribose and 2′-deoxyribose is also shown in the figure. Note that they differ only in the absence of a 2′-OH group in deoxyribose, a difference that makes DNA chemically more stable than RNA.

2. *A purine or pyrimidine base* attached to the 1′-carbon atom of the sugar by an *N*-glycosidic bond. The bases, which are shown in Figure 2-2, are the purines adenine (A) and guanine (G), and the pyrimidines

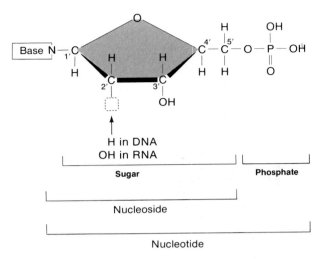

H in DNA
OH in RNA

Sugar   Phosphate

Nucleoside

Nucleotide

Figure 2-1   A typical nucleotide showing the three major components, the difference between DNA and RNA, and the distinction between a nucleoside and a nucleotide.

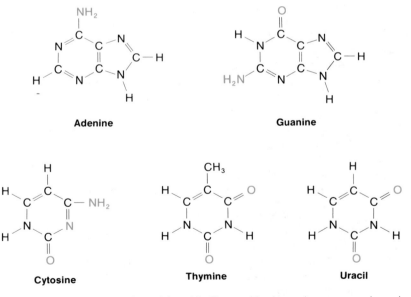

Adenine   Guanine

Cytosine   Thymine   Uracil

Figure 2-2   The bases found in nucleic acids. The weakly charged groups are shown in red. Adenine and guanine are derivatives of purines, the compounds in which all carbon atoms bear H atoms. Cytosine, thymine, and uracil are derivatives of pyrimidine, a compound with three double bonds in the rings and all C atoms bearing H atoms.

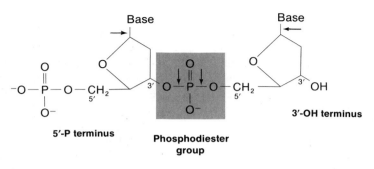

Figure 2-3 The structure of a dinucleotide. The vertical arrows show the bonds in the phosphodiester group about which there is free rotation. The horizontal arrows indicate the *N*-glycosidic bond about which the base can freely rotate. A polynucleotide would consist of many nucleotides linked together by phosphodiester bonds.

cytosine (C), thymine (T), and uracil (U). DNA and RNA both contain A, G, and C; however, T is found only in DNA and U is found only in RNA.

3. *A phosphate* attached to the 5′-carbon of the sugar by a phosphoester linkage. This phosphate is responsible for the strong negative charge of both nucleotides and nucleic acids.

A base linked to a sugar is called a **nucleoside;** thus *a nucleotide is a nucleoside phosphate*.

The nucleotides in nucleic acids are covalently linked by a second phosphoester bond that joins the 5′-phosphate of one nucleotide and the 3′-OH group of the adjacent nucleotide (Figure 2-3). This phosphate plus its bonds to the 3′- and 5′-carbon atoms is called a **phosphodiester.** The result of successive linkage of nucleotides in a polynucleotide is an alternating sugar-phosphate backbone having one 3′-OH terminus and one 5′-phosphate (5′-P) terminus.

A typical RNA molecule is the single-stranded polyribonucleotide that has just been described. However, except in unusual cases, DNA contains two polydeoxynucleotide strands wrapped around one another to form a double-stranded helix.

THE DOUBLE HELIX

Early physical studies of DNA indicated that the molecule is an extended chain having a highly ordered structure. The most significant observations, obtained by x-ray diffraction analysis, were that DNA is helical with the nucleotide bases stacked with their planes separated by 3.4 Å. Chemical analysis showed that a regularity exists in the molar content of the bases (generally called the **base composition**) adenine, thymine, guanine, and cytosine in DNA molecules isolated from many organisms. The regularity is that [A] = [T] and [G] = [C], in which the brackets denote mole fraction. James Watson and Francis Crick combined the physical and chemical data and showed that the two strands are coiled about one another to form a double-stranded helix (Figure 2-4). The sugar-phosphate backbones follow a helical path at the outer edge of the molecule, and the bases are in a helical array in the central core. The bases of one strand are hydrogen-bonded to those of the other strand to form the purine-pyrimidine base pairs AT and GC (Figure 2-5). Because each pair contains one two-ringed purine (A or G) and one single-ringed pyrimidine (T or C, respectively), the length of each pair is about the same and the helix can fit into a smooth cylinder.

The two bases in each base pair lie in the same plane and the plane of each pair is perpendicular to the helix axis. The base pairs are rotated with respect to each other to produce 10 pairs per helical turn. The diameter of the double helix is 20 Å and the molecular weight per unit length of the helix is approximately 2×10^6 per micrometer. Since the molecular weight of a

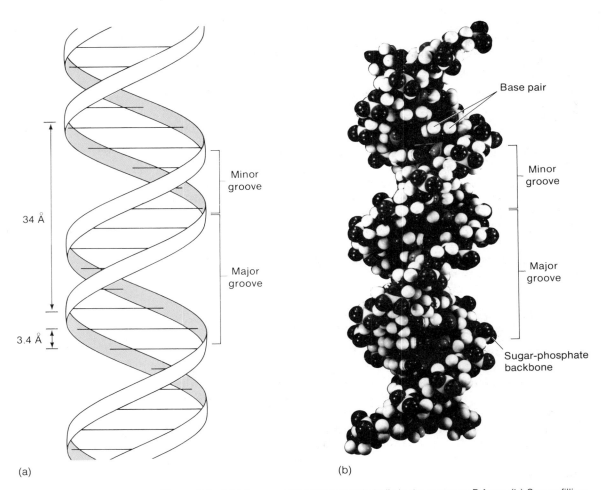

(a)   (b)

Figure 2-4   (a) Diagram of the DNA double helix in the common B form. (b) Space-filling model of DNA, again in the B form. [Courtesy of Sung-Hou Kim.]

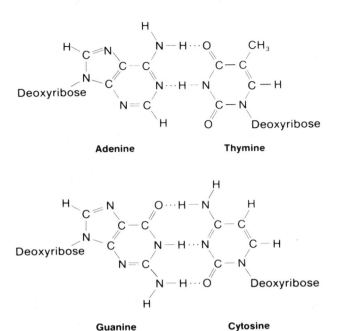

Figure 2-5 The two common base pairs of DNA.

Adenine   **Thymine**

Guanine   **Cytosine**

typical bacterial DNA molecule is about 2×10^9, such a molecule is 1 millimeter or 10^6 Å long and is very long and thin indeed.

DNA is a right-handed helix. This means that each strand will appear to follow a clockwise path moving away from a viewer looking down the helical axis.

The DNA helix has two external helical grooves, a deep wide one (the **major groove**) and a shallow narrow one (the **minor groove**), as indicated in Figure 2-4. The major groove is the site of binding of specific proteins.

The two polynucleotide strands of the DNA double helix are antiparallel—that is, the 3′-OH terminus of one strand is adjacent to the 5′-P terminus of the other (Figure 2-6). Thus, in a linear double helix there is one 3′-OH and one 5′-P terminus at each end of the helix. We will see in Chapter 8 that being antiparallel poses a constraint on the mechanism of DNA replication.

By convention the sequence of bases of a single chain is usually written with the 5′-P terminus at the left; for example, ATC denotes the trinucleotide P-5′-ATC-3′-OH, which may also be written pApTpC.

DENATURATION AND MELTING CURVES

The attractive forces that produce the three-dimensional structure of molecules, such as the hydrogen bonding between DNA base pairs, are fairly weak and easily disrupted at elevated temperatures. A macromolecule in a

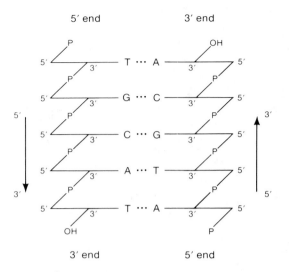

Figure 2-6 A stylized drawing of a segment of a DNA duplex showing the antiparallel orientation of the complementary chains. The arrows indicate the 5′ → 3′ direction of each strand. The phosphates (P) join the 3′ carbon of one deoxyribose (horizontal line) to the 5′ carbon of the adjacent deoxyribose.

disrupted state is said to be **denatured;** the ordered state, which is presumably that originally present in nature, is called **native.** The transition from the native to the denatured state is called **denaturation.** When double-stranded DNA, or native DNA, is heated, the hydrogen bonds between the strands are broken and the two strands separate; thus, *denatured DNA is single-stranded.*

A great deal of information about structure and stabilizing interactions in DNA has been obtained by studying denaturation. This is typically done by measuring some property of the molecule that changes as denaturation proceeds. The most common property is the absorption of ultraviolet light of wavelength 260 nanometers (nm). At this wavelength single-stranded DNA absorbs more strongly than double-stranded DNA. In early experiments denaturation was accomplished by heating a DNA solution, so a graph of a varying property as a function of temperature is called a **melting curve.** A convenient measure of the absorption is the absorbance (A) of a solution, which is defined as $-\log_{10}$[(intensity of light transmitted by a solution 1 cm thick)/(intensity of incident light)]. An example of a melting curve is shown in Figure 2-7. The state of a DNA molecule in different regions of the melting curve is also shown. Before the rise in A begins, the molecule is fully double-stranded. In the rise region, base pairs in various segments of the molecule are broken; the number of broken base pairs increases with temperature. In the initial part of the upper plateau a few base pairs remain to hold the two

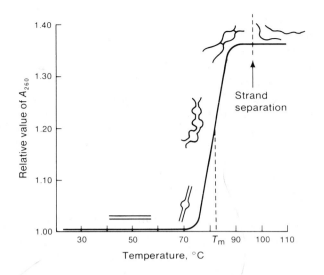

Figure 2-7 A melting curve of DNA showing T_m and possible molecular conformations for various degrees of melting.

strands together, until a critical temperature is reached at which the last base pair breaks and the strands separate completely.

A convenient parameter to characterize a melting transition is the temperature at which the rise in A_{260} is half complete. This temperature is called the **melting temperature** and is designated T_m. AT and GC pairs are held together by two and three hydrogen bonds, respectively, so a higher temperature is required to disrupt GC pairs. For this reason, the value of T_m is related to the base composition of the DNA, and in solutions standardized with respect to salt concentration and pH, T_m can be used to measure the base composition. At high pH the charge of several of the groups engaged in hydrogen-bonding is changed, so a base bearing such a group loses its ability to form these bonds. At a pH greater than 11.3 all hydrogen bonds are eliminated and DNA is completely denatured. At high temperatures at neutral pH, phosphodiester bonds are broken. These bonds are quite resistant to alkaline hydrolysis, so treatment at high pH is the method of choice for denaturing DNA without breaking covalent bonds.

When heated DNA is returned to room temperature or when alkali-denatured DNA is restored to neutral pH, the single strands remain separate. If the salt concentration of the solvent is low, the strong negative charge of the phosphate groups keep the strands extended and single-stranded throughout. However, if the salt concentration is high enough to neutralize the negative charge, the single strands fold back on themselves, forming compact molecules with much intrastrand base-pairing. However, since lengthy complementary sequences are rare, the paired segments rarely contain more than ten nucleotides.

RENATURATION

A solution of denatured DNA can be treated in such a way that native DNA re-forms. The process is called **renaturation** or **reannealing** and the re-formed DNA is called **renatured DNA.** Renaturation has proved to be a valuable tool in microbial genetics, since it can be used to demonstrate genetic relatedness between different organisms, to detect particular species of RNA, to determine whether certain sequences occur more than once in the DNA of a particular organism, and to locate specific base sequences in a DNA molecule.

Requirements for Renaturation

Two requirements must be met for renaturation to occur:

1. The salt concentration must be high enough that electrostatic repulsion between the phosphates in the two strands is eliminated. Usually 0.15 to 0.50 M NaCl is used.
2. The temperature must be high enough to disrupt the random, intrastrand hydrogen bonds described in the previous section. However, the temperature cannot be too high, or stable interstrand base-pairing will not occur. The optimal temperature for renaturation is 20–25° below the value of T_m.

Renaturation is a slow process compared to denaturation. The rate-limiting step is not the actual rewinding of the helix (which occurs in a few seconds) but the precise collision between complementary strands such that base pairs are formed at the correct positions. Since this is a result only of random motion, it is a concentration-dependent process; at concentrations normally encountered in the laboratory this takes several hours. Renaturation can be detected by the decrease in absorbance of a DNA solution.

Filter-binding Assays for Renaturation

Very thin filters (**membrane filters**) made of nitrocellulose are commercially available. These filters bind single-stranded DNA very tightly but fail to bind either double-stranded DNA or RNA. They provide an important method for measuring hybridization by the following technique (Figure 2-8). A sample of denatured DNA is filtered. The single strands bind tightly to the filter along the sugar-phosphate backbone, but the bases remain free. The filter is then placed in a vial with a solution containing a reagent that prevents additional binding of single-stranded DNA to the filter and a small amount of radioactive denatured DNA. After a period of renaturation, the filter is washed. Radioactivity is found on the filter only if renaturation has occurred.

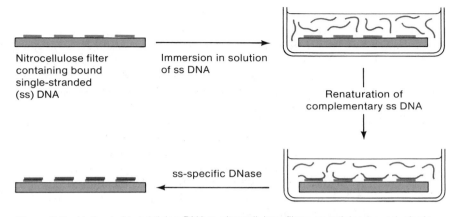

Figure 2-8 Method of hybridizing DNA to nitrocellulose filters containing bound, single-stranded DNA. In the final step the filter is treated with a single-strand-specific DNase, an enzyme that depolymerizes single-stranded, but not double-stranded DNA.

Filter binding can be used to determine whether two organisms have common base sequences. For example, if excess *E. coli* DNA is on the filter and a very small amount of denatured ^{14}C-labeled DNA isolated from *Salmonella* bacteria is applied, the fraction of the applied ^{14}C that is retained on the filter is the fraction of *Salmonella* DNA that has a sequence with sufficient complementarity with that of *E. coli* to hybridize under the annealing conditions used. The value of this fraction indicates the degree of evolutionary homology. This type of experiment has confirmed the basic expectations of evolutionary theory—that is, taxonomically related organisms have common sequences and the degree of overlap is connected to the relatedness determined by other criteria.

An important use of filter hybridization is the detection of sequence homology between a single strand of DNA and an RNA molecule. This is called **DNA-RNA hybridization,** and it is the method of choice to detect an RNA molecule that has been copied from a particular DNA molecule. In this procedure, a filter to which single strands of DNA have been bound, as above, is placed in a solution containing radioactive RNA. After renaturation the filter is washed and hybridization is detected by the presence of radioactive RNA on the filter.

Filter hybridization can also be used to assess the degree of homology between two segments. This is done by varying the pH, salt concentration, and renaturation temperature. Certain conditions (termed stringent) allow reannealing, and hence filter-binding, only when complementarity is nearly perfect. Less stringent conditions permit filter-binding with less complementarity.

DNA Heteroduplexes

Renaturation has been combined with electron microscopy in a procedure that allows the localization of common, distinct, and deleted sequences in DNA. This procedure is called **heteroduplexing.** Consider the DNA molecules #1 and #2 shown in Figure 2-9(a). These molecules differ in sequence only in one region. If a mixture of the two molecules is denatured and renatured, not only parental molecules but also hybrid molecules having unpaired single strands are produced, as shown in Figure 2-9(b). Figure 2-10 shows an actual electron micrograph of a heteroduplex. Measurement of the lengths of the single and double-stranded regions yields the endpoints of the regions of nonhomology. If the sequences by which the molecules differ have the same or nearly the same number of nucleotides, the two single strands of the bubble will have the same length.

Consider now molecule #3, shown in Figure 2-9(a). In this molecule, region A is deleted. If a hybrid is made between this molecule and molecule #2, the result is a molecule with a single loop, as shown in Figure 2-9(b).

Heteroduplexing is also possible between a double-stranded DNA molecule and an RNA molecule that is complementary to *part* of the DNA sequence. Denaturation of the DNA and reannealing with the RNA produces a molecule with a bubble, called an **R loop.** One branch of the bubble is a DNA-RNA hybrid, and the other is single-stranded DNA. This technique is useful to map DNA sequences from which particular RNA molecules are copied.

Figure 2-9 (a) Three DNA molecules to be heteroduplexed. Sequences A and A′ of molecules 1 and 2 differ. Neither sequence is present in the deletion molecule 3. The dashed lines indicate reference points. (b) Heteroduplexes resulting from renaturing molecules 1 and 2 or 2 and 3.

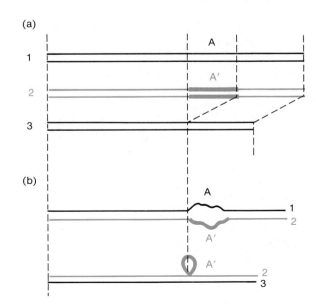

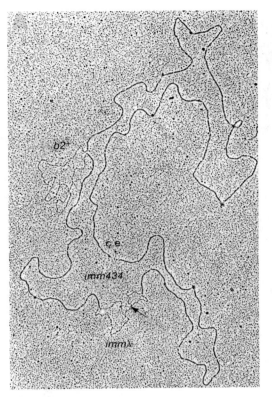

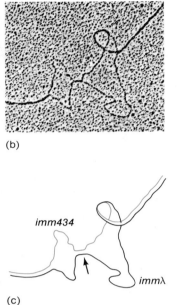

Figure 2-10 An electron micrograph of a heteroduplex between λ*imm*λ *b2* DNA, which carries the *b2* deletion, and λ*imm434 b2*⁺, in which the *imm*λ segment is replaced by the shorter, nonhomologous *imm434* segment. (a) Two bubbles of nonhomology are seen. The identity of each single-stranded segment is indicated. The arrow is explained in part (c). (b) An enlargement of the *imm434–imm*λ segment. (c) An interpretive drawing of panel (b). The arrow indicates a region of homology between the *imm434* and *imm*λ segments. The same region is indicated by the arrow in panel (a). [Courtesy of Barbara Westmoreland and W. Szybalski.]

CIRCULAR AND SUPERHELICAL DNA

The intact DNA molecules of most prokaryotes and viruses are circular. A circular molecule may be a **covalently closed circle,** which consists of two unbroken complementary single strands, or it may be a **nicked circle,** which has one or more interruptions (**nicks**) in one or both strands, as shown in Figure 2-11. With few exceptions, covalently closed circles are twisted, as shown in Figure 2-12. Such a circle is said to be a **superhelix** or a **supercoil.** Let us now examine what is meant by a circular DNA molecule being twisted.

Figure 2-11 A covalently closed circle and two kinds of nicked circles. Arrows indicate the nicks. A nicked circle is also called an open circle.

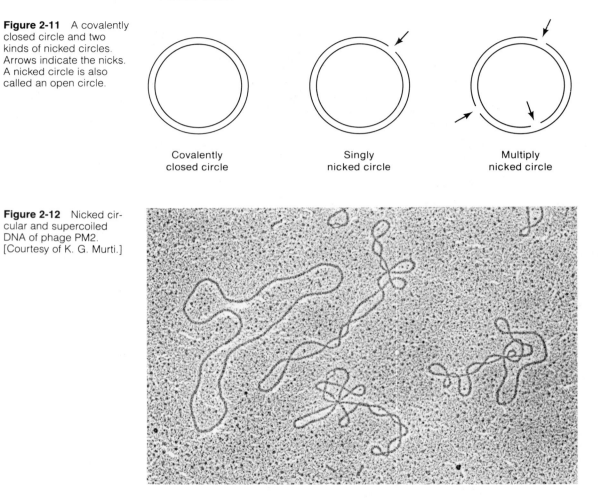

Covalently
closed circle

Singly
nicked circle

Multiply
nicked circle

Figure 2-12 Nicked circular and supercoiled DNA of phage PM2. [Courtesy of K. G. Murti.]

Twisted Circles

To begin this discussion it is worth reviewing what is meant by a right-handed or positive coil. Helical coiling is positive if, when looking down the helical axis, the coil follows a clockwise path and moves away from the viewer. If the path is counterclockwise, the coil is left-handed, or negative. Recall that DNA is a right-handed helix.

The two ends of a linear DNA helix can be brought together and joined in such a way that each strand is continuous. If, in so doing, one of the ends is rotated 360° with respect to the other to produce some unwinding of the double helix, and then the ends are joined, the resulting covalent circle will, if the hydrogen bonds re-form, twist in the opposite sense (here, opposite to the unwinding direction) to form a twisted circle, in order to relieve strain.

Such a molecule will look like a figure 8 (that is, have one crossover point or **node**). If it is instead twisted 720° prior to joining, the resulting superhelical molecule will have two nodes. The reason for the twisting is the following. In the case of a 720° unwinding of the helix, 20 base pairs must be broken (because the linear molecule has 10 base pairs per turn of the helix). However, a DNA molecule prefers to maintain a right-handed (positive) helical structure with 10 base pairs per turn and hence deforms in such a way that the underwinding is rewound and compensated for by negative (left-handed) twisting of the circle. Similarly, the initial rotation might instead be in the direction of overwinding, in which case the joined circle will twist in the opposite sense, forming a positive superhelix. *All naturally occurring superhelical DNA molecules are initially underwound and hence form negative superhelices.* Furthermore, the degree of twisting is about the same for all molecules; namely, one negative twist is produced per 200 base pairs, or, 0.05 twists per turn of the helix. In bacteria, the underwinding of superhelical DNA is not a result of unwinding prior to end-joining but is introduced into preexisting circles by an enzyme called **DNA gyrase,** which will be described in Chapter 8, when DNA replication is examined.

Single-Stranded Regions in Superhelices

We have just pointed out that the strain of underwinding can be accommodated by negative supercoiling. Three other arrangements that could counteract the strain of underwinding are possible: (1) The number of base pairs per turn of the helix could change. This does not happen, though, for thermodynamic reasons. (2) All of the underwinding could be taken up by having one or more large single-stranded bubbles (Figure 2-13). (3) The under-

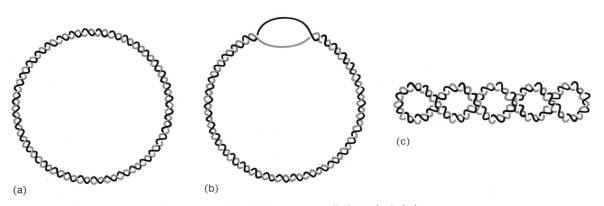

Figure 2-13 Different states of a covalent circle. (a) A nonsupercoiled covalent circle having 36 turns of the helix. (b) An underwound covalent circle having only 32 turns of the helix. (c) The molecule in part (b), but with four superhelical turns to eliminate the underwinding. In solution, (b) and (c) would be in equilibrium; the equilibrium would shift toward (b) with increasing temperature.

winding could be taken up in part by superhelicity and in part by bubbles. The real situation is alternative 3—that is, a state intermediate between forms (b) and (c) of the figure—because a DNA molecule is a dynamic structure that undergoes transient unwinding from thermal motion. If a circular molecule were made that was not underwound, transient breakage and remaking of hydrogen bonds (**breathing**) would introduce compensating transient negative twists. If the DNA is initially superhelical, the degree of supercoiling will fluctuate as breathing occurs: the strain produced by the underwinding is relieved in a superhelix both by the superhelicity and by an increase in the number and size of the bubbles and the duration of their existence. Thus, in a supercoil, the fraction of the molecule that is single-stranded at any moment is greater than in a nicked circle.

Experimental Detection of Covalently Closed Circles

In the life cycles of many organisms, the DNA molecules cycle through the various circular forms that have just been described. Two techniques are commonly used to distinguish these forms.

1. *Sedimentation at denaturing alkaline pH*. Above pH 11.3 a linear DNA molecule in a salt concentration of 0.3 *M* denatures to yield two single strands, each of which sediments about 30 percent faster than native DNA. However, the two strands of a covalently closed circle cannot separate (because there are no free ends to allow unwinding), so the molecules collapse in a tight tangle that sediments about three times faster than native DNA. If the circle has a single nick, one linear molecule and one single-stranded circle result, and the circle sediments 14 percent more rapidly than the former (Figure 2-14(a)). Figure 2-14(b) shows a sedimentation pattern for a mixture composed of linear molecules, nicked circles, and covalently closed circles in an alkaline solution.

2. *Equilibrium centrifugation in CsCl containing ethidium bromide*. **Ethidium bromide** (Figure 2-15) binds very tightly to DNA and, in so doing, decreases the density of the DNA by approximately 0.15 g/cm^3. It binds by intercalating between the DNA base pairs, thereby causing the DNA molecule to unwind as more ethidium bromide is bound. Because a covalently closed DNA molecule has no free ends, the entire molecule, as it unwinds, twists in the opposite direction; the degree of twisting increases as more molecules intercalate. Ultimately, the DNA molecule is unable to twist any more, so no more ethidium bromide molecules can be bound. On the contrary, a linear DNA molecule or a nicked circle does not have the topological constraint of reverse twisting and can therefore bind more of the ethidium bromide molecules. Because the density of the DNA and ethidium bromide complex decreases as more ethidium bromide is bound and because more ethidium bromide can be bound to a linear molecule or an open circle than

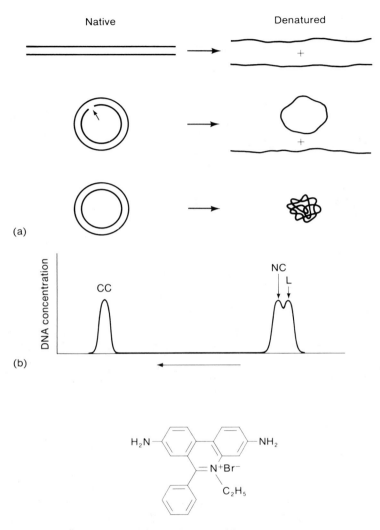

Figure 2-14 (a) Products of the denaturation of different forms of DNA. (b) Separation of covalently closed circles (CC), nicked circles (NC), and linear (L) molecules by sedimentation in alkali. The horizontal axis represents the length of a centrifuge tube. Sedimentation is from right to left.

Figure 2-15 Chemical formula for ethidium bromide.

to a covalent circle, the covalent circle has a higher density at saturating concentrations of ethidium bromide. Therefore, covalent circles can be separated from the other forms in an equilibrium density gradient, as shown in Figure 2-16. Equilibrium centrifugation is reviewed in a later section of this chapter, namely, Methods Used to Study Macromolecules.

Supercoiling is easily detected by electron microscopy, as already shown in Figure 2-12. Several methods, which we will not describe, are available for measuring the degree of supercoiling.

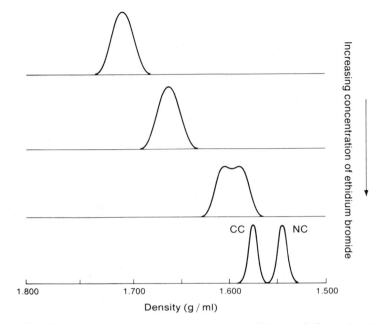

Figure 2-16   Effect of ethidium bromide on the density of DNA in a CsCl solution. A mixture of equal parts of nicked circles (NC) and covalently closed circles (CC) is centrifuged in CsCl containing various concentrations of ethidium bromide. The density of the DNA molecules decreases until, at saturation, the two components separate. The covalent circles bind less ethidium bromide and are therefore at a higher density.

SPECIAL BASE SEQUENCES AND THEIR STRUCTURAL CONSEQUENCES

An enormous variety of base sequences have been observed in DNA. Most do not have any special features that cause them to be recognized as unusual. Of particular interest are several types of repeated sequences. These sequences are common to regulatory regions and to sites of enzymatic activity and may in some cases impart special properties to either double- or single-stranded nucleic acids. Other sequences in which purines and pyrimidines alternate can cause DNA to form a left-handed helical region.

The nucleotide **palindrome** is a sequence of the general form

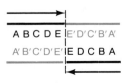

in which A and A′, B and B′, and so forth, are complementary bases able to pair. The dashed vertical line is an axis of symmetry: the double-stranded segment to the right of the axis can be superimposed on the one to the left by a 180° rotation in the plane of the page. Other terms used to describe palindromes are **inverted repeat, inverted repetition,** and region of **dyad symmetry,** a term used by crystallographers. Palindromes range in length up to about 50 base pairs. The two inverted sequences may be separated by a spacer: for example,

A B C D E U V W X Y Z E′D′C′B′A′

A′B′C′D′E′U′V′W′X′Y′Z′E D C B A

in which case only the term inverted repeat is used. Because DNA breathes, molecules containing palindromes and inverted repeats can in theory assume alternative structures (Figure 2-17). Once complementary strands have separated, intramolecular base pairing can cause a double-stranded branch to form between adjacent complementary sequences with, in rare cases, higher probability than the original base pairing. This structure is referred to as a **cruciform.** They have been produced in the laboratory but have not yet been detected in DNA isolated from cells.

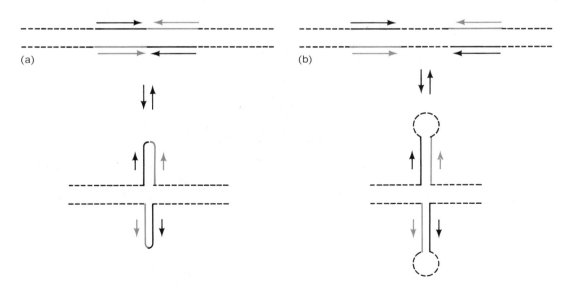

Figure 2-17 Possible alternative forms of a DNA molecule containing two inverted repeats that are (a) adjacent or (b) separated by a spacer. The horizontal arrows denote orientation of the sequences.

Figure 2-18 (a) Hairpin and (b) stem-and-loop structures that can form from two types of palindromes in single-stranded nucleic acids.

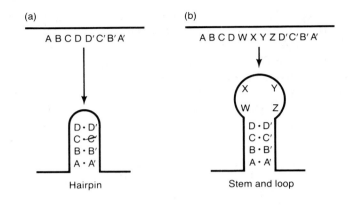

Both palindromes and interrupted inverted repeats have significant effects on the structure of single-stranded DNA (either obtained by denaturation or the natural single-stranded DNA found in certain phages) and on RNA. There is no simple mechanism with either single-stranded DNA or RNA to prevent intrastrand hydrogen bonding between adjacent or nearby complementary sequences. Thus, a palindrome produces an intrastrand double-stranded segment called a **hairpin** (Figure 2-18(a)), and an interrupted inverted repeat produces a structure consisting of a double-stranded segment with a terminal single-stranded loop, known as a **stem-and-loop** (Figure 2-18(b)). We will see in Chapters 6 and 7 that stem-and-loop structures are important in the regulation of RNA synthesis.

A sequence may also be repeated in the same orientation with or without a spacer.

$$\text{A B C D E U V W X Y Z A B C D E}$$
$$\text{A' B' C' D' E' U' V' W' X' Y' Z' A' B' C' D' E'}$$

These are called **direct repeats.** They typically consist of tens of base pairs and have large interruptions. They do not provide alternative structures to double-stranded DNA and have no effect on single-stranded molecules.

THE STRUCTURE OF RNA

A typical cell contains about ten times as much RNA as DNA. With the exception of the RNA of one known phage and a few viruses, RNA is a single-stranded polynucleotide. In bacteria there are three primary types of RNA—ribosomal RNA (of which there are three forms in bacteria), transfer RNA (of which there are about 50 different structures), and messenger RNA (of which there are almost as many different molecules as there are genes). All of these molecules superficially resemble single-stranded DNA in that single-stranded regions are interspersed with intramolecular double-stranded regions. Between 1/2 and 2/3 of the bases are paired. In single-stranded DNA the pairing is random and the paired regions tend to contain six or fewer pairs. Furthermore, if a sample of identical DNA molecules is denatured and intramolecular hydrogen bonds are allowed to form, the base-pairing pattern may differ from one molecule to the next. On the contrary, in RNA the double-stranded regions may contain up to 50 base pairs and a particular molecule has a definite base-pairing pattern. The structures of the different classes of RNA molecules will be discussed in Chapter 6.

NUCLEASES

Both DNA and RNA can be hydrolyzed to free nucleotides either chemically or enzymatically. For example, with both DNA and RNA, at pH 1 the phosphodiester bonds and the *N*-glycosidic bond between the base and the sugar are broken. Since free bases are produced, this procedure has been used in determining base composition.

Nucleases are enzymes that depolymerize nucleic acids. Most nucleases show chemical specificity and are either a deoxyribonuclease (DNase) or a ribonuclease (RNase). Many DNases act only on single-stranded or only on double-stranded DNA, though some degrade both kinds. Furthermore, some nucleases cleave only at the end of a nucleic acid, removing either a single nucleotide or a short oligonucleotide; these are called **exonucleases** and they may act specifically at either a 3′ or a 5′ terminus. **Endonucleases** act within the strand; some of these are specific in that they cleave only between particular bases.

Nucleases serve a variety of biological functions and have also been useful in the laboratory for removing unwanted nucleic acids and as an early step in determining the base sequence of DNA and RNA molecules.

Restriction endonucleases are a class of nucleases, each of which acts on a particular base sequence. They are exceedingly important in modern DNA technology and will be described in detail in Chapter 20.

METHODS USED TO STUDY MACROMOLECULES

Several methods of studying macromolecules are used repeatedly in molecular genetics. Some of these—e.g., spectrophotometry and chromatography—are usually covered in elementary chemistry courses and will not be discussed here. Three techniques—velocity sedimentation, equilibrium centrifugation in a density gradient, and gel electrophoresis—will be described briefly in this section. For additional details, the reader should consult the references at the end of this book.

Velocity Sedimentation

Several important properties of macromolecules can be determined from sedimentation data obtained with high-speed centrifuges.

The velocity with which a macromolecule moves is mainly a function of its molecular weight and its shape. The ratio of molecular velocity to centrifugal force is called the **sedimentation coefficient, s.** That is,

$$s = \text{velocity/centrifugal force}.$$

The unit of s, 10^{-13} seconds, is called one **svedberg** or one S. It is very common to refer to a molecule whose s value is 30 svedbergs as a 30S molecule.

The value of s for a particular molecule is often the same in many different solutions, so an s value is frequently considered to be a constant that characterizes a molecule. Furthermore, since the value of s depends on molecular weight and shape, changes in the s value, as experimental conditions are varied, can be used to monitor changes in molecular weight (such as aggregation or degradation) and in shape (such as unfolding an extended molecule).

The most common type of sedimentation in use today is **zonal centrifugation.** In this procedure (Figure 2-19), a centrifuge tube is filled with a sucrose solution whose concentration decreases continuously from the bottom of the centrifuge tube to the top of the tube. This density gradient stabilizes the liquid against mixing that might be caused by mechanical and convective disruptive forces. The density of the solution of molecules to be sedimented is adjusted to be lower than the density of the sucrose solution at the top of the tube, and the sample is layered on the surface, forming a band or zone. Because of the density gradient, this procedure is often called **sucrose gradient centrifugation.** After layering the sample, the tube is centrifuged in a swinging bucket rotor for a particular time. After centrifugation is completed, a tiny hole is punched in the bottom of the tube and drops of the solution are collected. These drops represent successive layers of the tube. The drops are assayed for a particular macromolecule to obtain the distribution of the concentration of this macromolecule along the tube. Some useful assays are radioactivity, optical absorbance, biological infectivity, and enzymatic activity.

Equilibrium Centrifugation

Another centrifugation technique, **equilibrium centrifugation in a density gradient,** is also widely used. In this procedure, the macromolecules (usually nucleic acids or virus particles) are suspended in a solution of CsCl whose concentration is chosen so that its density, ρ, is approximately equal to that of the macromolecules. During centrifugation the fairly heavy Cs^+ ions also sediment and, by a balance between sedimentation (tending to bring the

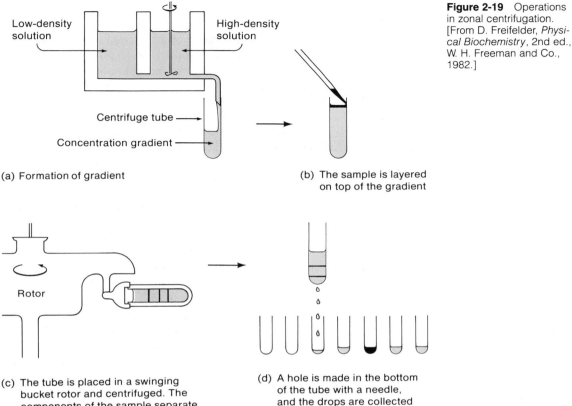

Figure 2-19 Operations in zonal centrifugation. [From D. Freifelder, *Physical Biochemistry*, 2nd ed., W. H. Freeman and Co., 1982.]

Low-density solution

High-density solution

Centrifuge tube

Concentration gradient

(a) Formation of gradient

(b) The sample is layered on top of the gradient

Rotor

(c) The tube is placed in a swinging bucket rotor and centrifuged. The components of the sample separate according to their *s* values.

(d) A hole is made in the bottom of the tube with a needle, and the drops are collected in a series of tubes.

Cs^+ ions to the bottom of the centrifuge tube) and diffusion (tending to produce equal concentrations everywhere), they form a nearly linear concentration (and hence, density) gradient in the centrifuge tube, with the maximum density at the bottom of the tube. The macromolecules move in the density gradient. Those in the upper reaches of the tube move toward the bottom, stopping at the position at which their density equals the solution density. Similarly, macromolecules in the lower part of the tube move upward to the same position. In this way the macromolecules form a narrow band in the tube. If the solution contains macromolecules having different densities, each macromolecule forms a band at the position in the gradient that matches its own density, and thus the macromolecules can be separated. The resolution of the technique is extraordinary. For example, DNA molecules with a density of 1.708 g/cm^3 can be separated from other DNA molecules in which the naturally occurring ^{14}N atoms have been substituted by ^{15}N atoms and which therefore have a density of 1.722 g/cm^3.

Gel Electrophoresis

Most biological macromolecules carry an electrical charge and thus can move in an electric field. For example, if the terminals of a battery are connected to the opposite ends of a horizontal tube containing a solution of positively charged protein molecules, the molecules will move from the positive end of the tube to the negative end. The direction of motion obviously depends on the sign of the charge but the rate of movement depends on the magnitude of the charge and, as in sedimentation, on the shape of the molecule (that is, its frictional resistance to movement). The mass of the molecule plays no direct role in the rate of migration (in contrast with sedimentation) and influences the rate only indirectly when the surface area of the molecule, which affects its frictional resistance, is a function of its mass.

The most common type of electrophoresis used in molecular genetics is zonal electrophoresis through a gel, or **gel electrophoresis.** This procedure can be performed such that the rate of movement depends only on the molecular weight of the molecule, as will be seen below.

An experimental arrangement for gel electrophoresis of DNA is shown in Figure 2-20. A thin slab of an agarose or polyacrylamide gel is prepared containing small slots ("wells") into which samples are placed. An electric field is applied and the negatively charged DNA molecules penetrate and move through the agarose. A gel is a complex network of molecules, and migrating macromolecules must squeeze through narrow, tortuous passages. The result is that smaller molecules pass through more easily and thus the

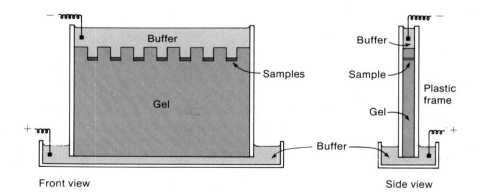

Front view   Side view

Figure 2-20 Apparatus for slab-gel electrophoresis capable of running seven samples simultaneously. The liquid gel is allowed to harden in place. An appropriately shaped mold is placed on top of the gel during the hardening in order to make wells for the samples (red). After electrophoresis, the slab is stained by removing the plastic frame and immersing the gel in a solution of the stain. Excess stain is removed by washing. The components of the sample appear as bands, which may be either visibly colored or fluorescent when irradiated with ultraviolet light. The region of the gel in which the components of one sample can move is called a *lane.* Thus, the gel shown has seven lanes. [After D. Freifelder, *Physical Biochemistry,* 2nd ed., W. H. Freeman and Co., 1982.]

migration rate increases as the molecular weight, M, decreases. For unknown reasons the distance moved, D, depends roughly logarithmically on M, obeying an equation of the form

$$D = a - b \log M,$$

in which a and b are empirically determined constants that depend on the buffer, the gel concentration, and the temperature. Figure 2-21 shows the result of gel electrophoresis of a collection of DNA molecules.

ISOLATION OF NUCLEIC ACIDS

Isolation of DNA is an essential step in many experiments. Although the methods are straightforward, the particular procedure must be tailored to the organism from which the DNA is to be obtained, because the structure and composition of organisms vary. The differences between the techniques follow from how the DNA is enclosed and the fraction of the total dry weight that is DNA (which varies from about 1 percent in mammalian cells to 10 percent in bacteria to about 50 percent in phages). The common feature in all procedures is that a cell or virus is first broken open, and then DNA is separated from such other components as protein, RNA, lipid, and carbohydrates. The basic procedures are the following:

1. *Phages*. The simplest procedure is employed with phages. An aqueous suspension of particles is shaken with phenol, a reagent that does not mix with water. A small amount of phenol enters the aqueous layer, breaking open the protein coat and denaturing the individual protein molecules. Most of the denatured protein either enters the phenol layer or precipitates at the phenol-water interface. The DNA remains in the aqueous layer. Addition of ethanol precipitates the DNA, which can be collected and redissolved in a solution having any desired composition.

2. *Bacteria*. The contents of bacterial cells are enclosed in a multilayered cell wall and an inner cell membrane. These structures cannot simply be removed by exposure to phenol but can be solubilized by successive treatment of a cell suspension with **lysozyme** (an enzyme isolated from chicken egg white) and one of several different detergents, the most common one being sodium dodecyl sulfate (SDS). The result of this treatment is release of the contents of the cells. RNA is then removed by digestion with specific enzymes (RNases), protein is removed both enzymatically and with phenol, and the DNA is collected by precipitation with ethanol.

With both phages and bacteria, covalently closed circular DNA is separated from linear DNA and nicked circles by equilibrium centrifugation in CsCl solutions containing ethidium bromide, as described in an earlier section (Experimental Detection of Covalently Closed Circles).

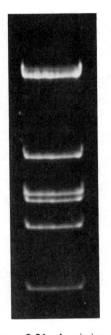

Figure 2-21 A gel electrophoregram of six fragments of *E. coli* phage λ DNA obtained by treating the DNA with the EcoRI restriction endonuclease. The direction of movement is from top to bottom. The DNA is made visible by the fluorescence of bound ethidium bromide. [Courtesy of Arthur Landy and Wilma Ross.]

Isolation of RNA follows the procedure used for DNA, with one major change. After the cells are broken open, DNA is degraded with a DNase. Protein is not usually removed with phenol since a large fraction of the RNA enters the phenol layer and becomes contaminated with protein. Treatment with certain enzymes that degrade proteins (**proteases**) can be used to remove protein. Alternatively, after DNase treatment the solution can be shaken vigorously with chloroform. This solvent is immiscible with water, and shaking creates a large surface area at the chloroform-water interface. Precipitated protein gathers at this interface and is easily removed.

Many experiments require that the complementary strands of a particular DNA molecule be separately purified. This can be accomplished in several ways. Details can be found in references given at the end of the book.

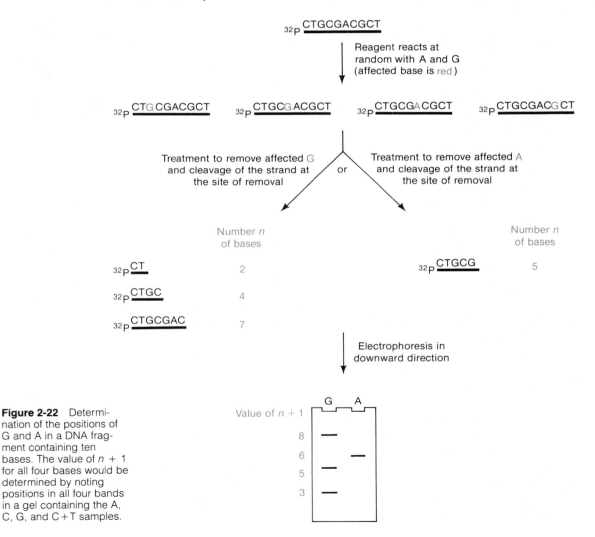

Figure 2-22 Determination of the positions of G and A in a DNA fragment containing ten bases. The value of $n + 1$ for all four bases would be determined by noting positions in all four bands in a gel containing the A, C, G, and C + T samples.

DETERMINATION OF THE BASE SEQUENCE OF DNA

A great deal of information about gene function and about other questions in microbial genetics has come from direct determination of the base sequence of DNA. Details of the two procedures in common use are beyond the scope of this book, but the rationale behind one procedure—the Maxam-Gilbert technique—is fairly straightforward (Figure 2-22). The procedure begins by cleavage of a DNA segment of interest into a set of overlapping fragments containing several hundred base pairs each. The technique described now applies to each strand. First, the two complementary strands are separated, and then each strand is made radioactive at the 5′ end. Next, four samples of a solution containing only one of the two complementary strands are subjected to four distinct chemical treatments that cause phosphodiester cleavage at the 5′ side of a particular nucleotide. Reaction conditions are chosen so that only one cleavage occurs in each strand. The number of molecules in the sample is so large that all potential cleavage sites are cut, which results in a set of fragments whose lengths are determined by the cleavage sites. The fragments are then separated by gel electrophoresis, using a gel system by which fragments differing in length by one nucleotide are resolved. The individual bands in the gel are identified by their radioactivity (that is, the nonradioactive fragment derived from each cleavage event is not seen). The result of this protocol is that the existence of a band corresponding to, say, 16 nucleotides among the set of fragments obtained by the treatment that cleaves next to guanine, for example, means that the 17th base from the 5′ end is guanine. Actually the four chemical treatments do not by themselves identify each base. Instead, they identify purines (A or G), G, pyrimidines (C or T), and C. An example of a sequencing gel is shown in Figure 2-23. G is identified by the presence of an intense band in the G column and a weak band in the A (actually the purine) column; A is represented by a band only in the A column; T is identified by the presence of a band in the C + T column; and C is found by noting a band in both the C + T and C columns. The base sequence is read directly from the gel; the bottom of the gel represents the 5′ terminus. Note that only one of the complementary strands has to be sequenced; however, usually both are examined, so the two sequences determined serve as a check on one another.

Figure 2-23 A portion of a sequencing gel. The sequence is read from the bottom to the top. Each horizontal row represents a single base. Each vertical column represents a sample treated to indicate the position of G, A or G, T or C, and C, respectively. The sequence from a portion of the gel is shown.

PROBLEMS

1. Which compound, a nucleoside or a nucleotide, contains a phosphate group?

2. Which bases in nucleic acids are purines and which are pyrimidines?

3. Which nucleic acid base is unique to DNA? to RNA?

4. How do ribose and 2′-deoxyribose differ?

5. What is the name of the bond formed between a N atom in a purine or pyrimidine and a C atom in ribose or deoxyribose?

6. In a nucleic acid which carbon atoms are connected by a phosphodiester group?

7. What chemical groups are at the end of a polynucleotide?

8. How many phosphate groups are there per base in DNA and in RNA?

9. How many hydrogen bonds are there in an AT and a GC base pair?

10. What chemical groups are at the end of a single polynucleotide strand?

11. How many turns of a double helix are there in a molecule consisting of 45 base pairs?

12. How many bases are in a molecule whose length is 2 micrometers?

13. In DNA, what is the relation between the sums ($[A] + [G]$) and ($[T] + [C]$), in which [ ] denotes mole fraction?

14. When referring to helices, what is meant by right-handed and left-handed? What is the handedness of a normal DNA helix?

15. In what sense are the two strands of DNA antiparallel?

16. What term is used to describe a DNA sequence (in one of the complementary strands) such as AGACGTCT?

17. What is meant by a nuclease, and how do endonucleases and exonucleases differ?

18. Can an exonuclease depolymerize a supercoiled DNA molecule?

19. Which of the following operations will induce formation of a supercoil: joining the ends of a linear DNA and doing nothing else; twisting the two ends of a linear DNA and then joining the ends together; joining the ends of a linear DNA and then twisting the circle?

20. A technique is used for determining base composition. Rather than giving mole fractions for individual bases, it yields the value of $[A]/[C]$. If this value is 1/3, what fraction of the bases are A?

21. Analysis of a DNA sample from a bacterium indicates that 18% of the bases are A. What fraction is C?

22. Could a single-stranded DNA molecule with base sequence 5'-GATTG-CCGGCAATC-3' fold back on itself to form a hairpin?

23. What would be the approximate molecular weight of a DNA molecule whose length is 16.4 μm?

24. One of the complementary strands of two DNA molecules is given below. Which DNA molecule would have the lower temperature for strand separation? Why?

    (1) AGTTGCGACCATGATCTG   (2) ATTGGCCCCGAATATCTG

25. ^{15}N-labeled DNA from phage T4 is mixed with T4 DNA of normal density. The solution is then heat-denatured and renatured. The resulting DNA is analyzed by centrifugation in a CsCl density gradient. How many bands will be observed and what will their relative proportions be?

26. Bacterial DNA can be density-labeled if the bacteria are grown in a medium containing a heavy isotope (such as ^{15}N or ^{13}C). If both strands are so labeled, the DNA is said to be heavy (HH) as opposed to normally light (LL) DNA. If equal amount of HH and LL DNA are mixed, heated to 100°C, and slowly cooled to allow renaturation to occur, 25% of the DNA will be HH, 25% will be LL and 50% will be HL (hybrid). Suppose 45 μg of HH DNA and 5 μg of LL DNA are mixed, heated to 100°C, and renatured, how many micrograms of HH, HL, and LL will result?

27. Consider a long linear DNA molecule, one end of which is rotated four times with respect to the other end, in the unwinding direction. The two ends are then joined.

    (a) If the molecule is to remain in the underwound state, how many base pairs will be broken?
    (b) If the molecule is allowed to form a supercoil, how many nodes will be present?

28. Suppose ten protein molecules are bound to a circular DNA molecule having a nick. Each bound protein molecule breaks one base pair. The nick is then sealed with DNA ligase (an enzyme that seals single-strand breaks having a free 3'-OH and 5'-P group) and the protein molecules are removed. What will be the shape of the DNA molecule after removal of the protein?

29. The sedimentation velocity properties of a supercoiled DNA molecule are being studied as a function of the concentration of added ethidium bromide. It is found that s decreases, reaches a minimum, and then increases. Explain. (Hint: Recall that s is a function of both molecular weight and shape.)

30. In gel electrophoresis what feature of the method causes DNA molecules to move at a rate that is dependent on their molecular weight? Do larger molecules move more slowly or faster than smaller molecules?

31. A DNA fragment containing 17 base pairs is sequenced by the Maxam-Gilbert procedure. The data are shown below; panels 1 and 2 correspond to the two complementary strands. Note that the 5'-terminal base does not appear in either lane, as it would have to be identified by a sugar-phosphate lacking a base; such molecules do not electrophorese with the nucleotides. What are the complete sequences of the two complementary strands, including the 5'-terminal bases?

Chapter 3

Proteins

All proteins are polymers of amino acids, yet each species of protein molecule has a unique three-dimensional structure determined principally by the amino acid sequence, in contrast with DNA, which has a universal structure—the double helix. This makes the study of proteins very complex, however, the diversity of protein structures enables these molecules to catalyze the thousands of different processes required by a cell.

The study of the detailed structure of proteins is beyond the scope of this book, and only a brief review will be given. For further information the reader should consult the references at the end of this book.

CHEMICAL STRUCTURE OF A POLYPEPTIDE CHAIN

A typical protein molecule consists of one or more polypeptide chains. The building blocks of a polypeptide are 20 different amino acids. Most of these amino acids have the basic structure shown in Figure 3-1—namely, a carbon

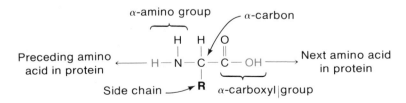

Figure 3-1 Basic structure of an α-amino acid. The NH_2 and COOH groups are used to connect amino acids to one another. The red OH of one amino acid and the red H of the next amino acid are removed when two amino acids are linked together (see Figure 3-3).

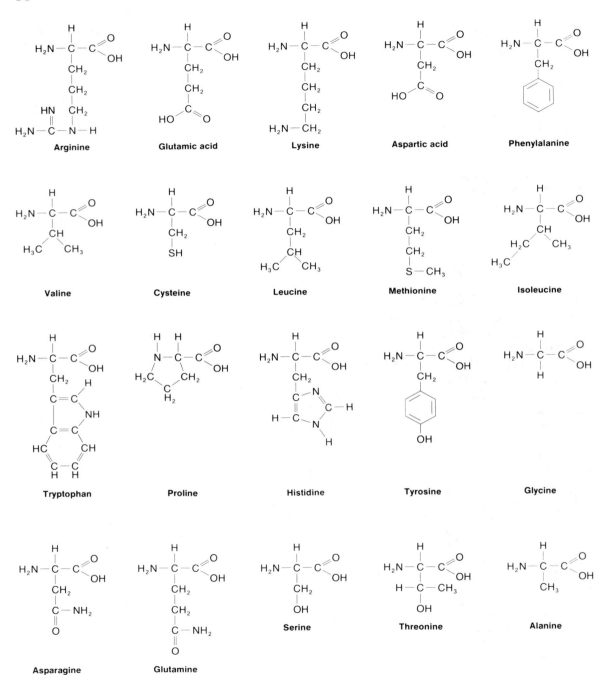

Figure 3-2 Chemical structures of the amino acids.

atom (the α carbon) to which is attached an amino group, a carboxyl group, and a **side chain,** usually denoted R (Figure 3-2). (Only proline does not have this structure.) The side chains are of several types—charged, uncharged, rings, sulfhydryl-containing, etc.—and the chemical and physical properties of each amino acid are determined mainly by these side chains. In a polypeptide chain amino acids are covalently joined via the amino group of one and the carboxyl group of the adjacent one, forming a **peptide bond** (Figure 3-3(a)). Thus, a polypeptide is a polymer of amino acids in which α-carbon atoms and peptide groups alternate to form a linear backbone with the side chains projecting from each α-carbon atom (Figure 3-4), without participating in the backbone structure. The term "linear" refers to the fact that the backbone is not branched. However, as will be seen in the following sections, a polypeptide chain is highly folded and can assume a variety of three-dimensional shapes. Each shape in turn consists of several standard elementary three-dimensional conformations and other conformations that may be unique to that molecule.

A typical polypeptide chain contains about 300–700 amino acids (molecular weight, 30,000–70,000), though both smaller and larger chains are known.

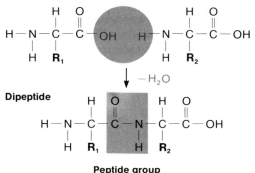

Figure 3-3 Formation of a dipeptide from two amino acids by elimination of water (shaded circle) to form a peptide group (shaded rectangle).

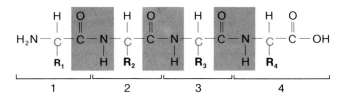

Figure 3-4 A tetrapeptide showing the alternation of α-carbon atoms (red) and peptide groups (shaded). The four amino acids are numbered below.

Polypeptides containing more than 1000 amino acids are rare. Many proteins contain several polypeptide chains, a phenomenon that will be discussed later. The largest proteins of this type have molecular weights of about 500,000. The largest polypeptides are small compared to typical nucleic acid molecules (molecular weight, 10^6–10^{10}).

PHYSICAL STRUCTURE OF A POLYPEPTIDE CHAIN

A fully extended polypeptide chain, if it were to exist, would have the conformation shown in Figure 3-5. (The chain is not perfectly straight because the C—N and C—C bonds in which the α-carbon atom participates are not colinear.) Such an extended zig-zagged molecule could not exist without stabilizing interactions to maintain the extension. In fact, a single polypeptide is never completely extended but is folded in a complex way, as will be seen in the following section.

Folding of a Polypeptide Chain

Three rules govern the manner of folding:

1. The peptide bond has a partial double-bond character and hence is constrained to be planar. Free rotation occurs only between the α-carbon atom and the peptide group. Thus, the polypeptide chain is

Figure 3-5 The conformation of a hypothetical fully extended polypeptide chain. The length of each amino acid residue is 3.61 Å; the repeat distance is 7.23 Å. The α-carbon atoms are shown in red. Side chains are denoted by R.

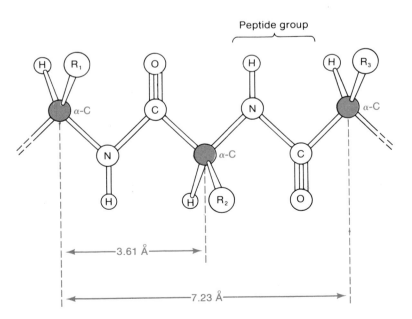

flexible but is not as flexible as would be the case if there were free rotation about all of the bonds.

2. The side chains of the amino acids cannot overlap. Thus, the path of the backbone can never be truly random because certain orientations are forbidden.

3. Two charged groups having the same sign will not be very near one another. Thus, like charges tend to cause extension of the chain.

In addition to these rules, folding behaves according to general tendencies, a few of which are the following:

1. Amino acids with polar side chains tend to be on the surface of the protein in contact with water.

2. Amino acids with nonpolar side chains tend to be internal. Very hydrophobic side chains tend to cluster.

3. Hydrogen bonds tend to form between the carbonyl oxygen of one peptide bond and the hydrogen attached to a nitrogen atom in another peptide bond. This hydrogen-bonding gives rise to two fundamental polypeptide structures called the α helix and the β structure, which will be described in the following section.

4. The sulfhydryl group of the amino acid cysteine (Figure 3-2) tends to react with an —SH group of a second cysteine to form a covalent S—S (**disulfide**) bond. Such bonds pose powerful constraints on the structure of a protein (Figure 3-6). Most proteins contain several cysteines and it is not yet generally possible to predict which cysteines will be paired.

These tendencies should not be considered invariant because there are many exceptions. Nevertheless they indicate that *the structure of a protein may be changed markedly by a single amino acid substitution*—for example, a polar amino acid for a nonpolar one; similarly, the change might be minimal if one nonpolar amino acid replaces another nonpolar one. This notion will be encountered again in Chapter 10 when mutations are considered.

The three-dimensional shape of a polypeptide chain is a result of a balance between each of the rules and tendencies just described and can be very complex. However, in examining many polypeptide chains, it has become apparent that certain geometrically regular arrays of the chain are found repeatedly in different polypeptide chains and in different regions of the same chain. These are the arrays resulting from hydrogen-bonding between different peptide groups. They are described in the next section.

Protein folding is exceedingly important since the properties of most proteins are determined by the folding. For example, the catalytic sites of enzymes are formed by bringing together distant regions of a polypeptide chain. Disruption of the precise pattern of folding invariably results in loss of biological activity of proteins.

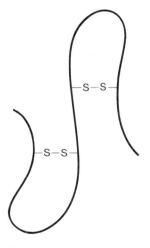

Figure 3-6 A polypeptide chain in which four cysteines are engaged in two disulfide bonds.

Hydrogen-Bonded Conformations:
The α Helix and the β Structure

In the absence of any interactions between different parts of a polypeptide chain, free rotation of each bond, except for the peptide bond, would occur continually, and the chain would assume a large number of changing conformations collectively called a **random coil.** However, interactions do occur; for example, hydrogen bonds easily form between the H of the N—H group of the peptide group and the O of the carbonyl of another peptide unit. In the absence of all side-chain interactions the most stable hydrogen-bonded structure of this sort is the **α helix.** In the α helix the polypeptide chain follows a helical path that is stabilized by hydrogen-bonding between peptide groups. Each peptide group is hydrogen-bonded to two other peptide groups, one three units ahead and three units behind the chain direction (Figure 3-7). The helix has a repeat distance of 5.4 Å, a diameter of 2.3 Å, and contains 3.6 amino acids per turn. Thus, it is a much tighter helix than the DNA helix. The α helix is the preferred form of a polypeptide chain because, in this structure, all monomers are in an identical orientation and each one forms the same hydrogen bonds as any other monomer. Thus, polyglycine, which

Figure 3-7 Properties of an α helix. (a) The two hydrogen bonds in which peptide group 4 (red) is engaged. The peptide groups are numbered below the chain. (b) An α helix drawn in three dimensions, showing how the hydrogen bonds stabilize the structure. The red dots represent the hydrogen bonds. The hydrogen atoms that are not in hydrogen bonds are omitted for the sake of clarity.

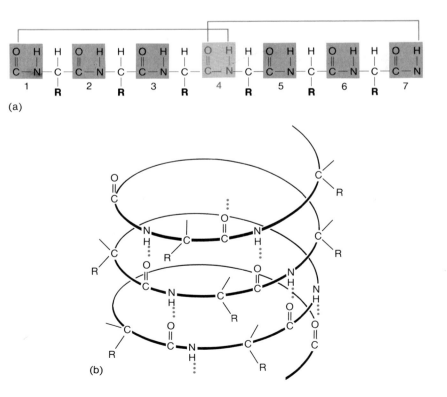

lacks side chains (the R group of glycine is a hydrogen atom) and hence cannot participate in any interactions other than those just described, is an α helix. Study of polylysine indicates how the composition of the medium can affect protein structure. Lysine, which also has an amino group in its side chain, is charged in a certain pH range and uncharged otherwise. When uncharged, it forms an α helix. However, if the pH is altered and the side chain becomes charged, the repulsion caused by the like charges destroys the helical structure and the molecule becomes highly extended.

If the amino acid composition of a real protein is such that the helical structure is extended a great distance along the polypeptide backbone, the protein will be somewhat rigid and fibrous (not all rigid fibrous proteins are α helical, though). This structure is common in many structural proteins, such as the α-keratin in hair.

Another common hydrogen-bonded conformation is the **β structure.** In this form, the molecule is almost completely extended (repeat distance = 7 Å) and hydrogen bonds form between peptide groups of polypeptide segments lying adjacent and parallel with one another (Figure 3-8(a)). The side chains lie alternately above and below the main chain.

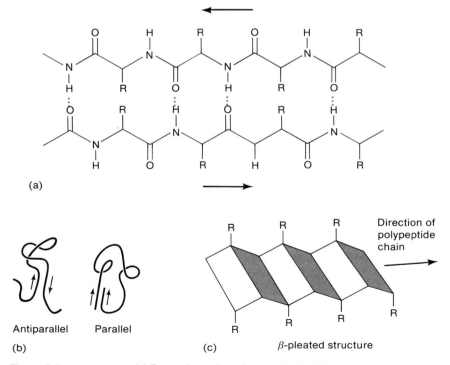

(a)

(b) Antiparallel   Parallel

(c)   β-pleated structure

Direction of polypeptide chain

Figure 3-8 β structures. (a) Two regions of nearly extended chains are hydrogen-bonded (red dots) in an antiparallel array (arrows). The side chains (R) are alternately up and down. (b) Antiparallel and parallel β structures in a single molecule. (c) A large number of adjacent chains forming a β pleat.

Two segments of a polypeptide chain (or two chains) can form two types of β structure, which depend on the relative orientations of the segments. If both segments are aligned in the N-terminal-to-C-terminal direction or in the C-terminal-to-N-terminal direction, the β structure is **parallel.** If one segment is N-terminal to C-terminal and the other is C-terminal to N-terminal, the β structure is **antiparallel.** Figure 3-8(b) shows how both parallel and antiparallel β structures can occur within a single polypeptide chain.

When many polypeptides interact in the way just described, a pleated structure results called the **β-pleated sheet** (Figure 3-8(c)). These sheets can be stacked and held together in rather large arrays by van der Waals forces and are often found in fibrous structures such as silk.

Fibrous and Globular Proteins

Few proteins are pure α helix or β structure; usually regions having each structure are found within a protein. Since these conformations are rigid, a protein in which most of the chain has one of these forms is usually long and thin and is called a **fibrous protein.** In contrast are the quasi-spherical proteins called the **globular proteins** in which α helices and β structures are short and interspersed with randomly coiled regions; numerous intrastrand interactions create a compact structure.

The fibrous proteins are typically responsible for the structure of cells, tissues, and organisms. Some examples of structural proteins are collagen (the protein of tendon, cartilage, and bone), and elastin (a skin protein). Some of the fibrous proteins are not soluble in water—examples are the proteins of hair and silk.

The catalytic and regulatory functions of cells are performed by proteins that have a well-defined but deformable structure. These are the globular proteins, of which the catalytic proteins, or enzymes, and the regulatory proteins are the largest group. Globular proteins are compact molecules having a generally spherical or ellipsoidal shape. Large segments of the polypeptide backbone of a typical globular protein are α-helical. However, the molecule is extensively bent and folded. Usually, the stiffer α-helical segments alternate with very flexible randomly coiled regions, which permit bending of the chain without excessive mechanical strain. Numerous segments of the chain, which might be quite distant along the backbone, form short parallel and antiparallel β structures; these also are responsible in part for the folding of the backbone. The α helix and β structures are called the **secondary structure** of the molecule. The extensive folding of the backbone is usually called the **tertiary structure** or **tertiary folding.** The amino acid sequence is the **primary structure.**

A very important distinction can be made between secondary and tertiary structures; namely, secondary structure results from hydrogen-bonding between peptide groups, whereas tertiary structure is formed from α helices and β structures and several different side-chain interactions.

The most prevalent interactions responsible for tertiary structure are the following:

1. Ionic bonds between oppositely charged groups in the side chains.
2. Hydrogen bonds between the hydroxyl group in tyrosine and a carboxyl group of aspartic or glutamic acids.
3. Hydrophobic clustering between the hydrocarbon side chains of phenylalanine, leucine, isoleucine, and valine.
4. Metal-ion coordination complexes between amino, hydroxyl, and carboxyl groups, ring nitrogens, and pairs of SH groups.

Hydrophobic clustering (item 3) is the most important stabilizing feature.

Figure 3-9 shows a schematic diagram of a hypothetical protein (in two dimensions) in which several of these interactions determine the structure.

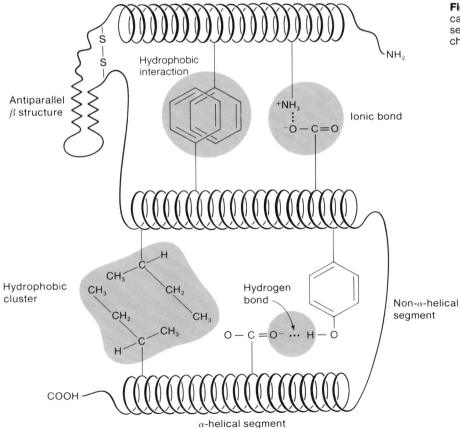

Figure 3-9 A hypothetical globular protein having several types of side-chain interactions.

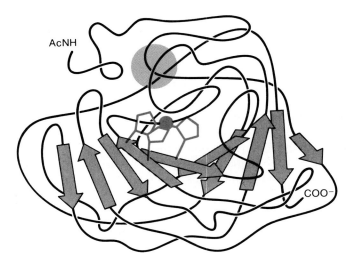

Figure 3-10 An idealized drawing of the tertiary structure of human carbonic anhydrase. Shown are the peptide chain and the three histidines (red) that coordinate to a zinc ion at the active site. Individual β-sheet strands are drawn as arrows from the amino to carboxyl ends. Note the twist of the β sheets. A hydrophobic cluster is shaded in red. [After K. K. Kannan et al. 1971. *Cold Spring Harbor Symp. Quant. Biol.*, 36: 221.]

This figure should be examined carefully for it indicates the role of different features of a protein molecule in determining the overall conformation of the molecule. One can see the following:

1. Most interactions, except for those between like charges, bring distant amino acids together.
2. Hydrogen bonds sometimes bring distant amino acids together, but usually a single hydrogen bond makes a more subtle change in position.
3. Both β structures and α helices make portions of the polypeptide backbone stiff and linear.

Not shown are van der Waals forces, which produce specific interactions between clusters of amino acids which may or may not be nearby in the polypeptide chain.

Figure 3-10 is an idealized drawing of the three-dimensional structure of the enzyme carbonic anhydrase, showing a β structure, a hydrophobic cluster, and three amino acids in a coordination bond with a metal ion.

PROTEINS WITH SUBUNITS

A polypeptide chain usually folds so that nonpolar side chains are internal—that is, out of contact with water. However, it is rarely possible for a polypeptide chain to fold in such a way that *all* nonpolar groups are internal.

Thus, nonpolar amino acids on the surface often form clusters in an effort to minimize contact with water. A polypeptide having a large hydrophobic patch can further reduce contact with water by pairing with a hydrophobic patch on another polypeptide. Similarly, if a molecule has several distantly located hydrophobic patches, a structure consisting of several polypeptides in contact effectively minimizes contact with water. The protein is then said to consist of identical **subunits;** this is in fact a very common phenomenon, with two, three, four, and six subunits occurring most frequently. (A multisubunit protein may also contain unlike subunits. This will be discussed shortly.)

One can take the point of view expressed above that multisubunit proteins exist because they dispose of hydrophobic groups in an efficient way. An alternative and probably more correct view is that many proteins have evolved so that folding forms hydrophobic patches that allow subunit assembly to occur. Such evolution has been valuable because multisubunit systems have three important advantages: (1) Subunits provide an economical way to utilize DNA. That is, if a large protein is built from several identical subunits, only a fairly small amount of DNA is needed to encode the protein. (2) The activity of multisubunit proteins can be very efficiently and rapidly switched on and off. How this occurs (allosteric effects) is described in references given at the end of the book. (3) Biologically active sites—for example, catalytic sites of enzymes—can often be produced between regions brought nearby by subunit formation; these same sites might not be formable within a single polypeptide chain.

The multisubunit proteins that have been described so far consist of several identical subunits. Different polypeptide chains can also aggregate and form proteins made up of nonidentical subunits, and in fact this is quite common. For example, hemoglobin, the oxygen carrier of blood, consists of four subunits, two each of two different types; likewise, RNA polymerase, which catalyzes synthesis of RNA, has five subunits of which four are different. The subunit structure of a protein is occasionally called **quaternary structure.**

ENZYMES

Enzymes are special proteins that catalyze chemical reactions. Their catalytic power exceeds all man-made catalysts. A typical enzyme accelerates a reaction 10^8- to 10^{10}-fold, though some highly active enzymes increase the reaction rate by a factor of 10^{15}. Enzymes are also exceedingly specific in that each catalyzes only a single reaction or set of closely related reactions. Furthermore, only a small number of reactants, often only one, can participate in a single catalyzed reaction. Since nearly every biological reaction is catalyzed by an enzyme, a very large number of distinct enzyme molecules are required.

The product of many genes is an enzyme. Hence the ability to measure enzyme activity is an important part of microbial genetics, since enzyme activity can be a reliable measure of gene expression. There are numerous ways to study enzyme reactions. Two of these procedures, optical assays and radioactivity assays, are very commonly used. In an optical assay of an enzyme, some component in the reaction mixture is detected by the ability of that component to absorb light of a particular wavelength. An important and commonly used optical assay is that for the enzyme β-galactosidase, which we will encounter frequently in this book.

The enzyme β-galactosidase catalyzes cleavage of lactose to form galactose and glucose:

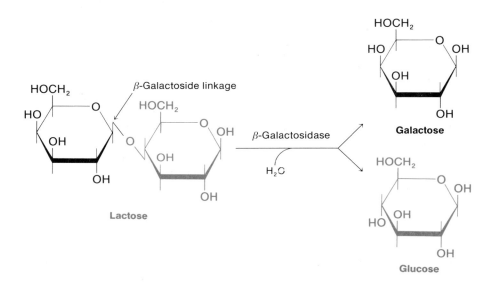

The bond broken is a β-galactoside linkage, as indicated by the arrow. To detect this cleavage the synthetic substrate (a **substrate** is a substance acted on by an enzyme) *o*-nitrophenyl galactoside (ONPG) is used. A β-galactoside linkage is also present in ONPG, so β-galactosidase can hydrolyze ONPG. The value of this substrate is that it is colorless but, when cleaved, yields galactose and *o*-nitrophenol, which is intensely yellow:

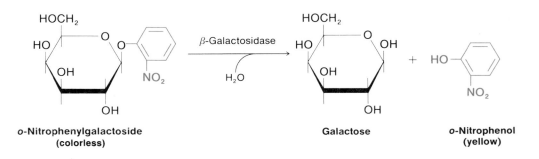

Thus, the activity of the enzyme can easily be followed by assaying the concentration of *o*-nitrophenol at a wavelength of 420 nm (blue light).

In a radioactivity assay a reactant that is radioactive is added to a reaction mixture and either the appearance of another radioactive substance or the loss of the radioactive reactant is measured. There is enormous variety in such assays. One of the most common assays, one used to measure polymerization, is based upon the fact that all proteins and nucleic acids are insoluble in 1 *M* trichloracetic acid (TCA), whereas all amino acids and nucleotides are TCA-soluble. This property allows one to measure protein synthesis by adding a radioactive (e.g., ^{14}C) amino acid to a reaction mixture containing the other 19 nonradioactive amino acids and the appropriate enzymes and factors. After a period of time, TCA is added and the mixture is filtered and washed with TCA. The ^{14}C-labeled amino acid is soluble and passes through the filter but if ^{14}C-protein has been made, it will be precipitated and ^{14}C will be retained on the filter. Counting the filter-bound radioactivity is then a measure of the extent of reaction.

PROBLEMS

1. What is meant by a side chain of an amino acid?

2. Which amino acid has no side chain, that is, only a hydrogen atom as a side chain?

3. What chemical groups are present at the termini of polypeptide chains?

4. Only one amino acid side chain engages in covalent bonds between different polypeptide segments. Which one is it?

5. Why is the structure of a random coil not constant in time?

6. In simple terms what is the nature of a hydrophobic interaction?

7. What term describes the tendency of a macromolecule to fold so that nonpolar groups cluster?

8. Name the kinds of bonds in proteins in which each of the following amino acids might participate: cysteine, lysine, isoleucine, glutamic acid.

9. Which atoms are covalently linked in a peptide bond?

10. Which amino acid, serine or alanine, will tend to be internal and which will be external?

11. About which bond in the polypeptide backbone is there no free rotation?

12. Which amino acids are linked in a disulfide bond?

13. Which of the following sets of three amino acids are probably clustered within a protein? (1) Asn, Gly, Lys; (2) Met, Asp, His; (3) Phe, Val, Ile; (4) Tyr, Ser, Lys; (5) Ala, Arg, Pro.

14. In an α helix, what fraction of the peptide bonds are engaged in hydrogen bonds?

15. Which of the following bonds or interactions are essential for forming an α helix and a β structure: hydrogen bonds, van der Waals forces, hydrophobic interactions, ionic bonds?

16. Which of the following functional classes of proteins is more likely to be fibrous rather than globular: enzymes, regulatory proteins, structural proteins?

17. What type of interaction is primarily responsible for the joining of subunits?

18. Most proteins completely unfold when organic solvents are added. Explain this phenomenon.

19. How would you expect the ratio of polar to nonpolar amino acids to change, on the average, with increasing size of a typical protein? (Think about the relative change in surface area and volume.)

20. Glycine is often found in regions of a polypeptide that are very tightly folded. Why might this be the case?

21. What would you guess to be the environment of a glutamine that is internal? What is the environment of an internal lysine?

Bacteria

Bacteria are free-living unicellular organisms. They have a single chromosome, which is not enclosed in a nucleus (they are prokaryotes), and compared to eukaryotes they are simple in their physical organization. Bacteria have many features that make them suitable objects for the study of fundamental biological processes. For example, they are grown easily and rapidly and, compared to cells in multicellular organisms, they are relatively simple in their needs. The bacterium that has served the field of molecular genetics best is *Escherichia coli* (usually referred to as *E. coli*), though other bacterial species have also been important. The principal features of these bacteria that make them useful are their lack of pathogenicity (disease-causing potential), simplicity of growth, and short life cycle (20–25 minutes under optimal laboratory conditions). How one grows these bacteria and counts their number and some of the essential properties of bacteria are described in this chapter.

GROWTH OF BACTERIA

Bacteria can be grown in a variety of media and by several techniques. Careful control of growth conditions, such as media composition, pH, and aeration, is necessary to achieve bacterial populations with reproducible characteristics. Some of these procedures and several parameters that are useful in describing the growth of a bacterial population comprise this section.

Growth Media

Bacteria can be grown in a **liquid growth medium** or on a solid surface. A population growing in a liquid medium is called a bacterial **culture.** A culture is initiated by placing a small amount of bacteria—an **inoculum**—into sterile medium in a flask or tube. If the liquid is a complex extract of biological material, it is called a **broth.** An example is tryptone broth, which is the milk protein casein hydrolyzed by the digestive enzyme trypsin to yield a mixture of amino acids and small peptides. Another common broth is prepared from an extract of beef. If the growth medium is a well-defined mixture containing no organic compounds other than a carbon source (such as a sugar), it is called a **minimal medium.** A typical minimal medium contains each of the ions, Na^+, K^+, Mg^{2+}, Ca^{2+}, Fe^{2+}, NH_4^+, Cl^-, phosphate buffered to neutral pH, and SO_4^{2-}, and a source of carbon atoms such as glucose or glycerol. Other metal ions are also required but in such small quantities that they are usually provided as contaminants in the salts used to prepare the medium. The best source of carbon is glucose; in minimal media containing glucose, bacteria grow more rapidly than in a minimal medium with any other carbon-containing compound.

   If a bacterium can grow in a minimal medium—that is, if it can synthesize *all* necessary organic substances, such as amino acids, vitamins, and lipids—the bacterium is said to be a **prototroph.** If any organic substances other than a carbon source must be added for growth to occur, the bacterium is termed an **auxotroph.** For example, if the amino acid leucine is required in the growth medium, the bacterium is a leucine auxotroph; the genetic (phenotypic) symbol for such a bacterium is Leu^-. A prototroph would be indicated Leu^+.

   Bacteria are frequently grown on solid surfaces. The earliest surface used for growing bacteria was a slice of raw potato. This was later replaced by media solidified by gelatin. Because many bacteria excrete enzymes that digest gelatin, an inert gelling agent was sought. **Agar,** which is a gelling agent obtained from a particular seaweed, is resistant to nearly all bacterial enzymes and is now universally used. A solid growth medium is called a **nutrient agar,** if the liquid medium is a broth, or a **minimal agar,** if a minimal medium is solidified. Solid media are typically prepared by pouring a hot molten agar solution into a **petri dish** and allowing the medium to cool and harden. In lab jargon a petri dish containing a solid medium is called a **plate,** and the act of depositing bacteria on the agar surface is called **plating.**

Some Parameters of a Bacterial Culture

When bacteria are inoculated in a liquid medium, they slowly start to grow and divide. After an initial period of slow growth called the **lag phase,** they begin a period of rapid growth in which they divide at a fixed time interval called the **doubling time.** The number of cells per milliliter, the **cell density,** doubles repeatedly, giving rise to a logarithmic increase in cell number; this stage of growth of the bacterial culture is called the **exponential** or **log phase.** This stage of growth continues until a cell density (for *E. coli*) of about 10^9 cells/ml is reached, at which point nutrients and O_2 become limiting and the growth rate decreases. (The maximum cell density may be five- to tenfold lower for many other bacterial species.) Ultimately, at a cell density of 2 to 3×10^9 cells/ml, no further growth is possible and the cell number becomes constant. This stage is the **stationary phase.** A typical growth curve for a bacterial culture is shown in Figure 4-1.

The doubling time of a culture varies with temperature and the composition of the growth. The optimal temperature depends on the bacterial species. For *E. coli* and for bacteria whose natural host is an animal, the optimal temperature is roughly 37°C. Those whose habitat is the soil may be within 1–2° of 30°C. Maximum growth rates are invariably achieved with broths, because the bacteria does not need to utilize energy to synthesize a large number of essential organic compounds. In fairly rich broths *E. coli* has a doubling time of 22–25 minutes. It is interesting to note how quickly a culture multiplies. For instance, one bacterium with a doubling time of 22 minutes passes through 30 generations in 11 hours and generates 10^9 cells. In a sufficiently large culture vessel a single bacterium could multiply in less than two days to form a number of bacteria equal in mass to that of the earth.

In minimal media the rate-limiting substance in the growth medium is usually the carbon source. With glucose, the optimal carbon source, the

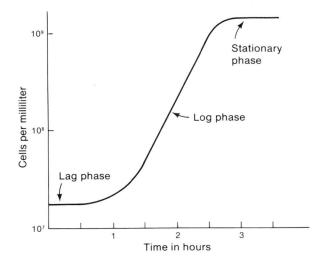

Figure 4-1   A typical growth curve for a bacterial culture. The *y* axis is logarithmic, so the curve is a straight line in log phase.

doubling time of *E. coli* at 37°C is about 45 minutes. With less efficiently utilized sugars the doubling time can be as much as 60 minutes, and with carbon compounds that are not sugars, several hours can be required for a culture to double in cell density. Most experiments in microbial genetics are carried out with media in which the doubling time is less than 60 minutes.

In order to maintain continual growth a culture must be repeatedly diluted. If a stationary-phase culture is diluted, it enters lag phase. However, if a log-phase culture is diluted into fresh medium, already warmed to the same temperature at which the cells have been growing, the culture remains in log phase. Cultures can be maintained in continuous growth by repeated dilution during log phase.

Chemostat Cultures

Some years ago a technique was developed for maintaining a culture at a constant cell density, using a modified culture vessel called a **chemostat.** The technique can only be applied to an auxotroph. With a chemostat fresh medium drips into the culture vessel at a carefully controlled and constant rate. Each drop causes a drop of the culture to overflow through a siphon. A nutrient required by the auxotroph is provided in the medium at a concentration sufficiently low that cell growth is limited by it. Furthermore, the rate of flow of the fresh medium is adjusted to be so low that the cell population can fully utilize all of the nutrient; that is, the nutrient is never present at a concentration that is not growth-limiting. In such a system the flow rate of the fresh medium determines the growth rate of the culture. Since the mass of the bacteria remains constant in a chemostat, growth of the culture is linear, not exponential. Hence, the time required to add a volume of liquid equal to the capacity of the chemostat is the doubling time; that is, if liquid flows into the vessel at a rate of 0.2 ml/minute and the capacity of the chemostat is 25 ml, the doubling time will be $25/0.2 = 125$ minutes. Chemostat cultures are useful in measuring mutation rates, as will be seen in Chapter 10.

Synchronous Cultures

In a growing culture the cells are distributed at all stages of the cell cycle. Thus, chemical analysis of a culture always yields values that are averaged over the cell cycle. Many properties of bacteria are related to the stage of growth of a single cell, so it is often desirable to be able to produce synchronous populations, that is, cultures in which all cells are at the same stage of the cell cycle. A variety of procedures have been used to synchronize cultures. The earliest methods utilized starvation for essential nutrients followed by restoration of these nutrients or inhibition of the synthesis of particular macromolecules (for example, DNA). These procedures yield an apparent synchrony in that most cells divide in a fairly small time interval, but introduce sufficient physiological stress that they have not been particularly useful. Mechanical fractionation of a culture has been fairly successful. In one procedure a culture is passed through several layers of filter paper. Small cells penetrate the filter, and the largest cells, which are usually on the verge of cell division remain on the surface. Hence, washing the filter provides a population of cells ready to divide. The most successful procedure utilizes the binding of cells from a growing culture to a nitrocellulose filter. If the filter is kept in warmed and aerated medium, the bound cells continue to grow and divide. If the flow of medium is reversed through the filter, already bound cells remain bound, but newborn cells are removed from the filter. Since the flow can be reversed at any time, the technique, which has been called the "baby machine," provides a supply of newborn cells whenever they are needed. Unfortunately, synchronous growth rarely persists for more than two to three generations because of the natural spread in the length of the life cycle of individual cells.

COUNTING BACTERIA

Knowledge of the number of cells per unit volume, the cell density, is important for most experiments. To obtain such values requires counting the number of bacteria in a known volume. Three methods are in common use—colony formation, counting a known volume with a microscope (direct methods), and optical absorbance (an indirect method). Electronic counters, which measure both cell number and cell size, are convenient, but they are sufficiently expensive that they are not often used.

A bacterium growing on an agar surface also divides. Since most bacteria are not very motile on a solid surface, the progeny bacteria remain very near the location of the original bacterium. The number of progeny increases so much that a visible cluster of bacteria appears. This cluster is called a **bacterial colony** (Figure 4-2). Colony formation allows one to determine the number of bacteria in a culture. For instance, if 100 cells are plated, 100

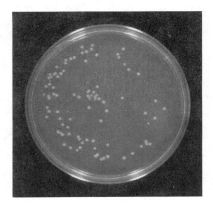

Figure 4-2 A petri dish with bacterial colonies that have formed on agar. [Courtesy of Gordon Edlin.]

colonies will be visible the next day. If 0.1 ml of a 10^6-fold dilution of a bacterial culture is plated and 200 colonies appear, the cell density in the original culture is $(200/0.1)(10^6) = 2 \times 10^9$ cells/ml.

Often it is necessary to know the cell density while a culture is growing. Counting colonies is clearly inadequate since the colonies take at least 10 hours to appear. Microscopic counting will work but is quite tedious and time-consuming. The most common method is to measure the optical absorbance. If a bacterial culture is placed in the path of a narrow beam of light, the intensity of the transmitted light will be less than that of the incident light. If the wavelength of the incident light is not absorbed by any of the intracellular molecules, the decreased transmission is caused entirely by *scattering* of the light by the bacteria. As long as the cell density is below about 5×10^8 cells/ml, the optical absorbance ($-\log_{10}$[intensity of transmitted light/intensity of incident light]) is proportional to the dry weight per ml and hence to the cell density. All that is needed to use this method is a standard curve relating absorbance and cell density. Such a curve is obtained by removing samples from a growing culture, measuring the optical absorbance, and determining the cell density by colony formation. From such a standard curve an absorbance measurement of any culture will henceforth yield the cell density. Cell size depends on the particular growth medium and on the bacterial species; thus, a standard curve is required for each medium and each species.

PREPARATION OF A PURE CULTURE

Genetic experiments must be done with a pure culture, that is, a population of bacteria obtained from the multiplication of a single cell—in other words, a clone. The need for a method to obtain such a population becomes important when one isolates either mutants or strains produced by genetic recombination. Often a colony containing mutants will be a mixture of mutant and wild-type cells (why this is true will become clear in Chapter 10 when mutagenesis is discussed), and colonies with a recombinant phenotype may consist of various recombinant cells. These problems can be eliminated by resuspending a colony in a buffer, diluting it, and replating to obtain single colonies. Such colonies are clones of cells from the original colony and will no longer contain a mixed population. Usually strains are not purified by resuspension and replating but by **streaking.** A sterile wire loop is first touched to a colony. The loop, which carries several million cells, is dragged across the surface of an agar plate ("streaked"). Cells are transferred to the agar, but the number decreases continually as the loop is dragged. After streaking for a few centimeters, the loop is sterilized by heating in a flame. It is allowed to cool and touched to the last part of the streak, and then streaked back and forth across the plate. The number of cells in the streak decreases as streaking proceeds until individual cells are well separated. The plate is then incubated

Figure 4-3   A streak plate. Streaking of a colony began at the top of the plate.

until growth occurs. A typical streak plate is shown in Figure 4-3. A colony is taken from the streak plate, a culture is prepared from the colony, and the culture is stored in a variety of ways. This pure master culture is then used for preparation of all other cultures.

IDENTIFICATION OF NUTRITIONAL REQUIREMENTS

Plating is a convenient way to determine if a bacterium is an auxotroph. This is done in the following way. Minimal agar and nutrient agar plates are prepared. Several hundred bacteria are plated on each plate and the plates are stored overnight in a constant-temperature incubator. Several hundred colonies are subsequently found on the nutrient agar because it contains so many substances that it can satisfy the requirements of nearly any bacterium. If colonies are also found on the minimal agar, the bacterium is a prototroph; if no colonies are found, it is an auxotroph and some required substance is not present in the minimal agar. Minimal plates are then prepared with various supplements. If the bacterium is a leucine auxotroph, the addition of leucine alone will enable a colony to form. If both leucine and histidine must be added, the bacterium is auxotrophic for both of these substances. In Chapter 10, where mutant selection is discussed, procedures for identifying auxotrophic mutants among a population of prototrophs will be described.

Table 4-1 shows the result of a plating experiment in which the nutritional requirements of a bacterium are deduced. The fact that there is no growth with supplement 1 shows that something other than histidine, leucine, and thymine is needed. Growth in the absence of thymine (supplement 2) and of leucine (supplement 3) indicates that neither thymine nor leucine is required. Note, however, that the presence of alanine in supplement 3 overcomes the deficiency of supplement 1, so alanine must be required. Finally, the lack of growth with supplement 4, in which alanine is present, shows that there is still another requirement. Since leucine is not needed, the only significant difference between supplements 3 and 4 is the presence of histidine in supplement 3. Therefore, histidine is also required, and the bacterium is auxotrophic for both alanine and histidine.

A variety of special growth media are available for determining whether a bacterium can utilize a particular substance as a carbon source. The most common media are the **color-indicator plates.** A popular medium, **EMB agar,** incorporates the dyes eosin and methylene blue into the medium. The color of these dyes is sensitive to pH. The medium also contains a sugar, for example, lactose, as a carbon source and a complete mixture of amino acids. Both Lac$^+$ and Lac$^-$ cells are able to form colonies on this medium. A Lac$^+$ cell utilizes the lactose and in so doing produces a local decrease in pH that causes the dyes to stain the colony an intense purple. A Lac$^-$ cell cannot utilize the lactose and instead uses some of the amino acids as carbon sources. One of the products of amino acid metabolism is ammonia, which increases the local pH, decolorizes the dyes, and causes the colonies to be white.

A medium on which all bacteria form colonies is called a **nonselective medium.** Mutants and wild type may or may not be distinguishable by growth on a nonselective medium. A color-indicator medium is an example of a nonselective medium on which mutant and wild-type bacteria can be distinguished. If the medium allows growth of only one type of cell (either mutant, wild type, or a combination of alleles), it is said to be **selective.** For example, a medium containing streptomycin is selective for the streptomycin-resis-

Table 4-1 Plating data enabling the determination of the nutritional requirements of a bacterium

Medium supplement	Growth	Conclusion
1. His, Leu, Thy	−	Needs some nutrient
2. Leu, Ala, His	+	Thy not needed
3. Ala, Thy, His	+	Ala needed (cf. plate 1)
		Leu not needed
4. Leu, Ala, Thy	−	His needed (cf. plates 1 and 3)

Note: Abbreviations of names of amino acids are given in Table 1-1, note 1.

tance (Str-r) phenotype and selects against the Str-s phenotype, and minimal medium containing lactose as the sole carbon source is selective for Lac$^+$ cells and against Lac$^-$ cells. Selective media are commonly used to identify genotypes.

THE PHYSICAL ORGANIZATION OF A BACTERIUM

The intracellular organization of bacteria is poorly understood, but some points are clear (Figure 4-4). The bacterial cell does not have its genetic material enclosed in a membrane-bounded nucleus and hence is a prokaryote. The intracellular material, which is found throughout the cell, is enclosed in a rigid multilayered cell envelope that gives it a defined shape—spherical, rodlike, and so forth (Figure 4-5). In this section we describe some of the properties of the nuclear material and the cell envelope.

A Complex DNA-Protein Structure: The *E. coli* Chromosome

The chromosome of *E. coli*, and presumably of most bacteria, is a single supercoiled double-stranded circular DNA molecule. For *E. coli* the total length of the circle is about 1300 μm, whereas the cylindrical bacterium has

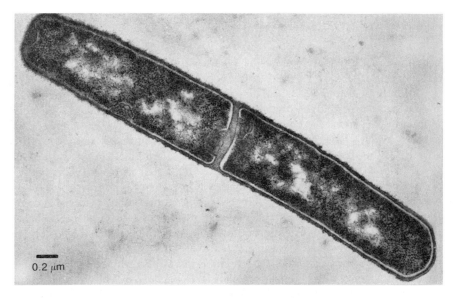

0.2 μm

Figure 4-4 An electron micrograph of a dividing bacterium. The layers of the cell wall can be seen. The light areas inside the cell are the DNA; note that the DNA is distributed throughout the cell. The fine dark particles are ribosomes, the units on which proteins are synthesized. [Courtesy of A. Benichou-Ryter.]

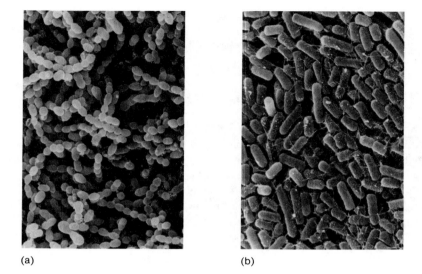

(a)   (b)

Figure 4-5 Two forms of bacteria. (a) Cocci (spheres that sometimes form chains). (b) Bacilli (rods). [From G. Shih and R. Kessel, *Living Images: Biological Microstructures Revealed by Scanning Electron Microscopy,* Jones and Bartlett, 1982.]

a diameter and a length of about 1 and 3 μm, respectively (Figure 4-6). Clearly the bacterial DNA must be highly folded when it is in a cell.

When *E. coli* DNA is isolated by a technique that avoids both DNA breakage and protein denaturation, a highly compact structure called a **nucleoid** is found. This structure contains a single DNA molecule, a fixed amount of protein, and variable amounts of RNA; the RNA is probably not part of the structure, but associates with it during isolation. An electron micrograph of the *E. coli* chromosome is shown in Figure 4-7. Two features of the structure should be noted: (1) The DNA is arranged in a series of loops and (2) each loop is supercoiled. A dense region containing membranous material is commonly seen in the central part; it is thought to be an artifact of the isolation procedure. The degree of condensation of the nucleoid is affected by a variety of factors, and questions exist about its intracellular state.

Introduction of a single-strand break into a supercoil by a DNase causes an abrupt transition to a nonsupercoiled form because of free rotation about the opposing sugar-phosphate bond (Chapter 2). However, if the nucleoid is nicked, it does not become an open circle but, with each nick, passes through 40–50 intermediate states in which the supercoiled loops are converted one by one to open loops. This clearly indicates that free rotation of the entire DNA molecule does not occur when one single-strand break is introduced.

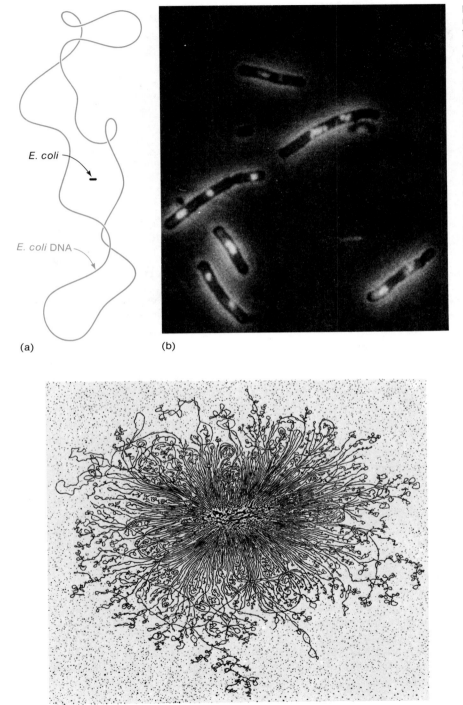

Figure 4-6 (a) Schematic diagram showing the relative sizes of *E. coli* and its DNA molecule, drawn to the same scale except for the width of the DNA molecule, which is enlarged approximately 10^6 times. (b) The localization of DNA in *E. coli*. Bacteria were exposed to a fluorescent dye that binds to DNA and then observed by fluorescence microscopy. The mode of sample preparation causes the DNA to condense slightly; in a living cell, the DNA occupies about twice as much space. [Courtesy of Todd Steck and Karl Drlica.]

E. coli

E. coli DNA

(a)　　　(b)

Figure 4-7 An electron micrograph of an *E. coli* chromosome showing the multiple loops emerging from a central region. [Bluegenes #1. © 1983. All rights reserved by Designergenes Posters Ltd., P.O. Box 100, Del Mar, CA 92014-0100, from which posters, postcards, and shirts are available.]

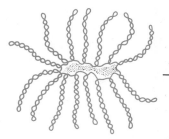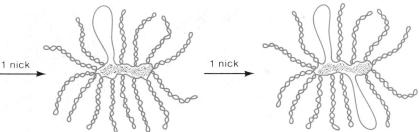

Figure 4-8   A schematic drawing of the highly folded supercoiled *E. coli* chromosome, showing only 15 of the 40–50 loops attached to proteins (stippled region) of unknown organization, and the opening of a loop by a single-strand break (nick).

The structure of the *E. coli* chromosome that has been deduced from these data is shown in Figure 4-8. The DNA is assumed to be fixed to proteins in a way (not yet understood) that causes the individual supercoiled loops to be isolated from one another and that prevents free rotation when a nick is introduced.

The enzyme DNA gyrase, which plays an important role in DNA replication (Chapter 8), is responsible for the supercoiling. If coumermycin, an inhibitor of *E. coli* DNA gyrase, is added to a culture of *E. coli* cells, the chromosome quickly loses its supercoiling. Furthermore, for unknown reasons *E. coli* has a system that controls the degree of supercoiling. In addition to DNA gyrase, which introduces negative superhelical twists, *E. coli* (and other bacteria) contains an enzyme (**topoisomerase I**) that removes superhelicity; *in vitro*, purified topoisomerase I converts a supercoil to a nonsupercoiled covalent circle. Mutants (*topA*) lacking topoisomerase I activity have been found. Nucleoids isolated from these mutants have increased supercoiling, about 32 percent greater than normal. Interestingly, although in the laboratory the mutants show no evidence for growth defects, they often acquire secondary mutations in or near the gyrase gene; these secondary mutations reduce gyrase activity slightly and cause a restoration of near-normal supercoiling to the nucleoid. The significance of these observations is unknown.

The Bacterial Cell Wall

Bacteria come in several shapes, each of which is determined by the arrangement of certain macromolecules in the bacterial cell walls. There are two distinct classes of cell envelopes known as **Gram-positive** and **Gram-negative.** Operationally these forms are distinguished by their ability to retain a crystal violet-iodine stain when treated with alcohol. Bacteria that retain the stain are Gram-positive; those that do not are Gram-negative. The cell wall of Gram-negative bacteria, such as *E. coli*, is much more complicated than that of Gram-positive cells. Figure 4-9(a) shows an electron micrograph of a section of *E. coli* in which the multiple layers of the cell wall are seen. Treatment of cells with various reagents, such as trypsin (which hydrolyzes protein), detergents (which remove lipids), and lysozyme (an enzyme obtained from egg white), causes dissolution of particular layers. Electron microscopy of samples so treated allows one to identify a sensitive layer as being the outer, middle, or inner layer. Moreover, by dissolving one or two layers, the remaining material can be purified and analyzed chemically. The result of such studies is the model shown in Figure 4-10. The cell wall consists of two

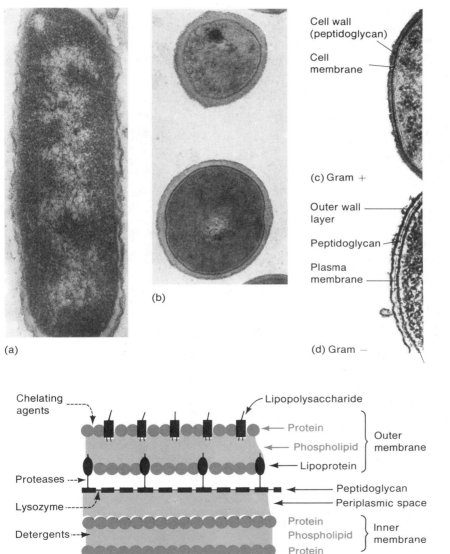

Cell wall (peptidoglycan)

Cell membrane

(c) Gram +

Outer wall layer

Peptidoglycan

Plasma membrane

(d) Gram −

(b)

(a)

Figure 4-9 Two types of bacterial cell walls. (a) An electron micrograph of a thin section of *E. coli* (a Gram-negative bacterium) showing the multiple layers. (Courtesy of Jack Pangborn.) (b) An electron micrograph of a thin section of a Gram-positive bacterium, *Staphylococcus stapholyticus*. The thick outer layer is the peptidoglycan. (Courtesy of Harriet Smith.) (c) An electron micrograph of the cell wall of a Gram-positive bacterium prepared in a way that shows the cell wall structure more clearly than in (b). (d) An electron micrograph of the cell wall of a Gram-negative bacterium prepared in a way that shows the three layers more clearly than in (a). The segments in (c) and (d) are enlarged about three times more than in (a) and (b).

Chelating agents

Lipopolysaccharide

Protein

Phospholipid

Lipoprotein

Outer membrane

Proteases

Lysozyme

Peptidoglycan

Periplasmic space

Detergents

Protein
Phospholipid
Protein

Inner membrane

Figure 4-10 Schematic diagram of the cell wall of a Gram-negative bacterium. The sites of attack by chelating agents, proteases, lysozyme, and detergents are indicated by broken arrows.

major segments, each called a membrane because of the presence of phospholipid. These two membranes, the inner and outer membrane, are separated by an aqueous region known as the **periplasmic space.** The outer membrane is a complex structure that is conveniently thought of as having two components—the protein-phospholipid membranous region and the **peptidoglycan layer.** Gram-positive bacteria differ from Gram-negative bacteria in that they lack the outer membrane, the peptidoglycan layer is thicker, and there is no periplasmic space (Figures 4-9(b) and 4-11).

The peptidoglycan layer, which can be seen most clearly as the thick outer layer in Figure 4-9(b), is an unusual substance. It is a polymer consisting of both sugar and peptide units. Individual polysaccharide chains are cross-linked by a pentaglycine peptide to form the large sheetlike structure (Figure 4-12). The remarkable feature of peptidoglycan is that the sheet ultimately closes on itself to form one enormous saclike macromolecule, which encloses

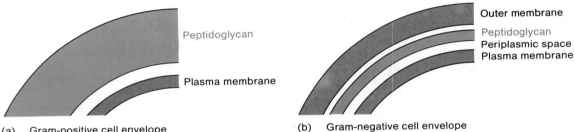

(a)   Gram-positive cell envelope   (b)   Gram-negative cell envelope

Figure 4-11   The differences between the cell walls of (a) Gram-positive and (b) Gram-negative bacteria.

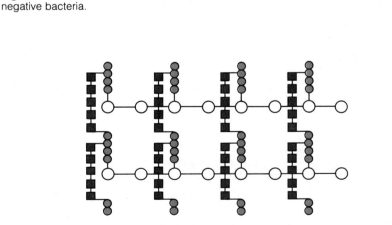

Figure 4-12   Schematic diagram of the peptidylglycan. The sugars are the large open circles, the tetrapeptides are small red circles, and the pentaglycine bridges are the five linked squares. The cell wall is a single, enormous macromolecular sac because of contiguous cross-linking.

the entire inner membrane and cytoplasm of the bacterium. The cross-linking is not always as regular as is shown in the figure; in fact, the particular variations of the cross-linking pattern determine the shape of each bacterial species.

Peptidoglycan is the site of attack of the enzyme lysozyme, which is commonly used as a stage in breaking cells open, or in rendering cells permeable to certain macromolecules. Thus, since peptidoglycan is the only rigid component of the cell wall, a lysozyme-treated bacterium assumes a spherical shape (whether the bacterium is initially spherical or rodlike, or has any other shape); a cell treated in this way is called a **spheroplast,** if some cell wall remains and a **protoplast** if none is present. The antibiotic penicillin interferes with the synthesis of peptidoglycan. Thus, if a bacterium is allowed to grow in the presence of penicillin, it enlarges without growth of the strong peptidoglycan layer; the volume of the cell then exceeds that which can be held by the weak membranous layers and the cell bursts, which accounts for the lethal effect of penicillin.

The inner or cytoplasmic membrane has a structure much like the basic membrane structure described in the previous section. This membrane provides the major osmotic barrier and determines, for the most part, which molecules can enter and leave the cytoplasm. *Uncharged* molecules having a molecular weight less than 100 are usually able to pass freely through the membrane. However, larger uncharged molecules and all charged molecules require various transport systems, which are more-or-less molecule-specific. Some charged molecules may never be able to enter a bacterium because of lack of the appropriate transport system. For example, no phosphorylated organic molecules (such as nucleotides) can enter a cell. Treatment of a cell with a good lipid solvent such as toluene or ether removes the lipid layer and thus also the permeability barrier. Cells treated in such a way are said to be **permeabilized.** They cannot grow and divide but are able to carry out many biochemical reactions. Permeabilized cells have been somewhat useful in the study of DNA synthesis since, in contrast with untreated cells, they are able to take up deoxynucleoside triphosphates, the immediate precursor of DNA. Permeabilizing is often also a convenient way to assay certain enzymes. For example, synthesis of the enzyme β-galactosidase has been widely studied as a means of understanding the regulation of gene expression (Chapter 7). It is assayed by permeabilizing cells with toluene prior to adding the charged substrate.

The outer membrane is a very complex structure. It is basically a mass of closely associated phospholipids between two layers of proteins. Interspersed in the outer protein layer are many units of lipopolysaccharide, a very unusual molecule. The lipopolysaccharide (LPS) units have several protective functions. First, they provide a second permeability barrier that keeps out many large molecules. Second, the LPS gives the cell a very hydrophilic surface, which decreases the ability of many protozoa, which feed on bacteria, to consume these cells.

Between the peptidoglycan layer and the inner membrane is an aqueous region called the **periplasmic space.** This region becomes evident when the LPS is removed or if the cell is subjected to a sudden decrease in osmotic pressure. (This often happens if cells are rapidly diluted from their growth medium into a dilute buffer.) These treatments cause a weakening of the outer membrane and the peptidoglycan layer, and subsequent release of certain proteins. Most of these proteins are nucleases and proteases and are thought to be used in digesting impermeable nutrients; for instance, in certain circumstances these proteins are excreted and enable a cell to use exogenous proteins and nucleic acids as food. Other molecules in the periplasmic space are proteins that bind specific ions, sugars, and amino acids, and are needed for transport of these substances across the inner membrane.

METABOLIC REGULATION IN BACTERIA

Bacteria are well-regulated and highly efficient organisms. For example, they rarely synthesize substances that are not needed. Thus, the enzymatic system for synthesizing the amino acid tryptophan is not formed if tryptophan is present in the growth medium; however, when the tryptophan in the medium is used up, the enzymatic system will be rapidly formed. The systems responsible for utilization of various energy sources are also efficiently regulated. A well-studied example (discussed in detail in Chapter 7) is the metabolism of the sugar lactose as an alternate carbon source to glucose.

Glucose is metabolized by all living cells by a series of chemical conversions in which the molecule is progressively broken down. Glucose is a fundamental carbon source in the sense that other sugars must be converted either to glucose or to a glucose-degradation product in order to be metabolized. Thus, bacteria use glucose more efficiently than lactose. The first step in the metabolism of lactose is its cleavage to form two sugars, glucose and galactose. However, the enzyme needed to catalyze this cleavage is not present in cells in significant quantities unless lactose is in the growth medium. If lactose is provided to a cell, through a complex multistep process the enzyme can then be formed, and glucose produced from lactose is broken down and used as a source of energy. The galactose that is also produced is converted to a component that enters the glucose-utilization pathway and is thereby also metabolized. That the system is regulated can be seen by the fact that if a cell is supplied with both glucose and lactose in the growth medium, there is no reason for the cell to synthesize the lactose-cleaving enzyme and, in fact, a simple system prevents the cell from wasting its energy synthesizing this enzyme until all the glucose is utilized.

The control of both tryptophan synthesis and lactose degradation are two examples of **metabolic regulation.** This very general phenomenon will be explored in Chapter 7.

PROBLEMS

1. What carbon source yields the highest growth rate of bacteria?

2. In what phase of a bacterial culture are all cells growing at the same rate?

3. If the doubling time of a bacterial culture is 20 min, by what factor does the cell density increase in 2 hr?

4. A bacterium fails to grow on minimal agar but grows if leucine is added. How many different amino acids are required for growth of this bacterium?

5. 0.1 ml of a bacterial culture is diluted into 9.9 ml of buffer; 0.1 ml of this dilution is again diluted in 9.9 ml of fresh buffer. Plating 0.1 ml from the second dilution tube yields an average of 72 colonies per plate on four plates. What is the cell density of the culture?

6. A microscopic count of a culture that has been in the refrigerator for two weeks shows that 453 cells are present, on the average, in each 0.001 ml of a 1000-fold dilution. Plating gives a cell density of 1.2×10^7 per ml. What information does the discrepancy in the two values give you?

7. A standard curve relating optical absorbance and cell density (determined by plating) has the following points: absorbance values of 0.2, 0.4, 0.6, and 0.8 correspond to 7.5×10^7, 1.5×10^8, 3×10^8, and 6×10^8 cells/ml, respectively. A cell sample taken from a growing culture has an absorbance of 0.7. What is its cell density?

8. A bacterial strain can grow on agar supplemented with arginine (Arg), tryptophan (Trp), and leucine (Leu). It fails to grow on agar containing (I) only Arg and Trp or (II) only Leu and Trp. It will grow if (III) only Arg and Leu are present. What is the genotype of the bacterium with respect to these three amino acids?

9. A bacterial mutant grows quite slowly in tryptone (acid-hydrolyzed milk protein) medium, but growth is normal in nutrient broth (an extract of beef). Since both media contain a very large number of substances, how might you explain the difference in growth?

10. What is the doubling time of a culture growing in a 75-ml chemostat if the flow rate for the growth medium is 0.3 min?

11. Recall that a Lac^+ cell will form a purple colony on an EMB-lactose plate. What will be the color of a $Mal^- Lac^+$ mutant on EMB containing both lactose and maltose?

12. What is the color of any bacterium growing on EMB-glucose agar?

13. In an effort to grow a Leu^- mutant you prepare minimal medium containing leucine. The cells grow to a maximum cell density of 2×10^8 per ml. You are surprised since Leu^+ cells in the same medium reach a cell density of 2×10^9 per ml. How might you explain this observation?

14. An A^- auxotrophic bacterium does not grow in minimal medium but does grow normally if A is added to the medium. Interestingly, it grows also if substance B is added. What information does this give you about the metabolic pathway for synthesis of B?

15. What molecules comprise the *E. coli* chromosome?

16. What role does RNA play in the structure of a bacterial chromosome?

17. What evidence supports the notion that the loops of a bacterial chromosome are in some way isolated from one another?

18. If you had never seen a chromosome with a microscope but had seen cells with nuclei, what information would nonetheless let you know that DNA must exist in a highly coiled form within cells?

Chapter 5

Bacteriophages

Bacteriophages, or phages, have played an important role in the development of molecular genetics. At the present time a few phages are the most completely understood of any organisms. Because phages are less complex than bacteria and higher cells and have available an enormous number of mutants, they have been extraordinarily useful in the study of replication, transcription, and regulation. In this chapter we present a survey of phages that will be sufficient to enable us to make use of their properties in subsequent discussions of gene expression and the genetics of bacteria. In Chapters 15–18 we will consider phage genetics in greater detail.

A word of curious terminology before beginning—the plural word *phages* refers to different species, whereas in common usage the word *phage* can be both singular and plural, referring in the plural sense to particles of the same species. Thus, P1 and P2 are both phages but a test tube might contain either 1 P1 phage or 100 P1 phage.

GENERAL PROPERTIES OF PHAGES

A bacteriophage is a bacterial parasite. By itself, a phage can persist, but it can not replicate except within a bacterial cell. Most phages possess genes encoding a variety of proteins. However, all known phages use the protein-synthesizing system, amino acids, and energy-generating systems of the host cell, and hence a phage can multiply only in a metabolizing bacterium.

Each phage must perform some minimal functions for continued survival. These are the following:

1. Protection of its nucleic acid from environmental chemicals that could alter the molecule (for example, break the molecule or cause a mutation).
2. Delivery of its nucleic acid to the inside of a bacterium.
3. Conversion of an infected bacterium to a phage-producing system which yields a large number of progeny phage.
4. Release of progeny phage from an infected bacterium.

These functions are carried out in a variety of ways by different phage species. All of these species have certain features in common, but differences in detail show the many ways in which specific biological functions can be accomplished.

An important observation that has been made is of the degree to which an individual phage particle uses parts of the machinery of the cell. Some phage species have fewer than 10 genes and depend almost entirely on cellular functions, whereas others have 30–100 genes and are more dependent on proteins encoded in their own genetic material. A few of the largest phage particles have so many of their own genes that, for certain functions such as DNA replication, they need no host genes. Surprisingly, in a few cases phage genes duplicate host genes.

Phage particles differ in their physical structures from species to species and often certain features of their life cycles are correlated with their structure. These structural differences are described in the following section.

STRUCTURES OF PHAGES

There are three basic phage structures: icosahedral tailless, icosahedral head with tail, and filamentous. Usually the phage particle consists of a single nucleic acid molecule—which may be single- or double-stranded, linear or circular DNA, or single-stranded, linear RNA—and one or more proteins. (The one known exception is phage $\phi6$, which contains three linear double-stranded RNA molecules, whose base sequences differ from one another.) The proteins form a shell, called either the **coat** or the **capsid,** around the nucleic acid; the nucleic acid is thereby protected from nucleases and harmful substances. Phages containing double-stranded DNA are typically 50 percent DNA by weight and hence are a useful source of DNA for physical studies. Figure 5-1 shows electron micrographs of the three basic structures; the components of a tailed phage are given in Figure 5-2. The following points are general:

1. In both icosahedral tailless and tailed phages the nucleic acid is contained in a hollow region formed by the capsid and is highly compact. In a filamentous phage the nucleic acid is embedded in the capsid and is present in an extended helical form.

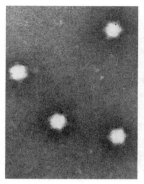 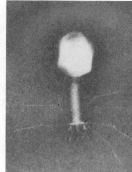 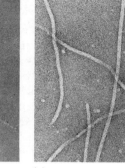

(a)   (b)   (c)

Figure 5-1 The three major morphological classes of phages. (a) Icosahedral, tailless: φX174; (b) Icosahedral, tailed: T4; (c) Filamentous: M13. [Courtesy of Robley Williams.]

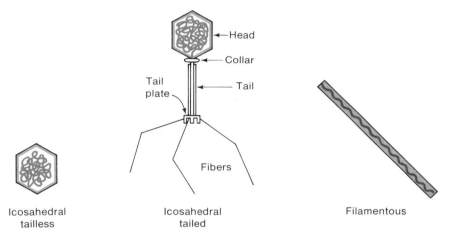

Icosahedral   Icosahedral   Filamentous
tailless   tailed

Figure 5-2 Diagrams of the three basic phage structures. The tailed phages do not always have a collar and can have from 0–6 tail fibers, the number depending on the phage type. The nucleic acid is shown in red.

2. The tail is a complex multicomponent structure often terminated by tail fibers.

3. In icosahedral phages the length of the DNA molecule is very much greater than any dimension of the head.

There are many variations on the basic structure of the tailed phages. For example, the length and width of the head may either be the same or the length may be greater than the width; however, short fat heads are not seen. The tail may be very short (barely visible in electron micrographs) or up to four times the length of the head, and it may be flexible or rigid. A complex baseplate may also be present on the tail; when present, it typically has from one to six tail fibers.

STAGES IN THE PHAGE LIFE CYCLE

Phage life cycles fit into two distinct categories—the **lytic** and the **lysogenic** cycles. A phage in the lytic cycle converts an infected cell to a phage factory, and many phage progeny are produced. A phage capable *only* of lytic growth is called **virulent.** The lysogenic cycle, which has been observed only with phages containing double-stranded DNA, is one in which no progeny particles are produced; the phage DNA usually becomes part of the bacterial chromosome. A phage capable of such a life cycle is called **temperate.** Most temperate phages also undergo a lytic cycle in certain circumstances. In this section only the lytic cycle is outlined. The lysogenic cycle will be described later.

There are many variations in the details of the life cycles of different virulent phages. There is, however, what may be called a basic lytic cycle, which applies to phages containing double-stranded DNA. This is the following (Figure 5-3):

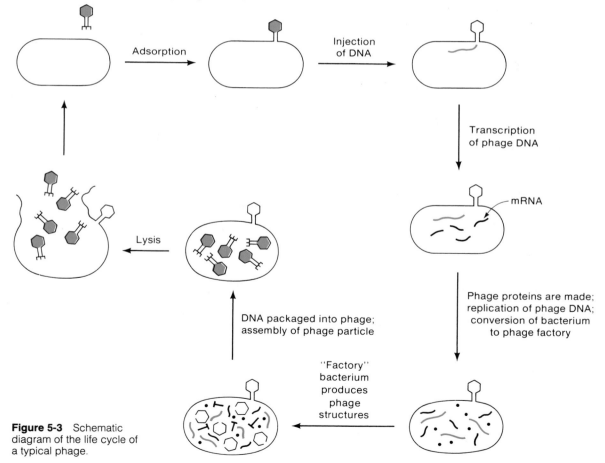

Figure 5-3 Schematic diagram of the life cycle of a typical phage.

1. *Adsorption of the phage to specific receptors on the bacterial surface* (Figure 5-4). Many different receptors exist and serve the bacteria for purposes other than phage adsorption.

2. *Passage of the DNA from the phage through the bacterial cell wall.* Some types of tailed phages use an injection sequence shown schematically in Figure 5-5. In this process the nucleic acid is transferred into the cell and is never exposed to the medium surrounding the recipient cell. Little is known about the transfer mechanisms for the other phage types. With tailless phages the nucleic acid is transiently susceptible to nuclease attack, so it is thought that the phage coat may break open and release its nucleic acid first onto the cell wall prior to entering the cell.

3. *Conversion of the infected bacterium to a phage-producing cell.* Following infection by most phages a bacterium loses the ability either to replicate or to transcribe its DNA; sometimes it loses both. This shutdown of host DNA or RNA synthesis is accomplished in many different ways (for example, degradation of host DNA) depending on the phage species. Shutdown is less common with phages containing single-stranded DNA or RNA.

4. *Production of phage nucleic acid and proteins.* By several mechanisms the phage directs the synthesis of a replicative system that specifically makes copies of phage nucleic acid. This programming is accomplished either by synthesis of phage-specific DNA and RNA polymerases or by addition of specificity elements to bacterial polymerases. In both cases many bacterial

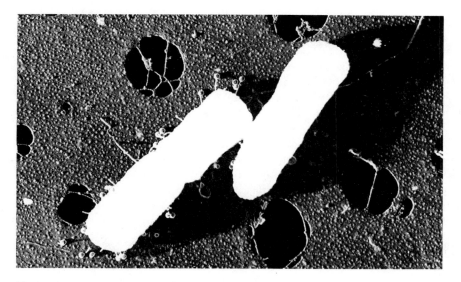

Figure 5-4 An electron micrograph of an *E. coli* cell to which numerous λ phage particles are adsorbed by their long tails. [Courtesy of T. F. Anderson.]

Figure 5-5 (a) Injection sequence of a tailed phage. In the injection stage the tail sheath contracts and drives a core protein tube through the cell wall like a hypodermic syringe. (b) Electron micrograph of T4 phage adsorbed to the cell wall of *E. coli*, observed in thin section. The tail sheath is contracted and the core is fixed firmly against the cell wall. The arrow shows a portion of the core projecting through the cell wall. DNA can be seen entering the cell from the two phage at the right. [Courtesy of Lee Simon.]

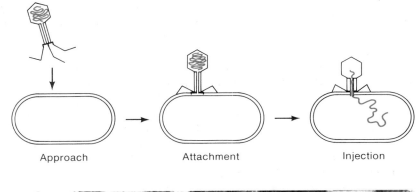

(a) Approach   Attachment   Injection

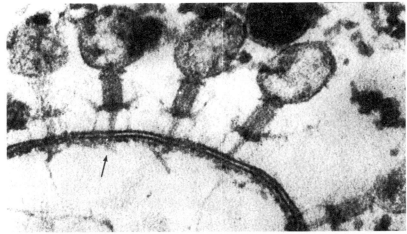

(b)

replication proteins are used. Synthesis of phage mRNA from the phage DNA is almost always initiated by the bacterial RNA polymerase; but after the first phage mRNA is made, either the bacterial polymerase is modified to recognize other start points for mRNA synthesis (**promoters**) or a phage-specific RNA polymerase is synthesized. RNA synthesis is regulated and phage proteins are synthesized sequentially in time as they are needed. Usually there is a fairly distinct difference in the time of synthesis of phage-specified enzymes (made early in the life cycle and called **early proteins**) and the structural proteins of the phage particle (made late in the life cycle). The temporal difference is accomplished by the timing of mRNA synthesis. The early proteins are encoded in **early mRNA** and the structural proteins are in **late mRNA**.

RNA-containing phages differ with respect to use of host replication enzymes—they must encode their own replication enzymes because bacteria do not contain enzymes that replicate RNA. Furthermore, the RNA molecules of phages containing single-stranded RNA serve as their own mRNA.

Phages containing single-stranded DNA also differ from the prototype in that they usually use unmodified RNA polymerase throughout the life cycle.

5. *Assembly of phage particles* (**morphogenesis**). Two types of proteins are needed for the assembly process: **structural proteins,** which are present in the phage particle, and **catalytic proteins,** which participate in the assembly process but do not become part of the phage particle. A subset of the latter class consists of the **maturation proteins,** which convert intracellular phage DNA to a form appropriate for packaging in the phage particle. With the icosahedral phages assembly occurs in several stages: (1) Aggregation of phage structural proteins to form a phage head and, when needed, to form a phage tail; at this point, the tail is not attached to the head. (2) Condensation of the nucleic acid and entry into a preformed head. (3) Attachment of the tail to a filled head. With filamentous phages the nucleic acid and the protein form a phage particle in a single step. The mechanism of nucleic acid condensation is not completely known. Usually 50–1000 phage particles are produced, the number depending on the particular phage species.

6. *Release of newly synthesized phage*. With most phages an enzyme called a **lysozyme** or an **endolysin** is synthesized late in the cycle of infection. The enzyme causes disruption of the cell wall and breakdown of the cell (**lysis**), so phage are released to the surrounding medium. The suspension of newly released phage is called a **lysate.** A few filamentous phages release progeny continuously by outfolding of the cell wall; this process is called **extrusion,** and it does not cause major damage to the cell. Cells infected with such filamentous phages can continue to produce virus particles for very long periods of time.

COUNTING PHAGE

Phage are easily counted by a technique known as the **plaque assay**; this technique is performed in the following way.

In Chapter 4 it was explained that plating 100 bacteria on agar yields 100 colonies. However, if 10^8 bacteria are plated, the 10^8 colonies that result are so near one another that they appear as a confluent, turbid layer of bacteria called a **lawn.** An alternative method of plating achieves maximal uniformity of the turbidity of the lawn. In this method the bacteria are mixed into a small volume of warm, slightly dilute, liquid agar which is then poured onto the surface of the solid medium. The liquid, known as **top agar** or **soft agar,** rapidly hardens, providing a very smooth surface, and the bacteria grow in this thin agar layer with much uniformity (Figure 5-6). If a phage is present in the hardened top agar, it can adsorb to one of the bacteria in the agar; shortly afterward, the infected bacterium will lyse and release about 100 phage, each of which will adsorb to nearby bacteria; these bacteria in turn will release a burst of phage which then can infect other bacteria in the

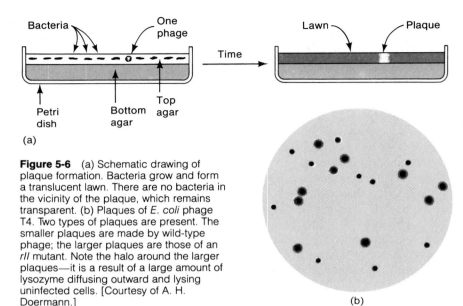

Figure 5-6 (a) Schematic drawing of plaque formation. Bacteria grow and form a translucent lawn. There are no bacteria in the vicinity of the plaque, which remains transparent. (b) Plaques of *E. coli* phage T4. Two types of plaques are present. The smaller plaques are made by wild-type phage; the larger plaques are those of an *rII* mutant. Note the halo around the larger plaques—it is a result of a large amount of lysozyme diffusing outward and lysing uninfected cells. [Courtesy of A. H. Doermann.]

vicinity. These multiple cycles of infection continue and after several hours, the phage will have destroyed all of the bacteria at a single localized area in the agar, giving rise to a clear, transparent circular region in the turbid, confluent layer. This region is called a **plaque.** Since one phage forms one plaque, the individual phage particles put on the plate can be counted.

The **efficiency of plating, EOP,** is defined as the fraction of phage particles that can form a plaque. The value of EOP is 1 or nearly 1 for many of the larger phages, but also can be less than 1 (0.1–0.5) for phages that make very small plaques. For most phages there is a range in plaque size, so for these phages many plaques may be too small to be visible. When phage lysates are stored, the EOP may decrease; when this occurs, it is usually a result of accumulated chemical damage or denaturation of the proteins in the tip of the tail. Storage of phage at 4°C or below -40°C in a glycerol solution reduces the loss of phage viability. Addition of high concentrations of proteins, such as serum albumin, are also protective.

PROPERTIES OF A PHAGE-INFECTED BACTERIAL CULTURE

In an earlier section the events following infection of one cell by a phage were described. In the laboratory one usually infects a bacterial culture with a large number of phage particles, and special techniques are needed to analyze the results of these interacting population. The first part of this section describes the measurement of the parameters needed to study an infected bacterial culture. Then, procedures that enable information to be obtained about a single infection will be described.

The Number of Participating Phage and Bacteria

The adsorption of phage to a bacterial culture is a random process and the variation in the distribution of phage among the cells is described by the Poisson distribution,

$$P(n) = \frac{m^n e^{-m}}{n!}$$

in which $P(n)$ is the fraction of the bacteria to which n phage have adsorbed when m is the average number of adsorbed phage per bacterium (the **multiplicity of infection** or **MOI**). That is, the fraction of bacteria infected with 0, 1, 2, 3, . . . , i phage is $P(0)$, $P(1)$, $P(2)$, $P(3)$, . . . , $P(i)$. Thus, if 3×10^8 phage adsorb to 10^8 bacteria ($m = 3$), the values of $P(0)$, $P(1)$, $P(2)$, $P(3)$, . . . are 0.05, 0.15, 0.22, 0.22, ... Since $P(0) = 0.05$, the sum of $P(1) + P(2) + . . . + P(i)$ must equal $1 - 0.05 = 0.95$. In other words, 95 percent of the bacteria are infected by at least one phage.

Note also that the value of $P(0)$ tells the fraction of the phage particles that have adsorbed. For example, in the infection just described, $P(0)$ should equal 0.05 for $m = 3$, if all of the added phage had adsorbed to the bacteria. However, in a particular experiment using 3×10^8 phage and 10^8 bacteria, if one observed that 12 percent of the bacteria remained uninfected, then the value of $P(0)$ would be 0.12; using this value one can calculate from the Poisson term $P(0) = e^{-m}$ that $m = 2.12$, which is the true value of the MOI. Thus, $2.12/3 = 0.71$ or 71 percent of the added phage particles actually adsorbed.

It is possible to measure $P(0)$ in a simple way (Figure 5-7). A known number of cells is used in an infection. After an adsorption period the bacterial suspension is diluted and plated, and the fraction of the bacteria able to form a colony is measured. Only uninfected cells can do so. Thus, $P(0)$ is the number of colonies formed divided by the total number of cells. As a check, the number of infected cells can be measured in the following way. First, antibodies that inactivate unadsorbed phage are added to the infected culture. (Antibodies are obtained from the blood of a rabbit that has been injected with a purified suspension of the phage.) Then, the phage suspension is plated on a lawn of phage-sensitive cells, where each infected cell, providing it is plated before lysis, will produce a *single* plaque. A cell that can form a plaque in this way is called an **infective center.**

The number of phage produced by an infected cell is called the **burst size.** This is an important parameter in many experiments because it is a

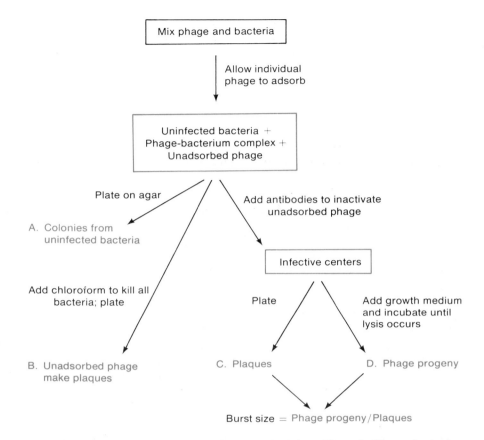

Figure 5-7   Scheme for determining the number of uninfected bacteria (A), unadsorbed phage (B), infective centers (C), progeny phage (D), and the burst size (D/C).

measure of the efficiency of phage production. It is measured by determining (by plaque counting) the number of phage produced after lysis of the culture and dividing by the number of infective centers.

Production of a Phage Lysate

Phage multiply much more rapidly than bacteria. That is, bacteria double in one generation time whereas, in one life cycle, the number of phage is increased by a factor equal to the burst size. This is easily seen in the example shown in Table 5-1, in which a single phage whose average burst size is 100 and whose life cycle lasts 25 minutes infects a one-milliliter bacterial culture

Table 5-1 Calculation of the phage and bacterial concentrations after various numbers of bacterial generations*

Number of generations	Phage / ml	Concentrations		
		Bacteria / ml		
		Approximate	Precise	
0	1	10^6	10^6	
1	10^2	2×10^6	$2 \times (10^6 - 1)$	
2	10^4	4×10^6	$(4 \times 10^6) - (2 \times 10^2) - 4$	
3	10^6	7.98×10^6	$(8 \times 10^6) - (2 \times 10^4) - (4 \times 10^2) - 8$	
4	10^8	1.4×10^7	$(1.6 \times 10^7) - (2 \times 10^6) - (4 \times 10^4) -$ $(8 \times 10^2) - 16$	
5	$1.4 \times 10^9 =$ $100 \times (1.4 \times 10^7)$	0	0	

*Initially a bacterial culture at a concentration of 10^6 cells/ml is infected with 1 phage. The doubling time of the bacteria and the life cycle of the phage are equal.

with a 25-minute doubling time. In this calculation, it is assumed that adsorption is always instantaneous and complete. Note that, in four generations, the number of bacteria has increased 14-fold, whereas the number of phage has increased 10^8-fold. At this time there are approximately eight times as many phage as bacteria; hence, all bacteria are infected. Thus, after 125 minutes the bacteria are gone and the original phage particle has produced 1.4×10^9 progeny.

By careful choice of media, growth conditions, and time of injection phage lysates containing very high concentrations of phage can be prepared. A common procedure uses a rich well-aerated medium and infection of the culture at a cell density of 5×10^8 cells/ml with an MOI of 0.1. One-tenth of these cells (5×10^7 per ml) are infected and release phage at a concentration of about 5×10^9 phage/ml at a time when the concentration of uninfected bacteria has reached 10^9 cells/ml. At this point the MOI is 5, all bacteria are infected, and after one more phage cycle the phage concentration is $1–2 \times 10^{11}$ per ml.

High concentrations of phage can also be prepared by growth on solid media using an amount of phage such that all bacteria are ultimately infected and lysed. For example, if 10^6 phage are placed in soft agar with about 2×10^8 bacteria, after about 6 hours the soft agar layer (which is completely clear) will contain about 10^{11} phage. Such a lysate is called a **plate lysate.** A plate lysate can be prepared from a single plaque by the following procedure. A sterile toothpick or wire is stabbed into a plaque and then stirred in liquid soft agar containing bacteria. The soft agar is then poured on a plate, where it hardens. Usually about 10^6 phage are transferred, so confluent lysis results.

The One-Step Growth Curve

Certain kinetic parameters of the phage infection can be determined by studying an infected culture. A classic experiment is the **one-step growth curve**. In this experiment a culture is infected with an MOI of about 0.1 (in order that no cell be infected with more than one phage). Phage antiserum is then added to inactivate any unadsorbed phage. The infected cells are diluted about 1000-fold into fresh warm medium (to prevent inactivation of progeny phage by the antibody), and at various times aliquots are taking for plating for plaques. At first the number of plaques is constant (Figure 5-8), because plaques are formed only by infected unlysed cells—that is, each infected cell produces one plaque. This period is called the latent period. At a particular time after infection (a time characteristic of each phage) the number of plaques increases. During this short time interval (the rise period) the infected cells are lysing. When all infected cells have lysed, the phage concentration remains constant. The ratio of phage produced to the initial number of infective centers is the burst size, and the number of minutes before the increase in plaque number occurs is the lysis time.

A modification of this experiment (also shown in Figure 5-8) enables one to determine the kinetics of phage production *within* the cell. In this procedure chloroform and lysozyme are added at various times after infection. Chloroform destroys the cell membranes, and lysozyme breaks down the

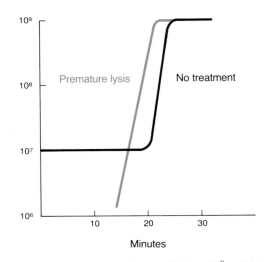

Figure 5-8 A hypothetical one-step growth curve. Cells at 10^8 cells/ml are infected at MOI = 0.1. Antiserum is added to inactivate unadsorbed phage, and then the culture is diluted 1000-fold. The black curve shows the number of plaques produced per ml of culture, without further treatment. The red curve shows the result of premature lysis by lysozyme and chloroform. The burst size is 100.

peptidoglycan layer (Chapter 4), which results in **premature lysis** of the infected cells. In this case the increase in the number of plaques represents intracellular production of phage.

The Single-Burst Technique

Studies of phage production in infected cultures yield information that has been averaged over a phage population. However, one would also like to know the events in a single infected cell. For example, the burst size of a particular phage is the average number of phage produced per cell, and one might be interested in knowing whether the burst size varies from cell to cell. A **single-burst experiment** can in some instances provide important information about cell-to-cell variation.

In a single-burst experiment a bacterial culture is infected (at an MOI appropriate to the phenomenon being studied—about 0.1 to study burst size), and after the phage are adsorbed, the infected cells are diluted in growth medium to an exceedingly low concentration—usually, about 0.05 infected cells/ml. Then, 1-ml aliquots are dispensed into hundreds (or thousands) of test tubes. According to the Poisson distribution, at this concentration 95.1 percent of the tubes will not contain infected cells, 4.8 percent will contain one infected cell, and 0.1 percent will contain more than one infected cell. Thus, 4.8/4.9, or 98 percent, of the tubes *containing infected cells* will contain one infected cell. The single cell in each tube is allowed to lyse, and the contents of each tube are plated in a single petri dish with indicator bacteria, so that plaques will form. Thus, the plaques on one plate are formed by the phage progeny of a single infected cell. The number of plaques observed on the various plates yields the distribution of burst sizes, which are found to range from less than 10 to several hundred. The wastefulness of this technique should be noted: to study 100 infected cells requires using about 2000 petri dishes of which roughly 1900 will contain no plaques.

The single-burst technique is used mostly to study the result of phage crosses in a single cell. This will be encountered again in Chapter 15.

SPECIFICITY IN PHAGE INFECTION

Several thousand phage species have been isolated. The ability of a particular phage to infect a bacterium is almost always limited to a single bacterial species and often to a few strains of that species. For example, no phage species that infects the genus *Pseudomonas* can also infect *E. coli*; furthermore, phages that grow in *Ps. fluorescens* generally fail to grow in *Ps. aeruginosa*. Among the *E. coli* phages, which are the most extensively studied phages, extraordinary specificity has been observed; for example, the phage φX174 grows well on *E. coli* strain C but fails to grow on most other laboratory strains. There are exceptions, though; for instance, *E. coli* phage T4 is capable of growing on many strains of *E. coli* and on certain species of the genus *Shigella*.

Several factors contribute to this specificity. One of these is the ability to adsorb. For example, *E. coli* phages will not adsorb to any known species of *Pseudomonas*. This is not unexpected inasmuch as there are specific receptors on the cell wall for phage adsorption. It is of course a distinct *dis*advantage to a bacterium to be able to adsorb phages, since this invariably kills the bacterial cell; thus, in the course of evolution cell wall components have evolved to be unrecognizable to most phage species. Phages have also evolved, in order to be able to recognize some bacterium; if a particular phage had not been able to do so, it would not exist at the present time.

If 10^8 *E. coli* B cells are infected with 10^{10} T6 phage particles and the infected cells are put on an agar surface, about 100 colonies (1 in 10^6) will form. Cultures prepared from these colonies invariably consist of cells that have lost the ability to be infected by T6; in particular, phage will no longer adsorb to the mutant cells. (These mutant cells have of course not been produced by the infection but existed in the culture and were selected by growth with an excess of phage.) These phage-resistant bacterial mutants are called T6-resistant (Figure 5-9), which is denoted either B/6 (read "B-bar-six") or Tsx-r. If 10^8 Tsx-r cells are used to form a bacterial lawn on agar and 10^8 T6 phage are added, about 10 plaques will result. The phage in these plaques carry a mutation in the tail fiber gene and have thereby regained the ability to adsorb to Tsx-r cells. These phage are called *h* mutants (for *h*ost range). They usually retain the ability to form plaques on Tsx-s bacteria and hence are said to have an "extended host range." Phage-resistant bacterial mutants arise in infections with all known phage species; also, *h* mutants can be obtained by infection of any of these strains.

An interesting phenomenon occurs if a bacterial culture is infected with both wild-type (T6h^+) and T6h phages at an MOI such that all bacteria are

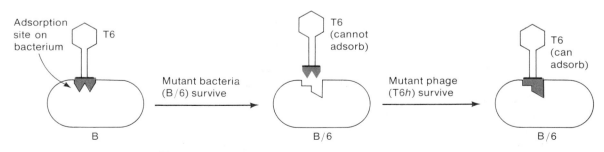

Figure 5-9 The generation of a phage-resistant bacterium, and a phage *h* mutant. Although the tail fiber is modified, for graphic purposes, the tail plate has been modified in the drawing.

infected with both phage types. One would expect that all progeny phage would form plaques on strain B (which plates both phage types) and half of the progeny (the h mutants) would plate on B/6. However, only one-fourth form plaques on B/6. The reason is that during the phage-assembly process no mechanism exists to match a DNA molecule with the h genotype with an h-type tail fiber. Thus, half of the phage with h DNA are in h^+ particles and cannot adsorb to the B/6 bacteria. Similarly, there are h^+ DNA molecules in h particles; they can adsorb to B/6 but produce h^+ progeny, which cannot go through further cycles of infection. This phenomenon is called **phenotypic mixing.** If the initial lysate resulting from the mixed infection is allowed to infect a B culture at an MOI well below 1, no phenotypically mixed progeny result.

HOST RESTRICTION AND MODIFICATION

Another specificity factor in phage infection is the inability of a "foreign" bacterial RNA polymerase to recognize phage DNA sequences for initiation of RNA synthesis. However, even when adsorption and mRNA synthesis are possible, with most bacteria there is usually another barrier called **host restriction** and **host modification.** This is a phenomenon in which a bacterium of a type X is able to distinguish a phage that has been grown in type-X bacterium from one grown in a different type such as Y and is able to prevent the phage grown in Y from carrying out a successful infection. The notation used to discuss this phenomenon is the following: a phage P grown in a bacterium X is denoted P·X. Host modification and restriction are illustrated by the data in Table 5-2. Note that λ·K, which has been grown in *E. coli* strain K, forms plaques at a low efficiency in strain B. Thus, λ·K is **restricted** by strain B. The phage population in these rare plaques (λ·B) has been modified by strain B, so the phage grow efficiently in strain B; however, λ·B now fails to grow in strain K—that is, it is restricted. The molecular

Table 5-2 The restriction and modification pattern of *E. coli* phage λ

Bacterial strain	Phage		
	λ·K	λ·B	λ·C
K	1	10^{-4}	10^{-4}
B	10^{-4}	1	10^{-4}
C	1	1	1

Note: Numbers indicate relative plating efficiency.

explanation for this is the following: *E. coli* B contains an enzyme called a restriction endonuclease (Chapter 2)—specifically, it is the EcoB nuclease—a site-specific nuclease that cuts DNA strands only near a specific base sequence (most restriction enzymes cut within a target sequence, but EcoB cuts near the sequence, see Chapter 20). Phage λ·K contains this sequence; when its DNA is injected into *E. coli* B, the phage DNA is broken. *E. coli* B also contains this sequence and would destroy its own DNA were the sequence not modified. A site-specific methylating enzyme (EcoB methylase) methylates an adenine in the sequence, thereby rendering the sequence resistant to the EcoB nuclease. When λ·K infects strain B, a few parental phage-DNA molecules in the large population of infected cells are methylated before they are restricted. All progeny DNA molecules are already methylated on one strand and the newly synthesized strands are also methylated rapidly (see Chapter 8), and restriction is avoided in these rare phage. Thus, a small population of phage having the B modification (λ·B) is produced. *E. coli* K also contains a restriction enzyme (EcoK). It attacks a base sequence that is different from the sequence recognized by EcoB. An EcoK methylase also protects *E. coli* K from self-destruction, producing the K modification. A λ phage that has always been grown in strain K—namely, λ·K—is methylated in the EcoK-specific sequence and is resistant to EcoK nuclease. However, λ·B has an unmethylated EcoK sequence, so λ·B DNA is usually broken when a strain K cell is infected. Occasionally, a λ·B DNA molecule escapes restriction and replicates, and its replicas have a methylated K-specific sequence. Thus, the rare progeny phage that result when λ·B successfully infect *E. coli* K are λ·K; they now lack the B modification and are restricted when infecting strain B.

Note in Table 5-2 that λ grown on strain C—i.e., λ·C—also fails to grow well in strains B and K, but neither λ·B nor λ·K is restricted by strain C. The reason for the lack of restriction is that strain C has no restriction nuclease active against any base sequence in λ DNA. The λ·C phage are restricted by both strains B and K because, of course, strain C does not have the EcoB and EcoK methylases.

Host restriction and modification is a widespread process, probably serving to destroy foreign DNA.

THE LYSOGENIC CYCLE

Lysogeny is an alternative reproductive pathway to the lytic cycle described earlier. It is best understood for *E. coli* phage λ.

Outline of the Lysogenic Life Cycle

There are two types of lysogenic cycles. The most common one, for which the *E. coli* phage λ pathway is the prototype, follows (Figure 5-10):

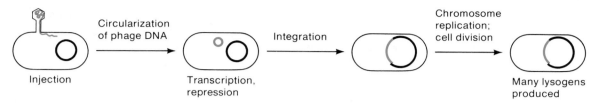

Figure 5-10 The general mode of lysogenization by insertion of phage DNA into a bacterial chromosome.

1. A DNA molecule is injected into a bacterium.
2. After a brief period of mRNA synthesis, which is needed to synthesize a repressor protein (which inhibits the synthesis of the mRNA species that encode the lytic functions) and a DNA-insertion enzyme, phage mRNA synthesis is turned off by the repressor.
3. A phage DNA molecule, typically a replica of the injected molecule, is inserted into the DNA of the bacterium.
4. The bacterium continues to grow and multiply, and the phage genes replicate as part of the bacterial chromosome.

The second and less common type, for which *E. coli* phage P1 is the prototype, differs from the preceding one in that there is no DNA-insertion system and the phage DNA becomes a plasmid (an independently replicating circular DNA molecule) rather than a segment of the host chromosome. In this chapter we will mainly consider the first type.

General Properties of Lysogens and Lysogenization

The following terms describe various aspects of lysogeny.

1. A phage capable of entering either a lytic or a lysogenic life cycle is a **temperate** phage.
2. A bacterium containing a complete set of phage genes is called a **lysogen.**
3. The process of forming a lysogen by infecting a bacterial culture with temperate phage is **lysogenization.**
4. If the phage DNA is contained within the bacterial DNA, the phage DNA is said to be **integrated.** The process by which this state of the DNA is achieved is called **integration** or **insertion.** Phage DNA in plasmid form is nonintegrated. Both integrated and plasmid phage DNA are called a **prophage.**

Two significant properties of lysogens are the following:

1. A lysogen cannot be reinfected by a phage of the type that first lysogenized the cell; this resistance to "superinfection" is called **immunity.**

2. Even after many cell generations, a lysogen can initiate a lytic cycle; in this process, which is called **induction,** the phage genes are excised as a single segment of DNA (Figure 5-11).

The molecular mechanism for immunity and the circumstances that give rise to induction will be discussed in Chapter 17.

More than 90 percent of the thousands of known phages are temperate. These phages are often unable to produce bursts as large as many highly virulent phages such as T4 and T7, but compensate by their ability to multiply in environmental conditions that are not suitable for rapid production of progeny. The meaning of this statement will become clear when we examine the possible outcomes of a single phage encountering a population of bacteria in nature.

Let us first consider a bacterial population that is actively dividing. If a phage can infect one cell and multiply (in a lytic cycle), the number of progeny phage will increase rapidly, as we saw in Table 5-1. However, if the bacteria were growing very slowly owing to exhaustion of nutrients in the surrounding medium, the phage infection might abort if the infected cell were to stop growing during the phage life cycle. (Remember that a phage can grow only in a bacterium that is actively metabolizing.)

When bacteria are starved of nutrients, they degrade their own mRNA and protein before they become dormant. Restoration of nutrients enables the bacteria to grow again. This is not true of a phage-infected cell in which the phage life cycle has been interrupted: usually the ability to produce phage is permanently lost, probably because the delicate balance of phage functions is destroyed by the protein and mRNA degradation. Furthermore, the cell dies. This deleterious outcome (permanent loss of both phage and bacteria) can be avoided if a lysogenic cycle is possible—that is, if the phage DNA can become dormant. When growth of the bacterium resumes, the phage genes replicate as part of the chromosome. Even though phage production has been suspended for the time being, because of the induction phenomenon, the potential for phage production remains.

Now let us return to an infection of an actively growing bacterial population in which phage are multiplying rapidly. When the number of phage

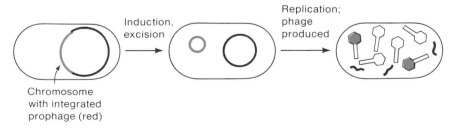

Chromosome
with integrated
prophage (red)

Figure 5-11   An outline of the events in prophage induction. The prophage DNA is in red. The bacterial DNA is omitted from the third panel for clarity.

exceeds the number of bacteria, the phage enter their final cycle of multiplication; after lysis occurs, no further multiplication is possible because there are no more sensitive bacteria. It is possible that years could pass before these phage particles might encounter another sensitive host bacterium and during this time various deleterious agents might damage the particles. Until a host cell appears, the phage particles have no chance to increase in number. However, if lysogenization could occur at a high multiplicity of infection (MOI), the phage genes could be maintained indefinitely, since the lysogen would grow whenever nutrients were available. Indeed, the two conditions that stimulate a lysogenic response of a temperate phage are (1) depletion of nutrients in the growth medium and (2) a high multiplicity of infection.

Phage species that are unable to lysogenize frequently possess exceedingly stable head and tail structures. For example, samples of phages T2 and T4 have maintained constant plaque-forming ability after more than 20 years in a refrigerator. In contrast, samples of λ usually begin to lose plaque-forming ability after several weeks.

Prophage Insertion with *E. coli* Phage λ

When λ DNA integrates, it is inserted at a preferred position in the *E. coli* chromosome, between the *gal* and *bio* (biotin) genes. The insertion site is called the λ attachment site and designated *att*, or, specifically, *att*λ. For most temperate phages there is only one integration site. Experiments to be described in Chapter 17 have shown that integration is a result of recombination between an attachment site in the phage DNA and one in the bacterial DNA. The attachment sites have a common base sequence, designated *O*, in which the exchange occurs, and are flanked by sequences that are specific to the bacterium and to the phage. The bacterial and phage attachment sites are written **BOB'** and **POP'**, respectively. The insertion process is shown schematically in Figure 5-12. The essence of the mechanism is circularization of λ DNA followed by physical breakage and rejoining of phage and host DNA—precisely in the two *O* regions. The exchange is catalyzed by a phage enzyme **integrase.** Since the attachment sites in the phage and bacteria are not the same, the **prophage attachment sites** differ from the bacterial and phage sites and from each other—namely, they are **BOP'** and **POB'**. Thus, the excision reaction is an exchange between these sites; another phage enzyme, **excisionase,** is required, along with integrase, to catalyze excision. Both insertion and excision are examples of site-specific recombination.

Nonintegrative Lysogeny

Most temperate phages form lysogens in the way described for λ—namely, a prophage is inserted at a unique site in the host chromosome. Phages have been observed for which there are several chromosomal *att* sites but this is

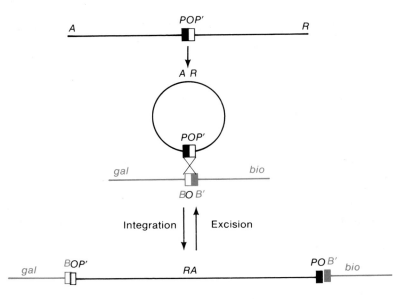

Figure 5-12   The mechanism of prophage integration and excision of phage λ. The phage attachment site has been denoted *POP'* in accord with subsequent findings. The bacterial attachment site is *BOB'*. The prophage is flanked by two new attachment sites denoted *BOP'* and *POB'*.

rare. We mentioned earlier that lysogeny with *E. coli* phage P1 is markedly different in that its prophage is not inserted into the chromosome. Following infection, P1 DNA circularizes and, like λ, is repressed. In the lysogenic mode it remains as a free supercoiled DNA molecule, roughly one or two per cell. Once per bacterial life cycle the P1 DNA replicates and somehow this replication is coupled to chromosomal replication (the coupling is controlled by a phage gene). When the bacterium divides, each daughter cell receives a P1 circle. How this orderly assortment is accomplished is unknown, but some suggestions will be given in Chapter 11 when segregation of plasmid DNA is discussed. The mechanism of prophage maintenance is not as foolproof in phage P1 as in temperate phages that insert their phage DNA into a chromosome; for example, in each round of cell division about 1 cell per 1000 fails to receive a P1 circle. It is not known whether this is due to occasional failure in replication or to imperfect segregation of plasmids into the daughter cells.

Plaques of Temperate Phages

When a virulent phage forms a plaque on a lawn of growing bacteria, the plaque is clear because all bacteria in the center of the plaque are killed and lysed. However, temperate phages, such as λ, form a plaque with a *turbid*

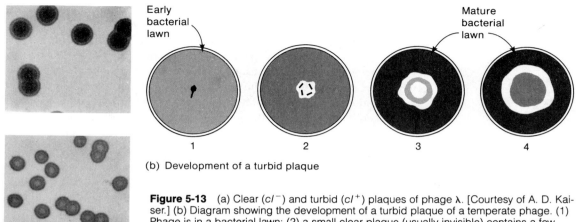

Early bacterial lawn

Mature bacterial lawn

1   2   3   4

(b) Development of a turbid plaque

Figure 5-13   (a) Clear (*cl⁻*) and turbid (*cl⁺*) plaques of phage λ. [Courtesy of A. D. Kaiser.] (b) Diagram showing the development of a turbid plaque of a temperate phage. (1) Phage is in a bacterial lawn; (2) a small clear plaque (usually invisible) contains a few lysogens (shown as rods); (3) the clear region enlarges but lysogens grow within the plaque; (4) clear region reaches maximum size and lysogens stop growing as nutrient is exhausted.

(a)

center (Figure 5-13). The turbidity is caused by the growth of phage-immune lysogenic cells in the plaque. When plating phage to obtain plaques, phage and bacteria are usually mixed in soft agar in a ratio of about 1 phage per 10^7 bacteria. The bacteria grow rapidly and the MOI is low, so the lytic cycle ensues. After several lytic cycles, the local MOI becomes high and a few cells are lysogenized; since availability of nutrients does not yet limit development of phage or cells, most cells are lysed. When the nutrients in the agar are depleted, the uninfected cells stop growing, and the plaque stops increasing in size. However, since there has been less bacterial growth within the plaque, nutrients are still present there. Therefore, the lysogenic cells, which are immune to subsequent infection by λ, continue to grow, forming a turbid center in the plaque.

Mutations in the immunity system are recognized by the clear plaques produced by the mutants. This phenomenon will be discussed further in Chapter 17.

COMPLEMENTATION WITH PHAGE

In Chapter 1 complementation in bacteria was described. The concept is also applicable to phages and is a common way to assign phage mutations to particular genes. The equivalent of the partial diploid test is the infection of a bacterium with two mutant phages, neither of which can grow alone; one then notes whether phage are produced by the doubly infected cells. This test is usually done with conditional mutants in the following way. Consider two temperature-sensitive mutations a^-(Ts) and b^-(Ts) in genes a and b of two respective phage; neither gene product is active at 42°C, so the mutants cannot grow at 42°C and phage production depends upon the complementation of separate Ts mutations. The phage are allowed to adsorb at 42°C to a host bacterium and then the infected cell is incubated at that temperature. If only a^-(Ts) or only b^-(Ts) phage have adsorbed, no phage will be produced. However, if both have adsorbed and the Ts$^-$ mutations are in different genes, the infected cell will contain a good copy of product A and a good copy of product B; complementation can then occur and progeny phage will be produced. Note that in addition to the parental types of a^-(Ts) and b^-(Ts) phage that are present in the resulting lysate, there are also some Ts$^+$ and a^-(Ts)b^-(Ts) recombinants.

ADDENDUM

This chapter has presented the basic features of the life cycles of temperate and virulent phages. Methods for counting phage by plating and various parameters of a phage-infected bacterium have been described. Little or no genetics has been discussed, and this chapter could well have been titled Phage Biology. We return to phages, and particularly phage genetics, later in the book. In Chapter 15 we will see the elementary aspects of phage genetics, examining genetic recombination and mapping. Surprising features of certain genetic maps and their molecular explanations will be given. Chapter 15 goes on to emphasize *E. coli* phage T4, which is the grandfather of phage genetics. Chapters 16 and 17 stresses phage λ, the best-understood phage. Chapter 16 examines the genetics and physiology of its lytic cycle, and Chapter 17 considers λ lysogeny. In Chapter 17 the role of genetic analysis in elucidating the phenomenon of lysogeny will be emphasized, and physical experiments that confirm many conclusions originally derived from genetic analysis will be presented briefly. In Chapter 18 the mechanism by which phages can transfer bacterial genes between different bacteria (transduction) will be examined. These chapters have been relegated to a later point in the book, because a good understanding of phage phenomena requires more knowledge of molecular biology and bacterial genetics than has yet been presented.

PROBLEMS

1. How many DNA molecules are contained in a typical phage particle?

2. What are the three morphological forms of phage particles?

3. How are phage particles usually counted?

4. How many plaques can be formed by a single phage particle?

5. What is the process by which an infected bacterium releases progeny phage?

6. A phage adsorbs to a bacterium in a liquid growth medium. Before lysis occurs, the infected cell is added to a large number of bacteria, and a lawn is allowed to form on a solid medium. How many plaques will result?

7. If 10^6 phage are mixed with 10^6 bacteria and all phage adsorb, what fraction of the bacteria will not have a phage?

8. List the main stages of a phage life cycle.

9. What is meant by term phage-host specificity, and what is the most frequent cause of this specificity?

10. Why can a particular phage not adsorb to any bacterial species?

11. A particular bacterial mutant cannot utilize lactose as a carbon source. If a phage adsorbs to such a bacterium and the infected cell is put in a growth medium in which lactose is the sole carbon source, can progeny phage be produced?

12. Why do phage plaques not enlarge indefinitely?

13. A sample of a wild-type virulent phage grown on strain A of *E. coli* plates on strain X with an efficiency of 10^{-4}. What is the most likely explanation for this low efficiency of plating?

14. Roughly speaking, what is a typical burst size of a phage whose nucleic acid is double-stranded DNA?

15. Since infection of a bacterium by a phage is usually lethal to the bacterium, why have bacteria not evolved to lose their phage receptors?

16. Bacteria are allowed to grow on an agar surface until a confluent turbid layer appears. Then, 10^3 T4 phage are spread on the surface. Six hours later (a time sufficient for plaque formation, if the phage had been added at the time the bacteria were placed on the agar) no plaques are evident. Explain.

17. Phage T4 normally forms small clear plaques on a lawn of *E. coli* strain B. A mutant of *E. coli* called B/4 is unable to adsorb T4 phage particles so that no plaques are formed. T4*h* is a host-range mutant phage capable of adsorbing to *E. coli* B and to B/4 and forms normal looking plaques. If *E. coli* B and the mutant B/4 are mixed in equal proportions and used to generate a lawn, what will be the appearance of plaques made by T4 and T4*h*?

18. In a broth containing glucose and yeast extract, *E. coli* grows with a doubling time of 30 minutes. Phage T7 has a life cycle of 20 minutes under these conditions and a burst size of 200 phage per infected cell. If a culture of 2×10^7 *E. coli* per ml is growing exponentially and 5000 plaque-forming units per ml of T7 phage are added, when will the culture lyse? (In working with this problem assume that phage adsorption is instantaneous and that multiply-infected bacteria give the same burst as singly infected bacteria.)

19. In an experiment in which bacteria and phage are mixed, how does one determine the actual MOI and the number of infective centers?

20. One ml of a bacterial culture at 5×10^8 cells/ml is infected with 10^9 phage. After sufficient time for greater than 99% adsorption, phage antiserum is added to

inactivate all unadsorbed phage. The infected cell is mixed with indicator cells in soft agar, and plaques are allowed to form. If 200 cells are put in each petri dish, how many plaques will be found?

21. P2 and P4 are bacteriophages of *E. coli*. They have the following properties: (1) When one P2 phage infects a bacterium, the bacterium usually bursts, giving about 100 P2 progeny. (2) When a P4 phage infects a bacterium, the bacterium survives because P4 is a defective phage. (3) When P2 phage and P4 phage coinfect the same bacterium, lysis of the bacterium gives 100 P4 progeny and no P2 progeny (because P4 inhibits the growth of P2). If 3×10^8 P2 and 2×10^8 P4 are added to 10^8 bacteria, then

    (a) How many bacteria will not be infected?
    (b) How many bacteria will survive?
    (c) How many bacteria will produce P2 progeny?
    (d) How many bacteria will produce P4 progeny?

22. How does a temperate phage differ from a virulent phage?

23. Does lysogeny occur at low or high multiplicity of infection?

24. At what stage of the growth cycle of a bacterial culture will lysogeny occur at high frequency, if the phage multiplicity is correct?

25. What plaque morphology is produced by a temperate phage?

26. What symbol designates the bacterial and phage *att* sites (i.e., *B, P*, etc.)?

MOLECULAR ASPECTS OF GENE EXPRESSION

Chapter 6

Gene Expression

In this chapter we review how genes are expressed, that is, how the information contained in genes is converted to molecules that determine the metabolism, structure, and form of bacteria and phages. Gene expression is accomplished through a sequence of events in which the information contained in the base sequence of DNA is first copied into an RNA molecule and then used to determine the amino acid sequence of a protein molecule. RNA molecules are synthesized by using the base sequence in a region of one of the DNA strands as a **template** in a polymerization reaction that is catalyzed by an enzyme called an **RNA polymerase.** The process by which the segment corresponding to a particular gene is selected and an RNA molecule is synthesized is called **transcription.** Protein molecules are then synthesized by using the base sequence of this RNA molecule to direct the sequential joining of amino acids in a particular order, so the amino acid sequence is a direct reflection of the DNA base sequence. The production of an amino acid sequence from an RNA base sequence is called **translation.** Some RNA molecules are not translated; these are used instead in several mechanical stages of translation. In this chapter we describe transcription and two features of translation—the genetic code and protein synthesis.

TRANSCRIPTION

The essential chemical characteristics of the enzymatic synthesis of RNA are the following:

1. The precursors in the synthesis of RNA are the four 5′-triphosphates of the ribonucleosides adenosine, guanosine, cytosine, and uridine. (A riboside is a base covalently linked to the sugar ribose.)
2. In forming an RNA molecule a 3′-OH group of the ribose at the 3′ end of the growing RNA molecule reacts with the innermost phosphate of a precursor nucleoside-5′-triphosphate. The two terminal phosphate groups are removed as inorganic pyrophosphate (PP_i), and a sugar-phosphate bond results, extending the RNA molecule by one nucleotide unit (Figure 6-1(a)).

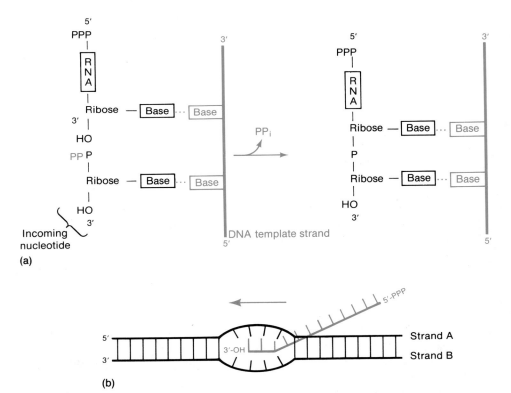

Figure 6-1 RNA synthesis. (a) The polymerization step in RNA synthesis. The incoming nucleotide forms hydrogen bonds (three dots) with a DNA base. Reaction occurs between the OH group in the upper nucleotide (the nucleotide at the 3′ end of the RNA molecule) and the black P in the triphosphate group, leading to removal of the red phosphates (PP_i). (b) Geometry of RNA synthesis. RNA is copied only from strand A of a segment of a DNA molecule. It is not usually copied from strand B in that region of the DNA. However, else-where—in a different gene, for example—strand B might be copied; in that case, strand A would not usually be copied in that region of the DNA. The RNA molecule is antiparallel to the DNA strand being copied.

3. The sequence of bases in an RNA molecule is determined by the base sequence of the DNA template. Each base added to the growing end of the RNA chain is chosen for its ability to base-pair with the DNA template strand; thus, the bases C, T, G, and A in a DNA strand cause G, A, C, and U, respectively, to be added to the growing end of an RNA molecule.

4. Nucleotides are added only to the 3′-OH end of the growing chain. Thus, the 5′ end of a growing RNA molecule is a triphosphate. The growing RNA strand and the DNA template strand are antiparallel to one another, like the two strands of a DNA molecule.

An important feature of RNA synthesis is that *the DNA molecule being copied is double-stranded, yet in any particular region of the DNA only one strand serves as a template*. The implications of this statement are shown in Figure 6-1(b).

The synthesis of RNA consists of five discrete stages: **promoter recognition, local unwinding, chain initiation, chain elongation,** and **chain termination.** These have the following characteristics.

1. RNA polymerase binds to DNA within a specific base sequence (20–200 bases long), called a **promoter.**

2. After the initial binding step local unwinding of the DNA occurs, and the RNA polymerase is said to have formed an **open promoter complex.**

3. The RNA polymerase becomes bound to a **polymerization start site,** which is very near the initial binding site. The first nucleoside triphosphate is placed at this site and synthesis begins.

4. RNA polymerase then moves along the DNA, adding nucleotides to the growing RNA chain.

5. RNA polymerase reaches a chain-termination sequence and both the newly synthesized RNA and the polymerase are released.

Promoter sequences have been isolated by mixing RNA polymerase with DNA and then treating the DNA with enzymes that digest DNA but leave intact the region bound to the RNA polymerase. After enzymatic digestion the RNA polymerase can be removed and the base sequence of the DNA determined. Examination of a large number of promoter sequences for different genes and from different bacteria has shown that they have many features in common. For example, the sequence TATAAT (or a nearly identical sequence)—often called a **Pribnow box** or **− 10 region**—is found as part of all prokaryotic promoters. A second sequence, the − 35 region, is also common and is located about 35 base pairs upstream (away from the direction of transcription) from the polymerization start site. These recognition sequences instruct RNA polymerase where to start synthesis. The strength of the binding of RNA polymerase to different promoters varies greatly; this variation is a fundamental mechanism for regulating gene expression. Some promoters have binding sites for proteins other than RNA polymerase, and for these promoters the site must be occupied by that protein for RNA polymerase to bind correctly. One of the most common sites of this type is one that binds a complex of cyclic AMP and the cyclic AMP receptor protein (CRP). Activity of these promoters is regulated by the concentration of cyclic AMP. An example of such regulation will be seen in Chapter 7.

Two kinds of termination events are known: those that are self-terminating (dependent on the DNA base sequence only) and those that require the presence of the termination protein **Rho.** Both types of events occur at specific but distinct base sequences. Self-termination usually occurs at base sequences in the template DNA strand that consist of a series of adjacent adenines preceded by a nucleotide palindrome that is usually interrupted by a few bases. It is believed that the RNA molecule folds back on itself to form a stem-and-loop configuration (Figure 6-2). Rho-dependent sequences have no distinguishing features that have yet been recognized.

Initiation of a second round of transcription need not await completion of the first, for the promoter becomes available once RNA polymerase has polymerized 50–60 nucleotides. For a rapidly transcribed gene such reinitiation occurs repeatedly, and a gene can be cloaked with numerous RNA molecules in various degrees of completion.

In a discussion in Chapter 1 of the mutant approach to solving biochemical problems it was pointed out that mutations can define a function. The

Figure 6-2 Base sequence of (a) the DNA of the *E. coli trp* operon at which transcription termination occurs and of (b) the 3′ terminus of the mRNA molecule. The inverted-repeat sequence is indicated by reversed red arrows. The mRNA molecule is folded to form a stem-and-loop structure. The relevant regions are labeled in red; the terminal sequence of U's in the mRNA is shaded in red.

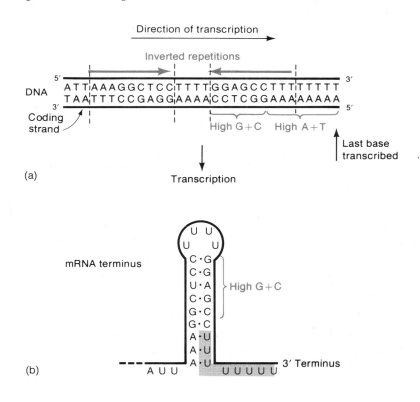

Table 6-1   Effect of promoter mutations on transcription of the *lacZ* gene

Genotype	DNA molecule contains both a p^- mutation and the *lacZ*$^+$ gene	Transcription of *lacZ*$^+$ gene
1. p^+lacZ^+	No p^-	Yes
2. p^-lacZ^+	Yes	No
3. p^+lacZ^+/p^+lacZ^-	No p^-	Yes
4. p^-lacZ^+/p^+lacZ^-	Yes	No
5. p^+lacZ^+/p^-lacZ^-	No	Yes

discovery of the promoter exemplifies this process since the existence of promoters was first demonstrated by the isolation of particular *E. coli* Lac$^-$ mutations, denoted p^-, that eliminate activity of the *lac* genes. These mutations did not map within any *lac* gene but were located adjacent to them. Furthermore, complementation tests showed that *lac*-gene expression was inhibited only with a mutation adjacent to the genes *in the same DNA molecule*. This feature can be seen by examining a partial diploid cell having two copies of the *lacZ* gene—for example, a cell containing a small DNA element called the F′*lacZ* plasmid. The presence of the *lacZ* gene enables the cell to synthesize the enzyme β-galactosidase. Table 6-1 shows that a wild-type *lacZ* gene is inactive when a p^- mutation is present on the same DNA molecule (either the chromosome or an F′ plasmid); this can be seen by comparing entries 4 and 5. A complete understanding of the nature of these mutations came from chemical analyses, which showed that in a cell with the genotype shown in entry 4 in the table, the *lacZ* gene is not transcribed, though transcription does occur if the genotype is that of entry 5. The p^- mutations are called **promoter mutations**.

The best-understood RNA polymerase is that of the bacterium *E. coli*. This enzyme, which consists of five protein subunits, is one of the largest enzymes known and can be easily seen by electron microscopy (Figure 6-3). The five subunits are organized in two groups. Four of the subunits comprise the **core enzyme**, which catalyzes the joining of the nucleoside triphosphates

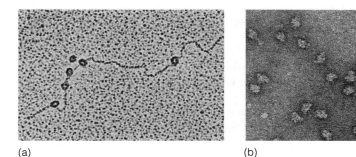

(a)   (b)

Figure 6-3   Electron micrographs of *E. coli* RNA polymerase. (a) Molecules bound to DNA. (b) The holoenzyme viewed by negative-contrast electron microscopy. ($\times$ 270,000). [Courtesy of Robley Williams.]

to the RNA; the fifth, the **σ subunit,** is required for promoter binding. Once polymerization begins, the σ subunit dissociates from the core enzyme. When transcription is completed, the core enzyme binds a different σ subunit and is then ready to bind to a promoter again.

MESSENGER RNA

Amino acids do not bind to DNA. Thus, intermediate steps are needed for arranging the amino acids in a polypeptide chain in the order determined by the DNA base sequence. This process begins with transcription of the base sequence of one of the DNA strands (the **coding** or **sense strand**) into the base sequence of an RNA molecule. In prokaryotes this RNA molecule, which is called **messenger RNA** or **mRNA,** is used directly in polypeptide synthesis. The amino acid sequence is obtained from mRNA by the protein-synthesizing machinery of the cell.

In prokaryotes mRNA molecules commonly contain information for the amino acid sequences of several different polypeptide chains; in this case, such a molecule is called **polycistronic mRNA.** (Cistron is a term commonly used to mean a base sequence encoding a single polypeptide chain.) The genes contained in a polycistronic mRNA molecule often encode the different proteins of a metabolic pathway. For example, in *E. coli* the ten enzymes needed to synthesize histidine are encoded in one mRNA molecule. The use of polycistronic mRNA is an economical way for a cell to regulate synthesis of related proteins in a coordinated way. For example, in prokaryotes the usual way to regulate synthesis of a particular protein is to control the synthesis of the mRNA molecule that encodes it (Chapter 7). With a polycistronic mRNA molecule the synthesis of several related proteins can be regulated by a single signal, so that appropriate quantities of each protein are made at the same time; this is termed **coordinate regulation.**

Not all base sequences in an mRNA molecule are translated into the amino acid sequences of polypeptides. For example, translation of an mRNA molecule rarely starts exactly at one end of the RNA molecule and proceeds to the other end; instead, initiation of polypeptide synthesis may begin hundreds of nucleotides from the 5′-P terminus of the RNA. The section of untranslated RNA before the region encoding the first polypeptide chain is called a **leader,** which in some cases contains regulatory sequences that influence the rate of protein synthesis (an example, the *trp*-gene system, will be described in Chapter 7). Untranslated sequences are found at both the 5′-P and the 3′-OH termini. Polycistronic mRNA molecules usually contain **spacer sequences** tens of bases long, which separate the **coding sequences;** each coding sequence corresponds to a polypeptide chain.

The coding sequence of each gene is obtained by transcription of only one DNA strand. It is quite rare for the two complementary base sequences in a particular gene to be transcribed, though a few exceptions are known in which two adjacent genes are transcribed from different strands, with a slight overlapping of the template segments. However, except for some small phages and some transposable elements (Chapter 12), *all mRNA molecules are not synthesized from the same DNA strand*, so in an extended segment of a DNA molecule, mRNA molecules would be seen growing in either of two directions (Figure 6-4), depending on which DNA strand functions as a template.

In bacteria most mRNA molecules are degraded within a few minutes after synthesis. This degradation enables cells to dispense with molecules that are no longer needed.

There are major differences between prokaryotes and eukaryotes in the relation between the transcript and the mRNA used for polypeptide synthesis. In prokaryotes the immediate product of transcription (called the **primary transcript**) is mRNA; in contrast, in eukaryotes the primary transcript must be converted to mRNA. This conversion, which is called **RNA processing,** consists of two types of events—modification of the termini and excision of untranslated sequences embedded *within* coding sequences. These events are illustrated diagramatically in Figure 6-5.

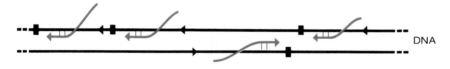

Figure 6-4 Schematic drawing showing that complementary DNA strands can be transcribed, but not from the same region of DNA. Promoters are indicated by black arrowheads, and termination sites by black bars. Promoters are present in both strands. Termination sites are usually located such that transcribed regions do not overlap, but this is not always the case.

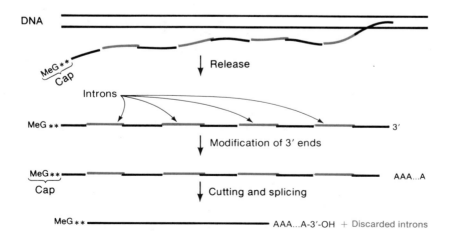

Figure 6-5 Schematic drawing showing production of eukaryotic mRNA. The primary transcript is capped before it is released. Then, its 3'-OH end is modified, and finally the introns are excised. MeG denotes 7-methylguanosine, and the two asterisks indicate the two nucleotides whose riboses are methylated.

The untranslated sequences within genes are called **intervening sequences** or **introns.** They are present in almost all eukaryotic transcripts but are rare in the free-living unicellular eukaryotes, such as yeast. Only one case is known of an intron in prokaryotic RNA. Introns are excised from the primary transcript and the remaining fragments are rejoined to form the mRNA molecule (Figure 6-5). Intron excision and the joining of coding sequences (**exons**) to form an mRNA molecule is called **RNA splicing.** This process is of no importance in classical microbial genetics, but we will see that it is something to be reckoned with in the cloning of eukaryotic genes by recombinant DNA techniques (Chapter 20).

TRANSLATION

The synthesis of every protein molecule in a cell is directed by an mRNA intermediate, which is copied from DNA (except for RNA viruses). Numerous steps follow mRNA production and these can conveniently be grouped into two categories: *information-transfer processes,* by which the RNA base sequence is converted to an amino acid sequence, and *chemical processes,* by which the amino acids are linked together. The complete series of events is called **translation.**

The translation system consists of four major components:

1. **Ribosomes.** These are particles on which the mechanics of protein synthesis is carried out. They contain the enzyme needed to form a peptide bond between amino acids, a site for binding one mRNA molecule, and sites for bringing in and aligning the amino acids in preparation for assembly into the finished polypeptide chain.

2. **Transfer RNA, or tRNA.** Amino acids do not bind to mRNA, but in the synthesis of a particular polypeptide, they must be ordered by the base sequence in the mRNA molecule. This ordering is accomplished by a set of adaptor molecules, the tRNA molecules. A tRNA molecule "reads" the base sequence of mRNA. The language read by the tRNA molecules is called the **genetic code,** which is a set of relations between sequences of three adjacent bases on an mRNA molecule and particular amino acids.

3. **Aminoacyl tRNA synthetases.** This set of enzymes catalyzes the attachment of an amino acid to its corresponding tRNA molecule.

4. **Initiation, elongation,** and **release factors.** These molecules are proteins needed at particular stages of polypeptide synthesis. They will not be discussed in this book.

In prokaryotes all of these components are present throughout the cell.

In outline, the mechanism of protein synthesis can be depicted as in Figure 6-6. An mRNA molecule binds to the surface of a protein-synthesizing particle, the ribosome. Appropriate tRNA-amino acid complexes, which have

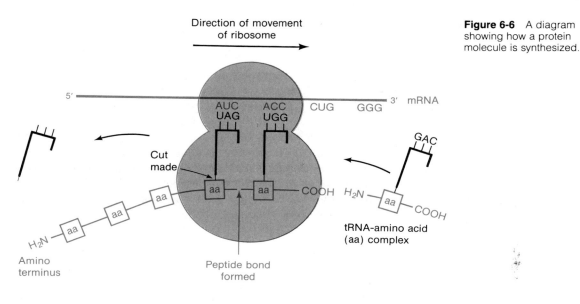

Figure 6-6 A diagram showing how a protein molecule is synthesized.

been made elsewhere by the aminoacyl tRNA synthetases, bind sequentially, one by one, to the mRNA molecule that is attached to the ribosome. Peptide bonds are made between successively aligned amino acids, each time joining the amino group of the incoming amino acid to the carboxyl group of the amino acid at the growing end. Finally, the chemical bond between the tRNA and its attached amino acid is broken and the completed protein is removed. The protein-synthetic mechanism will be described more fully later in the chapter.

An important feature of translation is that it proceeds in a particular direction, obeying the following rules:

1. RNA is translated from the 5′ end of the molecule toward the 3′ end—but not from the 5′ terminus itself nor all the way to the 3′ end.
2. Polypeptides are synthesized from the amino terminus toward the carboxyl terminus, by adding amino acids one by one to the carboxyl end. For example, a protein having the sequence NH$_2$-Met-Pro- . . . -Gly-Ser-COOH, would have started with methionine, and serine would be the last amino acid added to the chain.

These rules are illustrated schematically in Figure 6-7.

3′ ———————————— 5′	Coding strand of DNA
5′ ———————————▶ 3′	Synthesis of RNA
NH$_2$ ———————▶ COOH	Synthesis of protein

Figure 6-7 Direction of synthesis of RNA with respect to the coding strand of DNA, and synthesis of protein with respect to mRNA.

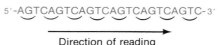

5′-AGTCAGTCAGTCAGTCAGTCAGTC-3′

Direction of reading

Figure 6-8 Bases in an RNA molecule are read sequentially in the 5′ → 3′ direction, in groups of three.

It is conventional when writing nucleotide sequences to place the 5′ terminus at the left, and with amino acid sequences, to place the NH_2 terminus at the left. Thus, polynucleotides are generally written to show both synthesis and translation from left to right, and polypeptides are also written to show synthesis from left to right. This convention is used in all of the following sections concerning the genetic code.

THE GENETIC CODE

Only four bases in DNA serve to specify 20 amino acids in proteins, so some combination of the bases is needed for each amino acid. Actually, more than 20 distinct combinations are needed for polypeptide synthesis, because signals are required for starting and stopping the synthesis of particular polypeptide chains. An RNA base sequence corresponding to a particular amino acid is called a **codon,** and the signal sequences are called **start codons** and **stop codons.** The genetic code is the set of all codons. Before the genetic code was elucidated, it was reasoned that if all codons were assumed to have the same number of bases, then each codon would have to contain at least three bases. Codons consisting of pairs of bases would be insufficient because four bases can form only $4^2 = 16$ pairs, and there are 20 amino acids; triplets of bases would suffice because these can form $4^3 = 64$ triplets. In fact, the genetic code is a **triplet code,** and all 64 possible codons carry information of some sort. Several different codons designate the same amino acid. Furthermore, in translating mRNA molecules the codons do not overlap but are used sequentially (Figure 6-8).

Codons and Features of the Code

The same genetic code is used by almost all biological systems and hence is said to be universal (the only known exceptions are mitochondria and a few unusual microorganisms). The complete code is shown in Table 6-2. The following features of the code should be noted:

1. Sixty-one codons correspond to amino acids.
2. Four codons are signals. These are the three stop codons—**UAA, UAG, UGA**—and the one start codon, **AUG.** The start codon also specifies the amino acid methionine. In rare cases, certain other codons (e.g., **GUG**) initiate translation.

Table 6-2 The "universal" genetic code

First position (5' end)	Second position				Third position (3' end)
	U	C	A	G	
U	Phe	Ser	Tyr	Cys	U
	Phe	Ser	Tyr	Cys	C
	Leu	Ser	Stop	Stop	A
	Leu	Ser	Stop	Trp	G
C	Leu	Pro	His	Arg	U
	Leu	Pro	His	Arg	C
	Leu	Pro	Gln	Arg	A
	Leu	Pro	Gln	Arg	G
A	Ile	Thr	Asn	Ser	U
	Ile	Thr	Asn	Ser	C
	Ile	Thr	Lys	Arg	A
	[Met]	Thr	Lys	Arg	G
G	Val	Ala	Asp	Gly	U
	Val	Ala	Asp	Gly	C
	Val	Ala	Glu	Gly	A
	[Val]	Ala	Glu	Gly	G

Note: The boxed codons are used for initiation.

3. In most cases several codons direct the insertion of the same amino acid into a protein chain; that is, the code is highly **redundant.** Only tryptophan and methionine are specified by one codon.

4. The redundancy is not random; except for Ser, Leu, and Arg, all codons corresponding to the same amino acid are in the same box. That is, *synonymous codons usually differ only in the third base*. For example, GGU, GGC, GGA, and GGG all code for glycine. The biochemical basis for redundancy will be discussed shortly when the phenomenon of wobble is described.

The codon assignments shown in Table 6-2 are completely consistent with all chemical observations and with the amino acid sequences of wild-type and mutant proteins. In every case in which a mutant protein differs by a single amino acid from the wild-type form, the amino acid substitution can be accounted for by a single base change between the codons corresponding to the two different amino acids. For example, substitution of proline by serine, which is a common mutational change, can be accounted for by the single base changes CCC → UCC, CCU → UCU, CCA → UCA, and CCG → UCG.

Transfer RNA and the Aminoacyl Synthetases

The decoding operation by which the base sequence within an mRNA molecule becomes translated to an amino acid sequence of a protein is accomplished by the tRNA molecules and a set of enzymes, the **aminoacyl tRNA synthetases.**

The tRNA molecules are small, single-stranded nucleic acids ranging in size from 73 to 93 nucleotides. Like all RNA molecules, they have a 3'-OH terminus, but the opposite end terminates with a 5'-monophosphate rather than a 5'-triphosphate (the reason will become clear in the final section of this chapter). Internal complementary base sequences form short double-stranded regions, causing the molecule to fold into a structure in which open loops are connected to one another by double-stranded stems (Figure 6-9). In two dimensions a tRNA molecule is drawn as a planar cloverleaf. Its three-dimensional structure is more complex.

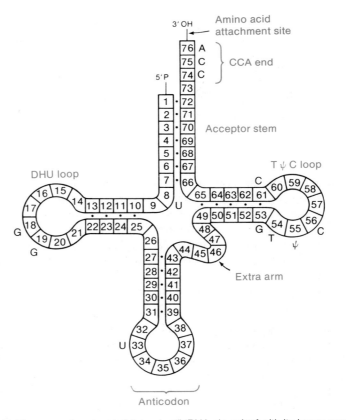

Figure 6-9 The currently accepted "standard" tRNA cloverleaf with its bases numbered. A few bases present in almost all tRNA molecules are indicated.

Three regions of each tRNA molecule are used in the decoding operation. One of these regions is the **anticodon,** a sequence of three bases that can base-pair with a codon sequence in the mRNA. *No normal tRNA molecule has an anticodon complementary to any of the stop codons UAG, UAA, or UGA, which is why these codons are stop signals.* A second site is the **amino acid attachment site,** the 3' terminus of the tRNA molecule; the amino acid corresponding to the particular mRNA codon that base-pairs with the tRNA anticodon is covalently linked to this terminus. These bound amino acids are joined together during polypeptide synthesis. A specific aminoacyl tRNA synthetase matches the amino acid with the anticodon; to do so, the enzyme must be able to distinguish one tRNA molecule from another. The necessary distinction is provided by an ill-defined region encompassing many parts of the tRNA molecule and called the **recognition region.**

The different tRNA molecules and synthetases are designated by stating the name of the amino acid that can be linked to a particular tRNA molecule by a specific synthetase; for example, leucyl-tRNA synthetase attaches leucine to tRNA^{Leu}. When an amino acid has become attached to a tRNA molecule, the tRNA is said to be **acylated** or **charged.** An acylated tRNA molecule is designated in several ways. For example, if the amino acid is glycine, the acylated tRNA would be written glycyl-tRNA or Gly-tRNA. The term **uncharged tRNA** refers to a tRNA molecule lacking an amino acid, and **mischarged tRNA** to one acylated with an incorrect amino acid.

Accurate protein synthesis, the placement of the "correct" amino acid at the appropriate position in a polypeptide chain, requires (1) attachment of the correct amino acid to a tRNA molecule by the synthetase and (2) fidelity in codon-anticodon binding.

Redundancy and the Wobble Hypothesis

Several features of the genetic code and of the decoding system suggest that something is missing in the explanation of codon-anticodon binding. First, the code is highly redundant. Second, the identity of the third base of a codon appears to be unimportant; that is, XYU, XYA, XYG, and XYC, where XY denotes any sequence of first and second bases, often correspond to the same amino acid. (Codons, like all RNA sequences, are by convention written with the 5' end at the left; thus, the first base in a codon is at the 5' end and the third base is at the 3' end.) Third, the number of distinct tRNA molecules that have been isolated from a single organism is less than the number of codons; since all codons are used, the anticodons of some tRNA molecules must be able to pair with more than one codon. Experiments with several purified tRNA molecules showed this to be the case.

These observations have been explained by the **wobble hypothesis,** which also provides insight into the pattern of redundancy of the code. Wobble refers to the less stringent requirement for base pairing at the third position

Table 6-3   Allowed pairings
according to the wobble hypothesis

Third position codon base	First position anticodon base
A	U,I
G	C,U
U	G,I
C	G,I

of the codon than at the first two positions. That is, the first two bases must form pairs of the usual type (A with U, or G with C), but the third base pair can be of a different type (for example, G with U). This observation was derived from the discovery that that the anticodon of yeast tRNA^{Ala} contains the nucleoside **inosine, I,** in the position that pairs with the third base of the codon (the first position of the anticodon). Later analyses of other tRNA molecules showed that inosine was common in this position, though not always present. Inosine can form hydrogen bonds with A, U, and C. In the wobble hypothesis all pairs of bases that can form hydrogen bonds are considered to be possible in the third position of the codon, except purine-purine base pairs, which would cause excessive distortion in the region of the pairing. These possible base pairs are listed in Table 6-3. The wobble hypothesis, later confirmed by direct sequencing of many tRNA molecules, explains the pattern of redundancy in the code in that certain anticodons (for example, those containing U, I, and G in the first position of the anticodon) can pair with several codons.

Unusual bases, of which inosine is one example, are found in several positions in tRNA. In all cases, they are not incorporated as such but are formed by modification of a standard base that is already in the tRNA. For example, an adenosine in the first position of the anticodon is always enzymatically converted to inosine. Adenosines in other positions are unaffected.

The Arrangement of Codons in a Typical Prokaryotic mRNA Molecule

Most prokaryotic mRNA molecules are polycistronic; that is, they contain sequences specifying the synthesis of several proteins. Thus, a polycistronic mRNA molecule must possess a series of start and stop codons for use during translation. If an mRNA molecule encodes three proteins, the minimal requirement would be the sequence

AUG (start),protein 1,stop–AUG,protein 2,stop–AUG,protein 3,stop

in which the stop codons might be UAA, UAG, or UGA. Actually, such an mRNA molecule is probably never so simple in that the leader sequence preceding the first start signal may be several hundred bases long, and spacer sequences that are 5–20 bases long are usually present between one stop codon and the next start codon.

OVERLAPPING GENES

When discussing coding and signal recognition earlier in the chapter, an implicit assumption has been that the mRNA molecule is scanned for start signals to establish the reading frame and that reading then proceeds in a single direction within the reading frame. The idea that several reading frames might exist in a single segment was not considered for many years. The notion of overlapping reading frames was rejected on the grounds that severe constraints would be placed on the amino acid sequences of two proteins translated from the same portion of mRNA. However, because the code is highly redundant, the constraints are actually not so rigid.

If multiple reading frames were used, a single DNA segment would be utilized with maximal efficiency. However, a disadvantage is that evolution could be slowed, because single-base-change mutations would be deleterious more often than if there were a unique reading frame. Nonetheless, some organisms—namely, small viruses and the smallest phages—have evolved having overlapping reading frames.

The *E. coli* phage φX174 contains a single strand of DNA consisting of 5386 nucleotides of known base sequence. If a single reading frame were used, at most 1795 amino acids could be encoded in the sequence and with an average protein size of about 400 amino acids, only 4–5 proteins could be made. However, φX174 makes eleven proteins containing a total of more than 2300 amino acids. This paradox was resolved when it was shown that translation occurs in several reading frames from three mRNA molecules (Figure 6-10). For example, the sequence for protein B is contained totally in the sequence for protein A but translated in a different reading frame. Similarly, the protein-E sequence is totally within the sequence for protein D. Protein K is initiated near the end of gene *A*, includes the base sequence of B, and terminates in gene *C*; synthesis is not in phase with either gene *A* or gene *C*. Of note is protein A′ (also called A*), which is formed by reinitiation within gene *A* and in the same reading frame, so that it terminates at the stop codon of gene *A*. Thus, the amino acid sequence of A′ is identical to a segment of protein A. In total, five different proteins obtain some or all of their primary structure from shared base sequences in φX174. This phenomenon, known as **overlapping genes,** has been observed in the related phage G4 and in the small animal virus SV40. It has not yet been seen in any cellular DNA. It should be realized that the single structural feature responsible for gene overlap is the location of each AUG initiation sequence, because these sequences establish the reading frame.

Figure 6-10 The map of *E. coli* phage φX174, showing the start and stop points for mRNA synthesis and the boundaries of the individual protein products. The solid regions are spacers.

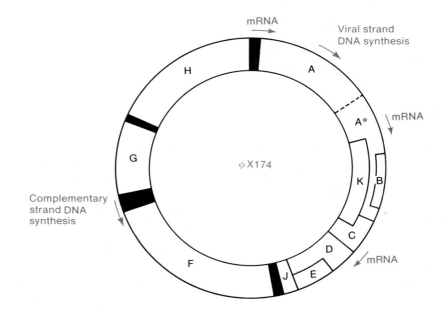

POLYPEPTIDE SYNTHESIS

In preceding sections how the information in an mRNA molecule is converted to an amino acid sequence having specific start and stop points has been discussed. In this section the chemical problem of attachment of the amino acids to one another is reviewed.

Polypeptide synthesis can be divided into three stages—**initiation, elongation,** and **termination.** The main features of the initiation step are the binding of mRNA to the ribosome particle, selection of the initiation codon, and the binding of acylated tRNA bearing the first amino acid. In the elongation stage there are two processes: joining together two amino acids by peptide bond formation, and moving the mRNA and the ribosome with respect to one another so that the codons can be translated successively. In the termination stage the completed protein is dissociated from the synthetic machinery and the ribosomes are released to begin another cycle of synthesis.

We begin our discussion with a description of the structure of the ribosome.

Ribosomes

A ribosome is a multicomponent ribonucleoprotein particle that contains several enzymes needed for protein synthesis and that brings together a single mRNA molecule and charged tRNA molecules in the proper position and orientation so that the base sequence of the mRNA molecule is translated into an amino acid sequence (Figure 6-6). The properties of the *E. coli* ribosome are best understood.

All ribosomes contain two subunits (Figure 6-11). For historical reasons, the intact ribosome and the subunits have been given numbers that describe how fast they sediment when centrifuged. For *E. coli* (and for all prokaryotes) the intact particle is called a **70S ribosome** and the subunits, which are unequal in size and composition, are termed **30S and 50S.** A 70S ribosome consists of one 30S subunit and one 50S subunit.

Both the 30S and the 50S particles contain RNA (called **rRNA** for *ribosomal RNA*) and >50 protein molecules (Figure 6-12). The 30S subunit contains one 16S rRNA molecule, and a 50S subunit contains two RNA molecules—one 5S rRNA molecule and one 23S rRNA molecule. In each particle usually only one copy of each protein molecule is present, though a few are duplicated or modified.

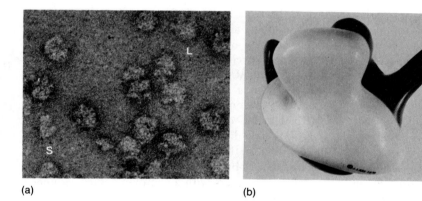

(a) (b)

Figure 6-11 Ribosomes. (a) An electron micrograph of 70S ribosomes from *E. coli*. Some of the ribosomes are oriented as in the model shown in panel (b). A few ribosomal subunits that also lie in the field are identified by the letters S and L, which stand for small and large. (b) A three-dimensional model of the 70S *E. coli* ribosome. The 30S particle is white and the 50S particle is red. [Courtesy of James Lake.]

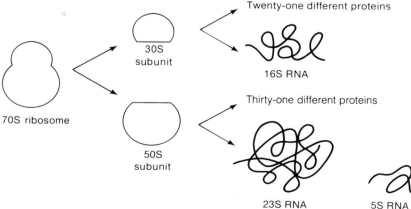

Twenty-one different proteins

16S RNA

70S ribosome

30S subunit

50S subunit

Thirty-one different proteins

23S RNA   5S RNA

Figure 6-12 Dissociation of a prokaryotic ribosome. The configuration of two overlapping circles will be used throughout this chapter, for the sake of simplicity. The correct configuration is shown in Figure 6-11.

Stages of Polypeptide Synthesis in Prokaryotes

An important feature of initiation of polypeptide synthesis in both prokaryotes and eukaryotes is the use of a specific initiating tRNA molecule. In prokaryotes this tRNA molecule is acylated with the modified amino acid N-formylmethionine (fMet); the tRNA is often designated tRNA^{fMet} (Figure 6-13). Both tRNA^{fMet} and tRNA^{Met} recognize the codon AUG, but only tRNA^{fMet} is used for initiation. The tRNA^{fMet} molecule is first acylated with *methionine*, and an enzyme (found only in prokaryotes) adds a formyl group to the amino group of the methionine. (In eukaryotes the initiating tRNA molecule is charged with methionine also, but formylation does not occur.) The use of these initiator tRNA molecules means that *while being synthesized*, all prokaryotic proteins have fMet at the amino terminus. However, this amino acid is frequently deformylated or removed later, and all amino acids have been observed at the amino termini of completed protein molecules isolated from cells.

Polypeptide synthesis in bacteria begins by the association of one 30S subunit (not the entire 70S ribosome), an mRNA molecule, fMet-tRNA, three proteins known as **initiation factors,** and guanosine 5'-triphosphate (GTP). These molecules constitute the **30S preinitiation complex** (Figure 6-14). Since polypeptide synthesis begins at an AUG start codon and AUG codons are found within coding sequences (that is, methionine occurs within a polypeptide chain), some signal must be present in the base sequence of the mRNA molecule to identify a particular AUG codon as a start signal. The means of selecting the correct AUG sequence differs in prokaryotes and eukaryotes. In prokaryotic mRNA molecules a particular base sequence (called the **ribosome binding site** or sometimes the **Shine-Dalgarno sequence**) near the AUG codon used for initiation forms base pairs with a complementary sequence in the 16S rRNA molecule of the ribosome.

Following formation of the 30S preinitiation complex, a 50S subunit joins with this complex to form a **70S initiation complex** (Figure 6-14).

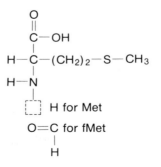

Figure 6-13 Comparison of the chemical structures of methionine and N-formylmethionine.

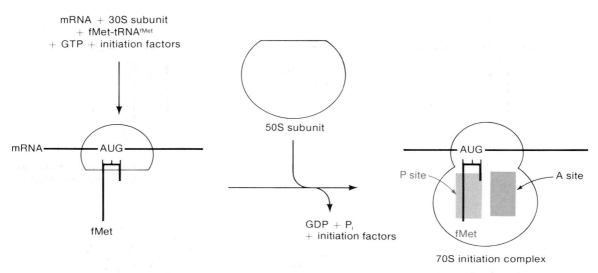

Figure 6-14 Early steps in protein synthesis: in prokaryotes: formation of the 30S preinitiation complex and of the 70S initiation complex.

The 50S subunit contains two tRNA binding sites. These sites are called the **A (aminoacyl) site** and the **P (peptidyl) site.** When joined with the 30S preinitiation complex, the position of the 50S subunit in the 70S initiation complex is such that the fMet-tRNA^{fMet}, which was previously bound to the 30S preinitiation complex, occupies the P site of the 50S subunit. Placement of fMet-tRNA^{fMet} in the P site fixes the position of the fMet-tRNA anticodon such that it pairs with the AUG initiator codon in the mRNA. Thus, *the reading frame is unambiguously defined upon completion of the 70S initiation complex*.

Once the P site is filled, the A site of the 70S initiation complex becomes available to any tRNA molecule whose anticodon can pair with the codon adjacent to the initiation codon. After occupation of the A site a peptide bond between fMet and the adjacent amino acid is formed by an enzyme complex called **peptidyl transferase.** As the bond is formed, the fMet is cleaved from the fMet-tRNA in the P site.

After the peptide bond forms, an uncharged tRNA molecule occupies the P site and a dipeptidyl-tRNA occupies the A site. At this point three movements occur: (1) the tRNA^{fMet} in the P site, now no longer linked to an amino acid, leaves this site, (2) the peptidyl-tRNA moves from the A site to the P site, and (3) the mRNA moves a distance of three bases in order to position the next codon at the A site (Figure 6-15). This step, called **translocation,** requires the presence of an elongation protein **EF-G** and GTP, and it is likely that mRNA movement is a consequence of the tRNA motion. After

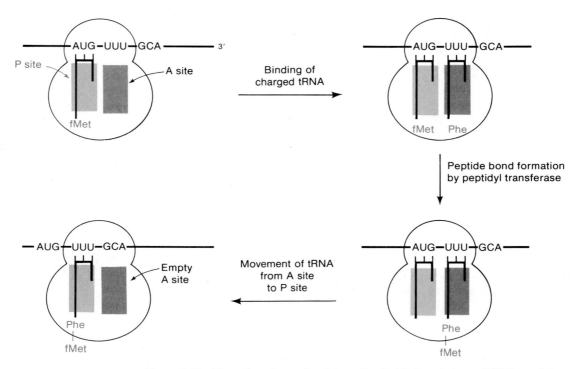

Figure 6-15 Elongation phase of protein synthesis: binding of charged tRNA, peptide bond formation, and movement of mRNA with respect to the ribosome.

mRNA movement has occurred, the A site is again available to accept a charged tRNA molecule having a correct anticodon.

When a chain termination codon (either UAA, UAG, or UGA) is reached, no tRNA exists that can fill the A site, so chain elongation stops. However, the polypeptide chain is still attached to the tRNA occupying the P site. Release of the protein is accomplished by proteins called **release factors,** which also causes dissociation of the 70S ribosome into its 30S and 50S subunits, completing the cycle.

If the mRNA molecule is polycistronic and the AUG codon initiating the second polypeptide is not too far from the stop codon of the first, the 70S ribosome will not always dissociate but will re-form an initiation complex with the second AUG codon. The probability of such an event decreases with increasing separation of the stop codon and the next AUG codon. In some genetic systems the separation is sufficiently great that more protein molecules are always translated from the first gene than from subsequent genes, and this is a mechanism for maintaining particular ratios of gene products (Chapter 7).

COMPLEX TRANSLATION UNITS

The unit of translation is almost never simply one ribosome traversing an mRNA molecule, but is a more complex structure, of which there are several forms. Two examples are given in this section.

After about 25 amino acids have been joined together in a polypeptide chain, an AUG initiation codon will be completely free of the ribosome and a second initiation complex can form. The overall configuration is that of two 70S ribosomes moving along the mRNA at the same speed. When the second ribosome has moved along a distance similar to that traversed by the first, a third ribosome can attach to the initiation site. The process of movement and reinitiation continues until the mRNA is covered with ribosomes at a density of about one ribosome per 80 nucleotides. This large translation unit is called a **polyribosome** or a **polysome,** and this is the usual form of the translation unit. An electron micrograph and a diagram of a polysome are shown in Figure 6-16.

The use of polysomes is particularly advantageous to a cell in that the overall rate of protein synthesis is increased compared to the rate that would occur without polysomes.

An mRNA molecule being synthesized has a free 5′ terminus. Since translation occurs in the 5′ → 3′ direction, the mRNA is synthesized in a direction appropriate for immediate translation. That is, the ribosome-binding site is transcribed first, followed in order by the initiating AUG codon, the region encoding the amino acid sequence, and finally the stop codon.

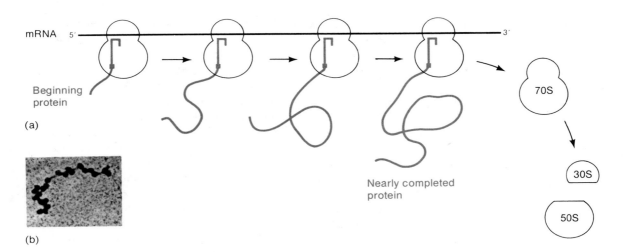

Figure 6-16 Polysomes. (a) Diagram showing relative movement of the 70S ribosome and the mRNA, and growth of the protein chain. (b) Electron micrograph of an *E. coli* polysome. [Courtesy of Barbara Hamkalo.]

Thus, the 70S initiation complex can re-form before the mRNA is released from the DNA. This process is called **coupled transcription-translation.** It does not occur in eukaryotes, because the mRNA is synthesized and processed in the nucleus and later transported through the nuclear membrane to the cytoplasm where the ribosomes are located. In prokaryotes, the coupling of transcription and translation is the rule.

ANTIBIOTICS AND ANTIBIOTIC RESISTANCE

Hundreds of antibiotics are known that inhibit gene expression. Some of these have been valuable tools in microbial genetics.

Rifampicin is a potent inhibitor of transcription in bacteria. It binds to one of the subunits of the core enzyme and prevents initiation of RNA synthesis. Thus, rifampicin has no effect on an RNA polymerase that has already initiated transcription. Other classes of antibiotics (e.g., streptolydigin) bind to RNA polymerase and inhibit all movement of RNA polymerase. Rifampicin has been used primarily in experiments in which specific inhibition of mRNA synthesis is desired.

Most of the known antibiotics inhibit polypeptide synthesis. Examples are streptomycin (Str) and kanamycin (Kan), which bind to the 30S ribosome and inhibit binding of mRNA; tetracycline (Tet), which inhibits binding of acylated tRNA to the 30S ribosome; chloramphenicol (Cam), which inhibits peptidyl transferase; and puromycin (Pur), which causes premature chain termination. These antibiotics have been used in biochemical studies to inhibit protein synthesis. In microbial genetics, alleles for sensitivity and resistance to these antibiotics are common genetic markers.

Penicillin, an antibiotic in common clinical use, does not affect protein synthesis, but interferes with particular steps in the production of the bacterial cell wall (Chapter 10).

Some bacteria are naturally resistant to certain antibiotics; others can acquire resistance by mutation. Natural resistance to penicillin is conferred by a gene that synthesizes β-lactamase, an enzyme that cleaves the lactam ring in penicillin, ampicillin, and in the cephalosporin antibiotics. This gene is carried on many plasmids (Chapter 11) and transposable elements (Chapter 12). (We have called this gene *amp* in this book, but in the plasmid and transposon literature the gene symbol *bla*, for *beta-lactamase*, will also be seen.) The most common mutational alteration that leads to antibiotic resistance is a change in the cell wall or cell membrane that prevents the antibiotic from entering the cell. Resistance to tetracycline is usually acquired in this way. Streptomycin acts by binding to one of the ribosomal proteins. Mutations can occur that alter the structure of the protein such that it cannot bind streptomycin, but without altering its ability to function in the ribosome. A cell carrying such a mutation is streptomycin-resistant. A curious mutation is one that confers streptomycin dependence. Here the ribosomal protein is altered in such a way that it can function *only* when streptomycin is bound to the mutant protein.

SYNTHESIS OF rRNA AND tRNA

Ribosomal RNA and tRNA are also transcribed from genes. The production of these molecules is not as direct as synthesis of bacterial mRNA. The main difference is that these RNA molecules are excised from large primary transcripts. Some of these transcripts include both rRNA and tRNA molecules. For example, one of the *E. coli* rRNA transcripts contains one copy each of 5S, 16S, and 23S rRNA, as well as four different tRNA molecules (Figure 6-17). Other transcripts include tRNA but not rRNA sequences. Highly specific RNases excise rRNA and tRNA from these large transcripts, and other enzymes produce the modified bases in tRNA. In no case do rRNA and tRNA transcripts include regions that are translated.

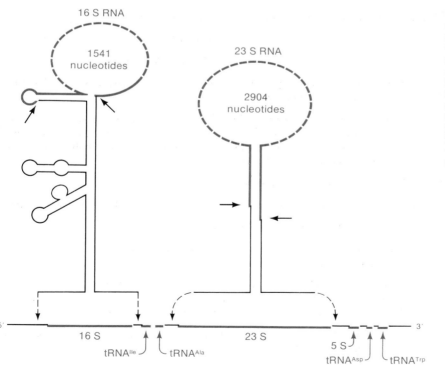

Figure 6-17 A schematic diagram of one of the *E. coli* rRNA transcripts from which 5S, 16S, and 23S rRNA molecules are excised. The regions containing the 16S rRNA and 23S rRNA molecules are shown in expanded form above the line designating the transcript. The arrows indicate the termini of the 16S rRNA and 23S rRNA molecules.

PROBLEMS

1. What are the chemical termini of an RNA molecule that has just been synthesized?

2. What common sequences are found in the 3′-terminal segment of prokaryotic mRNA?

3. In what way, if at all, does the chemistry of polymerization by RNA polymerases differ from that by DNA polymerases?

4. What is the direction of synthesis of RNA?

5. If a DNA-RNA hybrid were made, would the strands be parallel or antiparallel?

6. How must the statement, "both DNA strands cannot serve as a template for transcription," be modified to make it true?

7. Define coding strand and antisense strand.

8. What two regions are common to most prokaryotic promoters?

9. Write down the two RNA sequences which could conceivably result from complete transcription of the following DNA duplex.

$$5′ \ A \ G \ C \ T \ G \ C \ A \ A \ T \ G \ 3′$$
$$3′ \ T \ C \ G \ A \ C \ G \ T \ T \ A \ C \ 5′$$

Indicate the 5′ and 3′ ends of each transcript.

10. What parts of a mRNA molecule are not translated?

11. In terms of its role in translation, what is a ribosome?

12. How many codons could be contained in a four-letter code?

13. What amino acids correspond to the codons UUG and AAU? (Use the table; these do not have to be memorized.)

14. Poly(U) encodes polyphenylalanine. If a G is added to the 3′ terminus, the polyphenylalanine will have another amino acid at its terminus. What is the amino acid, and will the same amino acid be added at the terminus of polyphenylalanine if the G is added to the 5′ terminus?

15. What are the three stop codons?

16. What is the principal start codon and to what amino acid does it correspond?

17. What normal bases can pair with inosine?

18. If a codon were written PQR, would the anticodon be P′Q′R′ or R′Q′P′, in which a ′ denotes a complementary base?

19. Which of the following properties are essential for the function of a tRNA molecule: (1) recognition of a codon; (2) recognition of an anticodon; (3) ability to distinguish one amino acid from another; (4) recognition of DNA molecules?

20. Which chain-termination codon could be formed by a single base change from UCG, UUG, and UAU?

21. What is the minimum number of AUG sequences that would be in a tricistronic mRNA? Would it be likely that the actual number of AUGs would be larger?

22. Ribonuclease contains 124 amino acids. What is the least number of nucleotides you would expect to find in the gene encoding the protein? The first amino acid is not methionine.

23. The central region of a polypeptide containing a single lysine has the amino acid sequence Phe-Leu-Tyr-Ala-Lys-Gly-Glu . . . . A mutation is found that causes the polypeptide to terminate with Phe-Leu-Tyr-Ala. Which of the lysine codons was used in synthesis of the protein?

24. Which of the following amino acid changes can result from a single base change: (1) Met → Arg, (2) His → Glu, (3) Gly → Ala, (4) Pro → Ala, (5) Tyr → Val.

25. How many nucleoprotein subunits are there in a prokaryotic ribosome?

26. Do free (nontranslating) 70S ribosomes exist in bacteria?

27. With what polarity is mRNA read?

28. What is the direction of synthesis of a polypeptide chain?

29. At what stage of polypeptide synthesis do 70S ribosomes form?

30. Which site, A or P, can never be occupied unless the other site is also occupied?

31. Excluding the enzymes needed to synthesize the amino acids, make a rough count of the number of known enzymes required to synthesize a single protein molecule. What fraction of *E. coli* enzymes is involved in protein synthesis?

32. The synthetic polynucleotide 5'-A G G U U A U A G G A A A A A-3' is used as an mRNA in a system that does not require a start codon. What polypeptide is synthesized? Indicate the NH_2 and COOH termini.

33. Translation has evolved in a particular polarity with respect to the mRNA molecule. What would be the disadvantages of having the reverse polarity?

34. A polysome has a 3' and a 5' end. At which end will the polypeptide chain attached to the ribosome be longer?

Chapter 7

Regulation of
Gene Expression

The number of protein molecules produced per unit time by active genes varies from gene to gene, satisfying the needs of a cell and sometimes also avoiding wasteful synthesis. For many genes the different rates result mainly from different efficiencies of either recognition of a promoter by RNA polymerase or initiation of translation. However, bacteria need many gene products only on occasion, and regulatory mechanisms of an on-off type exist that enable such products to be present only when demanded by external conditions. In addition, more subtly regulated systems can adjust the intracellular concentration of a particular protein in response to needs imposed by the environment.

Prokaryotes are generally free-living unicellular organisms that grow and divide indefinitely as long as environmental conditions are suitable and the supply of nutrients is adequate. Thus, their regulatory systems are geared to provide the maximum growth rate in a particular environment, except when such growth would be detrimental. This strategy seems to apply also to the free-living unicellular eukaryotes such as yeast, algae, and protozoa, though less information is available about these organisms than for bacteria. Phages are less responsive to environmental fluctuations, probably because it is unlikely that a significant change would occur during the short life cycle of a typical phage. Instead, their systems are regulated on a time basis in order that particular proteins will be available at specific stages in the life cycle.

In this chapter we consider the basic mechanisms of metabolic regulation and present several examples of well-understood regulated systems. Since the number of regulatory mechanisms is quite large, only a few have been selected, each representing a strategy that has been observed in numerous systems.

COMMON MODES OF REGULATION

It is a general strategy in the living world that chemical changes are accomplished by a metabolic pathway, that is, by a sequence of reactions, each determined by a specific enzyme. In bacteria, regulation of the activity of such a pathway is often accomplished by synthesis (or lack of synthesis) of the entire set of enzymes and accessory proteins; that is, either all of the proteins are synthesized or none are made. This phenomenon, which is called **coordinate regulation,** results from control of the synthesis of a single polycistronic mRNA molecule that encodes all of the gene products (control of translation does occur, but it is much less common). Actually, few examples are known of turning transcription off completely. When transcription is in the off state, a basal level almost always remains, often consisting of only one or two transcriptional events per cell generation; hence, very little synthesis of the gene product occurs. For convenience, the term "off" will be used, but it should be kept in mind that usually what is meant is "very low." In only one case in bacteria, namely, in spores, is expression of most genes totally turned off.

Several mechanisms for regulation of transcription are common; the particular one used often depends on whether the enzymes being regulated act in degradative or synthetic metabolic pathways. For example, in a multistep degradative system the availability of the molecule to be degraded frequently determines whether the enzymes in the pathway will be synthesized. In contrast, in a biosynthetic pathway the final product is often the regulatory molecule. Even in a system in which a single protein molecule (not necessarily an enzyme) is translated from a monocistronic mRNA molecule, the protein may be **autoregulated**—that is, the protein itself may inhibit initiation of transcription, and high concentrations of the protein will result in less transcription of the mRNA that encodes the protein. The molecular mechanisms for each of the regulatory patterns vary quite widely but usually fall in one of two major categories—**negative regulation** and **positive regulation** (Figure 7-1). In a negatively regulated system, an inhibitor is present in the cell and prevents transcription. This inhibitor is called a **repressor.** An antagonist of the repressor, generally called an **inducer,** is needed to allow initiation of transcription. In a positively regulated system, an effector molecule (which may be a protein, a small molecule, or a molecular complex) activates a promoter; no inhibitor must be overridden. Negative and positive regulation are not mutually exclusive, and some systems

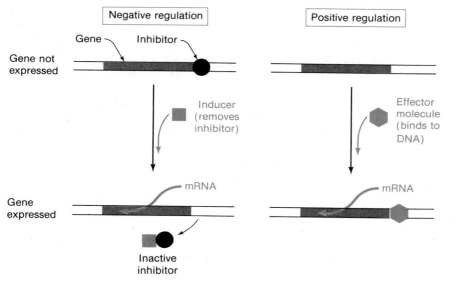

Figure 7-1 The distinction between negative and positive regulation. In negative regulation an inhibitor, bound to the DNA molecule, must be removed before transcription can occur. In positive regulation an effector molecule must bind to the DNA. A system may also be regulated both positively and negatively; in such a case, the system is "on" when the positive regulator is bound to the DNA and the negative regulator is not bound to the DNA.

are both positively and negatively regulated, utilizing two regulators to respond to different conditions in the cell.

A degradative system may be regulated either positively or negatively. In a biosynthetic pathway, the final product usually negatively regulates its own synthesis; in the simplest type of negative regulation, absence of the product increases its synthesis and presence of the product decreases its synthesis.

THE *E. coli* LACTOSE SYSTEM AND THE OPERON MODEL

Metabolic regulation was first studied in detail in the system in *E. coli* responsible for degradation of the sugar lactose, and most of the terminology used to describe regulation has come from genetic analysis of this system.

Lac⁻ Mutants

In *E. coli* two proteins are necessary for the metabolism of lactose—the enzyme **β-galactosidase**, which cleaves lactose (a β-galactoside) to yield galactose and glucose, and a carrier, **galactoside permease**, which is required for passage of lactose through the cell membrane into a cell. The existence of two different proteins in the lactose-utilization system was first shown by a combination of genetic experiments and biochemical analysis.

First, hundreds of mutants unable to use lactose as a carbon source—Lac⁻ mutants—were isolated. Some of the mutations were transferred by genetic recombination from the *E. coli* chromosome to an auxiliary DNA molecule (a plasmid called F′*lac*) carrying the genes for lactose utilization. Partial diploids having the genotypes F′*lac*⁻/*lac*⁺ or F′*lac*⁺/*lac*⁻ were constructed, using bacterial conjugation (Chapter 14). It was observed that these diploids always had a Lac⁺ phenotype (that is, they made β-galactosidase), so none produced an inhibitor of either enzymatic activity or gene function (or at least an inhibitor that could not be overcome by lactose). Other partial diploids were then constructed in which both the F′*lac* plasmid and the chromosome carried *lac*⁻ genes; these were tested for the Lac⁺ phenotype, with the result that all of the mutants initially isolated could be placed into two complementation groups, *lacZ* and *lacY*. The partial diploids F′*lacY*⁻*lacZ*⁺/*lacY*⁺*lacZ*⁻ and F′*lacY*⁺*lacZ*⁻/*lacY*⁻*lacZ*⁺ had a Lac⁺ phenotype, producing β-galactosidase, but the genotypes F′*lacY*⁻*lacZ*⁺/*lacY*⁻*lacZ*⁺ and F′*lacY*⁺*lacZ*⁻/*lacY*⁺*lacZ*⁻ had the Lac⁻ phenotype. The existence of two complementation groups was good evidence that the *lac* system consisted of at least two genes ("at least," because mutations had not yet been obtained in other genes).

Further experimentation was needed to establish the precise function of each gene. Experiments in which cells were placed in a medium containing ¹⁴C-labeled (radioactive) lactose showed that no [¹⁴C]lactose would enter a *lacY*⁻ cell, whereas it readily penetrated a *lacZ*⁻ mutant. This result indicated that the *lacY* gene is probably concerned with transport of lactose through the cell membrane into the cell and is the structural gene for *lac* permease. Enzymatic assays showed that β-galactosidase is present in *lac*⁺ but not *lacZ*⁻ cells. These results provided the initial evidence that the *lacZ* gene is the structural gene for β-galactosidase, a conclusion that was confirmed by immunological tests that demonstrated that an altered but inactive protein is present in *lacZ*⁻ cells. A final important result—that the *lacY* and *lacZ* genes are adjacent—was obtained by genetic mapping.

Regulation of the *lac* System: Inducible and Constitutive Synthesis and Repression

The on-off nature of the lactose-utilization system is evident in the following observations:

1. If a culture of Lac⁺ *E. coli* is growing in a medium lacking lactose or any other β-galactoside, the intracellular concentrations of β-galactosidase and permease are exceedingly low—roughly one or two molecules per bacterium. However, if lactose is present in the growth medium, the number of each of these molecules is about 10⁵-fold higher.

2. If lactose is added to a Lac⁺ culture growing in a lactose-free medium (also lacking glucose, a point that will be discussed shortly), both β-

galactosidase and permease are synthesized nearly simultaneously, as shown in Figure 7-2. Analysis of the total mRNA present in the cells before and after addition of lactose shows that no *lac* mRNA (the mRNA that encodes β-galactosidase and permease) is present before lactose is added and that the addition of lactose triggers synthesis of *lac* mRNA. The analysis is done by growing cells in a radioactive medium in which newly synthesized mRNA is radioactive, then isolating the mRNA, and finally allowing it to renature with the DNA of a λ phage that carries the *E. coli lac* genes (a λ*lac* transducing particle). Since the only radioactive mRNA that will renature to the *lac*-containing DNA is *lac* mRNA, the amount of radioactive DNA-RNA hybrid molecules is a measure of the amount of *lac* mRNA.

These two observations led to the view that the lactose system is **inducible** and that lactose is an **inducer.**

Lactose itself is rarely used in experiments to study the induction phenomenon for a variety of reasons; one important reason is that the β-galactosidase that is synthesized catalyzes the cleavage of lactose and results in a continual decrease in lactose concentration, which complicates the analysis of many types of experiments (for example, kinetic experiments). Instead, a

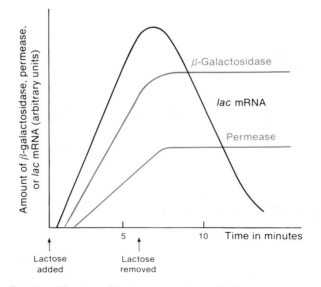

Figure 7-2 The "on-off" nature of the *lac* system. *Lac* mRNA appears very soon after lactose is added; β-galactosidase and permease appear at the same time but are delayed with respect to mRNA synthesis because of the time required for translation. When lactose is removed, no more *lac* mRNA is made and the amount of *lac* mRNA decreases owing to the usual degradation of mRNA. Both β-galactosidase and permease are stable. Their concentration remains constant even though no more can be synthesized. A third protein of the *lac* system, thiogalactoside transacetylase, is synthesized coordinately with β-galactosidase and permease. This protein, the product of the *lacA* gene, will be discussed later.

sulfur-containing analogue of lactose is used—usually isopropylthiogalacto-side (IPTG); this analogue is an inducer but not a substrate of β-galactosidase. Such a substance is said to be a **gratuitous inducer.**

Mutants have also been isolated in which *lac* mRNA is synthesized (hence also β-galactosidase and permease) in *both* the presence and the absence of an inducer. These mutants provided the key to understanding induction because they eliminated regulation; they were termed **constitutive.** (This term is now used to describe any system in which genes are continually transcribed.) Complementation tests—again with partial diploids carrying two constitutive mutations, one in the chromosome and the other in a plasmid—showed that the mutants fall into two groups termed *lacI* and *lacO^c*. The characteristics of the mutants are shown in Table 7-1. The *lacI^−* mutants are recessive (entries 3, 4). In the absence of an inducer a *lacI^+* cell fails to make *lac* mRNA, whereas this mRNA is made by a *lacI^−* mutant. Thus, the *lacI* gene is apparently a regulatory gene *whose product is an inhibitor that keeps the system turned off*. A *lacI^−* mutant lacks the inhibitor and hence is constitutive. Wild-type copies of the *lacI*-gene product are present in a *lacI^+/lacI^−* partial diploid, so the system is inhibited. The *lacI*-gene product, a protein molecule, is called the **lac repressor.** Genetic mapping experiments place the *lacI* gene adjacent to the *lacZ* gene and establish the gene order *lacI lacZ lacY*. How the *lacI* repressor prevents synthesis of *lac* mRNA will be explained shortly.

Dominance of *lacO^c* Mutants: The Operator

The *lacO^c* mutants are dominant (entries 1, 2, and 5 in Table 7-1), but the dominance is evident only in certain combinations of *lac* mutations, as can be seen by examining the partial diploids shown in entries 6 and 7. Both combinations are Lac^+, because a functional *lacZ* gene is present. However, in the combination shown in entry 6, synthesis of β-galactosidase is inducible

Table 7-1 Characteristics of partial diploids having several combinations of *lacI* and *lacO* alleles

Genotype	Constitutive or inducible synthesis of *lac* mRNA
1. F'*lacO^c lacZ^+ / lacO^+ lacZ^+*	Constitutive
2. F'*lacO^+ lacZ^+ / lacO^c lacZ^+*	Constitutive
3. F'*lacI^− lacZ^+ / lacI^+ lacZ^+*	Inducible
4. F'*lacI^+ lacZ^+ / lacI^− lacZ^+*	Inducible
5. F'*lacO^c lacZ^+ / lacI^− lacZ^+*	Constitutive
6. F'*lacO^c lacZ^− / lacO^+ lacZ^+*	Inducible
7. F'*lacO^c lacZ^+ / lacO^+ lacZ^−*	Constitutive

even though a $lacO^c$ mutation is present. The difference between the two combinations in entries 6 and 7 is that in entry 6 the $lacO^c$ mutation is carried on a DNA molecule that also has a $lacZ^-$ mutation, whereas in entry 7, $lacO^c$ and $lacZ^+$ are *carried on the same DNA molecule*. Thus, a $lacO^c$ mutation causes constitutive synthesis of β-galactosidase only when the $lacO^c$ and $lacZ^+$ alleles are located *on the same DNA molecule;* the $lacO^c$ mutation is said to be **cis-dominant,** since only genes *cis* to the mutation are expressed in dominant fashion. Confirmation of this conclusion comes from an important biochemical observation: the mutant enzyme (encoded in the $lacZ^-$ sequence) is synthesized constitutively in a $lacO^clacZ^-/lacO^+lacZ^+$ partial diploid (entry 6), whereas the wild-type enzyme (encoded in the $lacZ^+$ sequence) is synthesized only if an inducer is added. (The mutant enzyme is identified by an immunological test that detects the presence of a protein similar in structure to the active enzyme.) All $lacO^c$ mutations are located between the *lacI* and *lacZ* genes; thus, the gene order of the four elements of the *lac* system is

lacI lacO lacZ lacY

An important feature of all $lacO^c$ mutations is that they cannot be complemented (a feature of all *cis*-dominant mutations). That is, a $lacO^+$ allele cannot alter the constitutive activity of a $lacO^c$ mutation. Thus, *lacO* does not encode a diffusible product and must define a *site* or a noncoding region of the DNA rather than a gene. This site determines whether synthesis of the product of the adjacent *lacZ* gene is inducible or constitutive. The *lacO* region is called the **operator.**

The Operon Model

The regulatory mechanism of the *lac* system, which was elucidated by the elegant genetic analysis of François Jacob and Jacques Monod and termed by them the **operon model,** has the following features (Figure 7-3):

1. The lactose-utilization system consists of two kinds of components— *structural genes* needed for transport and metabolism of lactose, and *regulatory elements* (the *lacI* gene, the *lacO* operator, and the *lac* promoter). Together these components comprise the **lac operon.**
2. The products of the *lacZ* and *lacY* genes are encoded in a single polycistronic mRNA molecule. (This mRNA molecule contains a third gene, denoted *lacA*, which encodes the enzyme transacetylase. This enzyme is used in the metabolism of certain β-galactosides other than lactose and will not be of further concern.)
3. The promoter for the *lacZ lacY lacA* mRNA molecule is immediately adjacent to the *lacO* region. This location has been substantiated by the isolation and mapping of promoter mutants ($lacP^-$) that are completely incapable of making either β-galactosidase or permease, because no *lac* mRNA is made (these mutants were discussed in Chapter 6).

Figure 7-3 (a) Genetic map of the *lac* operon, not drawn to scale: the *p* and *o* sites are actually much smaller than the genes. (b) Diagram of the *lac* operon in (I) repressed and (II) induced states. The inducer alters the shape of the repressor, so the repressor can no longer bind to the operator.

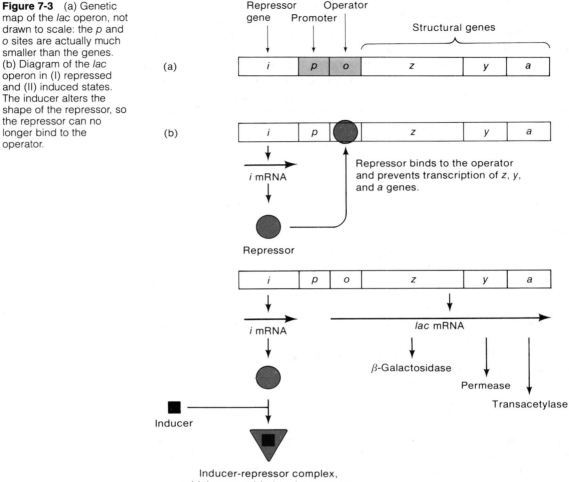

4. The *lacI*-gene product, the repressor, binds to a unique sequence of DNA bases, namely, the operator.
5. When the repressor is bound to the operator, initiation of transcription of *lac* mRNA by RNA polymerase is prevented.
6. Inducers stimulate mRNA synthesis by binding to and inactivating the repressor, a process called either **induction** or **derepression**. Thus, in the presence of an inducer the operator is unoccupied, and the promoter is available for initiation of mRNA synthesis.

Note that regulation of the operon requires that the *lacO* operator be adjacent to the structural genes of the operon (*lacZ, lacY, lacA*), but proximity of the *lacI* gene is not necessary, because the *lacI* repressor is a soluble protein and is therefore diffusible throughout the cell.

The operon model is supported by a wealth of experimental data and explains many of the features of the *lac* system as well as numerous other negatively regulated genetic systems. However, one aspect of the regulation of the *lac* operon—the effect of glucose—has not yet been discussed. Examination of this feature indicates that the *lac* operon is also subject to positive regulation, as will be seen in the next section.

Positive Regulation of the *lac* Operon

The function of β-galactosidase in lactose metabolism is to form glucose by cleaving lactose. (The other cleavage product, galactose, is also ultimately converted to glucose by the enzymes of the galactose operon.) Thus, if both glucose and lactose are present in the growth medium, activity of the *lac* operon is not needed, and indeed, no β-galactosidase is formed until virtually all of the glucose in the medium is consumed. The lack of synthesis of β-galactosidase is a result of lack of synthesis of *lac* mRNA. No *lac* mRNA is made in the presence of glucose, because in addition to an inducer to inactivate the *lacI* repressor, another element is needed for initiating *lac* mRNA synthesis; the activity of this element is regulated by the concentration of glucose. However, the inhibitory effect of glucose on expression of the *lac* operon is indirect.

The small molecule **cyclic AMP (cAMP)** is universally distributed in animal tissues, and in multicellular eukaryotic organisms it is important in regulating the action of many hormones (Figure 7-4). It is also present in *E. coli* and many other bacteria. Cyclic AMP is synthesized by the enzyme **adenyl cyclase,** and its concentration is regulated indirectly by glucose metabolism. When bacteria are growing in a medium containing glucose, the cAMP concentration in the cells is quite low. In a medium containing glycerol or any carbon source that cannot enter the biochemical pathway used to metabolize glucose (the glycolytic pathway), or when the bacteria

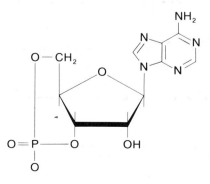

Figure 7-4 Structure of cyclic AMP.

are otherwise starved of an energy source, the cAMP concentration is high (Table 7-2). The mechanism by which glucose controls the cAMP concentration is poorly understood; the significant point is that *cAMP regulates the activity of the* lac *operon* (and several other operons as well).

E. coli (and many other bacterial species) contain a protein called the **cyclic AMP receptor protein (CRP)**, which is encoded in a gene called *crp*. A class of Lac⁻ mutations has been isolated that map quite far from the *lac* gene cluster. These mutations turn out to be in either the *crp* or the adenyl cyclase (*cya*) gene. Biochemical analysis showed that such mutants are unable to synthesize *lac* mRNA, indicating that both CRP function and cAMP are required for *lac* mRNA synthesis. CRP and cAMP bind to one another, forming a unit denoted **cAMP-CRP**, which has been shown in biochemical experiments with purified components to be required for transcription of the *lacZ–lacA* region. The requirement for cAMP-CRP is independent of the *lacI* repression system since *crp* and *cya* mutants are unable to make *lac* mRNA even if a *lacI⁻* or a *lacOᶜ* mutation is present. The reason is that the cAMP-CRP complex must be bound to a base sequence in the DNA in the promoter region in order for transcription to occur (Figure 7-5). Thus, *cAMP-*

Table 7-2 Concentration of cyclic AMP in cells growing in media having the indicated carbon sources

Carbon source	cAMP concentration
Glucose	Low
Glycerol	High
Lactose	High
Lactose + glucose	Low
Lactose + glycerol	High

Figure 7-5 Three states of the *lac* operon showing that *lac* mRNA is made only if cAMP-CRP is present and repressor is absent.

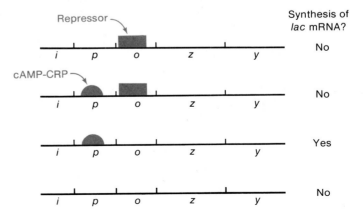

CRP is a positive regulator, in contrast with the repressor, and the *lac* operon is independently regulated both positively and negatively. The precise mechanism by which cAMP-CRP stimulates transcription is not known in detail and has recently (1986) become controversial.

Crp^- and cya^- mutants are not only Lac$^-$ but also defective in their ability to utilize maltose, galactose, arabinose, and several other sugars and polyhydric alcohols as a carbon source. This is because the operons responsible for utilization of these compounds (called **catabolite-sensitive operons**) all require bound cAMP-CRP. In fact, the easiest way to isolate a crp^- or cya^- mutant is to plate mutagenized wild-type cells (e.g., Lac$^+$Mal$^+$) on a color-indicator medium containing both lactose and maltose and isolate a Lac$^-$Mal$^-$ double mutant. These double mutants arise at roughly the same frequency as single Lac$^-$ or Mal$^-$ mutants and hence are likely to be the result of a single mutation. Confirmation of this conclusion comes from the observation that the double mutant is also Gal$^-$ and defective in the utilization of the other catabolite-sensitive carbon sources.

Differential Translation of the Genes in *lac* mRNA

The ratios of the number of copies of β-galactosidase, permease, and transacetylase (the third structural gene of the system), are 1.0 : 0.5 : 0.2. These differences, which are examples of **translational regulation,** are achieved in two ways:

1. The *lacZ* gene is translated first (Figure 7-6). Frequently, the *lac* mRNA molecule detaches from its translating ribosome following chain termination. The frequency with which this occurs is a function

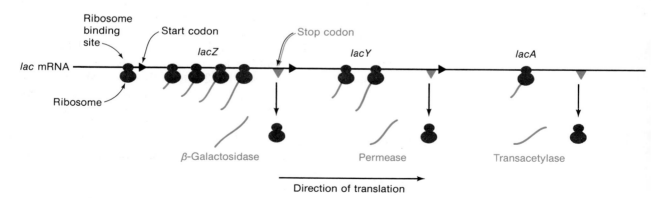

Figure 7-6 One explanation for polarity in the *lac* operon. All ribosomes attach to the mRNA molecule at the ribosome binding site. At each stop codon some ribosomes detach. Thus, the number of ribosomes translating each gene segment decreases for each subsequent gene.

of the probability of reinitiation at each subsequent AUG codon. Thus, there is a gradient in the amount of polypeptide synthesis from the 5′ terminus to the 3′ terminus of the mRNA molecule; this effect, which is called **polarity,** occurs with most polycistronic mRNA molecules.

2. All bacterial mRNA is degraded to nucleotides after several rounds of translation. Degradation is fairly slow but can occasionally begin after the first translation event. Degradation of *lac* mRNA is initiated more frequently in the *lacA* gene than in the *lacY* gene and more often in the *lacY* gene than in the *lacZ* gene. Hence, at any given instant, there are more complete copies of the *lacZ* gene than of the *lacY* gene, and more copies of the *lacY* gene than of the *lacA* gene.

In prokaryotes the following mode of regulation occurs repeatedly. The overall expression of activity of an operon is regulated by controlling transcription of a polycistronic mRNA, and the relative concentrations of the proteins encoded in the mRNA are determined by controlling the frequency of initiation of translation of each cistron. However, the mechanism by which transcription is regulated varies from one system to the next. An inducer-repressor system is common to many operons responsible for degradative metabolism, and cAMP-CRP is an element in many carbohydrate-degrading systems. However, particular features of the regulatory mechanisms differ; for example, two promoters are present in the galactose operon, and the arabinose operon is mainly positively regulated.

Isolation of Mutations in the *lac* Operon

In the preceding discussion it is clear that many of the conclusions have been drawn from the phenotypes of various mutants, as was pointed out in Chapter 1. It is of some value to the fledgling microbial geneticist to understand the source of these mutants. In initial studies of most operons mutants are collected merely by looking for an overall defective phenotype, in this case, Lac⁻, and the phenotype is identified by growth on standard color-indicator plates. For example, it was explained in Chapter 4 that on EMB agar plates a Lac⁺ colony is purple and a Lac⁻ colony is white. However, in later stages of analysis of an operon it is invariably necessary to collect large numbers of mutations that affect different aspects of the operon, for example, *lacI* mutations. Once the basic features are understood, it is usually possible (making use of a certain amount of cleverness) to design selection procedures for isolating mutations of particular types. As an example, we shall describe how some of the different *lac* mutations can be isolated.

1. *Constitutive mutants*. Xgal plates contain a colorless β-galactoside that produces an intense blue color when cleaved with β-galactosidase (a derivative of the dye indigo is released). However, the galac-

toside is not an inducer of the *lac* operon, so both Lac$^+$ and Lac$^-$ cells yield white colonies. However, if the cells are constitutive, induction is not needed, β-galactosidase is made, and the colonies are blue. Interestingly, *lacI$^-$* colonies are deep blue and *lacO^c* colonies are light blue; apparently, there is some residual repression that remains in most operator mutants.

2. *Permease-defective mutants*. The molecule *o*-nitrophenyl-β-D-thio-galactoside, TONPG, is a toxic compound to cells. It is transported into cells by *lac* permease but is not an inducer of the *lac* operon. If IPTG, a gratuitous inducer of the *lac* operon (see above), and TONPG are both included in agar, *lacY$^+$* cells will be killed; that is, survivors will be *lacY$^-$*. If the IPTG is left out, only constitutive cells will be sensitive; thus, survivors will be either *lacI$^-$ lacY$^-$* or *lacO^{c}lacY$^-$*. Since the *lacO* region is very short compared to the *lacI* gene, mutations will occur about fifty times more frequently in *lacI*.

3. *lacZ$^-$ mutants*. Mutants lacking one of the enzymes in the pathway for utilization of the sugar galactose, namely, *galE$^-$* mutants, lyse in the presence of galactose. Recall that the cleavage of lactose by β-galactosidase yields glucose and galactose, so *lac$^+$ galE$^-$* strains will not grow on a glycerol-lactose plate (glycerol is the carbon source), whereas *lac$^-$ galE$^-$* strains will. This procedure can yield *lacY$^-$* cells, since if the lactose does not enter the cell, a colony will form. However, by use of phenyl-β-D-galactoside, which does not require the permease for transport into the cell, survival depends on the absence of a functional *lacZ* gene. Note that this procedure would also yield promoter mutants, but these are quite rare, because of the small size of the promoter compared to the *lacZ* gene.

THE TRYPTOPHAN OPERON, A BIOSYNTHETIC SYSTEM

The tryptophan (*trp*) operon of *E. coli* is responsible for the synthesis of the amino acid tryptophan. Regulation of this operon occurs in such a way that when tryptophan is present in the growth medium, the *trp* operon is not active. That is, when adequate tryptophan is present, transcription of the operon is inhibited; however, when the supply is insufficient, transcription occurs. The *trp* operon is quite different from the *lac* operon in that tryptophan acts directly in the repression system rather than as an inducer, as will be seen shortly. Furthermore, since the *trp* operon encodes a set of biosynthetic rather than degradative enzymes, neither glucose nor cAMP-CRP functions in operon activity.

A simple on-off system, as in the *lac* operon, is not optimal for a biosynthetic pathway; a situation may arise in nature in which some tryptophan is available, but not enough to allow normal growth if synthesis of tryptophan

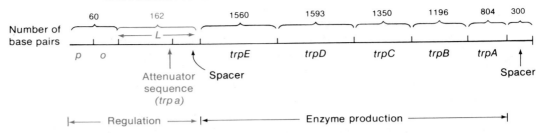

Figure 7-7 The *E. coli trp* operon. For clarity, the regulatory region is enlarged with respect to the coding region. The proper size of each region is indicated by the number of base pairs. *L* is the leader. The regulatory elements are shown in red.

were totally shut down. Tryptophan starvation when the supply of the amino acid is inadequate is prevented by a *modulating system* in which the amount of transcription in the derepressed state is determined by the concentration of tryptophan. This mechanism is found in many operons responsible for amino acid biosynthesis.

Tryptophan is synthesized in five steps, each requiring a particular enzyme. In the *E. coli* chromosome the genes encoding these enzymes are translated from a single polycistronic mRNA molecule. The *trpE* gene is the first one translated. Between the promoter and the operator are two regions called the **leader** and the **attenuator,** which are designated *trpL* and *trp a* (not *trpA*), respectively (Figure 7-7). The repressor gene *trpR* is located quite far from this gene cluster.

The regulatory protein of the repression system of the *trp* operon is the *trpR*-gene product. Mutations in either this gene or in the operator cause constitutive initiation of transcription of *trp* mRNA, as in the *lac* operon. This protein, which is called the *trp* **aporepressor,** does not bind to the operator unless tryptophan is present. The aporepressor and the tryptophan molecule join together to form the active *trp* repressor, which binds to the operator. The reaction scheme is:

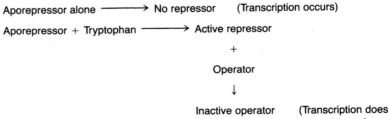

Thus, only when tryptophan is present does an active repressor molecule inhibit transcription. When the external supply of tryptophan is depleted (or reduced substantially), the equilibrium in the equation above shifts to the left, the operator is unoccupied, and transcription begins. This is the basic on-off regulatory mechanism.

In the on state a finer control, in which the enzyme concentration is varied by the amino acid concentration, is effected by (1) premature termination of transcription before the first structural gene is reached and (2) regulation of the frequency of this termination by the internal concentration of tryptophan. This modulation is accomplished in the following way.

A 162-base leader (noncoding) sequence is present at the 5′ end of the *trp* mRNA molecule. A mutant in which bases 123 through 150 are deleted synthesizes the *trp* enzymes in both derepressed cells and constitutive mutants at six times the normal rate, which indicates that bases 123–150 have regulatory activity. In nonmutant bacteria, after initiation of transcription most of the mRNA molecules terminate in this 28-base region, unless no tryptophan is present. The result of such termination is an RNA molecule that contains only 140 nucleotides and stops short of the genes encoding the *trp* enzymes. This 28-base region, in which termination occurs and is regulated, is called the **attenuator.** The base sequence (Figure 7-8) of the region in which termination occurs contains the usual features of a termination site— namely, a potential stem-and-loop configuration in the mRNA followed by a sequence of eight AT pairs (see Figure 6-2).

The leader sequence has several notable features:

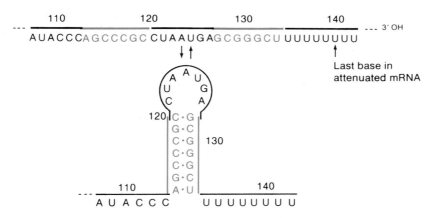

Figure 7-8 The terminal region of the *trp* leader mRNA (right end of *L* in Figure 7-7). The base sequence given is extended past the termination site at position 140 to show the long stretch of U's. The red bases form an inverted repeat sequence that could lead to the stem-and-loop configuration shown (segment 3–4, Figure 7-10).

1. It encodes a polypeptide containing 14 amino acids, the **leader polypeptide** (Figure 7-9).
2. Two adjacent tryptophan codons are located in the leader polypeptide at positions 10 and 11. We will see the significance of these repeated codons shortly.
3. Four segments of the leader RNA—denoted 1, 2, 3, and 4—are capable of base-pairing in two different ways—namely, forming either the base-paired regions 1–2 and 3–4 or just the region 2–3 (Figure 7-10). Two of these paired regions, 1–2 and 3–4, are also present in purified *trp* leader mRNA. The paired region 3–4 is in the terminator recognition region.

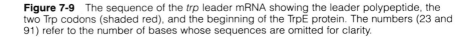

Figure 7-9 The sequence of the *trp* leader mRNA showing the leader polypeptide, the two Trp codons (shaded red), and the beginning of the TrpE protein. The numbers (23 and 91) refer to the number of bases whose sequences are omitted for clarity.

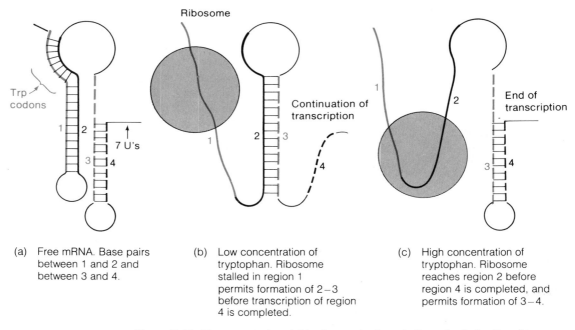

(a) Free mRNA. Base pairs between 1 and 2 and between 3 and 4.

(b) Low concentration of tryptophan. Ribosome stalled in region 1 permits formation of 2–3 before transcription of region 4 is completed.

(c) High concentration of tryptophan. Ribosome reaches region 2 before region 4 is completed, and permits formation of 3–4.

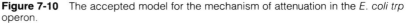

Figure 7-10 The accepted model for the mechanism of attenuation in the *E. coli trp* operon.

This arrangement enables premature termination to occur in the *trp* leader region by the following mechanism.

Termination of transcription is mediated through translation of the leader peptide. Because there are two tryptophan codons in this sequence, the translation of the sequence is sensitive to the concentration of charged tRNA^{Trp}. That is, if the supply of tryptophan is inadequate, translation will be slowed at the tryptophan codons. Three points should be noted. (1) Transcription and translation are coupled, as is usually the case in bacteria (Chapter 6). (2) Since sequences 2 and 3 are paired in the duplex segments 1–2 and 3–4, then the region 2–3 cannot be present simultaneously with regions 1–2 and 3–4. (3) All base pairing is eliminated in the segment of the mRNA that is in contact with the ribosome.

Figure 7-10 shows that the end of the *trp* leader peptide is in segment 1. A translating ribosome is in contact with about ten bases in the mRNA past the codons being translated, so when the final codons of the leader are being translated, segments 1 and 2 are not paired. In a coupled transcription-translation system, the leading ribosome is not far behind the RNA polymerase. Thus, if the ribosome is in contact with segment 2 when synthesis of segment 4 is being completed, then segments 3 and 4 are free to form the duplex region 3–4 without segment 2 competing for segment 3. The presence of the 3–4 stem-and-loop configuration allows termination to occur when the terminating sequence of seven uridines is reached. If there is no added tryptophan, the concentration of charged tRNA^{Trp} becomes inadequate and occasionally a translating ribosome is stalled for an instant at the Trp codons. These codons are located 16 bases before the beginning of segment 2. Thus, segment 2 is free before segment 4 has been synthesized and region 2–3 (the antiterminator) can form. In the absence of the 3–4 stem and loop, termination does not occur and the complete mRNA molecule is made, including the coding sequences for the *trp* genes. Hence, if tryptophan is present in excess, termination occurs and little enzyme is synthesized; if tryptophan is absent, termination does not occur and the enzymes are made. At intermediate concentrations the fraction of initiation events that result in completion of *trp* mRNA will depend on how often translation is stalled, which in turn depends on the concentration of tryptophan.

Repression versus Attenuation

The *trp* repressor-operator system does not operate as a simple on-off switch but can yield intermediate levels of operon expression. It is known that the synthesis of *trp* mRNA is partially repressed at all times in a cell growing in the absence of added tryptophan because the concentration of the *trp* enzymes is tenfold greater in a cell having a mutant (inactive) repressor than in a wild-type cell. This observation implies that in a wild-type cell, if the internal concentration of tryptophan were to fluctuate for any reason, the equilibrium between active and inactive repressor would shift to maintain a usable supply of tryptophan. Thus, it is not at all clear why the attenuation system is needed.

Repression has been studied independently of attenuation in a cell containing a gene fusion linking the *lacZ* gene to the *trp* promoter-operator region, lacking the attenuator. The activity of β-galactosidase as a function of concentration of tryptophan in the growth medium is a measure of the response of the repressor-operator system. Comparison of the behavior to that of an intact *trp* operon yields the contribution of the attenuation system. It was found that repression and attenuation are responsible for an 80-fold and 6–8-fold variation, respectively, of expression of the operon, for a total variation of 500–600-fold. Furthermore, repression is the dominant regulatory mechanism at higher concentrations of tryptophan; attenuation is not relaxed until starvation for tryptophan becomes severe (implying that charging of tRNA^{Trp} occurs when the concentration of tryptophan is quite low).

Many operons responsible for amino acid biosynthesis are regulated by attenuators equipped with the base-pairing mechanism for competition described for the *trp* operon. So far this has been described for the histidine, threonine, leucine, isoleucine-valine, and phenylalanine operons of the bacteria *E. coli, Salmonella typhimurium,* and *Serratia marcescens,* though in less detail than for the *trp* operon. Except for the phenylalanine operon, each lacks a repressor-operator system and is regulated solely by attenuation. Since these operons are regulated adequately (though the range of expression is not as great as that of the *trp* operon), one might ask why the *trp* operon (and the *phe* operon) has a dual regulatory system. This question has given rise to considerable speculation. The most obvious explanation is that the separate effects expand the range of tryptophan concentration in which regulation occurs. Whereas this is certainly true, there is more to it, because the *trp* repressor also regulates the activity of the *aroH* gene.

A common intermediate in the synthesis of the aromatic amino acids is made by three distinct enzymes, each having the same enzymatic activity. All three enzymes are needed only when all three aromatic amino acids must be synthesized. Thus, each enzyme is independently regulated by an amino acid-specific repressor. It has been suggested that in the distant past, the *trp* operon was regulated only by attenuation, and *aroH* was regulated as it is now, by the tryptophan-activated *aroH* repressor, which for historical reasons we have called the *trp* repressor. The existence of a tryptophan-sensitive repressor allowed the possibility of evolution of a sequence in the *trp* promoter into an operator that can bind the tryptophan-sensitive *aroH* repressor.

AUTOREGULATION

Many proteins are made from transcripts that are initiated at a constant rate. However, with some gene products the requirements of a cell vary greatly and the rate of transcription of the corresponding gene matches the need. One mechanism for the regulation of synthesis of monocistronic mRNA is **autoregulation.** In the simplest autoregulated systems the gene product is also a repressor: it binds to an operator site adjacent to the promoter. When the concentration of the gene product exceeds what the cell can use, a product molecule occupies the operator and transcription will be inhibited. At a later time the need may be greater, molecules will be consumed, and the concentration of unbound molecules will decrease. In these conditions, the molecule bound to the operator will leave the site, the promoter will be free, and transcription will occur. The synthesis of most repressor proteins—for example, the *lacI* product—is regulated in this way.

GENE FUSIONS

In studying operon function it is necessary to have not only a supply of mutations but also a means for quantitative assay of the gene products. In some cases the assays are so tedious and time-consuming that they seem to limit the rate of acquisition of information. Looking back at the early studies of the *lac* operon, one can see that the facility with which β-galactosidase can be measured quantitatively was of great significance. An early observation made while studying the *lac* operon made it clear that these assay problems in other systems could be circumvented.

The *pur* operon is responsible for the synthesis of purines. It is located "downstream" (in the direction of transcription) from the *lac* operon in the *E. coli* chromosome (Figure 7-11). Between these two operons is a gene *tsx*, which governs sensitivity to the phage T6. At one time a $lacZ^+$ *tsx-s* culture was taken and a *tsx-r* mutant (which is easy to obtain simply by adding T6 to the culture and picking survivors) was isolated. This mutant proved to be $lacZ^-$, and the mutation was found to be a deletion beginning within the *lacZ* gene, including all of the *tsx* gene, and extending into the *pur* operon past the *pur* operator and promoter. This deletion removes the RNA poly-

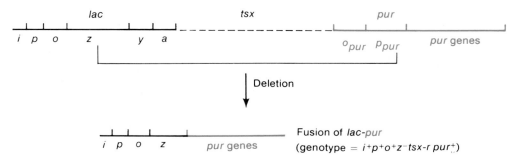

Figure 7-11 The *lac-pur* fusion and how it arose.

merase termination sequence in the *lac* operon; thus, when initiation begins from the *lac* promoter, an mRNA molecule is made that contains the proximal region of the *lacZ* gene and all of the *pur* genes. Thus, in the fused system transcription of the *pur* operon is induced by β-galactosides. The joining of operons or of genes is called **gene fusion.** Cells can survive a fusion as long as essential genes are not removed.

This type of gene fusion (transcriptional fusion) does not solve the assay problems described above but has nonetheless been quite valuable. In many systems regulation is accomplished by the concentration of particular proteins. For example, consider a system in which synthesis of a substance S is hypothesized to be controlled by the concentration of a protein P. The theory could be tested if there were some way to vary the amount of P in a controlled way. A fusion in which the gene encoding P is placed downstream of the *lac* promoter and operator could be used to create such variation. For example, addition of a *lac* inducer, for example, IPTG, would causes synthesis of P, and one could ask whether such induction leads to production of S. Similarly, synthesis of P could be turned off either by removal of the inducer or by addition of glucose, which would reduce the amount of cAMP-CRP. This fusion technique has been widely used. It is also useful as a means of producing a large amount of a desired biochemical. That is, a compound that exists in low concentration can be made in large quantities if the gene responsible for synthesis of the compound is fused to the *lac* promoter and a *lac* inducer is added. Examples of this technique will be seen in Chapter 20 in which genetic engineering is examined. In some cases the amount of a particular substance has been increased a thousandfold.

Fusions in which the *lacZ* gene is fused (downstream) to the promoter-operator region of other operons solve the assay problems discussed above. Such translational fusions have proved to be exceedingly valuable. For example, the *araC* gene product of the arabinose operon is very difficult to assay, but regulation of synthesis of the AraC protein has been elucidated by linking the *araC* promoter to the *lacZ* gene and measuring β-galactosidase activity.

In Chapter 19 a general method for generating translational gene fusions will be described.

REGULATION IN PHAGE SYSTEMS: EFFECTS ON RNA POLYMERASE

Phages are regulated in a large number of ways and an exhaustive description would be beyond the scope of this book. The temperate phages have a repressor-operator system (similar in principle to the *lac* system) for establishing and maintaining the prophage state, and this will be discussed in Chapter 17. Several mechanisms, which are rarely found other than in phage systems, involve direct effects on RNA polymerase.

Repressor-operator systems are regulated by controlling the accessibility of a promoter to RNA polymerase. In several phage systems promoters are not repressed, yet they are not recognized by the bacterial RNA polymerase. In other phage systems initiation of transcription is constitutive and regulation of *termination* at sites between particular genes controls gene expression. In the former cases, transcription of certain genes is controlled either by synthesizing specific RNA polymerases with different promoter-recognizing properties or by modifying the bacterial RNA polymerase, thereby changing the ability to recognize promoters. In the latter systems RNA polymerases are modified so that specific termination sites are ineffective.

E. coli phage T4 has a system for regulating the timing of synthesis of numerous classes of mRNA molecules. Early in the life cycle the bacterial RNA polymerase initiates transcription at a single class of hostlike promoter—the only class that it can recognize. Some of these transcripts encode proteins that induce modification of the host RNA polymerase. The modified polymerase can no longer bind to the original promoter but gains the ability to initiate at other promoters whose base sequences differ from that of the preceding class. A series of modifications occur in which either small molecules are covalently linked or phage protein molecules bind to a previously modified polymerase. These modifications cause the polymerase successively to ignore particular promoters and to initiate at new promoters. The net effect is an orderly control of the timing of synthesis of many species of mRNA. Note that not only are new mRNA species made when the time is right but species made at earlier times and not needed any more are no longer synthesized.

E. coli phage T7 makes three transcripts, each from the same DNA strand. Transcript I is initiated by *E. coli* RNA polymerase acting on promoter I. The other two promoters (II and III) are not recognized by the *E. coli* enzyme. Transcript I encodes two important proteins. One of these proteins is a new RNA polymerase that (1) does not recognize any promoters in the bacterial DNA, (2) fails to act at promoter I, and (3) acts at promoters II and III. The second protein inactivates the *E. coli* RNA polymerase. Thus, shortly after infection the phage has succeeded in preventing all synthesis of bacterial mRNA and has begun to synthesize important phage proteins (e.g., a T7 DNA polymerase) encoded in transcript II. The third transcript encodes structural proteins of the phage particle and a lysis enzyme, both of which, for the sake of efficiency, should be synthesized late in the life cycle. T7 has a unique mode of delaying synthesis of late mRNA, namely, the phage DNA that initiates the infection is injected so slowly that it takes about 12 minutes before promoter III has entered the cell. Thus, the overall sequence is the following: (1) make T7-specific RNA polymerase and inactivate host enzyme, (2) make phage DNA, (3) assemble phage particles, and (4) lyse the cell.

The life cycle of *E. coli* phage λ is regulated by controlling transcription termination, but the mechanism is quite different from that used in the *trp* attenuation system. In this case a gene product, N protein, which interacts with sites in the DNA, is responsible for *inhibiting* normal termination. This phenomenon is called **antitermination.** In λ antitermination occurs at several sites, one of which will be described.

Early in both the lytic and lysogenic pathways of λ, transcription is initiated from a promoter *pL* and yields a short transcript called L1 (Figure 7-12). This transcript includes a gene *N*, whose product is an antitermination protein. Transcription by the unmodified host RNA polymerase stops shortly after the terminus of the *N* gene, because the polymerase meets the normal transcription-termination sequence *tL1*. The *N*-gene product (aided by an *E. coli* protein) is able to bind to a short DNA sequence in the same region, called the N-utilization site (*nutL*). Following transcription of the *N* gene, translation of L1 occurs and the *N*-gene product is synthesized. As the concentration of the N protein increases, some N protein binds to *nutL;* when RNA polymerase makes its next transit across the *nutL* site, the polymerase acquires the N protein, and the altered form is then able to transcribe through the termination site. Transcription continues until a second termination site *tL2* is encountered. The altered RNA polymerase responds to this site and termination occurs, forming a longer transcript L2. By this sequence of events the synthesis of the proteins encoded in L2 is delayed by the amount of time required to synthesize a sufficient amount of the gene-*N* protein necessary for antitermination. Transcription of other λ genes is also regulated by antitermination. For example, after a series of antitermination events the prod-

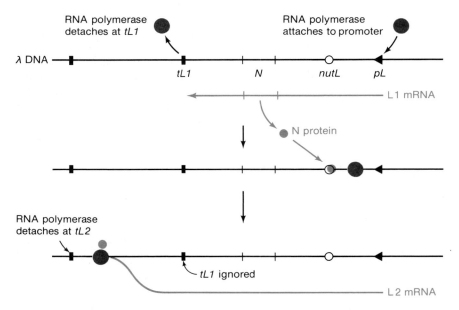

Figure 7-12 Antitermination of λ L1 mRNA induced by the binding of the λ gene-*N* protein to the *nutL* site in the DNA.

uct of the gene Q is made. The Q protein is also an antiterminator of a small constitutively synthesized RNA molecule R4. Downstream from the promoter for R4 is a site *qut* to which the Q protein binds. An RNA polymerase in transit picks up Q and thereby is able to ignore the termination signal for R4, which then becomes the leader for a large late mRNA molecule that encodes the head, tail, DNA packaging, and lysis proteins. In this way, by a series of delays—namely, the time required to make N, synthesize and extend mRNA, make Q, and synthesize the long mRNA, synthesis of the "late" proteins does not occur until ample DNA has been synthesized to produce a large number of progeny particles. This system will be examined further in Chapter 16.

PROBLEMS

1. What is meant by coordinate regulation?

2. Other than the ability to turn on and off a set of genes in an operon by a single regulatory element, what else is accomplished by having a set of genes contained in one polycistronic mRNA molecule?

3. Which type of regulation, positive or negative, involves removal of an inhibitor?

4. Would synthesis of an enzyme that is needed continually be regulated?

5. What is the biochemical action of an inducer?

6. Physically what is the consequence of binding of the *lac* repressor to the *lac* operator?

7. Which enzymes of the *lac* operon are regulated by the repressor?

8. What term describes a gene that is expressed continually, even though its transcription may be autoregulated?

9. Is the partial diploid F'*lacI*$^+$/*lacI*$^-$ inducible or constitutive?

10. Is the partial diploid F'*lacO*$^+$/*lacO*^{c} inducible or constitutive?

11. Why are all constitutively synthesized proteins not made at the same rate?

12. Is the *lac* repressor itself made constitutively or is it induced?

13. Is it necessary for a repressor gene to be adjacent to the operator?

14. Is it necessary for the operator to be very near the promoter?

15. When glucose is present, is the concentration of cyclic AMP high or low?

16. Can a mutant with either an inactive adenylate cyclase gene or an inactive *crp* gene synthesize β-galactosidase?

17. Does the binding of cAMP-CRP to DNA affect the binding of a repressor in any way?

18. Are all proteins translated from a single polycistronic mRNA necessarily made in the same quantity?

19. Repressors and aporepressors both bind molecules that are components of the metabolic pathway encoded in an operon. What is different about the positions in the pathway of the molecules bound, and what is different about the activity of the complex formed when binding occurs?

20. Is the attenuator, like the operator, a binding site?

21. Antitermination and attenuation are both concerned with termination of transcription. How do they differ, in principle, with respect to the role of RNA polymerase?

22. Explain how lactose molecules first enter an uninduced $lacI^+ lacZ^+ lacY^+$ cell to induce synthesis of β-galactosidase.

23. For each of the following diploid genotypes, indicate first whether β-galactosidase can be made, and second, whether synthesis of β-galactosidase is inducible (I) or constitutive (C) and finally, whether or not each cell could grow with lactose as sole carbon source. (i, p, o, z, y are used for $lacI, lacP, lacO, lacZ, lacY$, for simplicity.)

(a) $i^+ z^- y^+ / i^- z^+ y^+$
(b) $i^+ z^+ y^+ / o^c i^+ z^- y^+$
(c) $i^+ z^- y^+ / o^c z^+ y^+$
(d) $i^+ z^+ y^- / i^- z^- y^+$
(e) $i^- z^+ y^- / i^- z^+ y^+$
(f) $i^- z^+ y^+ / i^+ o^c z^- y^+$
(g) $i^+ p^- z^+ / i^- z^-$
(h) $i^+ o^c z^- y^+ / i^+ z^+ y^-$
(i) $i^+ p^- o^c z^- y^+ / i^+ z^+ y^-$
(j) $i^- p^- o^c z^+ y^+ / i^- z^+ y^-$

24. A cell that is wild-type with respect to the *lac* operon (+ for all alleles) is in a growth medium containing neither glucose nor lactose; that is, it is using another carbon source. How many proteins are bound to the DNA comprising the *lac* operon? How many if glucose is present?

25. You have isolated a Lac⁻ mutant and found by genetic analysis that its genotype is $lacZ^+ lacY^+$. The mutation, which you call $lacI^*$ is in the *lacI* gene. The partial diploid $lacI^* lacZ^+ lacY^+ / lacI^- lacZ^+ lacY^+$ is constructed and its phenotype is found to be Lac⁻—that is, $lacI^*$ is dominant. The diploid $lacI^* lacZ^+ lacY^+ / lacO^c lacI^+ lacZ^+ lacY^+$ is Lac⁺ (β-galactosidase is made). Suggest a property of the mutant repressor that would yield this phenotype. Would $lacI^* lacZ^+ lacY^+ / lacO^c lacI^+ lacZ^- lacY^+$ make β-galactosidase?

26. The purpose of regulating gene expression is to reduce wasted synthesis, because cells experiencing waste are at a definite disadvantage in the long run. Approximately 5% of the total protein made by $lacI^-$ cells, whose synthesis of β-galactosidase is unregulated, is β-galactosidase, far more than is needed.

(a) Suppose 10^6 wild type and 10^6 $lacI^-$ cells are placed in a growth medium lacking lactose. If the cells are grown for 50 generations (with continued dilution of the culture to keep the cells growing), what would you expect to happen to the ratio of wild-type and *lacI*⁻ cells?

(b) A gratuitous inducer is an inducer that is not a substrate of β-galactosidase. Many experiments are carried out with these inducers to avoid effects of changing concentration of lactose. Suppose a Lac⁺ culture is grown for hundreds of generations in the presence of a gratuitous inducer; what type of spontaneous mutant might accumulate in the culture?

27. A mutant strain of *E. coli* is found that produces both β-galactosidase and permease whether lactose is present or not.

(a) What are two possible genotypes for this mutant?

(b) Another mutant is isolated that produces no β-galactosidase at any time but produces permease if lactose is present in the medium. If a partial diploid is formed from these two mutants, in the absence of lactose neither β-galactosidase nor permease is made. When lactose is added, the partial diploid makes both enzymes. What are the genotypes of the two mutants?

28. How many proteins are bound to the *trp* operon when (1) tryptophan and glucose are present, (2) tryptophan and glucose are absent, or (3) tryptophan is present and glucose is absent?

29. An *E. coli* mutant is isolated that is simultaneously unable to utilize a large number of sugars as sources of carbon. However, genetic analysis shows that each of the operons responsible for metabolism of each sugar is free of mutation. What genotypes of this mutant are possible?

30. An operon has the gene sequence *A B C D E*. Neither the promoter nor the operator has been located. The repressor gene maps very far away from the structural genes. Various deletion mutants have been isolated. Some deletions of gene *E*, but none for any of the other genes, result in constitutive production of the mRNA of the operon. Where do you think the operator and the promoter are?

31. An operon responsible for utilizing a sugar Q is regulated by a gene called *kyu*. When Q is added to the growth medium, Qase is made; otherwise, the enzyme is not made. If the gene *kyu* is deleted (denoted Δ*kyu*), no Qase can be made. The partial diploid *kyu*⁺/Δ*kyu* is inducible. Two types of point mutants of *kyu* are found: *kyu1*, which never makes Qase, and *kyu2*, which makes the enzyme constitutively. The partial diploids *kyu*⁺/*kyu1* and *kyu*⁺/*kyu2* are inducible and constitutive, respectively. What is the likely mode of action of the protein encoded by the *kyu* gene?

32. The regulation of an operon responsible for synthesis of X is dependent on a repressor, a promoter, and an operator. In the presence of X, the system is turned off; an interaction between X and the repressor forms a complex which can bind to the operator.

(a) What kinds of mutations might occur in the repressor? Describe their phenotype (in terms of whether the operon is on or off).

(b) Describe the phenotype of a partial diploid with one wild-type and one mutant gene for each mutant gene.

PART 3

MAINTENANCE OF GENETIC INFORMATION

Chapter 8

DNA Replication

G enetic information is transferred from parent to progeny organisms by a faithful replication of the parental DNA molecules. Usually the information resides in one or more double-stranded DNA molecules. Some bacteriophage species contain single-stranded DNA instead of double-stranded DNA. In these systems replication consists of several stages in which single-stranded DNA is first converted to a double-stranded molecule, which then serves as a template for synthesis of single strands identical to the parent molecule. Viruses containing single- and double-stranded RNA molecules are also known; these organisms use several different modes of replication, some of which (in eukaryotes) include double-stranded DNA as an intermediate. The modes of replication of each of these types of molecules differ in detail, though certain fundamental features are common to each mode.

This chapter, which is an overview of DNA replication, will only examine a few general properties of the replication process.

THE BASIC RULE FOR REPLICATION OF ALL NUCLEIC ACIDS

The prime role of any mode of replication is to duplicate the base sequence of the parent molecule. The specificity of base pairing—adenine with thymine and guanine with cytosine—provides the mechanism used by all replication systems. Furthermore,

1. Nucleotide monomers are added one by one to the end of a growing strand by an enzyme called a **DNA polymerase.**

2. The sequence of bases in each new or **daughter strand** is complementary to the base sequence in the old or **parent strand** being copied—that is, if there is an adenine in the parent strand, a thymine nucleotide will be added to the end of the growing daughter strand when the adenine is being copied.

In the following section we consider how the two strands of a daughter molecule are physically related to the two strands of the parent molecule.

THE GEOMETRY OF DNA REPLICATION

The production of daughter DNA molecules from a single parental molecule gives rise to several topological problems, which result from the helical structure and great size of typical DNA molecules and the circularity of many DNA molecules. These problems and their solutions are described in this section.

Semiconservative Replication of Double-Stranded DNA

In the semiconservative mode of replication each parental DNA strand serves as a template for one new or daughter strand, and as each new strand is

Figure 8-1 The replication of DNA according to the mechanism proposed by Watson and Crick. The two replicas consist of one parental strand (black) plus one daughter strand (red). Each base in a daughter strand is selected by the requirement that it form a base pair with the parental base.

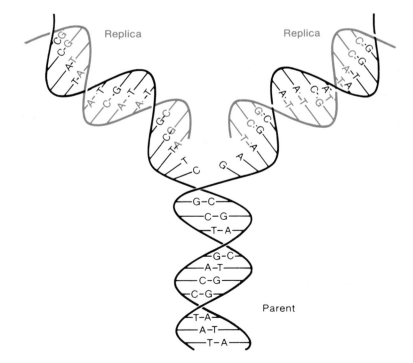

formed, it is hydrogen-bonded to its parental template (Figure 8-1). Thus, as replication proceeds, the parental double helix unwinds and then rewinds again into two new double helices, each of which contains one originally parental strand and one newly formed daughter strand.

That DNA replicates semiconservatively was proved by the now-classic experiment of Matthew Meselson and Franklin Stahl. They grew *E. coli* for many generations in a medium containing ^{15}N as the sole source of nitrogen and then transferred the cells to a medium containing the less dense isotope ^{14}N. DNA was isolated before the shift to the low-density medium and sedimented to equilibrium in CsCl (Chapter 2). The density of the DNA was about 1.722 g/cc, compared to 1.708, which is obtained for cells grown only in ^{14}N medium. After one generation of growth in the ^{14}N medium all of the DNA isolated from the cells had a density of $(1.708 + 1.722)/2 = 1.715$, exactly the density expected if one single strand contained ^{15}N and the complementary strand contained ^{14}N. Denaturation of this "hybrid" DNA yielded two components having the density of single-stranded [^{15}N]DNA and [^{14}N]DNA. In a second round of replication in ^{14}N medium the [^{14}N^{15}N]DNA was converted to equal amounts of two species, [^{14}N^{15}N]DNA and [^{14}N^{14}N]DNA, again the expectation of semiconservative replication.

The Geometry of Replication

Unwinding a double helix in semiconservative replication presents a mechanical problem. Either the two daughter branches at the Y-fork shown in Figure 8-2 must revolve around one another or the unreplicated portion must rotate. If the molecule were fully extended in solution, there would be no problem

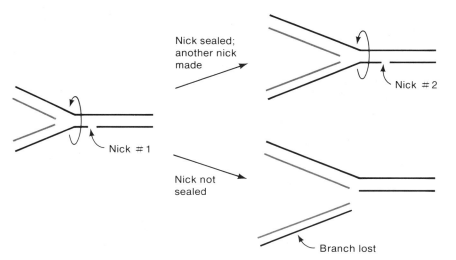

Nick sealed; another nick made

Nick not sealed

Nick #1

Nick #2

Branch lost

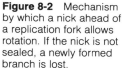

Figure 8-2 Mechanism by which a nick ahead of a replication fork allows rotation. If the nick is not sealed, a newly formed branch is lost.

and rotation of the unreplicated portion would be the simpler motion, as there would be less friction with the solvent. However, since in *E. coli* the molecule is 600 times longer than the cell that contains it (so it must be repeatedly folded, as was shown in Chapter 4), such rotation is unlikely. A simple solution would be to make a single-strand break in one parental strand ahead of the growing fork; this would enable a small segment of the unreplicated region to rotate, thereby eliminating the geometric problem (Figure 8-2). Then the only requirement would be to re-form the broken bond. However, this repair would have to occur before the replication fork had passed the nick; otherwise a daughter strand would be lost.

In bacteria most DNA molecules, both bacterial and phage, replicate as circular structures. This introduces a geometric problem that is more severe than that just described.

Geometry of the Replication of Circular DNA Molecules

The first proof that *E. coli* DNA replicates as a circle came from an autoradiographic experiment (genetic-mapping experiments to be described in Chapter 14 already had suggested that the chromosome is circular). Cells were grown in a medium containing [^{3}H]thymine so that all DNA synthesized would be radioactive. The DNA was isolated without fragmentation and placed on photographic film. Each ^{3}H-decay exposed one grain in the film, and after several months there were enough grains to visualize the DNA with a microscope; the pattern of black grains on the film located the molecule. One of the now-famous autoradiograms from this experiment is shown in Figure 8-3. Electron micrographs of replicating circular molecules of plasmids, phages,

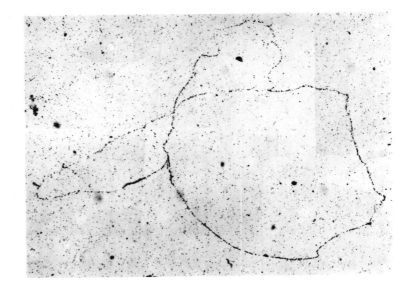

Figure 8-3 Autoradiogram of the intact replicating chromosome of an *E. coli* bacterium that has been allowed to incorporate [^{3}H]thymine into its DNA for slightly less than two generation periods. The continuous lines of dark grains were produced by electrons emitted during a two-month storage period by decaying ^{3}H atoms in the DNA molecule. [From J. Cairns *Cold Spring Harbor Symp. Quant. Biol.* 1963. 28: 44.]

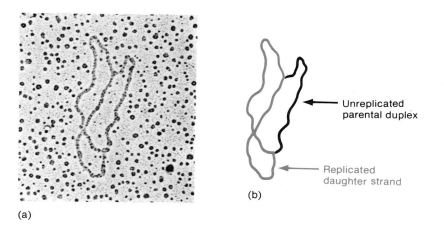

(a)

(b)

→ Unreplicated
parental duplex

← Replicated
daughter strand

Figure 8-4 θ replication. (a) Electron micrographs of a ColE1 DNA molecule (molecular weight = 4.2 × 10⁶) replicating by the θ mode. (b) Interpretive drawing showing parental and daughter segments. [Courtesy of Donald Helinski.]

and viruses have also been obtained (Figure 8-4). A replicating circle is schematically like the Greek letter θ (theta), so this mode of replication is usually called **θ replication.**

The unwinding problem in θ replication is formidable because lack of a free end makes rotation of the unreplicated portion impossible. An advancing nick, as in Figure 8-2, or a swivel at the replication origin (the point of initiation of replication) would solve the problem. These suggestions, though not quite correct, anticipate the correct mechanism. As replication of the two daughter strands proceeds along the helix, in the absence of some kind of swiveling the nongrowing ends of the daughter strands would cause the entire unreplicated portion of the molecule to become overwound (Figure 8-5). This in turn would cause *positive* supercoiling (Chapter 2) of the unreplicated portion. This supercoiling obviously cannot increase indefinitely because, if it were to do so, the unreplicated portion would become coiled so tightly that no further advance of the replication fork would be possible. This topological constraint could be avoided by the simple nicking-sealing cycle just described but the twists are removed in another way. As discussed

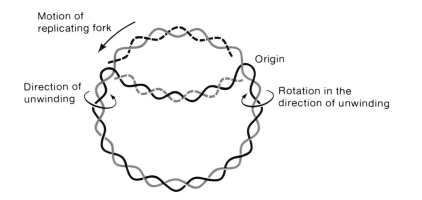

Motion of
replicating fork

Origin

Direction of
unwinding

Rotation in the
direction of unwinding

Figure 8-5 Drawing showing that the unwinding motion (curved arrows) of the daughter branches of a replicating circle lacking positions at which free rotation can occur causes overwinding of the unreplicated portion.

in Chapter 2, most naturally occurring circular DNA molecules are *negatively* supercoiled. Thus, initially the overwinding motion is no problem because it can be taken up by the underwinding already present in the negative supercoil. However, after about 5 percent of the circle is replicated, the negative superhelicity is used up and the topological problem arises.

Most organisms contain one or more enzymes called **topoisomerases.** These enzymes can produce a variety of topological changes in DNA; the most common are production of negative superhelicity and the removal of superhelicity. In *E. coli* the enzyme **DNA gyrase,** which is able to produce negative superhelicity, is responsible for removing the positive superhelicity generated during replication. That is, positive superhelicity is removed by gyrase introducing negative twists by binding ahead of the advancing replication fork.

DNA gyrase does have a DNA-breaking–sealing activity but this activity cannot by itself introduce negative superhelical turns. The mechanism of twisting is beyond the scope of this book but can be found in molecular biology references at the end of this book.

Termination poses a topological problem. When double-stranded circular DNA replicates semiconservatively, the result is a pair of circles that are linked as in a chain. Such a structure is called a **catenane.** Catenated molecules have been observed in numerous systems, and evidence is accumulating to indicate that they result from replication. Apparently, they are a precursor to the separated circles that ultimately result. Figure 8-6 shows that DNA gyrase is capable of decatenating two circles, and it is believed that gyrase is the enzyme responsible for separation of daughter molecules. Support for this hypothesis comes from a study of replication of the *E. coli* nucleoid (Figure 4-7) in a bacterial mutant that makes a temperature-sensitive DNA gyrase. Nucleoids isolated from these cells grown at temperatures at which gyrase was active (permissive temperature) appeared in a microscope as a spherical object. If the cells were instead grown at a temperature at which gyrase was nearly inactive, paired spheres accumulated, whose size

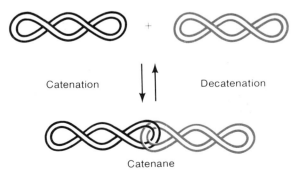

Catenation   Decatenation

Catenane

Figure 8-6   The processes of catenation and decatenation, catalyzed by DNA gyrase.

and appearance was consistent with the presence of two completely synthesized *E. coli* chromosomes. Treatment of isolated nucleoid pairs with gyrase caused them to separate into individual units.

ENZYMOLOGY OF DNA REPLICATION

The enzymatic synthesis of DNA is a complex process, primarily because of the need for high fidelity in copying the base sequence and for physical separation of the parental strands. The number of steps that must be completed is far too great to be accomplished by a single enzyme and, in fact, about twenty proteins are known at present to be necessary. The enzymes that form the sugar-phosphate bond (the phosphodiester bond) between adjacent nucleotides in a nucleic acid chain are called **DNA polymerases.**

The Polymerization Process

Three principal requirements must be fulfilled before DNA polymerases can catalyze synthesis of DNA:

1. The 5′-triphosphates of the four deoxyribonucleosides, deoxyadenosine, deoxyguanosine, deoxycytidine, and thymidine, are needed. Synthesis does not occur with the 3′-triphosphates or 5′-diphosphates or if one of the 5′-triphosphates is lacking.
2. A preexisting single-stranded template of DNA must be available.
3. A nucleic acid segment, hydrogen-bonded to a template DNA strand, must be present. Such a segment, which may be very short and either of the deoxyribo or ribo type, is called a **primer** (Figure 8-7). The need for a primer arises because *none of the known DNA polymerases is able to initiate a chain*. Thus, a primer chain with a free 3′-OH group is absolutely necessary for initiation of replication. (In this way, DNA polymerases differ from RNA polymerases, which can start chain growth.)

The reaction catalyzed by a DNA polymerase is the formation of a phosphodiester bond between the free 3′-OH group of the primer and the innermost phosphorus atom of the nucleoside triphosphate being incorporated at the new primer terminus (Figure 8-7). Thus, DNA synthesis occurs by the elongation of primer chains, *always in the 5′ → 3′ direction*. Recognition of the appropriate incoming nucleoside triphosphate during growth of the primer chain depends on base-pairing with the opposite nucleotide in the template chain. A DNA polymerase usually will catalyze the polymerization reaction, incorporating the new nucleotide at the primer terminus only when the correct base pair is present within the active site; in this reaction the two terminal phosphate groups of the nucleoside triphosphate are released as a pyrophosphate (PP_i) unit.

Figure 8-7 Addition of nucleotides to the 3'-OH terminus of a primer. The recognition step is shown as the formation of hydrogen bonds between the red A and the red T. The chemical reaction is between the red 3'-OH group and the red phosphate of the triphosphate.

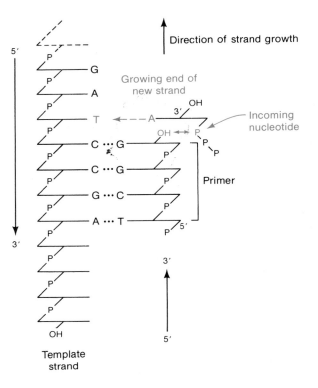

Figure 8-7 Addition of nucleotides to the 3'-OH terminus of a primer. The recognition step is shown as the formation of hydrogen bonds between the red A and the red T. The chemical reaction is between the red 3'-OH group and the red phosphate of the triphosphate.

Three DNA polymerases, numbered I–III, have been purified from *E. coli*. **Polymerase III (Pol III)** is the major replication enzyme. **Polymerase I (Pol I)** plays a secondary role and, as we will see, is responsible for removing RNA primers and replacing them with DNA. It is also required for repairing certain types of DNA damage (Chapter 9). **Polymerase II,** a minor enzyme, plays no role in either replication or repair and is of unknown function.

Error Correction

Pol I and Pol III both have the job of selecting a deoxynucleoside 5'-triphosphate that can hydrogen-bond to the template strand and of carrying out the polymerization reaction. Because of the need for faithful replication of a DNA base sequence, base selection must be extremely accurate. However, errors do occur on occasion, and systems have evolved for correcting these errors. A major error-correcting process is carried out by the polymerases themselves. DNA polymerases I and III of *E. coli* both have an exonuclease activity that acts only at the 3' terminus (a 3' → 5' exonuclease activity). This so-called **proofreading** or **editing function** excises a nucleotide from the 3'-OH end of the growing chain if it is not correctly base-paired to the corresponding nucleotide in the template chain.

The editing function of the polymerases is exceptionally efficient, but the integrity of the base sequence of DNA is so important that a second system exists for correcting the occasional error missed by the editing function. This correction system is called **mismatch repair.** In mismatch repair a pair of non-hydrogen-bonded bases that is not at the 3' end of a growing strand is recognized as incorrect and a polynucleotide segment is excised from one strand, thereby removing one member of the unmatched pair. The resulting gap is filled in by Pol I, which presumably uses this "third chance to get it right" to form only correct base pairs.

If it is to correct but not create errors, the mismatch repair system must be able to distinguish the correct base in the parental strand from the incorrect base in the daughter strand. If it were unable to do this, the correct base might sometimes be replaced by the complement of the incorrect base, thereby producing a mutation. The key to understanding the correction process came from the discovery of rare methylated adenines in DNA and from studies with *dam*⁻ (methylation-defective) mutants of *E. coli*. For any genetic locus the mutation frequency in a *dam*⁻ mutant is much higher than in a *dam*⁺ bacterium. This indicates that incorrectly incorporated bases are less frequently corrected in a *dam*⁻ mutant than in the wild type. The reason for this is that the mismatch repair system recognizes the degree of methylation of a strand and *preferentially excises nucleotides from the undermethylated strand* (Figure 8-8). The daughter strand is always the under-

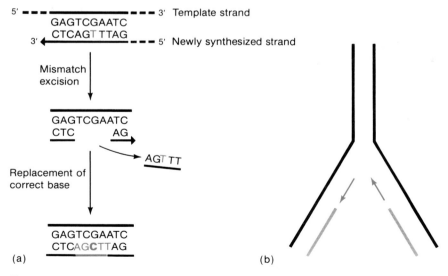

Figure 8-8 Mismatch repair. (a) Excision of a short segment of a newly synthesized strand, and repair synthesis. (b) Methylated bases in the template strand direct the excision mechanism to the newly synthesized strand containing the incorrect nucleotide. The regions in which methylation is complete are pink; the regions in which methylation may not be complete are red.

methylated strand, as its methylation lags somewhat behind the moving replication fork; the parental strand is fully methylated at the rare adenine sites, having been methylated in the previous round of replication.

DISCONTINUOUS REPLICATION

In the model of replication shown in Figure 8-1 both daughter strands are drawn as if replicating continuously. However, no known DNA molecule replicates in this way—instead, *one of the daughter strands is made in short fragments, which are then joined together.* The reason for this mechanism and the properties of these fragments are described in the following sections.

Fragments in the Replication Fork

All known DNA polymerases can add nucleotides only to a 3'-OH group. Examination of the growing fork indicates that if both daughter strands grew in the same overall direction, only one of these strands would have a free 3'-OH group; the other strand would have a free 5' end because the two strands of DNA are antiparallel (Figure 8-9). The solution to this geometric problem is that both strands grow in the 5' → 3' direction at a single growing fork and hence do not grow in the same direction along the parental molecule. Thus, one strand of the newly made DNA consists of fragments, as shown in Figure 8-10. This mechanism results in a single-stranded region of the parental strand on one side of the replication fork. This region results from the fact that synthesis of the discontinuous strand is initiated only periodically. (The signal for initiation is unknown, but it does not appear to be a particular base sequence.) In fact, the 3'-OH terminus of the continuously replicating strand is always ahead of the discontinuous strand. This had led to the use of the convenient terms **leading strand** and **lagging strand** for the continuously and discontinuously replicating strands, respectively (Figure 8-10). Such

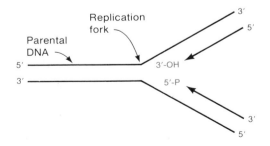

Figure 8-9   The termini (red) that would be present in a replication fork if both strands were to grow in the same overall direction.

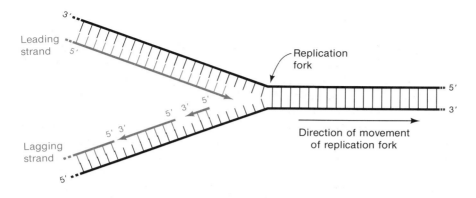

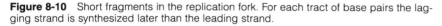

Figure 8-10 Short fragments in the replication fork. For each tract of base pairs the lagging strand is synthesized later than the leading strand.

regions have been seen in high-resolution electron micrographs of replicating DNA molecules (Figure 8-11).

How is synthesis of a precursor fragment initiated? To answer this, we must first consider the general question of initiation and priming of replication. This will also put us in a position to understand (1) how fragments are attached to one another, and (2) the role of polymerase I in replication of *E. coli* DNA.

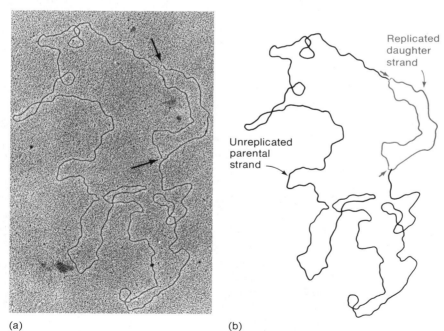

(a)   (b)

Figure 8-11 (a) A replicating θ molecule of phage λ DNA. The arrows show the two replicating forks. The segment between each pair of thick lines at the arrows is single-stranded; note that it appears thinner and lighter. (b) An interpretive drawing. [Courtesy of Manuel Valenzuela.]

The RNA Terminus of Precursor Fragments

As stated before, Pol III cannot lay down the first nucleotide to initiate chain growth but requires a primer. A polymerizing enzyme, distinct from Pol III, synthesizes a primer oligonucleotide, which is then extended by Pol III.

In *E. coli* initiation of synthesis of the leading strand and of the precursor fragments of the lagging strand occurs by somewhat different mechanisms, possibly because initiation of leading-strand synthesis requires priming of double-stranded DNA, whereas in initiation of synthesis of a precursor fragment, single-stranded DNA is primed (that is, the strand to be copied is already unwound). In both cases, the primer is a short *RNA* oligonucleotide whose size varies considerably, depending on whether the lagging or leading strand is being primed and on the particular organism. This RNA primer is synthesized by copying a particular base sequence from one DNA strand and differs from a typical RNA molecule in that after its synthesis *the primer remains hydrogen-bonded to the DNA template*. In bacteria two different enzymes synthesize primer RNA molecules. RNA polymerase, which is the same enzyme used for synthesis of most RNA molecules, primes the leading strand in some phage systems, once in each round of replication. **Primase,** the product of the *dnaG* gene, primes the precursor fragments of the lagging strand and may also prime leading-strand synthesis. In all cases, the growing end of the RNA primer is a 3'-OH group to which Pol III can easily add the first deoxynucleotide; the 5'-P end of the RNA chain, which remains free, has a 5'-triphosphate group. Thus, a precursor fragment has the following structure while it is being synthesized:

PPP-5' ——————————————————— 3'-OH
             RNA   DNA

The Joining of Precursor Fragments

The precursor fragments are ultimately joined to yield a continuous strand. This strand contains no ribonucleotides, so assembly of the lagging strand must require removal of the primer ribonucleotides, replacement with deoxynucleotides, and then joining. In *E. coli* the first two processes are accomplished by DNA polymerase I and joining is catalyzed by the enzyme DNA ligase, which can link adjacent 3'-OH and 5'-P groups at a nick. How this is done is shown in Figure 8-12. Pol III extends the growing strand until the RNA nucleotide of the primer of the previously synthesized precursor fragment is reached. It then moves away from the 3'-OH terminus, leaving a nick. *E. coli* DNA ligase cannot seal the nick because a triphosphate is present (it can only link a 3'-OH and a 5'-monophosphate). Polymerase I has an exonuclease activity that can remove a nucleotide from the 5' end of a base-paired fragment. It is effective with both a DNA or an RNA fragment.

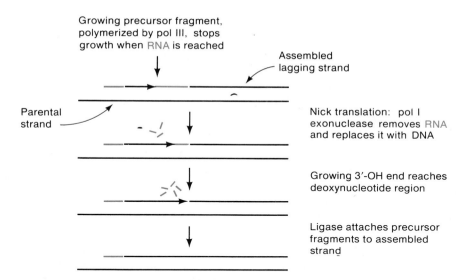

Growing precursor fragment, polymerized by pol III, stops growth when RNA is reached

Assembled lagging strand

Parental strand

Nick translation: pol I exonuclease removes RNA and replaces it with DNA

Growing 3'-OH end reaches deoxynucleotide region

Ligase attaches precursor fragments to assembled strand

Figure 8-12   Sequence of events in assembly of precursor fragments. RNA is indicated in red. The replication fork (not shown) is at the left.

This activity is called its **5′ → 3′ exonuclease** activity. Thus, Pol I acts at the 3′-OH terminus left by Pol III and moves in the 5′ → 3′ direction, removing ribonucleotides either one by one or as short oligonucleotides and adding deoxynucleotides to the 3′ end (this process is called **nick translation.** When all of the RNA nucleotides have been removed (plus probably some DNA nucleotides also), DNA ligase joins the 3′-OH group to the terminal 5′-P of the precursor fragment. By this sequence of events the precursor fragment is assimilated into the lagging strand. By this time, the next precursor fragment has reached the RNA primer of the fragment just joined and the sequence begins anew.

Synthesis of precursor fragments follows synthesis of the leading strand. In the next section, how the leading strand advances into the parental double helix is described.

Advance of the Replication Fork and the Unwinding of the Helix

DNA replication also requires a means of unwinding the parental double helix. Pol III is unable to unwind a helix. (Pol I can, but it is the only known polymerase that can do so.) Helix-unwinding is accomplished by enzymes called **helicases.** The helicase active in *E. coli* DNA replication is unknown but may be an enzyme called the **Rep protein.**

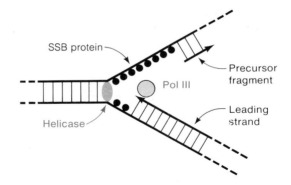

Figure 8-13   The unwinding events in a replication fork.

In *E. coli* the Pol III enzyme synthesizing the leading strand is not immediately behind the advancing Rep protein (Figure 8-13). Thus, behind the Rep protein are two single-stranded regions—a large one on the strand to be copied by precursor fragments and a smaller one just ahead of the leading strand. In order to prevent the single-stranded regions from annealing or from forming intrastrand hydrogen bonds, the single-stranded DNA is coated with *E. coli* single-strand binding protein (**SSB protein**). As Pol III advances, it must displace the SSB protein in order that base-pairing of the nucleotide being added can occur. It is not known whether Pol III carries out this displacement by itself or whether it is aided by another protein, that produced by the *dnaB* gene (and possibly other proteins as well).

BIDIRECTIONAL REPLICATION

Somewhat after initiation of synthesis of the leading strand at the replication origin, the first precursor fragment is synthesized. This is shown in part I of Figure 8-14, in which the overall direction in which the replicating fork moves is counterclockwise. In the discussion of lagging-strand replication just presented, it was noted that synthesis of each precursor fragment is terminated when the growing end reaches the primer of the previously synthesized fragment. However, in the case of the first precursor fragment, there is no earlier-made fragment. Thus, the precursor fragment becomes a leading strand for a second replication fork, moving clockwise, as shown in the figure. Clockwise replication requires the synthesis of precursor fragments in the second replication fork, but this can be achieved by the standard mechanism. The result of these events is that the DNA molecule will have two replication forks moving in opposite directions around the circle. This is called **bidirectional replication,** and it is an almost universal phenomenon. In a few systems, replication is unidirectional, presumably because in these

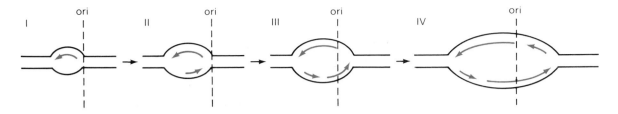

Figure 8-14 The formation of a bidirectionally enlarging replication bubble. I. The left-ward-leading strand starts at *ori*. II. The leading strand has progressed far enough that the first rightward precursor fragment begins. III. The leftward-leading strand has progressed far enough that the second rightward precursor fragment has begun. The first rightward precursor fragment has passed *ori* and has become the rightward-leading strand. IV. The rightward-leading strand has moved far enough that the first leftward precursor fragment has begun. There are now two complete replication forks.

systems a termination signal prevents movement of the first precursor fragment past the replication origin. Bidirectional replication is advantageous in that, compared to unidirectional replication, it halves the time required to replicate a circle.

ROLLING CIRCLE REPLICATION

There are numerous instances in which, in the course of replication, a circular phage DNA molecule gives rise to linear daughter molecules in which the base sequence of the DNA present in the phage is repeated numerous times, forming a concatemer (Figure 8-15). These concatemers are often an essential intermediate in phage production. Likewise, in bacterial mating, a linear DNA molecule is transferred by a replicative process from a donor cell to a recipient cell, as will be seen in Chapter 14. Both phenomena are consequences of initiation by covalent extension, an event that gives rise to a replication mode known as **rolling circle replication.**

Consider a duplex circle in which, by some initiation event, a nick is made having 3'-OH and 5'-P termini (Figure 8-16). Under the influence of a helicase and SSB protein a replication fork can be generated. Synthesis of

XYZABC   XYZABC   XYZABC   XYZABC

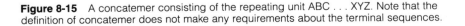

Figure 8-15 A concatemer consisting of the repeating unit ABC . . . XYZ. Note that the definition of concatemer does not make any requirements about the terminal sequences.

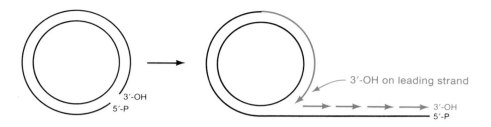

Figure 8-16   Rolling circle or σ replication. Newly synthesized DNA is shown in red.

a primer is unnecessary because of the 3'-OH group, so leading-strand synthesis can proceed by elongation from this terminus. At the same time, the parental template for lagging-strand synthesis is displaced. The polymerase used for this synthesis is apparently Pol III (though some phages use other enzymes). The displaced parental strand is replicated in the usual way by means of precursor fragments. The result of this mode of replication is a circle with a linear branch. This resembles the Greek letter σ, so rolling circle replication is sometimes called σ replication.

There are four significant features of rolling circle replication: (1) The leading strand is covalently linked to the parental template for the lagging strand. (2) Before precursor fragment synthesis begins, the linear branch has a free 5'-P terminus. (3) Rolling circle replication continues unabated, generating a concatemeric branch. (4) The circular template for leading-strand synthesis never leaves the circular part of the molecule.

A variant of the rolling circle mode, called **looped rolling circle replication,** generates a single-stranded circle from a double-stranded circular template. For *E. coli* phage ϕX174 this occurs in the following way (Figure 8-17). A phage protein (the A protein) nicks the viral-strand replication origin and becomes covalently linked to the newly formed 5'-P terminus. Using the Rep and SSB proteins and Pol III, chain growth occurs from the 3'-OH group, displacing the broken parental strand (which we call the (+) strand). This strand becomes coated with SSB protein and does not serve as a template for synthesis of precursor fragments. Synthesis continues until the origin is reached. At this point, the A protein binds to the 3'-OH group of the (+) strand and joins the 3'-OH and 5'-P groups of the (+) strand, dissociates, and attaches to the newly synthesized (+) strand. This process can continue indefinitely, generating numerous circular (+) strands. Note that in looped rolling circle replication, the displaced strand never exceeds the length of the circle, in contrast with ordinary rolling circle replication. We will encounter this mode again when transfer of F plasmids is examined (Chapter 11).

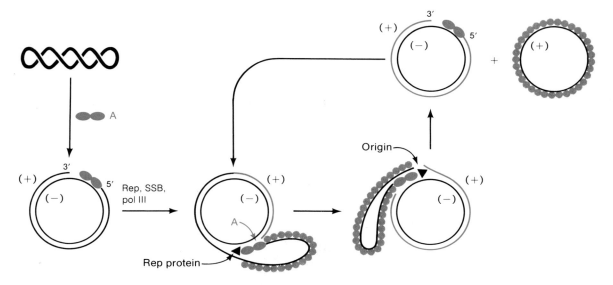

Figure 8-17 A diagram of looped rolling-circle replication of phage φX174. The gene-*A* protein nicks a supercoil and binds to the 5' terminus of a strand (known as the + strand) whose base sequence is the same as that of the DNA in the phage particle. Rolling circle replication ensues to generate a daughter strand (red) and a displaced + single strand that is coated with SSB protein and still covalently linked to the A protein. When the entire + strand is displaced, it is cleaved from the daughter + strand and circularized by the joining activity of the A protein. The cycle is ready to begin anew. Note that the − strand is never cleaved.

REGULATION OF BACTERIAL DNA SYNTHESIS AND REINITIATION OF A PARTIALLY REPLICATED DNA MOLECULE

In *E. coli* the time required for a single cell to double in size and divide depends upon the rate of production of useful energy and of precursor molecules. If glucose is provided as the sole carbon source and all other nutrients are simple inorganic compounds (that is, if the cells are in a glucose-minimal medium), it takes about 45 minutes at 37°C for a cell to replicate. If succinate is the sole carbon source, ATP is synthesized more slowly and the doubling time is 70 minutes. With even poorer carbon sources, the doubling time can be increased to 10 hours. In a glucose medium, the time required to replicate the bacterial DNA is 40 minutes; that is, initiation is delayed by a few minutes following completion of a round of replication. Surprisingly, in succinate medium, the replication time is still 40 minutes, so the time between successive rounds of replication is 30 minutes. In a medium in which the dou-

Figure 8-18 Stages of replicating DNA at various times in the *E. coli* life cycle when the doubling time is 22, 40, and 60 minutes. Colors alternate with round of replication.

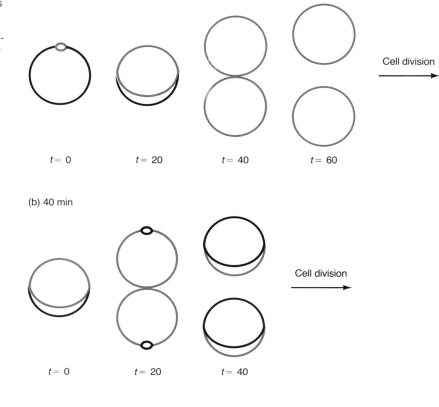

(a) 60 min

Cell division

$t = 0$ $t = 20$ $t = 40$ $t = 60$

(b) 40 min

Cell division

$t = 0$ $t = 20$ $t = 40$

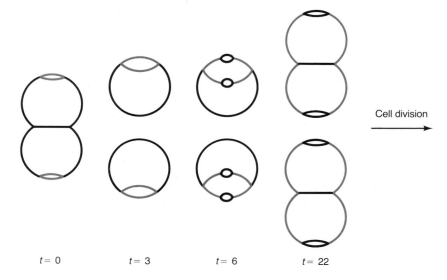

(c) 22 min

Cell division

$t = 0$ $t = 3$ $t = 6$ $t = 22$

bling time is 10 hours, the replication time is increased by only a few minutes. These facts indicate that the initiation of DNA replication must be regulated.

A good deal of evidence supports the idea that bacterial DNA replication is regulated by a repressor. A specific model (which is widely discussed but is exceedingly controversial) proposes that a pulse of repressor is synthesized just after the start of a new round of replication, triggered by doubling of a repressor gene that is presumably located near the replication origin. The repressor, which is not made after the initial pulse, inhibits further initiation until its concentration is lowered by increasing cell size. At a critical low concentration repression is relieved and initiation occurs.

Another significant aspect of bacterial DNA replication is the state of the DNA molecule at the time of cell division. An important observation made with synchronized populations of cells is that for doubling times greater than 40 minutes at 37°C, division occurs 20 minutes after termination of replication. The question then is how to reconcile a 40-minute replication time (see above) and a 20-minute preparation time for cell division with a 40-minute doubling time in glucose medium. The explanation, which is shown in Figure 8-18(b) is that newborn cells contain partially replicated DNA molecules. Note that with a 40-minute doubling time cell division is correlated with an initiation event that occurred two generations earlier. With a 60-minute doubling time (Figure 8-18(a)) a newborn cell has a completed chromosome. Another complexity occurs when *E. coli* is grown in a nutrient broth in which the doubling time may be as short as 21 minutes. In such media replication still takes 40 minutes—an apparent paradox. This observation is explained by the phenomenon of **premature initiation** (also called **dichotomous replication**). That is, in very rich media, following initiation and subsequent polymerization, a second initiation event occurs before replication is complete. Figure 8-18(c) shows how a second initiation event at the time replication is half complete allows segregation of two daughter molecules to occur at twice the normal rate. The mechanism underlying premature initiation is not known.

PROBLEMS

1. Fill in the blank word. A parent strand serves as a _____ for synthesis of a daughter strand.

2. In semiconservative replication what fraction of the DNA consists of one parental strand and one daughter strand after one, two, and three rounds of replication?

3. In which mode of replication does a parental circular DNA molecule yield two daughter circles?

4. In which mode of replication does a parental circle generate a circle with a linear branch?

5. Will a ^{15}N-labeled circle replicating in ^{14}N medium using the rolling circle mode ever achieve the density of ^{14}N^{14}N DNA?

6. How does rolling circle replication differ from θ replication with respect to the templates that are used?

7. What is the sense of the supercoiling produced by replicative movement of the growing fork in a circle?

8. What is the sense of supercoiling produced by DNA gyrase?

9. Can DNA gyrase convert a positive supercoil to a negative supercoil?

10. Which enzyme is able to separate two catenated circles?

11. Name three enzymatic activities of DNA polymerase I.

12. In what direction does a DNA polymerase move along a template strand?

13. What are the precursors for DNA synthesis?

14. DNA polymerization occurs by addition of a nucleotide to what chemical group?

15. What are the roles of the various exonuclease activities of the DNA polymerases in DNA replication?

16. How do Pol I and Pol III differ with respect to their ability to unwind the parental DNA in a replication fork?

17. Distinguish the roles of polymerases I and III in DNA replication.

18. What is the role of a helicase in DNA replication?

19. What is the function of SSB protein in advance of the replication fork?

20. Which enzymatic activity of Pol I and of Pol III is responsible for proofreading?

21. What is the chemical difference between the groups joined by a DNA polymerase and by DNA ligase?

22. How do organisms solve the problem that all DNA polymerases move in the same direction along a template strand, yet double-stranded DNA is antiparallel?

23. Are precursor fragments found in the leading or lagging strand?

24. Which strand, leading or lagging, is usually separated from the replication fork by a gap?

25. What is the chemical group (3'-P, 5'-P, 3'-OH, 5'-OH) at the sites indicated by the dots labeled a, b, and c in the drawing below?

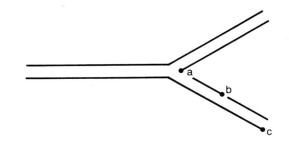

26. How does an RNA primer differ from a typical RNA molecule copied from DNA?

27. What must be done to two precursor fragments before they can be joined together?

28. Which enzymatic activity of Pol I is responsible for removal of RNA primers of precursor fragments?

29. What is the fundamental difference between the initiation of θ replication and of rolling circle replication?

30. Answer the following questions about rolling circle replication.

    (a) How many *de novo* initiation events (not events for initiating precursor fragments) are needed in a complete replication cycle?
    (b) If the circle has unit length, what is the maximum length of the linear branch that can be formed?
    (c) What is the primer?

31. If an $^{15}N^{15}N$ circle replicates in ^{14}N medium by the rolling circle mode, what will be the density distribution after three rounds of replication?

32. Consider a hypothetical phage whose DNA replicates exclusively by rolling circle replication. A phage whose DNA is radioactive in both strands infects a bacterium and is allowed to replicate in nonradioactive medium. Assuming that only daughter DNA from the branch ever gets packaged into progeny phage particles,

    (a) What fraction of the parental radioactivity will appear in progeny phage?
    (b) How many progeny phage will contain radioactive DNA? Will the occurrence of genetic recombination affect this value?

33. How long will it take to replicate a medium-sized phage DNA molecule whose molecular weight is 25 million (a typical size) and which replicates bidirectionally? You may use the fact that *E. coli* DNA replicates in 40 minutes.

34. The origin of the *E. coli* chromosome is on the genetic map shown below. What is the sequence (in time) of replication of the genes *A* through *G*? State any assumptions that you have made.

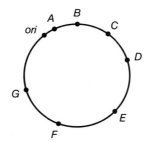

Chapter 9

DNA Damage
and Repair

Maintenance of the base sequence of DNA from one generation to the next is one of the prime goals of all biological systems. However, sequence alterations can arise in a variety of ways. For example, in Chapter 8 we saw that incorrect nucleotides are occasionally added to the growing terminus of a daughter DNA molecule. However, two levels of error-correcting systems—first, proofreading and, second, mismatch repair—serve to eliminate most of the misincorporated nucleotides. However, DNA is also subject to environmental damage from chemicals and radiation. In this chapter, we will see that bacteria possess systems to repair this type of damage.

BIOLOGICAL INDICATIONS OF DAMAGE TO DNA

When bacteria are exposed to radiation or various chemicals, they lose the ability to form colonies. Similarly, phage lose plaque-forming ability. This loss of viability can be expressed graphically by plotting the fraction of the initial population that survives various exposures to the radiation or the chemicals against some measure of exposure. The most detailed studies have been with radiation, in which the exposure is simply the total amount of radiation, or the radiation dose. Such a dose-response graph is called a **survival curve.** For bacteria such curves are obtained in the following way. Samples are drawn at intervals from a population of bacteria that is being irradiated, for example, with ultraviolet light or x radiation. The samples are

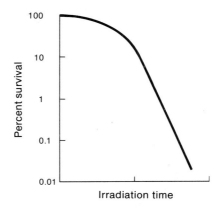

Figure 9-1 A typical ultraviolet-light survival curve for a bacterium. Initially, the curve is fairly flat because initial damage does not cause killing; this could either mean that several lethal hits are needed to kill a cell or that, at low doses, a great deal of the damage is repaired. Note that the y-axis is logarithmic.

plated, normally on a nutrient agar, and the colonies that form are counted. The fraction of the initial number of cells that remain able to produce colonies is plotted as a function of the dose. For phages the irradiated phage are plated on a lawn of bacteria so that plaques can form. An example of such a curve is shown in Figure 9-1.

Analysis of survival curves has given considerable insight into the various mechanisms of radiation damage and also provided the first suggestion that environmental damage to DNA is repaired. A simple mathematical theory, called hit theory, has been useful in analyzing survival curves. It is described in this section.

Hit Theory

In hit theory, equations are derived that enable one to calculate survival curves for various types of populations of N identical organisms exposed to a dose D of radiation (or some other external agent) that causes damage of some kind. In the simplest case we assume that each organism possesses only one sensitive site (a target) that, if damaged or "hit" by a radiation photon (a "particle" of light), will inactivate the organism. The number dN damaged by a dose dD is proportional to the number N that existed prior to receiving that dose; that is, $-dN/dD = kN$, in which the constant k is a measure of the effectiveness of the dose and is proportional to the fraction of incident photons that causes an inactivating hit—in other words, the probability that a single photon can cause such a hit. Integrating this equation from $N = N_0$ at $D = 0$ yields

$$N = N_0 e^{-kD} \qquad (1)$$

The surviving fraction $S = N/N_0$ is

$$S = N/N_0 = e^{-kD} \qquad (2)$$

so that a plot of $\ln S$ versus D gives a straight line with a slope of $-k$. Curves of this type are called **exponential** or **single-hit curves.** They are observed when phages are irradiated with ionizing radiation such as x rays (Figure 9-2).

Let us now consider a population of slightly different organisms in which each organism contains n sites, *each* of which must be hit (damaged) if the organism is to be inactivated. In this case, inactivation requires at least n hits ("at least" because statistically some sites will be hit twice, and we assume that two hits on one site are not more effective than one hit on that site). The probability of one unit being hit by a dose D is $1 - e^{-kD}$, so the probability P_n that all n units become inactivated is

$$P_n = (1 - e^{-kD})^n$$

The surviving fraction S of the population is $1 - P_n$ or

$$S = 1 - (1 - e^{-kD})^n \qquad (3)$$

Expansion of this equation yields

$$S = 1 - (1 - ne^{-kD} + \ldots + e^{-nkD})$$

At large values of D the higher order terms become negligible compared to ne^{-kD}, so that at high dose, $S = ne^{-kD}$, or

$$\ln S = \ln n - kD \qquad (4)$$

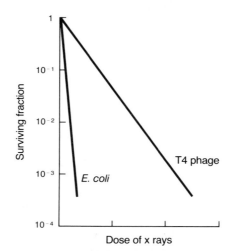

Figure 9-2 Survival curves for two x-irradiated populations, one of the bacterium *E. coli* and the other of the phage T4.

A plot of Equation 3 for $k = 1$ and various values of n shows that for small values of D, ln S changes slowly (Figure 9-3). At large D, Equation 4 predominates and the curve becomes linear. Extrapolation of the linear part (high-dose region) of a curve yields $S = n$ as the y intercept. Thus, if experimental data for sufficiently large values of D can be obtained, the number of targets n can be determined. Curves of this sort are called n-hit curves, and a system yielding such a curve is said to have n-hit inactivation kinetics. An example of a system that exhibits this effect of hit number is the inactivation of yeast by x rays. Yeast, like all microorganisms and probably all cells, is inactivated by radiation damage to DNA. Furthermore, yeast exist in both a haploid and a diploid state, and cell lines of higher ploidy (e.g., tetraploid, having four chromosome sets) have been constructed. The existence of an exponential (straight-line) curve for stationary-phase haploid cells (Figure 9-4) suggests that a single lesion in any chromosome is sufficient to prevent colony formation. This conclusion is supported by the data for stationary-phase diploid and tetraploid cells in which curves are obtained that extrapolate to approximately 2 and 4, respectively. As might be expected, straight lines are observed also for x-ray inactivation of all phages and for most bacteria. The data for bacteria are not always straightforward because most bacteria consist of mixed populations (at least from the point of view of radiosensitivity). For example, a fraction of the cells consist of incompletely separated daughters so that damage to one of the cells does not prevent colony formation by the other cell (that is, the cells behave like diploids). Secondly,

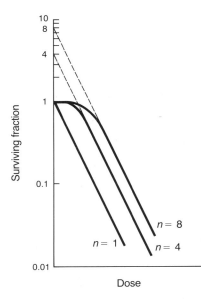

Figure 9-3 Survival curves for various values of n, showing that at high doses each curve becomes linear and that extrapolation to the y-axis yields n as the intercept.

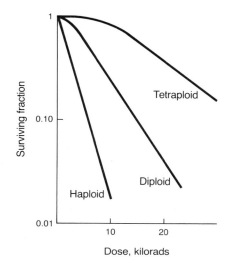

Figure 9-4 Survival curves for haploid, diploid, and tetraploid yeast cells, showing the effect of ploidy on radiosensitivity. The two multihit curves extrapolate to approximately 2 and 4, respectively, at zero dose.

as shown in Chapter 8, because of the relative timing of DNA replication and cell division many individual cells possess two partially replicated chromosomes.

The subject of the mechanism of inactivation of microorganisms by ionizing radiation is beyond the scope of this book but a few conclusions may be stated: (1) Ionizing radiation causes three types of damage to DNA— single-strand breaks, double-strand breaks, and alterations of bases. (2) Single-strand breaks are for the most part re-sealed by DNA ligase and do not contribute to lethality. (3) Double-strand breaks are lethal. With phages a double-strand break prevents injection of all of the phage DNA; that is, only a fragment is injected. With bacteria a double-strand break is lethal for two reasons: replication of the DNA cannot be carried to completion, and the free ends initiate degradation of the DNA by nucleases. (4) Damage to bases, which is an oxidative process requiring molecular O_2, is invariably lethal, probably because the damaged bases constitute a replication block.

It was pointed out in the derivation of Equation 1 that the constant k is in some way a measure of the probability of a hit. The relation between k and this probability can be seen by looking at the effect of radiation in a slightly different way. Radiation such as x rays produce ionizations in matter. The number of ionizations per unit volume is proportional to dose, and a fixed fraction of the ionizations produce a lethal hit. Thus, if V is the volume of the sensitive target molecule (V is called the **target volume**), the average number of hits within the target volume is cVD, in which c is a proportionality constant relating the number of ionizations and the number of hits. If the hits are random and independent, the probability $P(n)$ that n hits occur within the volume is given by the Poisson distribution, or

$$P(n) = \frac{e^{-cVD}(cVD)^n}{n!}$$

For a single-hit mechanism the surviving fraction $S = 1 - P(n)$ is

$$S = e^{-cVD}$$

Comparison to Equation 2 shows that $k = cV$, or k is proportional to the volume of the target. This is not hard to understand—if one DNA molecule A has twice the number of nucleotides as a second DNA molecule B, then the dose required to damage a nucleotide pair (for example, either a double-strand break or base damage) in A will be half that required to damage one in B. That is, A is twice as sensitive as B. This phenomenon can easily be seen by examining curves for x-ray inactivation of phages having DNA molecules of different sizes, as shown in Figure 9-5. Phage T5, which has the largest DNA molecule of the three, is most sensitive, and T7, which is the smallest, is the least sensitive. Furthermore, the ratio of the k values is the ratio of the DNA molecular weights.

The dose producing a survival of $1/e$ (37 percent) has particular significance for a system with single-hit kinetics. From Equation 2 we can see that *a survival of $1/e$ occurs when there is, on the average, one lethal hit per molecule or organism*. One can then calculate the rate of production of damages. For example, the curves shown in Figure 9-5 are for x irradiation in the absence of molecular O_2 (i.e., in a nitrogen atmosphere), so that all lethal hits are double-strand breaks, as pointed out in item (4) above. For phage T7, whose DNA has a molecular weight of 25×10^6, the dose yielding $1/e$ survival is 1.49×10^5 rads (rads are a common measure of exposure to ionizing radiation), so double-strand breaks are produced at a rate of 1 double-

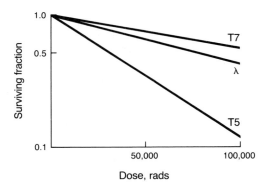

Figure 9-5 Loss of plaque-forming ability of the three phages irradiated with x rays. The molecular weights of the phage DNA molecules are: T7, 25×10^6; λ, 31×10^6; T5, 76×10^6.

strand break per 1.49×10^5 rads per 25×10^6, or 2.7×10^{-7} breaks/ rad/10^6 molecular weight units. This same value is obtained from the λ and T5 curves.

Ultraviolet Radiation

Ultraviolet radiation (UV) also causes inactivation of bacteria and phages. Nucleic acids and proteins absorb light in nearly the same range of wavelengths: 260 nm is the absorption maximum for nucleic acids and 280 nm is the maximum for proteins. However, analysis of UV survival curves for a variety of bacteria and phages makes it clear that the target molecule is DNA. The experiment consists of irradiating several identical phage samples with different wavelengths of UV. The survival curves all show single-hit kinetics, but the slopes depend on the wavelength used. The constant k is evaluated from each curve using Equation 4, and then k is plotted against wavelength, yielding a graph of radiosensitivity versus wavelength; this graph is called an **action spectrum.** (In general, an action spectrum is a graph of some parameter that characterizes a dose-response curve versus wavelength.) Such experiments show that the most effective wavelength is 260 nm, and that 280-nm radiation is quite ineffective. In fact, *the action spectrum for UV irradiation matches the absorption spectrum of DNA,* which suggests that DNA, not protein, is the target molecule. The absorption spectrum of RNA is quite similar to that of DNA, but because of the large number of RNA molecules in cells and because of the similarity in the action spectra of bacteria and phages that have no RNA, the possibility of an RNA target was never seriously considered.

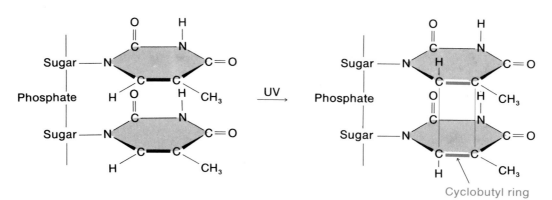

Figure 9-6 Structure of a cyclobutylthymine dimer. Following ultraviolet (UV) irradiation, adjacent thymine residues in a DNA strand are joined by formation of the bond, shown in red. Although not drawn to scale, these bonds are considerably shorter than the spacing between the planes of adjacent thymines, so the double-stranded structure becomes distorted. The shape of the thymine ring also changes as the C=C double bond (heavy horizontal line in left panel) of each thymine is converted to a C—C single bond (horizontal red lines in right panel) in each cyclobutyl ring.

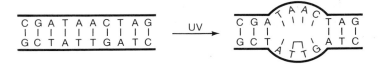

Figure 9-7   Distortion of the DNA helix caused by two thymines moving closer together when joined in a dimer. The dimer is shown as two joined lines.

Chemical analyses of UV-irradiated bacteria and phages, as well as of irradiated purified DNA, have shown that the major photoproduct is an intrastrand dimer formed by two adjacent pyrimidines as a result of UV. The most important dimer is apparently the thymine dimer, shown in Figure 9-6. The significant effects of the presence of thymine dimers are the following: (1) the DNA helix becomes distorted as the thymines, which are in the same strand, are pulled toward one another (Figure 9-7), and (2) as a result of the distortion, hydrogen-bonding to adenines in the opposing strand, though possible (because the hydrogen-bonding groups are still present), is significantly weakened; this structural distortion causes inhibition of advance of the replication fork, as will be discussed in a later section.

BIOLOGICAL INDICATION OF REPAIR

From the beginning, analysis of UV survival curves was paradoxical. First, it was found that bacteria almost never yielded a single-hit curve. One might have concluded that damage was required in both DNA strands, so that two hits are required. However, the survival curves were not two-hit either. Furthermore, when survival curves were extended to very high doses, a truly linear portion was never observed: the curve continued to bend downward. The significance of this phenomenon was not understood at first, though we now know that it is a result of repair of UV damage. That is, if fairly efficient repair occurs, small doses will not be effective in killing bacteria or phages. However, as the dose increases, either the putative repair system will become saturated, or some types of damage may be nonrepairable, or the repair system itself will become damaged. Any one of these effects would result in a survival curve that has an initial plateau region followed by continued downward curvature.

The most striking observations that led to the concept of DNA repair were two discoveries: (1) an increase in the surviving fraction of bacteria resulting from certain postirradiation treatments, and (2) mutant strains that were more sensitive to radiation than wild-type strains. Furthermore, the survival of plaque-forming ability of irradiated phage was also affected by use of these bacterial mutants as a lawn. These observations will be described in the following sections.

Postirradiation Effects on Survival Levels

Repair of UV damage was first recognized by a chance observation that the survival level of irradiated bacteria was increased when the cells were left in a window and exposed to sunlight before being allowed to form colonies. A quantitative analysis of this phenomenon, which is called **photoreactivation,** is shown in Figure 9-8(a,b). Panel (a) shows that exposure to visible light increases the survival to a level that is determined by the UV dose. In panel (b) the data of panel (a) are combined into two survival curves, with and without exposure to visible light. These curves clearly indicate that visible light eliminates some of the damage introduced by the UV light. Biochemical analysis has indicated that photoreactivation is an enzymatic reaction in which an enzyme, called **photolyase,** cleaves dimers, restoring them to the monomeric state (why all are not cleaved is unknown). The enzyme is inactive unless exposed to visible light, apparently using the energy of the light to perform the cleavage. Although photolyase was purified more than 20 years ago, the light-sensitive component has never been identified. Mutants (Pho$^-$) that cannot photoreactivate and lack photolyase have been isolated. They show no other phenotypic properties, so presumably photoreactivation is the sole function of photolyase.

UV-irradiated phages are not photoreactivated by exposure to light, since they do not possess photolyase. However, if examined in the correct way, irradiated phage can be photoreactivated. Phage are normally plated by mixing them with an excess number of bacteria and adding the mixture

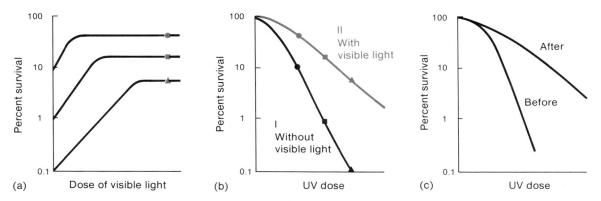

Figure 9-8   Types of repair. (a) Increase in survival of three different samples of ultraviolet-light-irradiated bacteria as a function of the dose of visible light. This is called photoreactivation. (b) A pair of survival curves showing the effect of postirradiation with visible light. Curve I consists of the points taken from the y-axis of part (a) and curve II is a plot of the red points taken from the plateaus in part (a). (c) Survival curves before and after incubation in buffer following ultraviolet-light irradiation. This is called liquid-holding recovery.

to soft agar. The same number of plaques will result if the phage are pre-adsorbed to bacteria in a *non*nutrient buffer (to inhibit phage development temporarily), and then these phage-bacterium complexes (infective centers) are placed in soft agar. However, if irradiated phage are preadsorbed to bacteria and a portion of the infective centers is exposed to intense visible light before plating, the infective centers that have been irradiated with visible light will produce more plaques. Clearly, this is a result of light activation of photolyase in the cells and subsequent photoreactivation of the UV-irradiated phage.

Another response to damage that points to the existence of a second repair system is **liquid-holding recovery.** If UV-irradiated cells are held in a nonnutrient buffer for several hours before plating, the surviving fraction for a particular dose is increased (Figure 9-8(c)). When first observed, it was suggested that delaying cell growth or perhaps merely DNA replication created additional time for some repair process to occur. Furthermore, since liquid-holding recovery takes place without light from any source, it was hypothesized that *E. coli* possesses two distinct repair processes—a light-dependent one (photoreactivation) and a light-independent one. Proof of the existence of two repair systems came from the isolation of an extraordinarily UV-sensitive *E. coli* mutant called B_s (Figure 9-9). This mutant does not show higher viability after UV irradiation when held in a buffer in the dark (that is, it does not exhibit dark repair), but it does photoreactivate normally. The Pho$^-$ mutant, described above, can undergo liquid-holding recovery, confirming the conclusion that there are two repair systems. Liquid-holding recovery is now known to be a manifestation of a general phenomenon called **dark repair,** which we will see is accomplished in several ways.

E. coli B_s exhibits loss of another repair phenomenon, called **host-cell reactivation** of phage. That is, the survival curve for UV-irradiated phage of

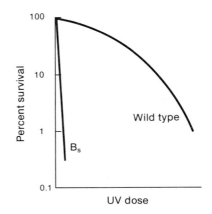

Figure 9-9 Survival curves showing the great sensitivity to ultraviolet light of the mutant *E. coli* B_s.

certain types (e.g., *E. coli* phage T1) is steeper if the phage are plated on B_s than on the wild-type strain B. The explanation is that strain B can repair some of the UV damage in the phage and B_s cannot.

A major mechanism for several types of dark repair is the elimination of thymine dimers from DNA. However, dimers are not cleaved and converted to monomers, as in photoreactivation, but instead *the dimer is completely excised from the DNA*. Evidence comes from the following experimental result. A population of bacteria is UV-irradiated and then incubated for various periods of time in a nonnutrient buffer (that is, subjected to liquid-holding recovery). During this period the number of thymine dimers present in the DNA, determined by direct biochemical analysis, continually decreases, and at the same time, thymine dimers appear both in the intracellular fluid and in the buffer. The biochemical mechanism for this excision will be described in a later section.

It might seem surprising that cells should have evolved systems for repairing damage caused by UV, which is not an agent commonly encountered in nature by most cells. To understand this, one must remember that the properties of cells reflect ancient history. During the period when life was evolving in the primordial seas, the earth was not enveloped by a stratospheric ozone layer. Without this layer a very high flux of UV would have reached the surface of the earth and would have been very damaging to primitive organisms. Thus, acquiring the ability to repair UV-induced damage had great survival value and this ability has persisted for eons because, as will be seen shortly, the dark repair system is also able to repair other types of damage. How photoreactivation has persisted is less clear since it seems to be active only against the pyrimidine dimers formed by ultraviolet irradiation. Furthermore, the enzymatic system is present also in animals whose exposure to UV light is very small, in barophyllic bacteria that grow only in oceanic deep trenches (where there is no light), and in humans.

Mutants of the Dark-Repair Systems

An effective approach to the analysis of dark repair came from a study of UV-sensitive mutants. In the preceding section the sensitive mutant B_s was mentioned. Unfortunately, at the time this mutant was isolated, techniques had not yet been developed for carrying out genetic analysis with *E. coli* strain B, which had certain properties that were not understood. Thus, a similar mutant in the more accessible K12 strain, which engages in sexual mating (Chapter 14) and genetic recombination, was sought. The search for mutants made use of the phenomenon of host-cell reactivation described above. Several hundred thousand cells of a mutagenized sample of *E. coli* were spread on agar along with about 10^7 UV-killed T1 phage. The concentration of phage on the plate was such that before colonies became visible, each microcolony had been infected with several UV-killed phage. If the microcolony consisted of wild-type cells, a fraction of the UV-killed phage was reactivated, and these went on to infect and lyse the microcolony. Colonies of mutants unable to engage in host-cell reactivation were also infected, but they did not release progeny phage and hence survived to produce visible colonies. These colonies were streaked on agar to isolate the mutant cell from free phage and wild-type cells (Chapter 4), bacterial cultures were prepared, and the radiosensitivity of the cultures was tested. Many of these colonies proved to be exceedingly sensitive to UV. Complementation tests showed that the mutations fell into three classes, which defined the genes *uvrA, uvrB,* and *uvrC.* Biochemical analysis showed that the *uvrA* and *uvrB* mutants are defective in excision of thymine dimers (as well as in repair of many types of chemical damage). The role of the *uvrC* gene is unknown. Survival curves for some of these mutants are shown in Figure 9-10.

In studies of genetic recombination in *E. coli* (described in Chapter 14), which were totally unrelated to the repair phenomenon, recombination-deficient mutants were isolated. These mutations mapped in three genes, designated *recA, recB,* and *recC.* Various properties of these mutants were examined; among the findings was the discovery that they are exceedingly sensitive to UV radiation (Figure 9-10). However, biochemical analysis showed that these mutants excised thymine dimers normally, indicating that a system that utilizes recombination, or at least the *rec* genes, is responsible for another class of repair. This type of repair will be described shortly.

In the course of studying DNA replication in *E. coli* a mutation was isolated in the gene *polA,* which encodes DNA polymerase I. The mutant was viable, which suggested that polymerase I is not the major replication enzyme. This finding provided the impetus for seeking other DNA polymerases in *E. coli,* and in fact polymerase III was isolated from the *polA*⁻

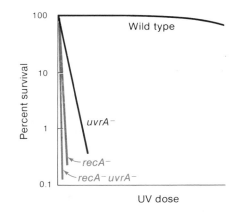

Figure 9-10 Survival curves of *E. coli* showing the sensitization to ultraviolet light resulting from the *uvrA*⁻, *recA*⁻, and *recA*⁻*uvrA*⁻ mutations.

mutant. Detailed examination of the *polA*⁻ mutant ultimately showed that it retained normal $5' \rightarrow 3'$ exonuclease activity and had a residual polymerizing activity of about 2 percent of the wild-type, which was sufficient for it to play its essential role in the removal of RNA primers and the joining of precursor fragments (Chapter 8). An important observation about the PolA⁻ phenotype was that the mutant had somewhat increased sensitivity to UV, which suggested that polymerase I might be a significant component of a repair system. Photoreactivation is normal, so the putative system must be dark repair. In the next section we will see that Pol I is active in a late step in the repair of DNA from which a thymine dimer has been excised.

BIOCHEMICAL MECHANISMS FOR REPAIR OF THYMINE DIMERS

There are four major pathways for dealing with thymine dimers in DNA, which can be subdivided into two classes—light-induced repair and light-independent pathways. The latter can be accomplished by three distinct mechanisms: (1) excision of the damaged bases (**excision repair**), (2) reconstruction of a functional DNA molecule from undamaged fragments (**recombination repair**), and (3) tolerance of the damage (**SOS repair**). The biochemical mechanisms for each of these repair processes are described in this section.

Excision Repair

Excision repair is a multistep enzymatic process (Figure 9-11). Two distinct mechanisms have been observed for the first step, an **incision** step. In *E. coli* a repair endonuclease recognizes the distortion produced by a thymine dimer and makes two cuts in the sugar-phosphate backbone: one is 8 nucleotides to the 5′ side of the dimer and another is 4–5 nucleotides to the 3′ side (panel (a)). At the 5′ incision site a 3′-OH group is produced, which DNA polymerase I uses as a primer and synthesizes a new strand while displacing the DNA segment that carries the thymine dimer. The final step of the repair process is joining of the newly synthesized segment to the original strand by DNA ligase. The excised fragment is ultimately degraded to single nucleotides plus a thymine dimer dinucleotide by the combined activity of numerous scavenging exo- and endonucleases. Note that the role of Pol I was anticipated by the observation described in the previous section that *polA*⁻ mutants are UV-sensitive.

In several other systems (e.g., the bacterium *Micrococcus luteus* and *E. coli* phage T4) the incision step occurs in two distinct stages (panel (b)). The first is an enzymatic cleavage of the *N*-glycosidic bond in the 5′ thymine nucleotide of the dimer. Incision of the strand is completed by an endonu-

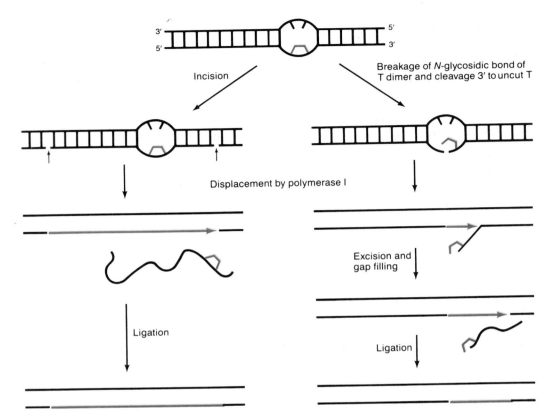

Figure 9-11 Two modes of excision repair. (a) The *E. coli* mechanism. Two incision steps are followed by gap-filling and displacement by polymerase I. (b) The *M. luteus* mechanism. A pyrimidine dimer glycosylase breaks an *N*-glycosidic bond and makes a single incision. Pol I displaces the strand, which is removed by an exonucleolytic event. In both mechanisms the final step is ligation.

clease activity that recognizes a deoxyribose lacking a base; the enzyme makes a single cut at the 5′ side of the remaining thymine in the dimer site. Then, the deoxyribose is removed, and Pol I acts at the new 3′-OH group, displacing the strand and filling the gap. The displaced strand is excised by one of several different enzymes.

The incision activity of *E. coli* is determined by a complex of the products of the three genes *uvrA, uvrB,* and *uvrC,* mentioned in the preceding section. The *uvrA-* and *uvrB*-gene products are two subunits of the excision endonuclease. The UvrC product is necessary for maximum endonuclease activity *in vivo,* but its precise role is not known.

The Uvr system is able to repair lesions other than thymine dimers. These lesions have in common either the displacement of bases, as in thymine dimer formation, or the addition of bulky substituents on the bases. It is assumed that the incision enzyme recognizes a helix distortion.

Recombination Repair

If excision repair accounted for all dark repair, one might expect that a UV dose yielding an average of one lethal event per $uvrA^-$ cell would produce one or a very small number of thymine dimers per cell. However, the number is about 300 unexcised dimers, which seems rather large and suggests that the cells possess another repair system. Evidence for such a system comes from the observation described earlier that cells mutant in the gene *recA*, a gene that is essential for genetic recombination in *E. coli*, are exceedingly UV-sensitive (Figure 9-10). Excision of thymine dimers occurs in a $recA^-$ mutant, so *recA*-mediated repair clearly differs from excision repair. The existence of two repair systems is confirmed by a quantitative analysis of the survival curve of a $uvrA^- recA^-$ double mutant, which is more UV-sensitive than either of the single mutants (Figure 9-10).

In order to discuss the mechanism of recombination repair, the effect of a thymine dimer on DNA replication must be understood. When polymerase III reaches a thymine dimer, the replication fork is temporarily stalled. A thymine dimer is still capable of forming hydrogen bonds with two adenines because the chemical change in dimerization does not alter the groups that engage in hydrogen bonding. However, the dimer introduces a distortion into the helix, and when an adenine is added to the growing chain, polymerase III reacts to the distorted region as if a mispaired base had been added; the editing function (Chapter 8) then removes the adenine. The cycle begins again—an adenine is added and then it is removed; the net result is that the polymerase is stalled at the site of the dimer. A cell in which DNA synthesis is permanently stalled cannot complete a round of replication, so a colony cannot form. However, stalling is only temporary, for there are two different ways by which DNA synthesis can restart—**postdimer initiation** and **transdimer synthesis.** These are responsible for **recombination repair** and **SOS repair,** respectively. In this section recombination repair is described; SOS repair is discussed in the succeeding section.

One way that a thymine-dimer block is dealt with is to pass it by and initiate chain growth beyond the block (Figure 9-12). This is called postdimer initiation; its mechanism, which appears to involve unprimed reinitiation, is unknown. The result of postdimer initiation is that the daughter strands have large gaps, one for each unexcised thymine dimer. There is no way to produce viable daughter cells by continued replication alone, because the strands

Figure 9-12 Blockage of replication by thymine dimers (represented by joined lines) followed by re-starts several bases beyond the dimer. The black region is a segment of ultraviolet-light-irradiated parental DNA. The red region represents synthesis of a daughter molecule from right to left. The daughter strand contains gaps.

having the thymine dimer will continue to turn out gapped daughter strands, and the first set of gapped daughter strands will be fragmented when the growing fork enters a gap. However, by a recombination mechanism called **sister-strand exchange** an intact double-stranded molecule can be made.

The essential idea in sister-strand exchange is that a single-stranded segment free of any defects is excised from a "good" strand on the homologous DNA segment at the replication fork and somehow inserted into the gap created by excision of a thymine dimer (Figure 9-13). How this occurs is unknown, but the RecA product and presumably genetic recombination are required. The combined action of polymerase I and DNA ligase joins this inserted piece to adjacent regions, thus filling in the gap. The gap formed in the donor molecule by excision is also filled in completely by polymerase I and ligase. If this exchange and gap filling are done for each thymine dimer, two complete daughter single strands can be formed and each can serve in the *next* round of replication as a template for synthesis of normal DNA molecules. Note that the system fails if two dimers in opposite strands are very near one another because then no undamaged sister strand segments

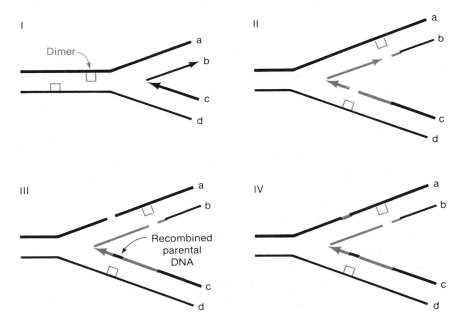

Figure 9-13 Recombination repair. I. A molecule containing two thymine dimers (red boxes) in strands a and d is being replicated. II. By postdimer initiation, a molecule is formed whose daughter strands b and c have gaps. If repair does not occur, in the next round of replication, strands a and d would yield gapped daughter strands, and strands b and c would again be fragmented. III. A segment of parental strand is excised and inserted into strand c. IV. The gap in strand b is similarly filled in by repair synthesis. Such a DNA molecule would probably engage in a second exchange in which a segment of c would fill the gap in b. DNA synthesized after irradiation is shown in red. Heavy and thin lines are used for purposes of identification only.

are available to be excised. The molecular details of recombinational repair are not known, and the model shown in Figure 9-13 must be considered to be a working hypothesis.

Recombination repair is an important mechanism because it eliminates the necessity for delaying replication for the many hours that would be needed for excision repair to remove all thymine dimers. It may also be the case that some kinds of damage cannot be eliminated by excision repair—for example, alterations that do not cause helix distortion but do stop DNA synthesis.

Recombination repair also occurs with UV-irradiated phages of some types; this is indicated by the fact that if a population of a phage that fails to engage in excision repair is UV-irradiated, fewer phage will plate on a $recA^-$ host than on a $recA^+$ host.

Since recombination repair occurs after DNA replication, in contrast with excision repair, it has been called postreplicational repair. Recently, the term **daughter-strand gap repair** has been introduced since only the gaps formed opposite dimers, rather than the dimers themselves, are repaired.

SOS Repair

SOS repair includes a **bypass system** that allows DNA chain growth across damaged segments at the cost of fidelity of replication. It is an error-prone process; that is, even though intact DNA strands are formed, the strands often contain incorrect bases. (The principle involved is that survival with mutations is better than no survival at all.) SOS repair is not yet thoroughly understood but is thought to invoke a relaxation of the editing system in order to allow polymerization to proceed across a dimer (transdimer synthesis) despite the distortion of the helix. A great deal of evidence now indicates that SOS repair is the major cause of UV-induced mutagenesis.

An important feature of SOS repair is an absolute requirement for a functional *recA* gene (that is, mutagenesis caused by UV irradiation does not occur in a $recA^-$ bacterium). The RecA protein has several functions in SOS repair; the most important one for our purposes is that it directly inhibits the editing function of polymerase III by binding tightly in the region of the distortion resulting from a pyrimidine dimer. When DNA polymerase III encounters a dimer site to which RecA is bound, RecA interacts with the Pol III subunit responsible for editing and inhibits the editing function. Thus, the replication fork advances. Most of the time Pol III can utilize the residual base-pairing ability of the thymine dimer and put two adenines in the daughter strand. However, mispairing is enhanced by the distortion, which normally would activate the editing response. The presence of RecA at the dimer site inhibits editing and causes the mispaired base to remain in the daughter strand as a mutation.

A mutation in two other genes, *umuC* and *umuD*, causes UV sensitivity. Since UV-induced mutagenesis does not occur in *umu* mutants, one assumes that these genes participate in SOS repair. However, their role is not known.

Another important feature of SOS repair is that the system is *induced* as a result of damage to the DNA. In particular, sufficient amounts of RecA protein and UmuC and UmuD proteins are not available until some time after damage has occurred. The best evidence for this point comes from an analysis of the survival of UV-inactivated phage on UV-irradiated bacterial host cells.

Most phages do not need a fully functional (or even viable) host to produce progeny phage particles. For instance, a dose of UV yielding 10 percent survival of the ability to form colonies does not prevent *E. coli* from producing an infective center with phage λ. However, if the λ particles are also UV-irradiated, the irradiated *E. coli* cells are *better* able to support growth of the irradiated λ than are unirradiated cells. That is, UV-irradiated λ will produce more infective centers with irradiated *E. coli* than with an unirradiated host (Figure 9-14). This phenomenon, which is called **UV-reactivation** or **W reactivation** (for Jean Weigle, who discovered it), clearly involves a repair process and occurs only if the host is Rec$^+$; the phenomenon does not require the *uvr* genes. Although more phage survive, the surviving population contains a higher proportion of phage mutants when the irradiated host is used; that the mutation frequency increases suggests that this is an example of SOS repair. The SOS system has been turned on by the UV irradiation of the host.

Since SOS repair allows the frequency of replication errors to increase when necessary, it must be regulated in an on-off fashion in order to keep the normal error frequency low. Since repair is only needed following certain types of DNA damage, it seems reasonable that some feature of the damage

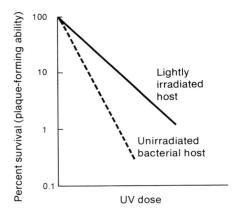

Figure 9-14 W reactivation of ultraviolet-light-irradiated phage λ. The dashed line shows the survival curve (for plaque-forming ability) obtained when λ phage irradiated with various doses of ultraviolet light are plated on unirradiated bacteria. The solid line represents survival of plaque-forming ability, when ultraviolet-light-irradiated λ are plated on lightly irradiated bacteria.

would be the inducer. The same might be said of other types of UV repair, such as recombination repair and excision repair, and indeed these systems are inducible and are controlled by the same elements that regulate SOS repair. These repair systems plus several other operons comprise the SOS regulon (the term **regulon** refers to a set of operons controlled by a common element).

The SOS regulatory system has two components; the genes *lexA* and *recA*. These genes have the following features:

1. The *lexA* gene encodes a common repressor of all SOS operons. The LexA repressor binds to a common operator sequence adjacent to each gene or operon; this arrangement includes the *lexA* gene, so LexA is autoregulated.
2. The RecA protein is an inducer. The RecA product has several activities; in addition to the effect on editing described above, it is a protease. The proteolytic activity is very weak against most proteins, but quite strong against the LexA repressor. (In Chapter 17 we will see that it is also active against phage repressors.)
3. Following DNA damage of the sort inflicted by UV irradiation, there is a burst of synthesis of *recA* mRNA. However, this does not occur in *recA*⁻ mutants, because functional RecA protein is required to turn the operon on.

Let us consider how the SOS operons should be regulated. In the absence of DNA damage its gene products are unnecessary. Following the occurrence of DNA damage the system must be active, but once repair is complete, it should optimally be shut down rapidly. The three features of the SOS regulon that enable these regulatory events to occur are (1) a damage-induced activation of the protease activity of the RecA protein, (2) a vulnerable proteolytic site in the LexA protein, and (3) the self-repression of transcription of the *lexA* gene.

The RecA protein in an undamaged cell lacks protease activity. In a UV-irradiated cell RecA binds to thymine dimer sites, probably to the short segment of single-stranded DNA generated by the distortion (Chapter 9). Binding of RecA protein to DNA causes a shape change in the protein, which thereby acquires protease activity. The activated RecA molecules cleave the LexA repressor, which allows transcription of all operons of the SOS regulon to increase about 50-fold. This increase produces an adequate supply of the Uvr enzymes and of RecA proteins. Since *lexA* is autoregulated, the LexA protein is also made in abundance. However, the large amount of RecA protein, much of which has been activated to a protease, continues to cleave LexA, so all proteins of the SOS regulon continue to be made. Once all repair is completed, RecA loses its proteolytic activity and LexA is no longer cleaved. During the period in which RecA was active, LexA was made at a rapid rate; thus, without RecA protease activity, LexA rapidly accumulates, binds to the SOS operators, turns off the SOS operons, and the state of the cell existing before DNA damage occurred is reestablished.

WHY DO DAMAGED CELLS DIE?

We have discussed a variety of repair systems that seem to be able to fix everything, yet we have not explained why, with increasing dose, cells are still killed by various forms of radiation and toxic chemicals. There are many possible explanations. One trivial reason is that there are probably some alterations that are not repairable—for instance, a base might be altered to yield a replication block that cannot be overcome. Also, damaged regions may be opposite one another or so clustered that an undamaged template is not available for repair. A second possibility is that the repair systems themselves can be damaged; this is not unlikely since it takes a very large amount of damage before lethality ensues in a wild-type organism. A third and reasonable possibility is that a lethal event may often take place before repair occurs.

PROBLEMS

1. Are thymine dimers formed between adjacent thymines in the same strand or between opposite thymines in complementary strands?

2. If UV-irradiated bacteria are incubated in a buffer prior to plating, will the number of colonies formed be increased or decreased by the treatment?

3. If UV-irradiated phage are incubated in a buffer or exposed to light prior to plating, how will the efficiency of plaque formation change?

4. Which repair system cleaves thymine dimers?

5. Which enzymes are required for excision repair in *E. coli*?

6. What is the difference between incision and excision?

7. Name two features of SOS repair that distinguish it from all other repair systems.

8. Would you expect photoreactivation to occur with equal efficiency over a wide range of wavelengths or will it exhibit a spectrum with maximum effectiveness in a small range of wavelengths?

9. A cell possesses a repair system capable of removing half of the damage produced by some agent. Which of the following statements would be true of a survival curve: (1) a dose yielding x percent survival when repair does not occur would yield $2x$ percent survival if repair occurs; (2) the dose required for a particular percentage of survival in the absence of repair is doubled when repair occurs.

10. The ability of UV-irradiated T4 phage to form plaques is almost the same on both Uvr^+ and Uvr^- bacteria, with the survival curve on the Uvr^- bacteria being only slightly steeper than the one obtained with Uvr^+ bacteria.

    (a) How might you explain this fact?
    (b) A T4 mutant is UV-irradiated and plated on Uvr^+ and Uvr^- bacteria. The survival curves are much steeper than with wild-type phage and considerably steeper with Uvr^- bacteria than with Uvr^+ bacteria. Suggest a defect for the mutant.

11. A bacterial repair system called X removes thymine dimers. You have in your bacterial collection the wild-type (X^+) and an X^- mutant. Phage λ, when UV-irradiated and then plated, gives a larger number of plaques on X^+ than on X^- bacteria. It has been proposed on the basis of survival curve analysis that the X enzyme is inducible. To test this proposal, UV-irradiated λ phage are adsorbed to both X^+ and X^- bacteria in the presence of the antibiotic chloramphenicol (which inhibits protein synthesis). No thymine dimers are removed in the X^- cell and 50% are removed in the X^+ cell. In the absence of chloramphenicol, the same results were obtained.

(a) Is X an inducible system?

(b) Suppose 5% of the thymine dimers are removed in the presence of chlor-amphenicol and 50% in its absence; how would your conclusion be changed?

12. A Uvr^- bacterial culture that has been grown for many generations in medium containing $[^3H]$thymidine is transferred to nonradioactive medium. The culture is UV-irradiated; since the cells are Uvr^-, no radioactive material appears in the medium. A phage sample is used to infect the irradiated culture, and several minutes later radioactive thymine dimers are found in the medium. The phage does not cause degradation of bacterial DNA from unirradiated bacteria. Explain the appearance of the dimers.

13. It is generally observed that if the UV survival curves of two different phages are compared, the phage with the larger DNA molecule has the steeper curve.

(a) Explain this phenomenon.

(b) You have isolated a new phage whose DNA is very large but which has the radiosensitivity of a phage with a small DNA. Suggest a possible explanation.

14. The radiosensitivity (measured by colony formation) of bacteria obtained from a log-phase culture and from late stationary phase is measured. Both give curves with the general shape shown in Figure 9-1 of the text, but the cells in log phase are more resistant to the radiation. Suggest two possible explanations, one of which includes DNA repair and one of which does not.

15. What is the significance of the dose yielding $1/e$ survival?

16. Four double-stranded DNA phages are x-irradiated. Exponential survival curves are seen. The doses yielding $1/e$ survival are 1.55×10^4, 2.60×10^4, 4.14×10^4, and 8.45×10^4 rads for phages I–IV, respectively. The molecular weights of the DNA molecules of three of the phages are known: I, 120×10^6; II, 72×10^6; IV, 22×10^6. What is likely to be the molecular weight of the DNA of phage III?

Chapter 10

Mutagenesis, Mutations and Mutants

I n previous chapters the use of mutants has been encountered repeatedly. In this chapter we explain what a mutant actually is, how it is formed, and how it is detected.

BIOCHEMICAL BASIS OF MUTANTS

The term mutant refers to an organism in which either the base sequence of DNA or the phenotype has been changed. These definitions are actually the same since the base sequence of DNA determines the amino acid sequence of a protein. The chemical and physical properties of each protein are determined by its amino acid sequence, so a single amino acid change is capable of altering the activity of, or even completely inactivating, a protein. It is easy to understand how an amino acid substitution can change the structure and, hence, the biological activity of a protein. For instance, consider a hypothetical protein whose three-dimensional structure is determined entirely by an interaction between one positively charged amino acid (for example, lysine) and one negatively charged amino acid (aspartic acid). A substitution of methionine, which is uncharged, for the lysine would clearly destroy the three-dimensional structure. Similarly, a protein might be stabilized by a hydrophobic cluster, in which case substitution of a polar amino acid for a nonpolar one in the cluster would also be disruptive.

An amino acid substitution does not always lead to a mutant phenotype. For instance, a hydrophobic cluster might be virtually unaffected by a replacement of a leucine by another nonpolar amino acid such as isoleucine. When an amino acid substitution has no detectable effect on phenotype, it is called a **silent mutation.** A base change without an amino acid alteration (for example, in the third position of a codon), is also a silent mutation.

The shapes of proteins are determined by such a variety of interactions that sometimes an amino acid substitution is only partially disruptive. This could be manifested as a reduction, rather than a loss, of activity of an enzyme. For example, a bacterium carrying such a mutation in the enzyme that synthesizes an essential substance might grow very slowly (but it would grow) unless the substance is provided in the growth medium. Such a mutation is called a **leaky mutation;** one tries to avoid these mutations in genetic studies.

Several common alterations other than amino acid substitutions also eliminate activity of a protein:

1. Deletion of $3n$ bases (n a positive integer), which causes one or more amino acids to be absent in the completed protein.
2. A base deletion or addition that causes shift in the reading frame such that all amino acids starting from the site of the mutation are different.
3. A chain termination mutation, in which a base change generates a stop codon. Such mutations cause the mutant protein to be a fragment of the wild-type protein.

SPONTANEOUS MUTATIONS

Mutations are random events and there is no way of knowing when or in which cell a mutation will occur. However, every gene mutates spontaneously at a characteristic rate, making it possible to assign probabilities to particular mutational events. Thus, there is a definite probability that a given gene will mutate in a particular cell, and likewise a definite probability that a mutant allele of the gene will occur in a population of a particular size. Mutations are also random in the sense that their occurrence is not related to any adaptive advantage they may confer on the organism in its environment; in this section the basis for this conclusion will be presented.

The Random and Nonadaptive Nature of Mutation

The idea that mutations are spontaneous random events unrelated to adaptation was not accepted by many microbiologists until the late 1940s. Prior to that time it was believed that mutations occur in bacterial populations *in response to* particular selective conditions. The basis for this belief was the observation that when antibiotic-sensitive bacteria are spread on an agar containing the antibiotic, some colonies form that consist of cells having an inherited resistance to the drug. The initial interpretation of this observation (and similar ones) was that these adaptive variations were *induced* by the selective agent itself. However, in 1943 this interpretation was shown to be incorrect by an experiment that demonstrated the spontaneous and non-adaptive nature of mutation, an experiment that marked the birth of microbial genetics. In this experiment the origin of mutations in *E. coli* that confer resistance to phage T1 (T1-r mutations) was investigated. The approach was to compare the number of T1-r mutant cells arising in different cultures of T1-s (T1-sensitive) cells with the number found in repeated samples of the same size taken from a single culture. A statistical test called a **fluctuation test** was used to analyze the results of the experiment. The data shown in Table 10-1 were obtained from one experiment in which twenty 0.2-ml cul-

Table 10-1   The number of T1 phage-resistant *E. coli* mutants in small individual cultures and in samples from a large bulk culture.

Small individual cultures		Samples from large culture	
Culture	T1-r colonies per plate	Sample	T1-r colonies per plate
1	1	1	14
2	0	2	15
3	3	3	13
4	0	4	21
5	0	5	15
6	5	6	14
7	0	7	26
8	5	8	16
9	0	9	20
10	6	10	13
11	107		
12	0	Mean	16.7
13	0		
14	0	Variance	15
15	1		
16	0	Variance/Mean	0.9
17	0		
18	64		
19	0		
20	35		
Mean	11.4		
Variance	694		
Variance/Mean	60.8		

Source: Data from S. E. Luria and M. Delbrück. *Genetics* 28 (1943): 491.

tures and one 10-ml culture, each containing a T1-s bacterial strain at an initial concentration of 10^3 cells/ml, were grown to a concentration of 2.8×10^9 cells/ml (21 generations). Each of the small cultures and ten 0.2-ml samples from the large culture were plated on individual plates covered uniformly with about 10^{10} T1 phage (enough to kill all T1-s cells), and the number of colonies that formed (which would be T1-r) were counted. Each plate received the same number of bacterial cells (5.6×10^8), but the number of T1-r colonies formed depended on whether the cells had been grown in the small individual cultures or in the large bulk culture. No T1-r cells were detected in 11 of the 20 small cultures and the numbers in the other 9 of these cultures ranged from 1 to 107; in contrast, the 10 samples from the large culture each had about the same number. The alternatives expected in such an experiment were the following: (1) if T1-r bacteria arise in response to the phage, there should be about equal numbers in all populations of the same size; (2) if T1-r cells arise by spontaneous mutation *at different times* in the growth of the cultures in the absence of the phage, the numbers in different culture should vary greatly. The results of the experiment shown in Table 10-1 and others like it were consistent with the second alternative, which is depicted in Figure 10-1. Panel (a) shows four separate bacterial cultures and the pedigrees by which the cells could be derived from a single ancestor. A mutation to phage resistance is shown to have occurred during a different generation in each lineage (line of descent). The two extreme

Figure 10-1 The fluctuation test. (a) Pedigrees showing the source of the mutants (red colonies) found in each of four different samples. In each pedigree, the occurrence of a mutation is indicated by a shift of the path to red. (b) Typical results of sampling from a single mixed culture. Black and red rods indicate wild type and mutant bacteria, respectively; they are not drawn to scale, being roughly 10^8 times too large compared to the colonies.

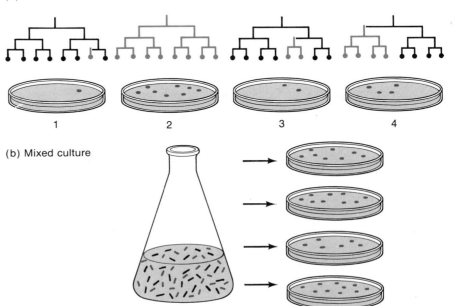

(a) Individual cultures

1   2   3   4

(b) Mixed culture

cases are cultures 1 and 2. Culture 1 depicts a mutation occurring just prior to sampling, resulting in only a single mutant colony. Culture 2 depicts a mutation occurring soon after addition of bacteria to the medium, yielding an entirely mutant population. Cultures 3 and 4 are intermediate cases. Thus, separate cultures will contain significantly different numbers of mutant cells. Panel (b) shows the result of growing the four lineages in panel (a) in a single large culture. With this arrangement, the mutants from each lineage become uniformly dispersed in the medium, so sample-to-sample variation is greatly reduced.

Direct evidence for the spontaneous origin of T1-r mutations in *E. coli* cells not exposed to the phage was obtained in the early 1950s by a procedure called **replica plating** (Figure 10-2). In this procedure bacteria are plated, and after colonies have formed, a piece of sterile velvet mounted on a solid support is pressed onto the surface of the plate. Some bacteria from each colony stick to the fibers, as shown in panel (a). Then, the velvet is pressed onto a fresh plate, transferring the cells, which give rise to new colonies

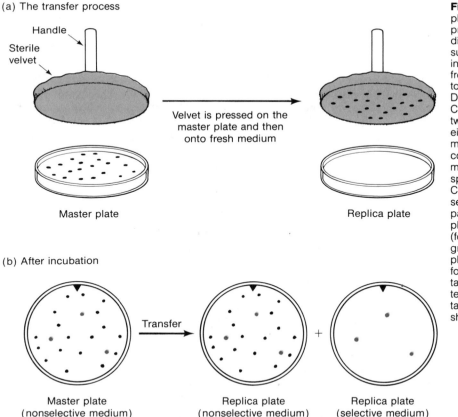

(a) The transfer process

Handle

Sterile velvet

Velvet is pressed on the master plate and then onto fresh medium

Master plate

Replica plate

(b) After incubation

Transfer

+

Master plate (nonselective medium)

Replica plate (nonselective medium)

Replica plate (selective medium)

Figure 10-2 Replica plating. (a) The transfer process. A velvet-covered disk is pressed onto the surface of a master plate in order to transfer cells from colonies on that plate to a second medium. (b) Detection of mutants. Cells are transferred onto two plates containing either a nonselective medium (on which all form colonies) or a selective medium (for example, one spread with T1 phage). Colonies form on the nonselective plate in the same pattern as on the master plate. Only mutant cells (for example, T1-r) can grow on the selective plate; the colonies that form correspond to certain positions on the master plate. Colonies containing mutant cells are shown in red.

having positions identical to those on the first plate. Panel (b) shows how the method was used to demonstrate the spontaneous origin of T1-r mutants. Master plates containing 10^7 colonies growing on nonselective medium (lacking T1) were replica-plated onto a series of plates that had been spread with T1 phage. After incubation for a time sufficient for colony formation, a few colonies of T1-r bacteria appeared in the same positions on each of the replica plates. This meant that the T1-r cells that formed the colonies must have been transferred from corresponding colonies on the master plate. Since the colonies on the master plate had never been exposed to the phage, the mutations to resistance occurred by chance in cells not exposed to the phage.

Mutation Rates

The **mutation rate** is the probability that a gene undergoes mutation in a single generation. Measurement of mutation rates is important in population genetics, studies of evolution, and in analyzing the effect of environmental mutagens.

Estimation of mutation rates in bacteria is complicated by the fact that a mutation can occur at any time during the growth of a culture, and division of the mutant bacteria usually will result in an increase in their number at a rate the same as that of the increase of the population as a whole. The result, shown by the occurrence of phage-resistant mutants in different cultures (Table 10-1, column 2) is that a culture may contain either many mutants, some mutants, or none. With a mutation rate *per generation* of μ the probability P_0 of obtaining no mutants in a culture of N cells is, by the Poisson distribution, $e^{-\mu N}$. (This follows from the fact that the total number of divisions of individual cells required to produce N cells from 1 cell is $N - 1$, and N is a large number.) Thus, $-\mu N = \ln P_0$, or

$$\mu = -(1/N) \ln P_0$$

In the fluctuation test data in Table 10-1, 11 of the 20 small cultures contained no phage-resistant mutants, and the average number N of cells per culture was 5.6×10^8. Thus, the mutation rate could be estimated as

$$\mu = -(1/5.6 \times 10^8) \ln (11/20)$$
$$= 1.1 \times 10^{-9} \text{ per cell per round of replication}$$

The fluctuation test is one important method for estimating mutation rates in bacteria. Another method is the following.

Consider a culture of wild-type *E. coli* obtained by inoculating a single colony into liquid medium. At various times samples are removed and tested for a particular phenotype. In the experiment to be described the wild-type phenotype is sensitive to phage T5 and the mutant is a T5-r cell. Data for this experiment are shown in Figure 10-3. Note that the culture was grown in the absence of any phage, so that neither T5-r nor T5-s bacteria were selected, and changes in allele frequency were determined entirely by muta-

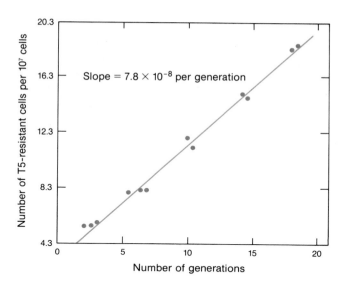

Figure 10-3 Accumulation of cells resistant to phage T5 resulting from recurrent mutation in a population of *Escherichia coli* growing in conditions that allow continuous growth at constant cell density. [After Kubitschek, H. E. 1970. *Introduction to Research with Continuous Cultures*, p. 33. Reprinted by permission of Prentice-Hall, Inc., Englewood Cliffs, NJ.]

tion. The figure shows that the frequency of T5-r cells increased linearly over the course of the experiment. The reason for the increase was that almost all of the cells in the initial population were T5-s; in each generation, some of these T5-s cells underwent spontaneous mutation to become T5-r, and thus T5-r cells accumulated in the population as time progressed. Note that the T5-r cells are so rare in the experiment that reverse mutation from T5-r to T5-s may be ignored.

The linearity of the curve can be deduced from a simple equation. We denote the frequencies of T5-s and T5-r cells in the population by p and q, respectively, with subscripts indicating the generation. Since reverse mutation can be neglected, the frequency of T5-s cells in the nth generation will equal the proportion of T5-s bacterial cell lineages that escaped mutation for n consecutive generations. All other lineages will have mutated to T5-r at some prior time. If mutation from T5-s to T5-r occurs at the rate μ per generation, then the probability that a particular T5-s lineage will escape mutation in each of n generations must equal $(1 - \mu)^n$. Consequently,

$$p_n = (1 - \mu)^n p_0$$

Making the approximation $(1 - \mu)^n = 1 - n\mu$, which is valid when μ is very small, and substituting $1 - q = p$ yields

$$1 - q_n = (1 - n\mu)(1 - q_0) = 1 - n\mu - q_0 + n\mu q_0$$

The term $n\mu q_0$, which is the product of two small numbers, can be ignored, so rearrangement of the equation yields

$$q_n = q_0 + n\mu$$

This is the equation of a straight line, and it explains why the curve in Figure

10-3 is linear. Furthermore, the *slope* of the line is μ, the mutation rate. Thus, Figure 10-3 shows that the mutation rate of T5-s to T5-r is 7.8×10^{-8} mutations per generation. It should be noted that the mutation rates for T5-s to T5-r and T1-s to T1-r are quite different (by a factor of 70). This should not be surprising, because the mutation rates do not depend on the class of phenotypic changes (e.g., phage sensitivity to resistance) but on the size and nucleotide sequence of the gene and on the amino acid sequence and three-dimensional structure of the gene product.

The Origin of Spontaneous Mutations

Mutation requires modification of DNA, and several mechanisms for such modification are known. The two mechanisms that are most important for spontaneous mutagenesis are (1) errors occurring during replication and (2) spontaneous alteration of a base.

Errors in nucleotide incorporation at the end of a growing strand during DNA replication occur with sufficiently high frequency that the information content of a daughter DNA molecule would differ significantly from that of the parent were it not for two mechanisms for correcting such errors—editing and mismatch repair. These have already been described in Chapter 8.

A source of incorporation errors not described earlier is **tautomerism** of the nucleotide bases. Some of the bases exist in alternative forms having different base-pairing properties. Examples are shown in Figure 10-4. A rare form of adenine can form a stable hydrogen bond with cytosine, and the enol form of thymine can pair with guanine. Thus, at the time of incorporation, an incorrect base will be correctly hydrogen-bonded to the template strand, so it is not recognized by the editing function as incorrect. However, when the base later assumes its normal structure, a mismatched base pair will be present, which will ordinarily be corrected by the mismatch repair system. However, if the elapsed time is so great that the region of the daughter strand containing the incorrect base has become methylated, the mismatch repair system will be unable to distinguish parental and daughter strands and mutation can occur. Both editing and mismatch repair are exceedingly efficient; however, they are not perfect, and mutations do occur.

Another source of spontaneous mutation is an alteration of 5-methyl-cytosine (MeC), a methylated form of cytosine that pairs with guanine and is present to the extent of about 5 percent of the cytosines in the DNA of many bacteria and viruses. Both cytosine and 5-methylcytosine are subject to occasional loss of an amino group. For cytosine, this loss yields uracil (Figure 10-5). Since uracil pairs with adenine instead of guanine, replication of a molecule containing a GU base pair will ultimately lead to substitution of an AT pair for the original GC pair (by the process GU → AU → AT in successive rounds of replication). However, cells possess an enzyme (*uracil glycosylase*) that specifically removes uracil from DNA, so the C-to-U con-

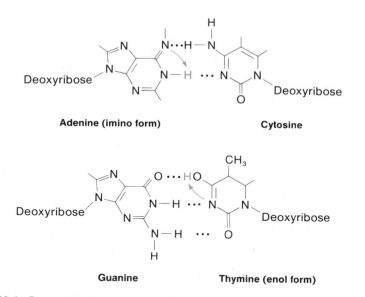

Adenine (imino form)   **Cytosine**

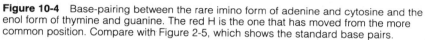

Guanine   **Thymine (enol form)**

Figure 10-4 Base-pairing between the rare imino form of adenine and cytosine and the enol form of thymine and guanine. The red H is the one that has moved from the more common position. Compare with Figure 2-5, which shows the standard base pairs.

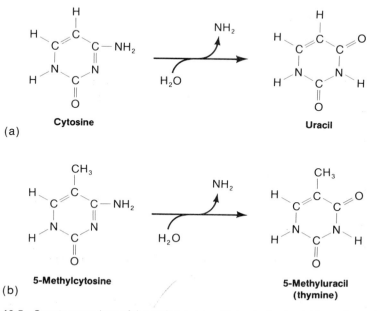

Figure 10-5 Spontaneous loss of the amino group of (a) cytosine to yield uracil and (b) 5-methylcytosine to yield thymine.

version rarely leads to mutation. Loss of the amino group of 5-methylcytosine yields 5-methyluracil, which is the same as thymine (Figure 10-5). Since thymine is a normal DNA base, no thymine glycosylase exists, so the GMeC pair becomes a GT pair. A GT pair is subject to correction by mismatch repair. However, since MeC is a methylated base and therefore is present in a methylated strand, the mismatch repair system does not recognize the thymine as incorrect. The direction of correction will be random, sometimes yielding the correct GC pair and sometimes an incorrect AT pair. Thus, MeC sites, which exist at only a few locations in a gene, constitute highly mutable sites, and the mutations are always GMeC → AT changes.

Sites within a gene at which mutations occur with much higher frequency than at other sites were first observed in the fine-structure mapping experiments of the T4 *rII* locus to be described in Chapter 15, in which several thousand independently isolated *rII* mutations were mapped (Figure 10-6). These sites were called **hot spots.** In later years determination of the base sequence of several bacterial genes and of mutant alleles has shown that most hot spots are GMeC base pairs.

Some hot spots yield large deletions rather than point mutations. The base sequences of these regions indicate that the deletion is often bounded by a repeated sequence. Two possible mechanisms for the production of a deletion in molecules having repeated nucleotide sequences are illustrated in Figure 10-7. These mechanisms are recombinational excision and a particular type of replication error.

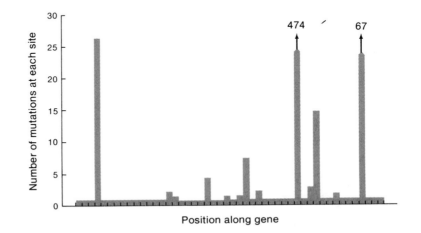

Figure 10-6 A portion of the T4 *rII* gene showing the number of mutations isolated at each site. [From S. Benzer. 1961. *Proc. Nat. Acad. Sci.*, 47, 410.]

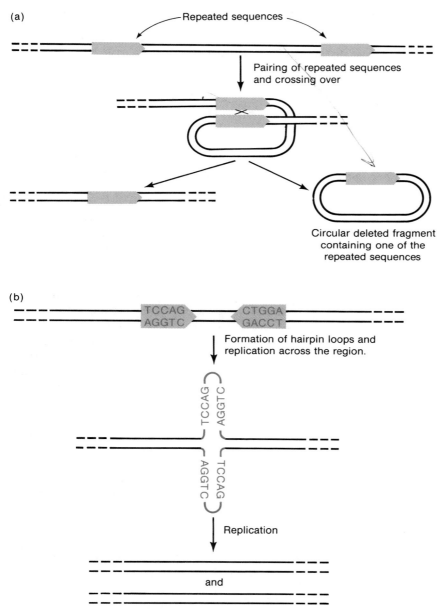

Figure 10-7 Two mechanisms for the production of deletions in molecules with repeated nucleotide sequences. (a) Crossing over (homologous recombination) between paired direct repeats. (b) Replication across an aberrant form of a molecule with an inverted repeat.

ISOLATION OF MUTANTS

Mutants are usually isolated from large populations of organisms, and some means is needed to distinguish a mutant and a wild-type organism. Furthermore, the mutants may constitute an extremely small fraction of the total population, so that prior to searching for a mutant it is necessary to increase the mutant fraction. These two procedures—detection, or screening, and enrichment—are described in this section.

Mutants can be detected in several ways. Let us first examine two simple procedures used to detect bacterial mutants that either are resistant to an antibiotic or unable to utilize a particular sugar. These procedures are the following:

1. An antibiotic-resistant mutant is found simply by plating a large number of bacteria on agar containing the antibiotic. Only a resistant cell can form a colony. This method applies to the isolation of mutants resistant to any chemical compound that can be incorporated into agar and of phage-resistant mutants. In the latter case, about 10^9 phage are spread on the surface of a plate before the bacteria are plated. Sensitive cells are lysed and resistant mutants form colonies.

2. Sugar-utilization mutants are isolated by means of color indicator plates, as described in Chapters 4 and 7. If the mutant fraction is high enough, for example, 10^{-5}, the mutants can be found by direct inspection: on MacConkey-lactose plates a single white (Lac$^-$) colony can usually be detected among about 10^4 small red (Lac$^+$) colonies.

A sugar-utilization mutant can grow on a color-indicator plate lacking the sugar, since other carbon sources are present. However, if we seek a leucine-requiring (Leu$^-$) mutant, we cannot use an agar lacking leucine, because the desired mutant will not grow. Although it is more tedious, replica-plating solves this problem. A portion of a population of Leu$^+$ cells containing some Leu$^-$ mutants is plated on nutrient agar such that there are several hundred colonies on a plate. This "master plate" is then replicated onto two minimal agar plates, one containing leucine and the other lacking leucine. The leucine-containing plate is a reference plate and indicates that cells have been transferred from each colony of the master plate. Leu$^-$ mutants will fail to grow on the leucine-deficient medium (the "test plate"). Thus, a Leu$^-$ mutant is identified by its growth on the reference plate and lack of growth on the test plate; it can then be isolated from the reference plate.

Replica-plating can also be used to isolate temperature-sensitive mutants. This is done by forming colonies at 30°C and then transferring the colonies to two plates that are incubated at either 30°C or 42°C. A colony that forms on the 30°C plate but is absent from the 42°C plate contains a temperature-sensitive mutation.

An essential difference exists between isolation of a streptomycin-resistant (Str-r) mutant and a Leu$^-$ mutant. Isolation of a Str-r colony entails a *positive selection* in that only Str-r cells can grow on a plate containing streptomycin. Thus, as many as 10^{10} cells can be put on a plate. However, isolating a Leu$^-$ mutant entails *negative* selection: no more than a few hundred colonies can be examined on a single plate to screen for *lack* of growth by replica-plating. Thus, thousands of plates would be needed to find a single mutant if the mutant fraction was 10^{-5} (compared to the ability to detect 1 in 10^{10} for a positive selection. This problem is minimized by two techniques. First, the starting culture can be treated with a mutagen in order to increase the mutation frequency (to be discussed in the next section), and, second, an enrichment procedure can be used to kill a significant fraction of the wild-type cells. For some genes in some organisms mutagenesis can increase the mutant fraction to 10^{-3}, so the mutant can be found by replica-plating of a few plates. However, more commonly, induced mutagenesis does not increase the mutant fraction to more than one mutant per 10^5 cells, which necessitates the use of an enrichment procedure.

The most commonly used enrichment procedure is the **penicillin selection technique**. This method, which is applicable only to bacteria, is based upon the fact that penicillin kills bacteria by interfering with the synthesis of the bacterial cell wall. When penicillin is added to a culture of growing bacteria, cell wall synthesis stops but growth (enlargement) of the cell continues until the cell explodes. A Leu$^-$ cell in a growth medium lacking leucine cannot synthesize protein and thus cannot enlarge; hence, it is not killed by penicillin. Consider a culture containing one Leu$^-$ cell per 10^6 Leu$^+$ cells growing in medium containing leucine. The culture is filtered and the cells collected on the filter are resuspended, this time in a growth medium containing penicillin but *lacking leucine*. The Leu$^+$ cells continue to grow, and after one hour about 99 percent of them begin to explode; in contrast, the Leu$^-$ cells, which cannot grow, are unharmed. Incubation of the culture is stopped at this point, because the exploded cells release their contents to the medium and the pool of leucine released from these cells would allow the Leu$^-$ cells to grow and therefore also to be killed. The treated culture at this time consists of one Leu$^-$ cell per 10^4 cells and the mutant could be found with 50 replica plates. The cells could also be passed through another enrichment cycle; there would then be one Leu$^-$ cell per 100 Leu$^+$ cells, and the mutant cell might be found with a single replica-plating.

Phage mutants are detected in two ways. "Plaque-morphology" mutants are found because they look different from wild-type plaques. For example, some mutants make plaques that are turbid, or clear, or have a halo, or are a combination of these. Conditional mutants—that is, mutants that are either temperature-sensitive or host-dependent (forming plaques on one host but not on another)—are commonly used and can also be found by replica-plating. With a temperature-sensitive mutant, plaques are formed at 30°C and then transferred to two plates already seeded with bacteria, and incubated at either 30°C or 42°C. A temperature-sensitive mutant forms a plaque on the 30°C plate but not on the 42°C plate. Similarly, a host-dependent mutant (a type of conditional mutant) can be detected by plating on a permissive host and replica-plating onto either a permissive or a nonpermissive host. These host-dependent mutations, called suppressor-sensitive mutations, are described later in this chapter.

MUTAGENESIS

The production of a mutation requires that a change occur in the base sequence. This change can be stimulated to occur in five main ways: (1) removal of an incorrectly inserted base is prevented; (2) a base is inserted that tautomerizes and allows a substitution to occur in subsequent replication; (3) a previously

Table 10-2   Types of mutagens

Mutagen	Mode of action	Example	Consequence
Base analogue	Substitutes for a standard base during replication and causes a new base pair to appear in daughter cells in a later generation	5-Bromouracil 2-Aminopurine	$A \cdot T \rightarrow G \cdot C$, and $G \cdot C \rightarrow A \cdot T$ $A \cdot T \rightarrow G \cdot C$
Chemical mutagen	Chemically alters a base so that a new base pair appears in daughter cells in a later generation	Nitrous acid Hydroxylamine Ethyl methane sulfonate (EMS) Ultraviolet light	$G \cdot C \rightarrow A \cdot T$, and $A \cdot T \rightarrow G \cdot C$ $G \cdot C \rightarrow A \cdot T$ $G \cdot C \rightarrow A \cdot T$, $G \cdot C \rightarrow C \cdot G$, and $G \cdot C \rightarrow T \cdot A$ All single base-pair changes are possible.
Intercalating agents	Addition or deletion of one or more base pairs	Acridines	Frameshifts
Mutator genes	Excessive insertion of incorrect bases or lack of repair of incorrectly inserted bases	———	All single base-pair changes are possible.
None	Spontaneous deamination of 5-methylcytosine (MeC)	———	$G \cdot MeC \rightarrow A \cdot T$

Note: Italicized changes in base pairs are transversions; those that are not italicized are transitions.

inserted base is chemically altered to a base having different base-pairing specificity; (4) one or more bases are skipped during replication; or (5) one or more extra bases are inserted during replication. In the following sections we describe mutagens that act by one or more of these mechanisms and address the question of spontaneous mutagenesis. The information that follows is also presented in Table 10-2.

Methods for site-specific mutagenesis will be presented in Chapter 20.

Base-Analogue Mutagens

A base analogue is a compound sufficiently similar to one of the four DNA bases that it can be built into a DNA molecule during normal replication. Such a substance must be able to pair with a base in the template strand or the editing function will remove it. However, if the base has two modes of hydrogen-bonding, it will be mutagenic.

The substituted base 5-bromouracil (BU) is an analogue of thymine inasmuch as the bromine has about the same size as the methyl group of thymine (Figure 10-8). In subsequent rounds of replication BU functions like thymine and primarily pairs with adenine. In discussing tautomerism earlier in this

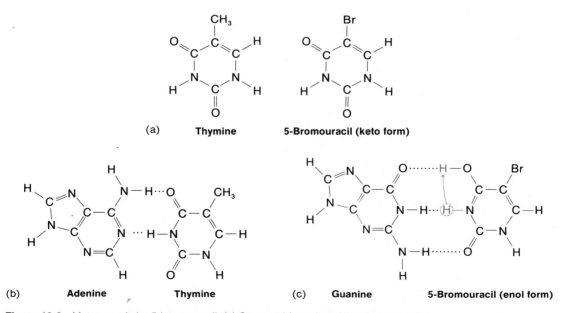

Figure 10-8 Mutagenesis by 5-bromouracil. (a) Structural formulas of thymine and 5-bromouracil. (b) A standard adenine-thymine base pair. (c) A base pair between between guanine and the enol form of 5-bromouracil. The red H in the dashed circle shows the position of the H in the keto form. When tautomerization occurs, the red double bond forms.

chapter it was pointed out that thymine occasionally assumes a form that can pair with guanine. The mutagenic activity of 5-bromouracil stems in part from a shift in the equilibrium caused by the bromine atom; that is, the enol form exists for a greater fraction of time for BU than for thymine. Thus, if BU replaces a thymine, in subsequent rounds of replication, it occasionally generates a guanine, which in turn specifies cytosine, resulting in formation of a GC pair (Figure 10-9). BU can also induce a change from GC to AT. The enol form is actually sufficiently prevalent that BU is sometimes (but infrequently) incorporated into DNA in that form. When that occurs, BU is acting as an analogue of cytosine rather than thymine. However, even though it may become part of the DNA by temporarily having the base-pairing properties of cytosine, the keto form is the predominant form; hence, in subsequent rounds of replication BU will usually pair like thymine. Thus, a GC pair, which, as a result of an incorporation error, is converted to a GBU pair, ultimately becomes an AT pair, as shown in Figure 10-9(b).

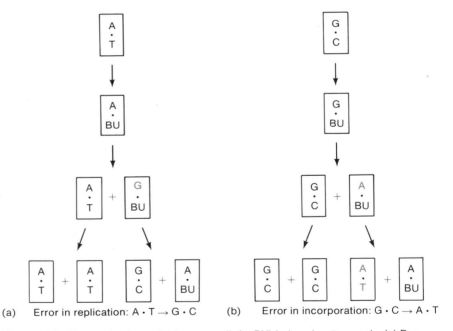

(a)   Error in replication: A · T → G · C   (b)   Error in incorporation: G · C → A · T

Figure 10-9   Two mechanisms of 5-bromouracil- (or BU)-induced mutagenesis. (a) During replication, BU, in its usual keto form, substitutes for T and the replica of an initial AT pair becomes an ABU pair. In the first mutagenic round of replication the BU, in its rare enol form, pairs with G. In the next round of replication, the G pairs with a C, completing the transition from an AT pair to a GC pair. (b) During replication of a GC pair a BU, in its rare enol form, pairs with a G. In the next round of replication the BU is again in the common keto form and it pairs with A, so the initial GC pair becomes an AT pair. The replica of the ABU pair produced in the next round of replication is another AT pair.

Recent experiments suggest that BU is also mutagenic in another way. Because of complex regulatory pathways for nucleotide synthesis, the BU nucleoside triphosphate inhibits production of dCTP. The ratio of TTP to dCTP then becomes quite high and the frequency of misincorporation of T opposite G increases. The rate of misincorporation then exceeds the capacity of the editing and mismatch repair systems, and a persistent incorrectly incorporated thymine will pair with adenine in the next round of DNA replication, yielding a GC → AT change in one of the daughter molecules.

Both base-pair changes induced by BU maintain the original purine (Pu)-pyrimidine (Py) orientation. That is, the original and the altered base pairs both have the orientation PuPy—namely, AT and GC. If the original pair was TA, the altered pair would be CG—that is, PyPu for both the original and the altered pairs. A base change that does not change the PyPu orientation is called a **transition.** Base-analogue mutations are always transitions. Later we will see changes from PuPy to PyPu and from PyPu to PuPy; when such a change of orientation occurs, the mutation is called a **transversion.** Note that BU induces transitions in both directions—AT → GC by the tautomerization route, and GC → AT by the misincorporation route.

Chemical Mutagens

A chemical mutagen is a substance that can alter a base that is already incorporated in DNA and thereby change its hydrogen-bonding specificity. Three commonly used chemical mutagens are nitrous acid (HNO_2), hydroxylamine (HA), and ethylmethane sulfonate (EMS). The chemical structures of these are shown in Figure 10-10.

Nitrous acid primarily converts amino groups to keto groups by oxidative deamination. For example, cytosine and adenine are converted to uracil (U) and hypoxanthine (H), which form the base pairs UA and HC. Therefore, the changes are GC → AT and AT → GC as cytosine and adenine respectively are deaminated.

Hydroxylamine reacts specifically with cytosine and converts it to a modified base that pairs only with adenine, so a GC pair ultimately becomes an AT pair.

Ethylmethane sulfonate (EMS), an **alkylating agent,** is a potent mutagen. Many sites in DNA are alkylated by these agents; of prime importance

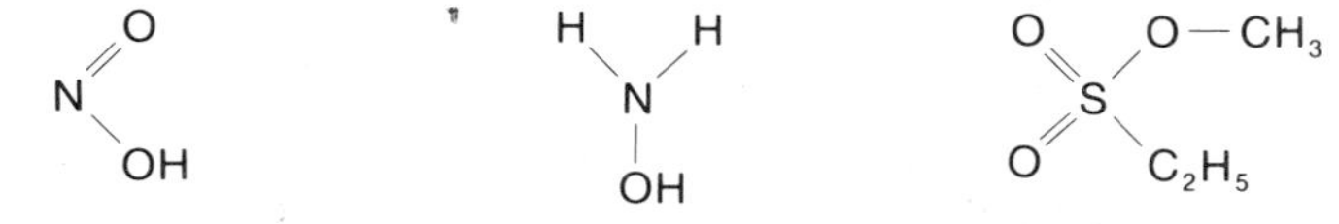

Nitrous acid   **Hydroxylamine**   **Ethyl methane sulfonate**

Figure 10-10 Structures of three chemical mutagens.

for the induction of mutations is the addition of an alkyl group to the hydro-gen-bonding oxygen of guanine and thymine. These alkylations impair the normal hydrogen-bonding of the bases and cause mispairing of G with T, leading to the transitions AT → GC and GC → AT (the latter markedly predominates). EMS also reacts with adenine and cytosine.

Another phenomenon resulting from alkylation of guanine is **depurination**, or loss of the alkylated base from the DNA molecule by breakage of the bond joining the purine nitrogen and deoxyribose. Depurination is not always mutagenic, since the gap left by loss of the purine can be efficiently repaired. However, sometimes the replication fork may reach the apurinic site before repair has occurred. When this happens, replication stops just before the apurinic site, the SOS system is activated, and replication proceeds, almost always putting an adenine nucleotide in the daughter strand opposite the apurinic site. Since the original parental base (which was removed) was a purine, the base pair at that site will be a mismatch (PuA) and after replication the base pair at that site will change orientation from PuPy to PyPu, the first example we have seen of a transversion. Treatment of phages with buffers at pH 4 also produces depurinations. On replication of phages treated in this way, numerous transversions occur, in agreement with the adenine-insertion mechanism just suggested.

Mutagenesis by Intercalating Substances

Acridine orange, proflavine, and acriflavine (Figure 10-11), which are substituted acridines, are planar, three-ringed molecules whose dimensions are roughly the same as those of a purine-pyrimidine pair. In aqueous solution, these substances can insert in DNA between the bases in adjacent pairs, a process called **intercalation.** When DNA containing intercalated acridines is replicated, additional bases appear in the sequence (Figure 10-12). The usual addition is a single base, though occasionally two bases are added. Deletion

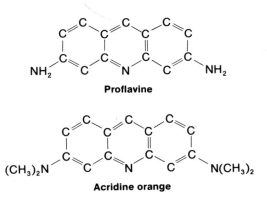

Proflavine

Acridine orange

Figure 10-11 Structures of two mutagenic acridine derivatives.

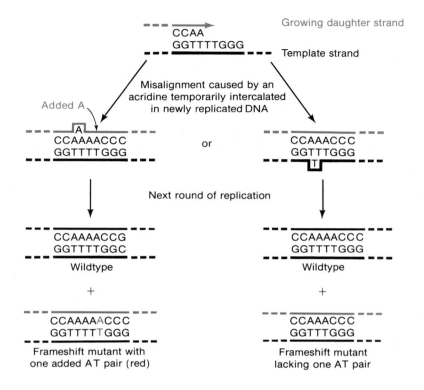

Figure 10-12 Proposed mechanism for misalignment mutagenesis generated by intercalation of an acridine molecule in the replication fork. Acridine is present only in the first round of replication; the second round serves to produce a true-breeding mutant. The left and right paths generate a base addition and deletion, respectively. The growing daughter strand is shown in red.

of a single base also occurs but this is far less common than base addition. Mutations of this sort are called **frameshift mutations.** This is because the base sequence is read in groups of three bases when it is being translated into an amino acid sequence and the addition of a base changes the reading frame (Figure 10-13).

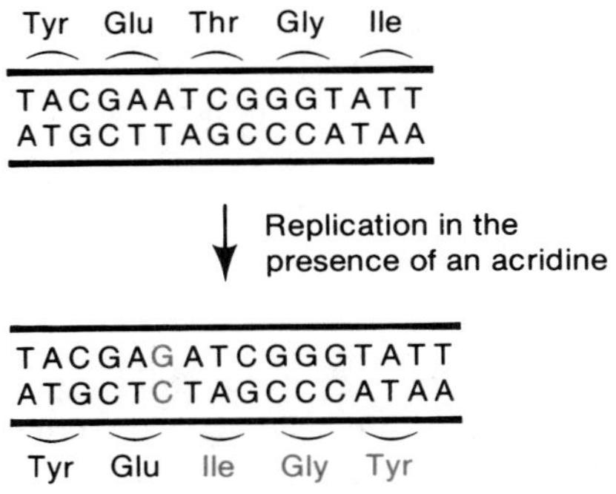

```
Tyr  Glu  Thr  Gly  Ile
⌒    ⌒    ⌒    ⌒    ⌒
TACGAATCGGGTATT
ATGCTTAGCCCATAA
```

↓ Replication in the
presence of an acridine

```
TACGAGATCGGGTATT
ATGCTCTAGCCCATAA
⌒    ⌒    ⌒    ⌒    ⌒
Tyr  Glu  Ile  Gly  Tyr
```

Figure 10-13 A base addition (red) resulting from replication in the presence of an acridine. The change in amino acid sequence read from the upper strand in groups of three bases is also shown in red.

Mutator Genes

In *E. coli* there are genes that in a mutant state cause mutations to appear very frequently in other genes throughout the chromosome. These genes are called **mutator genes.** This is a misnomer, because the function of each gene is probably to keep the mutation frequency low; that is, it is only when the product of a mutator gene is itself defective that there is widespread production of mutations. Some mutators are mutations in the editing function of DNA polymerases. Another is a mutant *dam* gene, the gene responsible for the methylation of DNA and directing the mismatch repair system to the correct template strand (Chapter 8, Figure 8-8). In a *dam⁻* mutant there is little or no methylation of the parent (or any other) strand, and the mismatch repair system frequently excises the parental (correct) base and inserts a base that can pair with the daughter (incorrect) base. The products of the mutator genes *mutH*, *mutL*, and *mutS* participate in an as-yet-unknown way in mismatch repair. The mechanism of action of these mutators should be known in the near future. Other mutators are poorly understood.

Expression of a Mutation After Mutagenesis

If a culture of Lac⁺ cells is exposed to a mutagen and then plated on a color-indicator plate on which Lac⁺ cells are red and Lac⁻ cells are white, no white colonies will be found. Most colonies will, of course, be red, but a few will have a red half and a white half. These **sectored colonies** consist of a mixture of Lac⁺ and Lac⁻ cells. They arise for the following reason. If a mutagen changes one member of a base pair, the daughter DNA molecule that was copied from the unaltered template will have the parental sequence of base pairs. The daughter that was copied from the template with the altered base will have a mutant sequence. Thus, in order to isolate a colony that consists only of mutant cells, the mutagenized culture must, prior to plating, be allowed to go through at least one doubling to eliminate cells with heteroduplex templates. This doubling is said to allow the mutation to be expressed. Since cell growth is often delayed considerably by the conditions used for chemical mutagenesis, in practice a mutagenized culture is allowed to grow for many hours, usually overnight, before plating is done.

REVERSION

So far, we have discussed changes from the wild-type to the mutant state. The reverse process, in which the wild-type phenotype is regained, also occurs. This process is called **back mutation, reverse mutation,** or, most commonly, **reversion,** and a reverse mutant is called a **revertant.** How revertants are detected was described in Chapter 1.

Reversion is a result of spontaneous mutagenesis and, like the formation of all spontaneous mutations, is a nearly random process. Thus, revertants are not formed by any adaptive pressure on a population of organisms but preexist in populations and are merely selected by various screening techniques. It is, of course, possible to stimulate reversion by mutagenesis.

Second-Site Revertants

One way that the wild-type phenotype may be restored is to regain the wild-type genotype completely (that is, the wild-type base sequence). However, this is not always what happens because reversion can occur in several ways.

The somewhat large value of the frequency of spontaneous reversion of a typical point mutation suggests that spontaneous reversion rarely results in a restoration of the wild-type base sequence. In a population of about 10^9 leu^+ cells, only about 100 leu^- mutants are typically present, and the mutations creating these mutants will in general be distributed over roughly 100 different sites in the DNA. In a second population there may also be 100 mutants; some may be mutant at sites present in the first population but most will be located at other sites. In many populations there will be, roughly speaking, mutations at about 500 different sites. Thus, if we were to insist that a reversion were to occur at one particular site (in order to revert a particular mutation to the wild-type base sequence), then on the average only $100/(500 \times 10^9)$ or one cell in 5×10^9 would be mutated at that site. If base changes occur at random and a change from any one base pair to one of the other three is equally probable, then the original base sequence would be restored in about one cell in 1.5×10^{10} cells. However, it is usually found that in a population derived from a single Leu$^-$ mutant, about one in 10^8 cells is Leu$^+$. For any particular mutation, the number just stated can vary widely, but it is frequently observed that the fraction of the mutant population with the revertant phenotype is much too high to be explained by a return to the wild-type base sequence. The explanation is simply that reversion events occurring at many different sites can produce the same phenotype. A reversion event other than a precise reversal at the original mutant site is called a **second-site mutation** or a **suppressor mutation.** This has been confirmed by biochemical data in which it has been shown that the amino acid sequence in a revertant is rarely the wild-type sequence and that the original mutational amino acid substitution is usually still present. Some suppressor mutations are a base change at the original site, but with insertion of a non-wild-type amino acid at that site.

Consider a hypothetical protein containing 97 amino acids whose structure is determined entirely by an ionic interaction between a positively charged (+) amino acid at position 18 and a negative one (−) at position 64 (Figure 10-14). If the (+) amino acid were replaced by a (−) amino acid, the protein would clearly be inactive. Three kinds of reversion events would restore

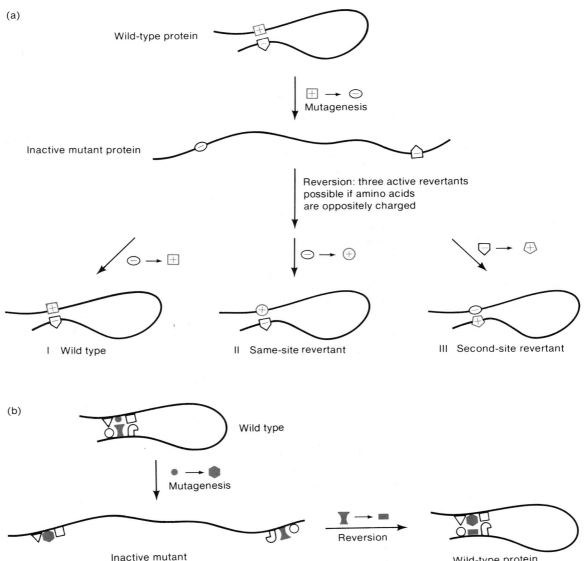

Figure 10-14 Several mechanisms of reversion. In panel (a) the charge of one amino acid is changed and the protein loses activity. The activity is returned by (I) restoring the original amino acid, or (II) by replacing the (−) amino acid by another (+) amino acid, or (III) by reversing the charge of the original (−) amino acid. In each case the attraction of opposite charges is restored. In panel (b) the structure of the protein is determined by interactions between six hydrophobic amino acids. Activity is lost when the small circular amino acid is replaced by the bulky hexagonal one and is restored when space is made by replacing the concave amino acid by the small rectangle.

activity (Figure 10-14(a)): (1) The original (+) amino acid could be put back. (2) A different (+) amino acid could be put at position 18. (3) The (−) amino acid at position 64 could be replaced by a (+) amino acid; this second-site mutation would restore the activity of the protein. A possibility which would not generally work but which might be effective in a specific case is to insert a (+) amino acid at position 17 or 19.

Figure 10-14(b) shows another, but more complicated, example of intragenic reversion. In this case the structure of a protein is maintained by a hydrophobic interaction. The replacement of an amino acid with a small side chain by a bulky phenylalanine changes the shape of that region of the protein. A second amino acid substitution providing space for the phenylalanine could restore the protein structure.

Second-Site Revertants of Frameshift Mutations

Reversion of frameshift mutations almost always occurs at a second site. It is of course possible that a particular added base could be removed or a particular deleted base could be replaced by a spontaneous event, but this is fairly improbable. Second-site reversion of a frameshift mutation has two requirements illustrated in Figure 10-15: (1) the reversion event must be very near the original site of mutation, so very few amino acids are altered between the two sites; and (2) the segment of the polypeptide chain in which both changes occur must be able to withstand substantial alterations.

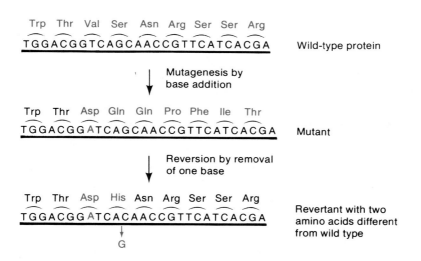

Figure 10-15 Reversion by base deletion from an acridine-induced base-addition mutant.

Reversion as a Test of Isogenicity

Mutations are usually introduced to enable some biological system to be understood. As pointed out in Chapter 1, if the properties of a mutant are to be compared to those of a wild-type bacterium or phage, it is essential that only a single mutation be introduced. That is, the mutant and the wild-type organisms must be isogenic, for otherwise one cannot know what really caused a change in properties. This problem is important in mutagenesis since some mutagens tend to produce mutations in clusters. Reversion is a useful test for isogenicity, since if an observed mutant phenotype were the result of two mutations, reversion of the phenotype would occur at an exceedingly low frequency.

Consider a tryptophan-requiring mutant, isolated from a mutagenized population, that grew exceedingly slowly even when tryptophan was added to the medium. One might hypothesize that the mutation affected a protein that was common to the tryptophan biosynthetic pathway and to some other important pathway. Alternatively, the mutant might carry mutations in two distinct genes. The hypothesis could be distinguished by plating a large number of mutants on medium lacking tryptophan and selecting a Trp^+ revertant. If the revertant grew at a normal rate, one would conclude that both the tryptophan requirement and the slow growth rate resulted from the same mutation. However, if the revertant grew very slowly, two mutations in distinct genes were probably present in the original mutant.

Reversion can also indicate the existence of a mutation that had not been recognized earlier. The following experiment was once carried out in the author's laboratory. Two phage λ mutants, one red^- and one P^-, each of which was capable of normal integration in the lysogenic pathway, were crossed, and a double mutant, $\lambda red^- P^-$, was isolated. This double mutant could not engage in a lysogenic response. The most obvious hypothesis was that pairing the two mutations causes a defect in lysogeny. However, it was also possible that one of the initial mutants contained a silent mutation (call it x^-) that had a phenotypic effect when combined with the other mutation. For example, suppose the P^- mutant was actually $x^- P^-$ and the recombinant was $red^- x^- P^-$, and it was the $red^- x^-$ combination that prevented lysogenization. To test for the present of the hypothetical x mutation, both $red^+ P^-$ and $red^- P^+$ revertants were selected from a lysate of the $red^- P^-$ recombinant. Both revertants were able to lysogenize. These revertants arose at a reasonable frequency, so it was very unlikely that the hypothetical x mutation, if it had been present at all, would have reverted along with the red and P reversion events. Thus, it was clear that the combination $red^- P^-$ alone prevented lysogenization.

Reversion as a Means of Detecting Mutagens and Carcinogens

In view of the increased number of chemicals used and present as environmental contaminants, tests for the mutagenicity of these substances have become important. Furthermore, most carcinogens are also mutagens, so that mutagenicity provides an initial screening for these hazardous agents. One simple method for screening large numbers of substances for mutagenicity is a reversion test using auxotrophic mutants of bacteria. In the simplest type of reversion test a compound that is a potential mutagen is added to solid growth media, known numbers of a mutant bacterium are plated, and the number of revertant colonies that arise is counted. A significant increase in the reversion frequency above that obtained in the absence of the compound tested would identify the substance as a mutagen. However, simple tests of this type fail to demonstrate the mutagenicity of a number of potent carcinogens. The explanation for this failure is that some substances are not directly mutagenic (or carcinogenic), but are converted to active compounds by enzymatic reactions that occur in the liver of animals and have no counterpart in bacteria. The normal function of these enzymes is to protect the organism from various noxious substances that occur naturally by chemically converting them to nontoxic substances. However, when the enzymes encounter certain manmade and natural compounds, they convert these substances, which may not be themselves directly harmful, to mutagens or carcinogens. The enzymes are contained in a component of liver cells called the **microsomal fraction.** Addition of the microsomal fraction of the rat liver to the growth medium as an activation system has been used to extend the sensitivity and usefulness of the reversion test system. The use of the microsomal fraction is the basis of the **Ames test** for carcinogens.

In the Ames test histidine-requiring (His$^-$) mutants of the bacterium *Salmonella typhimurium,* containing either a base substitution or a frameshift mutation, are used to test for reversion to His$^+$. In addition, the bacterial strains have been made more sensitive to mutagenesis by the incorporation of mutations that inactivate the excision-repair system and that make the cells more permeable to foreign molecules. Since some mutagens act only on replicating DNA, the solid medium used contains enough histidine to support a few rounds of replication but not enough to permit formation of a colony. The procedure is the following. Rat-liver microsomal fraction is spread on the agar surface and bacteria are plated. Then, a paper disc saturated either with distilled water (as a control) or a solution of the compound being tested is placed in the center of the plate. The test compound diffuses outward from the disc, forming a concentration gradient. If the substance is a mutagen or is converted to a mutagen, colonies form. With a highly effective mutagen colonies will be present all over the surface of the medium, including far from the disc where the concentration is low; in contrast, with a weak mutagen colonies will form only very near the disc, where the concentration is high. The procedure is highly sensitive and permits the detection of very weak mutagens. A quantitative analysis of reversion frequency can also be carried out by incorporating known amounts of the potential mutagen in the medium. The reversion frequency depends on the concentration of the substance being tested and, for a known carcinogen or mutagen, correlates roughly with its known effectiveness.

The Ames test has now been used with thousands of substances and mixtures (such as industrial chemicals, food additives, pesticides, hair dyes, and cosmetics), and numerous unsuspected substances have been found to stimulate reversion in this test. A high frequency of reversion does not mean that the substance is definitely a carcinogen but only that it has a high probability of being so. As a result of these tests, many industries have reformulated their products: for example, the cosmetic industry has changed the formulation of many hair dyes and cosmetics to render them nonmutagenic. Ultimate proof of carcinogenicity is determined from testing for tumor formation in laboratory animals. The Ames test and several other microbiological tests (Devoret test, Chapter 17) are used to reduce the number of substances that have to be tested in animals since to date only a few percent of more than 300 substances known from animal experiments to be carcinogens failed to increase the reversion frequency in the Ames test.

SUPPRESSION

Suppression or **intergenic reversion** refers to a mutational change in a second gene that eliminates or suppresses a mutant phenotype. The most common type of intergenic reversion is one in which the second-site mutation not only eliminates the effect of the original mutation but also suppresses mutations in many other genes as well. This mode of suppression has been studied most carefully with conditional mutations, the class of mutation that has the wild-type phenotype in certain circumstances (genetic background, temperature, etc.) and produces a mutant phenotype in other conditions. One type of conditional mutation is the temperature-sensitive mutation. In this section we shall consider another major class—**suppressor-sensitive mutations,** which behave like wild type when a suppressor molecule is present. An example of this phenomenon is a phage mutant that can grow in one strain of bacteria (denoted Sup$^+$) but fails to grow in other strains (Sup0). We will see that this type of suppression is based on changes in the decoding system.

Suppressor-sensitive mutations are of two main types: **nonsense** or **chain termination** mutations, and **missense** or **amino acid substitution** mutations. The nonsense type will be considered first.

Genetic Detection of a Nonsense Suppressor

Chain termination mutations are very common, for they can arise in many ways. For example, a single base substitution in any of the codons AAG, CAG, GAG, UCG, UUG, UGG, UAC, and UAU can give rise to the chain termination codon UAG. If such a mutation occurs within a gene, a mutant protein with little or no function will result because there is no tRNA molecule whose anticodon is complementary to UAG (Chapter 6). Thus, only a fragment of the wild-type protein is produced, and this fragment will usually have little or no biological function unless the mutation is very near the carboxyl terminus of the wild-type protein.

In certain bacterial mutants a chain termination mutation does not cause termination; this can be seen in the following simple system. Consider a bacterium B that has mutated such that chain growth terminates at a Tyr site in the *lacZ* gene, making the bacterium genotypically and phenotypically Lac$^-$ (Figure 10-16). We now seek a Lac$^+$ revertant of strain B by mutagenizing a population of B, allowing several generations of growth to express the reversion, and plating the culture on lactose-minimal agar, which is selective for Lac$^+$ colonies. Three classes of revertants will be found, as shown in the figure. In class I, a rare class, the chain termination mutation is reversed by a base substitution mutation that converts UAG back to UAC. In this class the complete protein chain will be present, the tyrosine will be restored at the correct position, and the protein will have the wild-type amino acid sequence. In class II a new amino acid, serine, is present in the now-complete protein chain, and the base sequence is likewise altered from the original wild-type sequence. In this case the chain termination mutation has

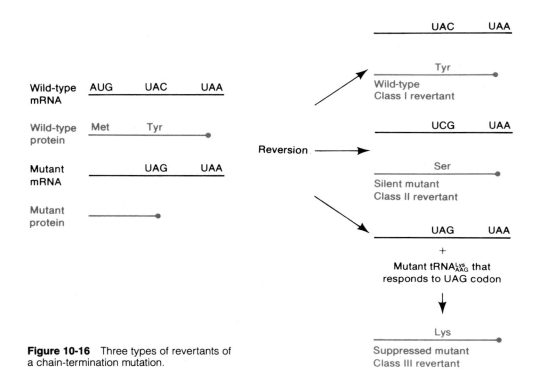

Figure 10-16 Three types of revertants of a chain-termination mutation.

been converted to a silent mutation that has a wild-type or near wild-type phenotype. In the figure this has happened as a result of a base substitution mutation converting UAG to UCG, the serine codon, and the substitution of serine for tyrosine does not markedly alter function; that is, serine is an acceptable amino acid at this position in the protein. The class III revertants are those of interest for this discussion. These revertants produce a complete polypeptide chain, usually by supplying some other amino acid at the mutant site without in any way altering the original mutant base sequence—that is, the codon remains UAG. Class III revertants also typically have another property: a mutant phage with a chain termination mutation in an essential protein can often form a plaque on a class III revertant. The revertant cell has gained the ability to ignore, or **suppress,** UAG-type chain termination mutations by translating the mutant UAG codon into an amino acid.

The molecular explanation for this type of suppression is that the class III revertant, which is called a suppressor-containing bacterium or a **suppressor mutant,** contains an altered tRNA molecule—one that has the anticodon CUA, which can pair with UAG. This mutant tRNA is called a **suppressor tRNA.**

How can such a mutant tRNA molecule arise? Since a tRNA molecule is the product of a tRNA gene, these mutants occur in exactly the same way as any other mutant—by a replication error. Note that, like any other mutation, a suppressor tRNA is not formed in response to a chain termination mutation; rather, it has been selected from a population of bacteria containing both it and a chain termination mutation by growth in a medium in which only cells having a wild-type phenotype can form a colony.

What has been mutated in the production of a suppressor tRNA? Clearly it must be a normal tRNA gene. Thus, in the example given previously, a tRNA^{Lys} molecule whose anticodon is CUU, has been altered to have the anticodon CUA, which can hydrogen-bond to the codon UAG.

Inasmuch as a single base change is sufficient to alter the complementarity of an anticodon and a codon, there are (at most) eight tRNA molecules having a complementary anticodon that, with a single changed base, will also suppress a UAG codon. Thus, the following amino acids (whose codons are also indicated) can be put at the site of a chain termination codon: Lys (AAG), Gln (CAG), Glu (GAG), Ser (UCG), Trp (UGG), Leu (UUG), and Tyr (UAC and UAU). Note that these are the same amino acid codons that can be altered by mutation to form a UAG site.

Suppressors also exist for chain termination mutations of the UAA and UGA type. These too are mutant tRNA molecules whose anticodons are altered by a single base change.

In conventional notation, suppressors are given the genetic symbol *sup* followed by a number (or occasionally a letter) that distinguishes one suppressor from another. A cell lacking a suppressor is designated *sup0* or *sup⁻*.

Several features of nonsense suppression should be recognized:

1. A particular UAG suppressor will not suppress all UAG chain termination mutations, because the amino acid it inserts will not always produce an active protein.
2. Suppression may be incomplete in that the new proteins may have activity sufficient for colony formation, but the activity may be subnormal.
3. A cell can survive the presence of a suppressor only if the cell also contains two or more copies of the tRNA gene. Clearly if a tRNA^{Ser} molecule that reads the UCG codon is mutated, then all normal UCG codons would be read as a stop codon. However, there are multiple copies of most tRNA genes, so in any living cell containing a suppressor tRNA, there will always be an additional copy of a wild-type tRNA that can function in normal translation.

Normal Termination in the Presence of a Suppressor tRNA

If a cell contains a UAG suppressor, then proteins terminated by a single UAG codon will not be completed and the existence of a suppressor tRNA should be lethal. Two features of the normal chain termination process are responsible for the survival of a suppressor-containing cell:

1. Protein factors active in translation termination respond to chain termination codons even though a tRNA molecule that recognizes the codon is present, that is, suppression is weak. For example, if the probability of recognition is only x percent, then only x percent of the prematurely terminated mutant proteins would be completed and only x percent of normally terminated proteins would be ruined by lack of termination.
2. Normal chain termination often utilizes pairs of distinct termination codons such as the sequence UAG-UAA. Thus, the existence of a UAG suppressor would not prevent termination of a doubly terminated protein.

It is likely that both of these mechanisms are responsible for the existence of viable suppressor-containing organisms. Suppression of potentially lethal UAG and UGA mutations is very efficient; termination is prevented more than 50 percent of the time. (Note that the *activity* of the suppressed mutant may not always be as high as 50 percent; this is because the particular amino acid insertion is not always well tolerated by the protein rather than because of inefficient suppression of chain termination.) A partial explanation for the viability of organisms containing UAG and UGA suppressors is that when either UAG or UGA is used as a natural stop signal, it usually occurs adjacent to or either 3 or 6 base pairs away from a second and different stop codon in the same reading frame.

Natural chain termination is accomplished most commonly by a single UAA codon. This is consistent with the observation that UAA suppressors are typically very inefficient, preventing termination of mutationally induced UAA codons between 1 and 5 percent of the time. Furthermore, cells containing UAA suppressors are generally unhealthy, growing rather slowly compared to cells lacking any suppressor or containing either a UAG or UGA suppressor. Presumably, several percent of the normal proteins, perhaps those needed critically, are damaged by not being terminated at the right time.

Suppression of Missense Mutations

Suppression can also occur for missense mutations. For example, a protein in which valine (nonpolar) has been mutated to aspartic acid (polar), resulting in loss of activity, can be restored to the wild-type phenotype by a missense suppressor that substitutes alanine (nonpolar) for aspartic acid. Such a substitution can occur in three ways: (1) a mutant tRNA molecule may recognize two codons, possibly by a change in the anticodon loop; (2) a mutant tRNA molecule can be recognized by an incorrect aminoacyl synthetase and be misacylated; and (3) a mutant synthetase can charge an incorrect tRNA molecule. Suppression of missense mutations is necessarily inefficient. If a suppressor that substitutes alanine for aspartic acid worked with, say, 20 percent efficiency, then in virtually every protein molecule synthesized by the cell at least one aspartic acid would be replaced, which is a situation that a cell could not possibly survive. The usual frequency of missense suppression is about 1 percent. In this way a small amount of a functional, essential protein is made and thereby a mutant cell is able to survive. However, missense suppression still introduces a significant number of defective proteins of all other types and, as a result, a cell carrying a missense suppressor usually grows slowly and is generally unhealthy.

PROBLEMS

1. Is a base-pair change always a mutation?

2. Does a base-pair change necessarily change the phenotype?

3. Which of the following base-pair changes are transitions and which are transversions? (1) AT → TA, (2) AT → GC, (3) GC → TA.

4. Do base analogues produce transitions or transversions?

5. Assuming that all possible base changes can occur with equal frequency, what would be the ratio of transversions to transitions in a large collection of mutants?

6. Distinguish a missense and a nonsense mutation.

7. What makes a particular mutation temperature-sensitive?

8. How can tautomerization cause mutation?

9. By what mechanisms does 5-bromouracil induce mutations?

10. Does a frameshift always cause a phenotypic change?

11. A deletion occurs that eliminates a single amino acid in a protein. How many base pairs were deleted?

12. A mutant is isolated that cannot be reverted. What biochemical type(s) of mutation might it carry?

13. One class of point mutation cannot be reverted by intragenic reversion at a second site? What is this class?

14. Why is a liver microsomal fraction included in the Ames test for mutagens?

15. One variety of a temperature-sensitive mutation is the cold-sensitive (Cs) mutation, which has a mutant phenotype below a particular temperature. Bacteria containing the mutation *ess2*(Ts) can form colonies at 32°C, but not at 37°C and 42°C, whereas those containing the mutation *ess5*(Cs) form colonies at 42°C, but not at 32°C and 37°C. What would be the phenotype of an *ess2*(Ts)*ess5*(Cs) double mutant?

	32°C	37°C	42°C
ess2(Ts)	+	−	−
ess5(Ts)	−	−	+

16. *E. coli* polymerase I possesses several enzymatic activities. Two important activities are the polymerizing function and the 3'-5' exonuclease. Mutant polymerases have been found which either increase or decrease mutation rates in an organism containing the mutant enzyme. A mutant that increases the mutation rate is called a mutator; a mutant that decreases the mutation rate is called an antimutator. The mutator and antimutator activities are usually a result of changes in the ratio of the two enzymatic activities described above. How do you think the ratios change in a mutator and in an antimutator?

17. An enzyme has the property that if amino acid 28, which is glutamic acid, is replaced by asparagine in a mutant, all activity is lost. If in this mutant protein, amino acid 76, which is asparagine, is then replaced by glutamic acid, full activity of the enzyme is restored. What can you say about amino acids 28 and 76 in the normal protein?

18. If a cell contains 2000 genes and if the average mutation rate per gene is 1 $\times$ 10^{-5} per generation, what is the average number of *new* mutations per cell per generation?

19. A fluctuation test was carried out to estimate the mutation rate of an *E. coli* locus conferring resistance to phage T1. If 5 of the 12 small independent cultures contained no phage-resistant mutants after growth of the cultures was completed

and the average number of bacterial cells per culture was 5×10^8, what is the estimated rate of mutations to T1 resistance?

20. A fluctuation test is carried out for two different genes A and B. The following data are obtained. For gene A, 22 of 40 cultures had no mutants, with $N = 5.6 \times 10^8$. For gene B, 15 of 37 cultures had no mutants, with $N = 5 \times 10^8$. What are the mutation frequencies for the two genes?

21. In microorganisms which mutation rates are more easily measured: ability to inability to synthesize proline (pro^+ to pro^-) or the reverse (pro^- to pro^+)?

22. Which of the following amino acid substitutions would be likely to yield a mutant phenotype if the change occurred in a fairly critical part of a protein: (1) Pro → His; (2) Arg → Lys; (3) Thr → Ile; (4) Val → Ile; (5) Gly → Ala; (6) His → Tyr; (7) Gly → Phe?

23. Several hundred independent missense mutants have been isolated in the A protein of $E. coli$ tryptophan synthetase, a protein having 268 amino acids. Fewer than 30 of the positions were represented with one or more mutant. Why do you think that the number of different positions represented by amino acid changes is so limited?

24. The molecule 2-aminopurine is an analogue of adenine, pairing with thymine. It also pairs on occasion with cytosine. What types of mutations will be induced by 2-aminopurine?

25. Nitrogen mustard reacts efficiently with guanine causing ring cleavage and subsequent hydrolysis of the N-glycosidic bond. What base-pair change does this cause?

26. Can a mutation induced by nitrous acid be induced to revert at the same site by treatment with nitrous acid?

27. Two hundred Leu$^-$ mutants of a bacterial strain are examined separately to determine reversion frequencies. Of these, 90 revert at a frequency of 10^{-5}, 98 at 3×10^{-6}, 6 at 3×10^{-11}, and 6 at 10^{-10}.

    (a) What type of mutant is probably contained in the class whose reversion frequency is 10^{-11}: single point mutations, double point mutations, or deletions?
    (b) Can you say anything from these frequencies about whether any of the classes of mutations are chain termination mutations?

28. Revertants of temperature-insensitive mutations often prove to be temperature-sensitive. That is, they exhibit a wild-type phenotype at low temperature and mutant at a higher temperature. Explain this phenomenon.

29. A mutation of a bacterial Lac strain yielding a Lac$^-$ colony has been isolated. Several lines of experiments indicate that the mutation resulted from production of a UGA codon. Spontaneous revertants are sought and found at a frequency of 10^{-8} per cell per generation and 9 of 10 of them were caused by suppressor tRNA molecules. What do think is the rate of production of suppressor mutations in the original Lac culture?

part 4

GENETICS OF BACTERIA AND PHAGES

Chapter 11

Plasmids

Plasmids are circular, supercoiled DNA molecules present in most species, but not all strains, of bacteria. They are small, ranging in size from about 0.2 to 4 percent that of the bacterial chromosome. Under most conditions of growth, plasmids are dispensable to their host cells. However, many plasmids contain genes that have value in particular environments, and often these genes are the main indication that a plasmid is present. For example, R plasmids render their host cells resistant to certain antibiotics, so in nature a cell containing such a plasmid can better survive in environments in which a fungal antibiotic is present.

In many bacterial species plasmids are responsible for a particular type of gene transfer between individual cells, a property that accounted for the initial interest in the 1950s. Furthermore, like phages, plasmids are heavily dependent on the metabolic functions of the host cell for their reproduction. They normally use most of the replication machinery of the host and hence have been useful models for understanding certain features of bacterial DNA replication. Finally, they have been exceedingly valuable to the microbial geneticist in constructing partial diploids (Chapter 7) and as a gene-cloning vehicle in genetic engineering (Chapter 20).

TYPES OF PLASMIDS

This chapter will be concerned primarily with plasmids of *E. coli* except when otherwise noted. Many types of plasmids have been detected in various

E. coli strains, but the greatest amount of information has been obtained about three main types—the F, R, and Col plasmids—which share some properties but which are, for the most part, quite different. The presence of an F, R, or Col plasmid in a cell is indicated mainly by the following traits:

1. *F, the sex, or fertility, plasmids*. These plasmids mediate the ability to transfer chromosomal genes (that is, genes not carried on the plasmid) from a cell containing an F plasmid to one that does not. F itself can also be transferred to a cell lacking the plasmid.
2. *R, the drug-resistance plasmids*. These plasmids make the host cell resistant to one or more antibiotics, and many R plasmids can transfer the resistance to cells lacking R.
3. *Col, the colicinogenic factors*. Col plasmids synthesize proteins, collectively called **colicins**, that can kill closely related bacterial strains that lack a Col plasmid of the same type. The mechanism of killing is different for different Col plasmids,

Further discussion of each of these plasmid types will be presented throughout the chapter.

With only a single exception (the killer-plasmid of yeast, which is an RNA molecule) all known plasmids are supercoiled circular DNA molecules (Figure 11-1). The molecular weights of the DNA range from about 10^6 for the smallest plasmid to slightly more than 10^8 for the largest one. Table 11-1 lists the molecular weights for several plasmids that are actively being studied.

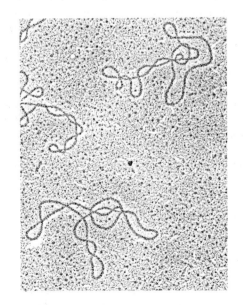

Figure 11-1 Two supercoiled plasmid DNA molecules.

Table 11-1 Several plasmids and selected properties

Plasmid	Mass, $\times 10^6$	No. copies/ chromosome	Self-transmissible	Phenotypic features
Col plasmids				
ColE1	4.2	10–15	No	Colicin E1 (membrane changes)
ColE2 (*Shigella*)	5.0	10–15	No	Colicin E2 (DNase)
ColE3	5.0	10–15	No	Colicin E3 (ribosomal RNase)
Sex plasmids				
F	62	1–2	Yes	F pilus
F'*lac*	95	1–2	Yes	F pilus; *lac* operon
R plasmids				
R100	70	1–2	Yes	Cam-r Str-r Sul-r Tet-r
R64	78	(limited)	Yes	Tet-r Str-r
R6K	25	12	Yes	Amp-r Str-r
pSC101	5.8	1–2	No	Tet-r
Phage plasmid				
λdv	4.2	≈50	No	λ genes *cro, cI, O, P*
Recombinant plasmids				
pBR322	2.9	≈20	No	High-copy-number
pBR345	0.7	≈20	No	ColE1-type replication

DETECTION OF PLASMIDS

Plasmids can be detected by both genetic and physical experiments. The first plasmid that was discovered was F. An *E. coli* strain (A) with phenotype Met$^-$Bio$^-$Thr$^+$Leu$^+$ was mixed with a second strain (B) with phenotype Met$^+$Bio$^+$Thr$^-$Leu$^-$, and the mixture was plated on minimal agar. At a frequency of about 10^{-7}, colonies formed on the minimal agar; these had the phenotype Met$^+$Bio$^+$Thr$^+$Leu$^+$ and hence were recombinants. In a second experiment strain A was treated with streptomycin (and then washed free of the antibiotic) prior to mixing the cells; recombinant colonies still formed. However, if strain B was first treated with streptomycin, no recombinants were found. This experiment indicated that the recombinants were derived from strain B, and that mating somehow involved a one-way transfer of genetic information. Another experiment involved a third strain C, which could not transfer any genetic information to B. However, if A and C bacteria were mixed and allowed to grow together for a long time and then colonies of C were isolated, these colonies (call them C') could transfer genes to strain B. The explanation was that strain A contained a fertility element, called F, which mediated transfer of chromosomal genes from bacterium A to bacterium B. Strain C lacked F, so no recombinants formed. However, when strains A and C were grown together, F was transferred to C, generating strain C', which could then (because it contained F) transfer genetic markers to B cells. Thus, the initial cross was

$$F^+ met^- bio^- thr^+ leu^+ \times F^- met^+ bio^+ thr^- leu^-$$

Because of the one-way nature of the transfer the cell containing F is said to be a **donor** or a **male,** and the cell without F is the **recipient** or **female.**

Variants of F, called F', are known that possess genetic markers. One that was studied in great detail carried the *lac* operon and was used in constructing the partial diploids described in Chapter 7. It is designated F'*lac*. Transfer of an F' is recognized easily with color-indicator plates containing an antibiotic (for example, streptomycin) to which the donor cells are sensitive and the recipients are resistant. For example, in a cross

$$F' lac^+/lac^- str\text{-}s \times lac^- str\text{-}r$$

in which the donor cell carries the *lac*⁺ marker on the plasmid and a *lac*⁻ marker in the chromosome (note the use of the / to distinguish plasmid and chromosomal markers), plating the mixture on EMB-lactose agar containing streptomycin yields white colonies (*lac*⁻ *str-r*) and purple colonies (F'*lac*⁺/ *lac*⁻ *str-r* recombinants). If the cultures had not been mixed, plating the donor alone would yield no colonies (because all cells are Str-s) and plating the recipient alone would yield only white colonies (all would be Lac⁻). Note that the function of the streptomycin marker is to prevent growth of donor cells; such a marker is termed the **counterselection,** or **counterselective marker.** Antibiotics are commonly used for counterselective purposes but other agents serve as well. For example, F' transfer can be detected in the cross F'*lac*⁺/*met*⁻ × *lac*⁻ *met*⁺ with the mixture plated on minimal agar containing lactose. Donors will not form colonies because methionine is lacking in the agar (methionine is the counterselective agent), and females lacking a transferred F'*lac*⁺ will not grow because they are Lac⁻ and cannot use the lactose in the agar. In this case, the **selected marker** is *lac*⁺.

The experiments with the F-containing strain (that is, the cross between strain A and strain B) and the F' strains show significant quantitative differences. If the cells are mixed for about 30 minutes prior to plating, recombinants arise at a frequency of about 10^{-7} per donor cell in the A × B experiment and about 50 percent in the F' experiment. The difference is that in the F'*lac* experiment the plasmid carries the genetic marker, whereas the genetic markers in the other experiment are chromosomal. Thus, in the A × B experiment even though about half of the recipients receive a copy of F, only a tiny fraction receive chromosomal markers. This process of rare **chromosomal mobilization** will be discussed later.

Another important experimental result was obtained in a study of the properties of a donor cell after a plasmid has been transferred. The mating

$$F' lac^+/lac^- str\text{-}s\ tsx\text{-}r \times F^- lac^- str\text{-}r\ tsx\text{-}s$$

was carried out, in which *tsx-r* indicates inability to absorb phage T6. After a period of mating a portion of the bacterial mixture was plated on lactose color-indicator plates containing streptomycin, and more than 90 percent of the colonies were Lac$^+$. Thus, most donors had transferred F$'lac^+$. In another part of the experiment the question of whether donors that had transferred F$'lac$ still had a copy of the plasmid was investigated. A portion of the mating mixture was treated with an excess of T6 phage; the number of phage was so great that the recipients, which were all Tsx-s, were lysed within a few minutes by the simple action of thousands of phage particles punching holes in the cell wall. The surviving cells, which were the Tsx-r donors, were then plated on lactose-indicator plates, and Lac$^+$ colonies formed. From several other experiments it was known that most of the original donor cells contained only one copy of F$'lac^+$. Thus, cells that had transferred the plasmid still retained a copy of the plasmid, which means that transfer is accompanied by DNA replication. Indeed, further tests showed that cultures derived from the colonies remained able to transfer the lac^+ marker.

Since physical experiments are frequently carried out with bacterial extracts, a plasmid is often discovered incidentally as a nonchromosomal circular DNA molecule present in a DNA sample isolated from a culture. This has led to the development of the following screening technique by which the presence of a plasmid in the cells of a single bacterial colony can be detected electrophoretically. A single colony is taken from a plate, lysed by the lysozyme-detergent procedure described in Chapter 2 (or sometimes lysed directly in the well), and then subjected to gel electrophoresis (Figure 11-2). The bacterial chromosome is very large and cannot penetrate the gel, but the plasmid DNA can do so. Since the rate of electrophoretic movement of DNA molecules through a gel increases with decreasing molecular weight, plasmid DNA, if present, will form a narrow band at a position in the gel characteristic of its molecular weight. The band can be visualized by the usual treatment of staining the gel with ethidium bromide (Chapter 2, Figure 2-21), which binds tightly to the DNA and fluoresces upon irradiation with ultraviolet light. From the distance moved in a particular time interval relative to that for plasmids of known molecular weight, the molecular weight of the plasmid DNA can be calculated, as shown in Figure 11-3.

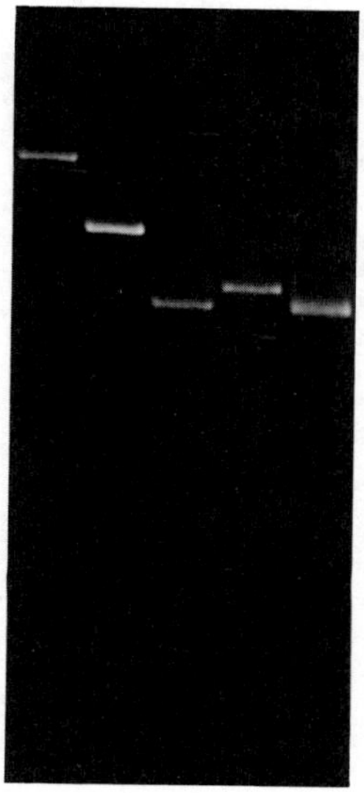

Figure 11-2 A gel electrophoregram showing the migration of five different plasmids. Movement is from top to bottom. Each vertical column ("lane") represents a single plasmid. After migration the gel is soaked in a solution of ethidium bromide, washed, and then illuminated with near-ultraviolet light. The ethidium bromide, which is bound to the DNA, fluoresces. The single intense band in each lane contains supercoiled DNA molecules. [Courtesy of Elaine Cocuzzo and Pieter Wensink.]

PURIFICATION OF PLASMID DNA

Plasmid DNA can be isolated from bacteria in a simple way. Plasmid-containing bacteria are opened by the lysozyme-detergent treatment (Chapter 2), and the resulting translucent solution, called a **cell lysate,** is centrifuged. The bacterial chromosome complex, which contains protein and RNA, is very large and compact and sediments to the bottom of the centrifuge tube; the smaller plasmid DNA remains in the clear supernatant, which is called a **cleared lysate.** Some chromosomal DNA is usually present in the cleared lysate, but since most of the plasmid DNA is covalently circular, this contaminating material can be easily removed by the following procedure. CsCl

Figure 11-3 Determination of the molecular weights of DNA molecules (red points) by gel electrophoresis with DNA molecules whose molecular weights are known (black points). The line is a plot of the black points; the molecular weights of the molecules being studied are then determined from the positions of the red points.

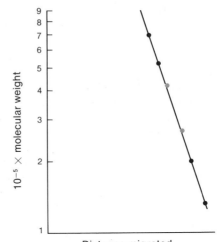

plus ethidium bromide (Chapter 2, Figure 2-15) is added to the cleared lysate, and the lysate is centrifuged to equilibrium (Chapter 2, Figure 2-16). When ethidium bromide is present, the covalently circular DNA has a higher density than the linear chromosomal fragments, so the plasmid DNA is purified. The only disadvantage of this technique, which is the one most commonly used, is that both nicked circles (which often form accidentally during isolation of plasmid DNA) and nonsupercoiled replicating molecules are discarded, since they come to equilibrium within the chromosomal-DNA band.

The isolation of DNA usually entails a deproteinization step (for example, treatment with phenol, as described in Chapter 2). When the DNA molecules of some *E. coli* plasmids are isolated without such a step, about half of the supercoiled DNA molecules contain three tightly bound protein molecules. This DNA-protein complex is called a **relaxation complex.** If this complex is heated or treated with alkali, proteolytic enzymes, or detergents, one of these proteins, which is a nuclease, nicks one DNA strand, thereby "relaxing" the supercoil to the nicked circular form (Figure 11-4). This nick occurs in only one strand and at a unique site. During relaxation the two

Figure 11-4 Nicking of one strand of a supercoiled DNA relaxation complex.

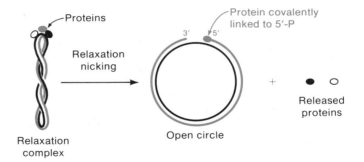

smallest proteins are released, but the largest protein becomes covalently linked to the 5′-P terminus of the nick. If, prior to relaxation, the supercoiled DNA is nicked by any of a number of laboratory techniques, the relaxation nuclease is unable to make its site-specific nick, indicating that the nuclease is active only on supercoiled DNA. This nicking, which presumably also occurs within cells, is an early step in plasmid transfer.

TRANSFER OF PLASMID DNA

In an earlier section we described the detection of F and F′ plasmids by virtue of their ability to be transferred from a donor cell to a recipient. In this section we examine the transfer process in some detail.

Stages in Transfer Process

One of the earliest discoveries about recombination between donor and recipient cells is that it requires cell-to-cell contact. This was demonstrated by experiments in which F^+ and F^- cultures were separated either by a porous membrane or a porous plug. Recombinants were not produced in this arrangement, so recombination could not result from movement of genetic material through the growth medium. This led to the use of the term **bacterial conjugation** for the cell-mating event, and some years later, striking electron micrographs were obtained of conjugating bacteria (Figure 11-5). A variety of experiments have shown that plasmid transfer can be divided into four stages:

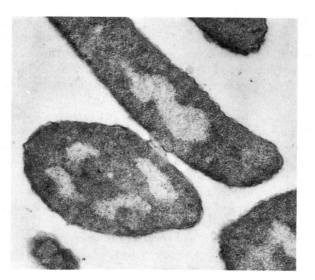

Figure 11-5 Electron micrograph of two *E. coli* cells during conjugation. The small cell is an *F⁻* cell; the larger cell contains F′*lac*. [With permission, from Caro, L. 1966. *J. Mol. Biol.*, 16, 269. Copyright: Academic Press Inc. (London) Ltd.]

1. Formation of specific donor-recipient pairs (**effective contact**).
2. Preparation for DNA transfer (**mobilization**).
3. DNA transfer.
4. Formation of a replicative functional plasmid in the recipient.

When cells containing F or an F′ come in contact with a recipient, all four steps occur. However, many types of plasmids are genetically unable to carry out all of these processes. Thus, the following terminology has been developed to describe different types of plasmids:

A **conjugative plasmid** is a plasmid carrying genes that determine the effective contact function.
A **mobilizable plasmid** can prepare its DNA for transfer.
A **self-transmissible plasmid**, such as F, is both conjugative and mobilizable.

The conjugative functions are, in many cases, not plasmid-specific, and hence one plasmid can assist transfer of a second. For example, a single cell may contain both F and ColE1. F is both conjugative and mobilizable (it is self-transmissible), and ColE1 is mobilizable but nonconjugative—that is, a cell containing only ColE1 cannot transfer the plasmid. In a cell containing both plasmids, F can provide the missing conjugative function to ColE1, and ColE1 can thereby be transferred to a recipient that lacks both plasmids. This process—namely, one in which a nonconjugative plasmid is transferred via the effective contact provided by a conjugative plasmid—is called **donation;** its hallmark is efficient transfer of the nonconjugative plasmid (10–100 percent). In contrast, the mobilization functions are usually plasmid-specific, so a self-transmissible plasmid cannot enable a nonmobilizable plasmid to be transferred by simple complementation. However, helped transfer can occur in some cases, though at low frequency. A requirement is base-sequence homology between the plasmids so that genetic recombination between the two plasmids to form a single transferable DNA molecule can occur. That is, the self-transmissible plasmid just carries along the nonmobilizable plasmid. This process is called **conduction,** in contrast with donation. The frequency (high versus low) of helped transfer of a non-self-transmissible plasmid is a useful criterion for determining whether a plasmid is mobilizable by donation or conduction.

In the historic experiment described earlier in which F mediated transfer of the chromosome, the chromosome was mobilized by conduction. That is, genetic recombination occurred between sequences in F and in the chromosome, and F carried the chromosome along into the recipient. That the chromosomal transfer occurred by conduction (which requires a low-frequency recombination event between F and the chromosome) accounts for the 10^7-fold lower frequency of transfer of chromosomal genes than of the lac^+ marker from the F′lac^+-containing cell.

In subsequent sections F will serve as the primary model for transfer.

Effective Contact and Pili

The first step in effective contact is pair formation between a donor and a recipient cell. This requires a hairlike protein appendage, called a **sex pilus,** on the donor cell (Figure 11-6). The pili on F- and R-containing cells are called **F pili** and **R pili,** respectively. The average number of F pili per cell is 1.4 to 2.7, the number depending on the growth medium.

The F pilus has been purified. It consists of a single, very hydrophobic protein, called **pilin,** which forms a hollow tubular structure. Four experimental results indicate that the F pilus is necessary for conjugation. (1) Plasmid mutants that cannot synthesize wild-type pilin cannot be transferred. (2) Pili can be stripped from cells by violent agitation of a culture. Pili reform over a period of about one-half hour. F is unable to be transferred from such stripped cells until the pili have regrown. (3) Purified pili can bind to recipient cells. (4) Several phages are known that bind to F pili (these are called **male-specific phages**). These phages are of two types, binding either to the tip of the pilus or to the shaft. Addition of those that bind to the pilus tip inhibit pair formation and transfer, whereas those that bind to the shaft do not interfere in a major way with F transfer.

The fact that pili are hollow tubes suggested some time ago that plasmid DNA is transferred through the core of the pilus. However, no experiment has ever given evidence for DNA within a pilus. Other experiments indicate that pili retract into the donor cell after pairing. Thus, it is widely believed that the pilus serves first to bring the pair into initial contact and then to draw the cells together into close contact. The nature and identity of the actual conjugation tube that occurs at the cell-cell junction is unknown.

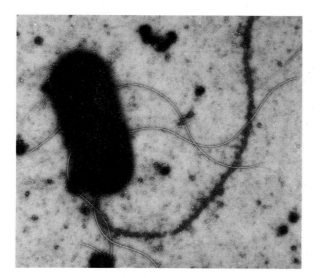

Figure 11-6 An *E. coli* cell showing a single sex pilus, which is coated with the F-specific phage R17 to make the very thin pilus visible as a rough dark appendage. The five heavy bright fibers are flagella. The very faint thin hairs are called fimbrae. [Courtesy of Barry Eisenstein.]

All mating systems do not entail the use of pili. For example, some strains of the bacterium *Streptococcus faecalis* carry a self-transmissible plasmid. Plasmid-free recipients produce a mating protein (analogous to the pheromones of female insects) that is not made in plasmid-containing (donor) cells. The pheromone causes the donor to synthesize a protein called adhesin that coats the donor cells and causes donor-recipient pairs to form. Once the plasmid is transferred, synthesis of the pheromone is inhibited.

Mobilization and Transfer

Mobilization begins when a plasmid-encoded protein, which is probably the nicking protein of the relaxation complex, makes a single-strand break in a unique base sequence called the **transfer origin** or in F, *oriT*. (Relaxation complexes have not been detected for all transmissible plasmids but presumably there is a protein serving the function just described.) This nick initiates rolling circle replication (Figure 8-16) and the linear branch of the rolling circle is transferred. It is thought that the nicking protein remains bound to the 5' terminus and that the replication mode is like the looped rolling-circle mechanism used by phage φX174 (Figure 8-17). The sequence of events during transfer is shown schematically in Figure 11-7.

Note that DNA synthesis occurs both in donor and recipient cells. The synthesis in the donor, called **donor conjugal DNA synthesis,** serves to replace the single strand that is transferred. Synthesis in the recipient cell (**recipient conjugal DNA synthesis**) converts the transferred single strand to double-stranded DNA.

In the usual situation during transfer the transferred strand is simultaneously replaced by donor conjugal synthesis and converted to double-stranded DNA in the recipient. This would indicate that DNA synthesis and transfer are coupled were it not for the following observations: (1) Transfer occurs even if donor synthesis is inhibited by appropriate host mutations, such as a temperature-sensitive mutation in the polymerase III gene; (2) donor synthesis occurs even if transfer is prevented by appropriate plasmid mutations; (3) transfer occurs even though recipient conjugal synthesis is inhibited by mutations in the recipient cell. These observations raise the question of the identity of the motive force for transfer since clearly it is not DNA replication. This question has not yet been answered.

Fertility Inhibition

If a single *E. coli* cell containing the F plasmid is added to a culture of growing F^- cells, after 15–20 generations of cell growth, a large fraction of the cells contain F. This is a result of transfer of a replica of F from an F^+ cell to an F^- cell without loss of F by the F^+ cell. After transfer, the F replica remaining in the original cell can replicate again, and another replica can be trans-

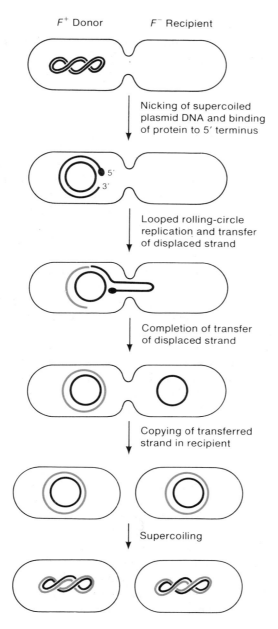

F^+ Donor F^- Recipient

Nicking of supercoiled
plasmid DNA and binding
of protein to 5' terminus

5'
3'

Looped rolling-circle
replication and transfer
of displaced strand

Completion of transfer
of displaced strand

Copying of transferred
strand in recipient

Supercoiling

Figure 11-7 A model for transfer of F plasmid DNA from an F^+ cell by a looped rolling-circle mechanism. The displaced single strand is transferred to the F^- recipient cell, where it is converted to double-stranded DNA. Chromosomal DNA and proteins of the relaxation complex in completed DNA molecules are omitted for clarity.

ferred to a second F^- cell. Every recipient acquires F and therefore becomes a donor from which F can be transferred to other F^- cells. Transfer can occur once or twice per cell generation, so F quickly spreads throughout a bacterial population.

Such rapid spread is not the case for most transmissible plasmids. For example, only about 0.02 percent of a population of cells containing most R plasmids are competent donors. That is, if an R^+/lac^- culture is mixed with an R^-lac^+ culture and the culture is diluted 1000-fold shortly after mixing in order that no further pair formation can occur, the number of R^+/lac^+ recombinants is only 0.02 percent the number of initial R^+ cells. However, if after transfer has occurred an excess of R^- recipients is added at a concentration that allows rapid pair formation, the R plasmid quickly spreads through the recipient culture. The kinetics of transfer show that an R^- cell that has just received an R plasmid is able to transfer the R plasmid almost immediately. This phenomenon, which is called **fertility inhibition,** is a result of the activity of a multisubunit repressor (encoded in two *fin* genes), which acts on an operator and prevents transcription of the genes required for transfer. Most R plasmids have an active *fin* repressor, which accounts for the low transfer frequency. F lacks one of the *fin* genes and hence the transfer system is always transcribed constitutively. With most R plasmids, once pair formation occurs, derepression occurs because the repressor is absent in the recipient, so recent recipients can transfer again almost immediately (in time, repressor is made and fertility decreases). Interestingly, if the culture is rediluted during the stage of epidemic transfer, the transfer genes are repressed again and donor competence drops again to 0.02 percent. The operator in F apparently can bind the R repressors because transfer of F is markedly reduced by the presence of an R plasmid; as might be expected, if the R plasmid contains a defective *fin* gene, transfer of F is not inhibited.

It should be noted that whereas fertility inhibition is a common feature in R plasmids, it is not a property of all R plasmids.

The *tra* Genes of F

In the preceding section we saw that transfer is modulated by the *fin* gene. Here we examine the genes whose activity is essential for transfer to occur. These genes constitute the *tra* operon.

Most of the *tra* genes are involved in pili synthesis. For example, the *traJ* gene encodes pilin, and the *traA, traB, C, E, F, G, H, K, L, U, V, Z* genes are needed for assembly of a functional pilus. Other genes are required for pairing, triggering transfer, actual transfer, nicking at the transfer origin *oriT*, DNA replication from the normal (nontransfer) replication origin *oriV*, surface exclusion (the inability of an F^+ cell to pair with another F^+ cell), and plasmid incompatibility (see later section). Those genes involved specifically with transfer are encoded in a single mRNA molecule; it is transcribed constitutively in F but not in R and Col plasmids. These and several genes are listed in Table 11-2 and shown on a map of F given in Figure 11-8. The complementation tests and the mapping procedures that enabled one to locate the various genes are beyond the scope of this book; the

Table 11-2 Some genes and sites of F plasmid and their functions

Gene	Function
traA, traB, traC, traE, traF, traG, traH, traJ, traK, traL, traU, traV, traW	Pili formation
traJ	Structural genes for pilin
traG, traN	Mating aggregation
traI, traM	Initiation of transfer
traM	Transfer
traO	Operator for finO gene
traY, traZ	Subunits of endonuclease that nicks oriT
oriT	Origin of transfer DNA synthesis
oriV	DNA replication origin
traS, traT	Surface exclusion (inhibition of mating between F-containing cells); encode membrane proteins
ilzA, ilzB	Lethal zygosis (killing of females by excess Hfr cells)
frp	F replication
inc	Incompatibility of IncF group
finO, finP	Fertility inhibition (found in Col, R, and F-like plasmids, but not F)

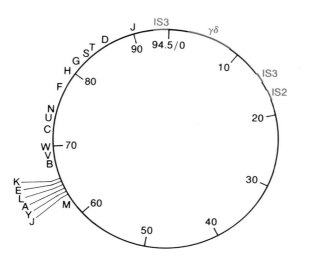

Figure 11-8 A map of the sex plasmid F. Points are given in kilobase pairs. The single capital letters refer to the midpoints of the locations of the corresponding *tra* genes. The sequences γδ, IS2, and IS3 are shown in red.

procedures can be found in the book *Plasmids* by Broda given in the references at the end of this book.

Host Restriction in Transfer

In Chapter 5 host restriction and modification of phages was described. This same phenomenon occurs in plasmid transfer since a plasmid carries the modification pattern of the host cell in which it resides. Thus, F'*lac* does not appear to be transferred efficiently from *E. coli* strain K12 to strain B and vice versa. In fact, transfer does occur but the DNA transferred from strain K12 is destroyed by B-specific restriction nucleases when it enters strain B. Just as with phages, the rare B recipient to which successful transfer of F'*lac* from strain K12 has occurred can then transfer F'*lac* to a strain-B recipient, but not back to K. Restriction in bacterial conjugation is exercised to a lesser extent than that shown in Table 5-2 (Chapter 5) for phages. This may have to do with the fact that a single strand is transferred and the single strand is not attacked efficiently by the restriction nuclease. Also, when the completed strand is synthesized, it rapidly becomes methylated. It is known that DNA that is methylated in one strand is not restricted.

PLASMID REPLICATION

A plasmid can replicate only within a host cell, so we might expect that all plasmids native to the same host species would have the same mode of replication. However, as with phages, there is enormous variation in both the enzymology and mechanics of plasmid DNA replication, as will be seen in the following sections.

Variations in the Use of Host Proteins in Replication Mechanisms

Plasmids rely heavily on the host replication proteins for their replication. However, even though polymerase III is the major replication protein in *E. coli*, the particular DNA polymerase responsible for chain growth is not the same for all *E. coli* plasmids. For example, an $F^+ polA^-$ culture, which has much reduced activity of DNA polymerase I, continues to produce F^+ progeny, whereas ColE1 fails to replicate unless the host possesses active polymerase I. This is because chain growth of F utilizes Pol III, but ColE1 uses Pol I for the initial stages of replication. Some plasmids use host gene products exclusively. For example, ColE1 can be replicated *in vitro* by adding purified ColE1 DNA to a cell extract prepared from cells that do not contain ColE1 or any other plasmid; clearly, such an extract cannot contain any plasmid-encoded gene product. Other plasmids require some plasmid gene products. For instance, F'*lac* plasmids carrying a temperature-sensitive mutation in F have been isolated that fail to replicate at 42°C; that is, at this temperature, $F'(Ts)lac^+/lac^-$ cells produce F^-lac^- daughter cells after several generations of growth at 42°C.

All plasmids examined to date replicate semiconservatively and maintain circularity throughout the replication cycle. However, there are significant differences in the replication pattern from one plasmid to the next. For example, some plasmids replicate unidirectionally and others bidirectionally. The plasmid RK6 replicates first in one direction and then later in the opposite direction from the same origin. The bidirectionally replicating plasmids terminate replication in two ways. In one type, termination occurs when the growing forks collide. Others have a fixed termination site that is sometimes reached by one growing fork before the other fork reaches it. In the most carefully studied plasmids, replication occurs by the so-called **butterfly** or **rabbit's ears** mode (Figure 11-9) in which a partially replicated molecule contains untwisted replicated portions, as is usually the case in θ replication and a supercoiled unreplicated portion. When the replication cycle is completed, one of the circles must be cleaved in order to separate the daughters. The result after one round of replication is one nicked molecule and one supercoiled molecule. The nicked molecule is then sealed and, somewhat later, supercoiled. Whether this is a general mechanism for plasmid replication is not known.

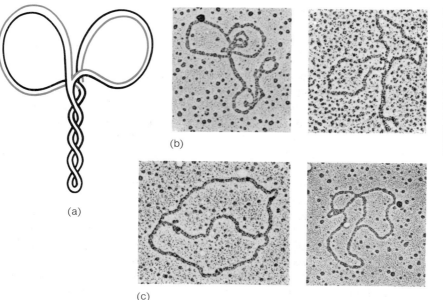

(a)

(b)

(c)

Figure 11-9 Replication of ColE1 DNA. (a) Diagram of a butterfly molecule. The newly synthesized strands are red. (b) Electron micrographs of butterfly molecules. (c) Nicked butterfly molecules, showing that a nick converts a butterfly molecule to a θ molecule. [Courtesy of Donald Helinski.]

In Vitro Plasmid Transfer

Often it is desirable to transfer a nontransmissible plasmid to a specific host cell. This is possible by transfer of purified DNA, as long as some genetic test is available to show that the recipient possesses the plasmid. Except for a few species (Chapter 13) bacteria are generally unable to take up DNA from the surrounding medium. However, some years ago it was found that many bacteria can be altered such that for a brief period of time DNA uptake is possible. The most commonly used general technique is called **CaCl$_2$ transformation.** In this procedure bacteria are placed in a solution of CaCl$_2$ and then subjected to several temperature shifts before DNA is added. By an unknown mechanism the cells become permeable to DNA and the intracellular nucleases are temporarily inactivated. If the DNA is a complete replicating entity, such as a plasmid, it can become permanently established in the recipient cell. If it is a DNA fragment, most of the DNA is digested by exonuclease V in *E. coli;* in some bacterial genera infrequent recombination allows a small fraction of the cells to maintain the DNA. A typical experiment would be to treat tetracycline-sensitive (Tet-s) cells with an R plasmid that is *tet-r* and to plate the cells on nutrient agar containing tetracycline. All Tet-r cells will contain the plasmid. Cells that acquire the Tet-r trait as a spontaneous mutation will not be found since they would normally be present in a Tet-s culture at a frequency of about 10^{-8}. If an amount of plasmid DNA is added such that all cells take up the DNA, a few percent of the treated cells will be Tet-r.

Control of Copy Number

We have just seen that plasmids normally use most of the replication machinery of the host. However, each type of plasmid possesses its own genes for regulating the rate of initiation of replication and, hence, the number of plasmid copies per cell. Plasmids may, on this basis, be classified as **stringent (low-copy-number)** plasmids, for which the number per cell is 1 to 2, or **relaxed (high-copy-number)** plasmids, for which there are 10–100 per cell. The *in vitro* plasmid-transfer technique (CaCl$_2$ transformation) has been used to show that the copy number is established and regulated by controlling the rate of initiation of DNA synthesis. If a plasmid-free culture is transformed with low-copy-number plasmid DNA at a DNA concentration such that no cell receives more than a single copy of the plasmid, the plasmid DNA replicates only once or twice before cell division. However, if a cell is transformed with a single DNA molecule of a high-copy-number plasmid, the plasmid DNA replicates repeatedly until the proper copy number is reached.

The generally accepted explanation for the regulation of copy number is that there is a plasmid-encoded repressor that is a negative regulator of the initiation of replication. Let us first see how this idea explains maintenance of copy number from generation to generation. In the current theory it is assumed that the repressor activity depends on concentration. Thus, as a cell grows (enlarges), the repressor concentration drops, replication is not inhibited, and the number of plasmid DNA molecules doubles. At this point there will exist twice the initial number of repressor genes; therefore, as a result of protein synthesis the repressor concentration also doubles, causing replication to stop. A similar sequence of events would occur if there were initially only one copy of a high-copy-number plasmid—that is, replication would continue until there is sufficient repressor to turn off synthesis. How can this explanation account for differences in copy number? The most likely possibility is that for the high-copy-number plasmids complete repression requires a higher concentration of repressor than does a low-copy-number plasmid. Thus, only when the number of plasmids per cell is high is the "gene dosage" high enough to cause inhibition. The following experiment supports this view by providing evidence that each plasmid controls its own copy number.

A hybrid plasmid, pSC134, was constructed (by recombinant DNA techniques; Chapter 20) that consists of a complete copy of each of two plasmids, ColE1 and pSC101 (an R plasmid). The copy numbers for these plasmids are roughly 18 and 6, respectively. The following three facts about pSC134 were obtained (Figure 11-10):

1. In wild-type cells, plasmid pSC134 replicates from the ColE1 replication origin and has a copy number of 16—roughly equal to that for ColE1. Thus, the higher-copy-number origin dominates.

2. If pSC134 is put into a *polA*⁻ cell by CaCl$_2$ transformation (recall that ColE1 cannot replicate in a *polA*⁻ cell), the pSC101 origin is used and the copy number becomes 6, the value for pSC101. These two results show that the copy number correlates with the replication origin that is being used.

3. If pSC101 DNA is taken into a bacterium containing 16 copies of pSC134 (again using CaCl$_2$ transformation), the pSC101 cannot replicate. This lack of replication shows that the pSC101 inhibitor is being made by pSC134.

The interpretation of these three results is the following. If the copy number were less than 6, both pSC101 and ColE1 origins would be active and replication from both origins would increase the number. If the number were greater than 6, the pSC101 origin would be inactive because the pSC101 inhibitor concentration would be above its inhibitory level. Synthesis of pSC101 would continue, though if it were self-regulated, the concentration would not exceed that produced by a copy number of 6. The ColE1 origin would remain active, for it would be shut off only if the copy number were to exceed 16. In a *polA*⁻ cell ColE1 cannot replicate; consequently, the copy number is totally controlled by the concentration of pSC101 inhibitor, and thus it will not exceed its normal value. This interpretation is completely consistent with

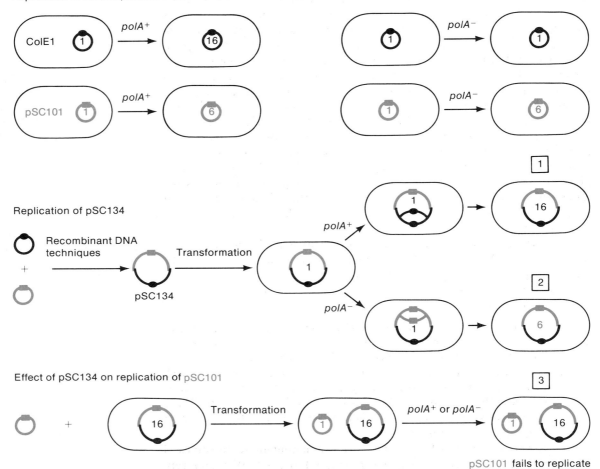

Figure 11-10 Diagram depicting the replication of ColE1, pSC101, and the hybrid plasmid pSC134, starting with one copy per cell. The black solid circle and the red solid square designate the replication origins of ColE1 and pSC101, respectively. The numbers (1, 6, and 16) in the DNA molecules indicate the number of copies in the cell. The boxed numbers (1, 2, and 3) refer to similarly numbered items in the text.

the facts, and with several other types of experiments that also support the repressor model of copy-number regulation.

The following important point must be understood about the repressor model. In a strain containing a single plasmid type, all of the plasmids in a particular cell are identical, so a repressor molecule cannot (for most plasmids) distinguish one DNA molecule from another. Thus, when the concentration of repressor is low enough that all DNA molecules are not repressed, the few that are free are drawn randomly from the population. This means that if one plasmid replicates to form two daughter plasmids, an individual daughter plasmid has the same probability of replicating a second time as any other molecule has of replicating (a first time). That is, at any instant, a DNA molecule is chosen for replication by random selection from the entire population of plasmids.

The molecular mechanism of inhibition is fairly well understood for ColE1, R1, and F. Details can be found in references given at the end of this book.

Plasmid Amplification

Some high-copy-number plasmids exhibit a phenomenon called **amplification.** If chloramphenicol, an inhibitor of protein synthesis, is added to a culture of plasmid-containing bacteria, initiation of replication of chromosomal DNA, but not plasmid DNA, is inhibited, and the number of plasmids per cell increases to 1000 or more. The repressor model explains this phenomenon. Bacterial DNA synthesis is inhibited because of the inability to synthesize specific initiation proteins. However, if the plasmid utilizes only *stable* bacterial replication proteins or stable plasmid-encoded proteins, plasmid DNA synthesis can continue. In fact, since the copy-number regulator is a concentration-dependent inhibitor, the amount of repressor will be insufficient (in the absence of protein synthesis) and plasmid replication will be initiated repeatedly, without regulation. Increased availability of bacterial replication proteins, owing to lack of bacterial DNA synthesis, plus metabolic instability of RNA molecules contribute to the excessive synthesis. (The argument just given assumes a protein repressor; it is likely that some plasmids have an RNA inhibitor.)

Amplification is a convenient way to increase the amount of plasmid DNA that can be isolated from a culture and accounts for the extensive use of high-copy-number plasmids in genetic engineering.

Incompatibility

Pairs of closely related plasmids usually cannot be stably maintained in the progeny of a single cell; such plasmids are said to be **incompatible.** The repressor model for initiation of plasmid replication also explains this phenomenon.

Let us first consider a cell that contains two plasmids, say, F and ColE1, that have different repressors. Replication of each type of plasmid will proceed independently of one another since the repressor of one type (e.g., F) does not regulate the replication of the other type (e.g., ColE1). Thus, F and ColE1 are compatible. Common terminology is to say that they belong to different **incompatibility groups.**

The situation is quite different with two plasmids A and B whose repressors are either identical or are similar enough that the repressor of A can regulate replication of B and vice versa. Let us consider a cell having one copy of A and one of B and which has enlarged sufficiently that initiation occurs (Figure 11-11). Since the two plasmid copies are selected at random for replication, the result of the first replication event is a cell having either one copy of A and two of B (1A,2B) or one of B and two of A (2A,1B). When the second replication event occurs, each cell will have four plasmids, but, depending on the plasmid that is replicated, the plasmid composition may be (1A,3B), (2A,2B), or (3A,1B), as shown in the figure. At this point the cell, which has twice the initial number of plasmids, can divide. The plasmid composition of the two daughter cells will be one of the following:

(1A,3B) becomes (1A,1B) + (1B,1B).
(2A,2B) becomes (1A,1A) + (1B,1B) or (1A,1B) + (1A,1B).
(3A,1B) becomes (1A,1A) + (1A,1B).

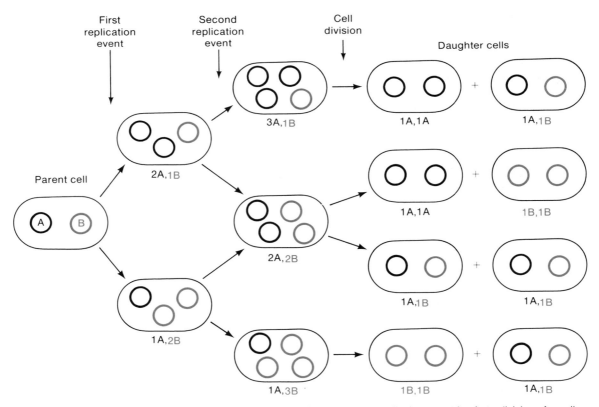

Figure 11-11 Four possible pairs of daughter cells that can arise from division of a cell containing one copy each of the incompatible plasmids A and B. See text for details.

Note that two possible types of cells, namely, (1A,1A) and (1B,1B), contain only one of the two kinds of plasmids; daughter cells obtained from these cells will of course continue to have only one kind of plasmid. In subsequent cell divisions each cell still containing one A and one B will, with 50 percent probability, produce daughter cells lacking one of the plasmids. Thus, as a single cell initially containing two incompatible plasmids divides, the percentage of the progeny population containing only one plasmid type will increase with each generation. Thus, incompatibility is a result of (1) two plasmids having a similar repressor and (2) the random selection of plasmids for DNA replication.

Hundreds of plasmids have been sorted into incompatibility (Inc) groups. This classification is of some value since members of a single group are evolutionarily related with respect to replication functions and also, to some extent, to features of the pili. The following test is a standard one for classification:

1. Transfer a second plasmid B into cells of a culture already containing a resident plasmid A, and select for the presence of plasmid B.
2. Pick at least ten colonies, and test these for the traits that identify the two plasmids.
3. If plasmid A is absent from all colonies, carry out a mating in which plasmid A is in the donor and B is in the recipient. If introduction of either plasmid always eliminates the other, the two plasmids are incompatible.
4. If the resident plasmid is still present in the progeny colonies, test the colonies carrying both plasmids for stability and independent replication of the two plasmids. Stability is shown by the continual presence of markers from both plasmids after many generations of growth in a nonselective medium. Separate replication is implied if the two plasmids are separately transferred in a mating.

If two plasmids coexist stably and are separably transferable, they are compatible, and hence belong to different incompatibility groups.

It has been found that many plasmids can be transferred between various enterobacteria (e.g., *E. coli*, *Salmonella typhimurium*, and *Shigella dysenteriae*), so these plasmids have, as a group, been classified into about 25 incompatibility groups. Plasmids can also be sorted according to the pili they produce (usually immunological relatedness of pili is detected). The different pili classes are: F, F-like, I, I-like, N, etc. A typical Inc group consists of plasmids that not only have a common replication system, but a common type of pili, and hence the pili class is used, to some extent, in naming the Inc groups. This can be seen in Table 11-3, which describes eight enterobacterial Inc groups.

Incompatibility is usually tested by determining whether two different plasmids can cohabit a cell. In one documented case seven different plasmid types were maintained in a single cell. The limitation may be the amount of

Table 11-3 Some incompatibility groups of plasmids that are transmissible to *E. coli**

Group	Sex pili	Group	Sex pili
FI	F	I1	I
FII	F-like	I2	I-like
FIII	F-like	N	N
FIV	F-like	P	RP4

*The F plasmids are all in group FI; R1 and R100, which are mentioned in the text are in group FII; Col plasmids are mainly in FII, FIII, and I1.

space available for excess DNA. For example, the cells with seven different plasmid types had a 25 percent excess of DNA over that of the chromosome.

Replication Inhibition by Acridines

With many plasmids, replication is inhibited by various agents that can inter-calate between the bases of DNA, particularly acridines (e.g., proflavin, acridine orange), without inhibiting chromosomal DNA replication. Such inhibition can lead to loss of the plasmid (**acridine curing**). The phenomenon, which is detected most easily when the plasmid contains host genes, is illus-trated in Figure 11-12. If a culture of a bacterium whose genotype is $F'lac^+/lac^-$ is grown in medium containing acridine orange, cells continue to grow and divide. However, after one generation the number of Lac^+ cells remains constant and the number of Lac^- cells increases. The explanation is the following. At the time of adding acridine orange, most cells contain two copies of $F'lac$. After one generation each cell contains only one $F'lac$ because plasmid replication is inhibited. Thus, in the next cell division only one plasmid is available for the two daughter cells; as a consequence, plasmid-free bacteria are produced. Acridine curing is a convenient aid in identifying genetic markers that reside in the chromosome.

The mechanism of acridine curing has been elusive. It works only in a fairly narrow range of acridine concentrations—too much inhibits chromo-some replication and too little fails to inhibit plasmid replication. Why plas-mid DNA replication is more sensitive to acridines than replication of the chromosome is unclear. Furthermore, curing has an extraordinary depen-dence on the pH of the growth medium, being optimal at pH 7.6 and quite inefficient at pH 7 and pH 8. Why some plasmids are not inhibited by acridines is unknown.

Some plasmid-containing strains, of which the F^+ strains are an exam-ple, are cured by growth at 45°C, exposure to low concentrations of deter-gents, and certain antibiotics at concentrations too low to exert their anti-biotic activity.

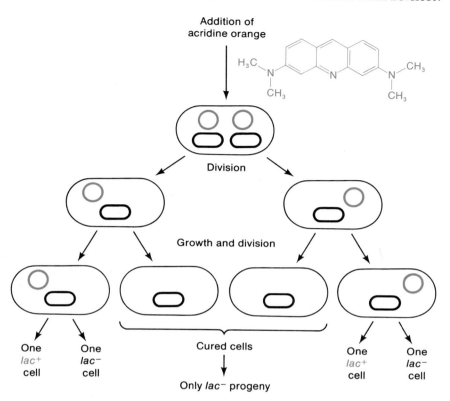

Addition of
acridine orange

Division

Growth and division

One
lac+
cell

One
lac−
cell

Cured cells

Only *lac−* progeny

One
lac+
cell

One
lac−
cell

Figure 11-12 Curing of a cell containing F'*lac+* (red circle) by growth in a medium containing acridine orange.

PARTITIONING OF PLASMID REPLICAS AT CELL DIVISION

A variety of experiments indicate that for low-copy-number plasmids partitioning of replicas between daughter bacterial cells during cell division is carefully regulated. That is, if a cell that is ready to divide contains two copies of plasmid DNA molecules (as is the case with F plasmids), each daughter cell receives one molecule. If partitioning were a random process, the fraction of cells lacking a plasmid after division would be determined by the Poisson distribution, and hence roughly 37 percent of the cells would be plasmid-free. However, for an F'*lac+*/*lac−* cell, segregation of Lac− cells occurs at a frequency of only about 1 per 10^4 cells per generation, which is a fairly typical value for the low-copy-number plasmids. Evidence that a genetically defined system regulates partition of F comes from the isolation of mutations in F that increase the segregation of plasmid-free cells to that expected for random partition. Very little is known about the mechanism that prevents segregation of plasmid-free cells, but it is clear that two distinct systems are active. One is a gene or a gene system called *par,* whose activity is a complete mystery. The other is a system that inhibits cell division in cells carrying only one copy of F; this is called the *ccd* (control of cell division) system.

It seems clear that the low-copy-number plasmids would have a partition function to avoid segregation of large numbers of plasmid-free cells. However, it is not obvious that such a system is needed for the high-copy-number plasmids. Indeed, partitioning may be random with some high-copy-number plasmids. A more-or-less standard test for random partitioning is the following. First, the rate of segregation of plasmid-free cells is measured. Usual values are about 10^{-6}–10^{-5} per generation, which is comparable to the expected value for random segregation from a cell with 20–30 plasmids; however, values observed for low-copy-number plasmids are in the same range, so no conclusion can be drawn from these data alone. Then, cells are treated in some way (for example, a small temperature increase with a Ts replication mutant) that reduces the number of plasmid copies to 2–3. In the case of some Col factors, segregation of Col$^-$ cells occurs at a rate of about 0.2 per generation, indicating that these plasmids do not have a Par function. However, for other plasmids the segregation rate is still a small number indicating that these plasmids have a Par function.

PROPERTIES OF PARTICULAR BACTERIAL PLASMIDS

In this section we examine certain plasmids with respect to plasmid-specific properties that have not yet been described.

F Plasmids

An important property of F is its ability to integrate into the bacterial chromosome to generate an Hfr cell. The properties of these cells will be discussed in detail in Chapter 14. Integration is a reciprocal exchange much like that occurring when phage λ lysogenizes a bacterium. However, integration of F into the *E. coli* chromosome differs from λ prophage formation in that there are several possible exchange sites in F and many sites in the chromosome at which integration can occur. More than 20 major sites and nearly 100 minor sites in the chromosome are known. The affinity of F for each site is not the same.

Excision of F also occurs, though this is quite rare (at some locations, excision occurs more frequently than the average). Often excision is imperfect and one cut is made at one of the two termini within the integrated F and a second cut is in the adjacent chromosomal DNA. In such a case the new F plasmids contains chromosomal genes, which may or may not be identifiable. This aberrant excision is the origin of the F' plasmids. In Chapter 14 the genetic procedures used to isolate a strain containing an F' plasmid will be described.

Integration of F into certain bacterial mutants that have defects in DNA replication gives rise to a phenomenon called **integrative suppression.** F uses many *E. coli* replication proteins but may not always use the bacterial *dnaA*-gene product, which is necessary to initiate chromosomal DNA replication. As stated earlier, plasmids generally carry their own genes for replication initiation. Thus, if F, as a free plasmid, is contained in an *E. coli* *dnaA*(Ts) mutant and the temperature is raised to 42°C (which inactivates the mutant DnaA protein), initiation of chromosomal replication is no longer possible; however, F can still replicate. In contrast, in a *dnaA*(Ts) mutant strain in which F is integrated (an Hfr strain) chromosomal replication can occur indefinitely at the high temperature. However, replication is not initiated at the *E. coli* replication origin but instead is initiated at the *oriV* site for F replication. Thus, integration of F suppresses the DnaA(Ts) phenotype by providing a DnaA-independent replication origin.

Screening for integrative suppression is one way to isolate a strain in which F is integrated. For example, an F'*lac*⁺/*str-s* male can be mated with a *lac*⁻ *dnaA*(Ts)*str-r* strain and Lac⁺ Str-r cells can be selected by plating at 42°C on a color-indicator plate containing streptomycin. All surviving colonies should contain an integrated F'*lac*. As a test, cultures can be derived from a few colonies and grown in the presence of acridine orange. Lac⁻ segregants should not appear because the plasmid is integrated.

The Drug-resistance (R) Plasmids

The drug-resistance, or R, plasmids, were originally isolated from the bacterium *Shigella dysenteriae* during an outbreak of dysentery in Japan and have since been found in *E. coli* and other bacteria. Their defining characteristics are that they confer resistance on their host cell to a variety of fungal antibiotics and are usually self-transmissible. Most R plasmids consist of two contiguous segments of DNA (Figure 11-13). One of these segments is called **RTF (resistance transfer factor);** it carries genes regulating DNA replication

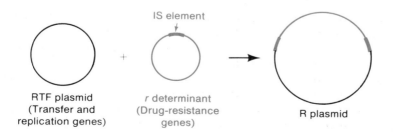

RTF plasmid
(Transfer and
replication genes)

r determinant
(Drug-resistance
genes)

R plasmid

Figure 11-13 The components of an infectious R plasmid. In Chapter 12 we will describe the mechanism for formation of two IS elements in the R plasmid from the single one present in the *r*-determinant.

and copy number, the transfer genes, and sometimes the gene for tetracycline resistance (*tet*), and has a molecular weight of 11×10^6. The other segment, sometimes called the **r determinant,** is variable in size (from a few million to more than 100×10^6 molecular weight units) and carries other genes for antibiotic resistance. Resistance to the drugs ampicillin (Amp), chloramphenicol (Cam), streptomycin (Str), kanamycin (Kan), and sulfonamide (Sul), in combinations of one or more, appears commonly. Small drug-resistance plasmids, lacking the ability to transfer but still containing the *tet* gene, are also known; one of these, pSC101, whose molecular weight is 5.8 $\times 10^6$, is commonly used in genetic engineering (Chapter 20). The two-component R plasmids are reminiscent of F′ plasmids, but the drug-resistance genes are not acquired by integration of RTF followed by aberrant excision; instead, they result from acquisition of transposons that carry antibiotic-resistance genes, as will be seen in Chapter 12.

Evidence for an RTF and an r component of the larger R plasmids is provided by both physical and genetic experiments. Examination of the DNA isolated from cultures of certain R-containing strains showed the presence of three sizes of circular DNA molecules. The length of the largest was the sum of the lengths of the other two molecules. Genetic studies of R plasmids carrying markers for resistance to chloramphenicol (Cam), streptomycin (Str), and tetracycline (Tet) indicated that strains often arose that were Tet-r Cam-s Str-s, Tet-s Cam-r Str-r, or sensitive to all three antibiotics. The Cam-r and Str-r markers were never dissociated—either both or none were present. Furthermore, Tet-r Cam-s Str-s cultures could transfer the Tet-r marker, but Cam-r Str-r Tet-s strains could not transfer any of the markers. Finally, Tet-r Cam-r Str-r strains could transfer either all three markers or just Tet-r. Physical studies showed that a mid-sized circle was present in Tet-r Cam-s Str-s cells and a small circle in Tet-s Cam-r Str-r cells. Thus, it was clear that the Tet-r marker resided on a self-transmissible plasmid (RTF), and another plasmid that was not self-transmissible (an r unit) carried the Cam-r and Str-r markers.

A critical observation was that in some strains and with some plasmids a cell line could not be maintained that stably carried only an RTF-r composite plasmid: RTF and r segregants continually arose. Furthermore, cell lines that initially contain RTF and r units usually produce some cells that contain a composite plasmid. Thus, these plasmids engage in some type of association-dissociation system. The existence of such an apparent equilibrium gives rise to a gene-amplification phenomenon that occurs when a cell carrying the Tet-r Cam-r Str-r plasmid is challenged with such large amounts of either Cam or Str that even these resistant cells are partially inhibited. With continued growth the culture undergoes a transition to increased resistance to these antibiotics. Associated with the transition is an increase in the size of the plasmid, which results from acquisition of multiple copies of the r determinant. Transition does not occur if large amounts of Tet are added, since there appears to be no mechanism to increase the number of copies of

RTF. It is unlikely that the antibiotic challenge stimulates production of these multi-r plasmids. More likely, a culture of R^+ cells contains rare cells in which the number of r determinants has somehow increased, and these mutant plasmids are simply selected for by the antibiotic challenge because of the increased growth rate of cells containing these plasmids. Apparently these multi-r plasmids are unstable, since if the antibiotic is removed from the growth medium and growth is allowed for many generations, the plasmid size gradually drops. Association and dissociation and the related transition phenomenon presumably are a result of recombination between common sequences, known as insertion sequences or IS elements (Chapter 12), present in both RTF and r. In association recombination occurs between IS elements in different circles, and in dissociation recombination between two elements in a single circle causes segregation of the two circles.

An interesting group of R plasmids are the F-like R plasmids, of which R1, R6, and R100 are representative. A relation between these plasmids and F was first noticed in cells containing both F and one of these plasmids: the plasmids caused fertility inhibition of F. Complementation analysis indicated that the F-like plasmids contained most of the F *tra* genes. Electron microscopic examination of heteroduplexes formed between F and an F-like R showed extensive homology in the transfer regions. Detailed analysis of these heteroduplexes shows that the F-like R plasmids consist of a large segment of F linked to a typical r determinant. Numerous inversions, deletions, and substitutions are also present in the F region, which suggests that these plasmids may have arisen a long time ago from a common ancestor.

R plasmids are of considerable medical interest since they can be transferred to bacteria that cause major epidemics, such as *S. typhimurium* and *Shigella dysenteriae,* and to strains that cause infections in hospitals (various enterobacteria, *Pseudomonas aeruginosa,* and *Staphylococcus aureus).* In fact, it has become clear that since the beginning of the "antibiotic era," R plasmids have been on the increase in nature. For example, penicillin was introduced to general use in the early 1940s. By 1946 14 percent of strains of *S. aureus* isolated in hospitals were Pen-r. The fraction was 38 percent in 1947, 59 percent in 1969, and nearly 100 percent by the 1970s. The majority of these resistant strains either carry an R plasmid or a *pen-r* gene of the type found in R plasmids. A particular problem has been the transfer of R^+ bacteria from animals to man. For example, animal carcasses, especially poultry, are frequently contaminated with R^+ *E. coli,* which can colonize the human intestine, and *S. typhimurium.* Studies by the U.S. government have shown that handling of raw meat is the usual route for transmission to man; bacteria from the meat get on utensils and kitchen surfaces, and later to humans. Numerous drug-resistant epidemics of salmonellosis in farm animals occurred between 1960 and 1980 in Great Britain, and the causative organisms were quickly transmitted to man. The infection could not easily be combatted because of resistance to standard antibiotics. The major cause of the infection of animals is apparently the extensive use of penicillin and tetracycline in animal feed; these substances lead to more rapid growth of the animals and hence are economically valuable. Nonetheless, in the early 1980s the United Kingdom Government Committee recommended that all antibiotics be eliminated from animal feed.

The Colicinogenic or Col Plasmids

Col plasmids are *E. coli* plasmids able to produce colicins, proteins that prevent growth of susceptible bacterial strains that do not contain a Col plasmid. They are one class of a general type of plasmid called a bacteriocinogenic plasmid, which produce bacteriocins in many bacterial species. Bacteriocins, of which colicins are one example, are proteins that can bind to the cell wall of a sensitive bacterium and inhibit one or more essential processes such as replication, transcription, translation, or energy metabolism. There are many types of colicins, each designated by a letter (e.g., colicin B) and each having a particular mode of inhibition of sensitive cells (Table 11-4). Colicin production is detected by an assay similar to that for detecting phage. A colicin-producing cell is placed on a lawn of sensitive cells; the colicin inhibits growth of nearby bacteria, producing a clear area, known as a **lacuna,** in the turbid layer of bacteria.

Colicin production is normally repressed but can be induced by a variety of agents. One that is commonly used is ultraviolet light. A small fraction of every population of Col$^+$ cells also produces colicin constitutively. Thus, the presence of the Col plasmid has survival value for the population, which is thereby able to compete more effectively in the wild with colicin-sensitive cells.

Colicins are probably of two types—true colicins and defective phage particles. The latter class is inferred from studies of many purified colicins. Only a few colicins are simple proteins; others look like phage tails when examined by electron microscopy and are thought to be gene products transcribed from remnants of ancient prophages. The hypothesis is that repeated mutation has resulted in loss of the genes for replication and head production, but genes encoding a repressor system, the lysis enzyme, and the tail pro-

Table 11-4   Properties of several *E. coli* colicins

Colicin	Action of colicin
Colicin B, Colicin Ib	Damages cytoplasmic membrane
Colicin E1, Colicin K	Uncouples energy-dependent processes by an unknown effect on the cell membrane.
Colicin E2	Degrades DNA
Colicin E3	Cleaves 16S rRNA

teins have survived intact. Presumably these phages share with phage T4 the property that adsorption without DNA injection causes an inhibition of macromolecular synthesis. True colicins also bind to specific recognition sites on the cell wall. As with phage adsorption, the sites have specific cell functions. Table 11-4 shows that colicins act in a variety of ways. A surprising feature of colicin activity is that in some cases *one* colicin molecule is sufficient to kill a target cell.

In most cases colicins are inactive against a cell that contains a related Col plasmid. This is called immunity. Interestingly, immunity is conferred by an excreted small protein that binds to the larger colicin protein. The immunity protein not only confers immunity on Col^+ cells but is also necessary for killing of Col^- cells. In the case of the colicin cloacin DF13 the role of the immunity protein is understood somewhat. The DF13 protein consists of three regions (Figure 11-14): a receptor-binding region, an RNase, and a segment that binds the immunity protein. The receptor-binding terminus is very hydrophobic and may be able to adsorb to the cell membrane. The immunity-binding segment has a strong negative charge that is neutralized by the positively charged immunity protein. After binding to the receptor, the colicin is cleaved. The N-terminal region remains outside the cell and the RNase segment enters the cell, leaving the immunity protein on the cell surface. The RNase acts on the RNA in ribosomes and thereby kills the sensitive cells.

The Col plasmids range in size from a molecular weight of a few million, for those that are not self-transmissible, to more than 60×10^6, for the self-transmissible plasmids. The best-studied Col plasmid is ColE1, whose molecular weight is 4.4×10^6. Its complete sequence of 6646 base pairs is known. It is used extensively in recombinant DNA research (Chapter 20). Many of the large self-transmissible Col plasmids are hybrids between a small Col plasmid and F or F′ plasmids.

ColE1 is a mobilizable but nonconjugative plasmid and has provided the most definitive evidence that transfer requires a plasmid-encoded nuclease (for ColE1 the nuclease is encoded by the *mob* gene), and a specific base sequence called *bom* (*basis of mobility*), which contains the cutting site. The critical experiment is shown in Figure 11-15, in which the compatible plasmids F and ColE1 cohabit a single bacterium and F donates the conjugal functions that ColE1 lacks, thereby enabling ColE1 to be transferred. Panel (a) shows the state of ColE1 in an F^- cell. The *mob* gene is transcribed, the

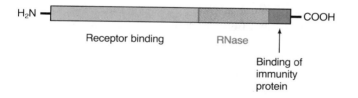

H_2N — Receptor binding — RNase — Binding of immunity protein — COOH

Figure 11-14 Regions of the polypeptide chain of cloacin DF13.

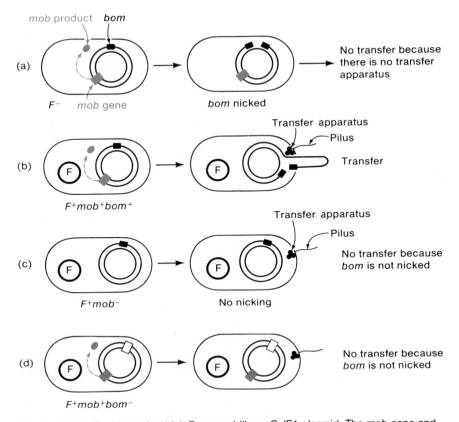

Figure 11-15   Conditions in which F can mobilize a ColE1 plasmid. The *mob* gene and the *mob* product are drawn in red. The *bom* site in the DNA is drawn as a solid box, when functional, and as an open box, when mutant or deleted. Transfer only occurs when ColE1 makes an active *mob* product that acts on a functional *bom* site and F provides the transfer apparatus.

mob product nicks in the *bom* site (the actual cutting site is called *nic*), and the ColE1 supercoil is converted to a nicked circle. Transfer cannot occur because ColE1 lacks the ability to form pili and hence to form conjugate pairs. In panel (b) the cell contains both F and ColE1. F causes synthesis of the pilus and the transfer apparatus, and ColE1 is transferred. Mutants (*mob⁻*) of ColE1 that fail to be transferred when F donates the transfer apparatus have been isolated. Genetic analysis shows that the *mob⁻* mutation is recessive, which indicates that the *mob* gene encodes a diffusible molecule, presumably a protein. Furthermore, the mutant plasmid, when isolated, is not in the form of a relaxation complex. The failure of F to help the *mob⁻* mutant to transfer is shown in panel (c), in which the F pilus and transfer apparatus are formed but nicking of the ColE1 DNA does not occur. Another ColE1 mutant, having the Mob$^-$ phenotype, but which has a *cis*-dominant mutation, is a deletion that removes the *bom* site. The Mob protein is made but nicking fails to occur and there is no transfer, as shown in panel (d).

The *Agrobacterium* Plasmid Ti

A crown gall tumor found in many dicotyledonous plants is caused by the bacterium *Agrobacterium tumefaciens*. The tumor-causing ability resides in a plasmid called Ti. When a plant is infected, some of the bacteria enter and grow within the plant cells and lyse there, releasing their DNA in the cell. From this point on, the bacteria are no longer necessary for tumor formation. By an unknown mechanism a small fragment of the Ti plasmid (the T DNA), containing the genes for replication, becomes integrated into the plant cell chromosomes. The integrated fragment breaks down the hormonally regulated system that controls cell division and the cell is converted to a tumor cell. This plasmid has recently become very important in plant breeding because specific genes can be inserted into the Ti plasmid by recombinant DNA techniques, and sometimes these genes can become integrated into the plant chromosome, thereby permanently changing the genotype and phenotype of the plant. New plant varieties having desirable and economically valuable characteristics derived from unrelated species can be developed in this way. This will be discussed further in Chapter 21.

Broad Host Range Plasmids

Most plasmids can exist in only a limited number of closely related bacteria; these are called narrow host range plasmids. However, the self-transmissible R plasmids of incompatibility group IncP of *E. coli* and of group IncP1 of *Pseudomonas aeruginosa* are notable in that they can be transferred to and maintained in bacteria of a large number of species. These are called **broad host range plasmids**. Why some plasmids have a narrow host range and others a broad host range is unknown. The broad host range plasmids have recently become exceedingly useful since they integrate (albeit at low frequency) into the host chromosome of numerous species and have thereby enabled the establishment of genetic systems in species in which mapping had not previously been possible. In this way, genetic maps have been obtained for several economically important bacteria; such mapping facilitates the construction of strains with desirable characteristics.

Most of these plasmids are able to mobilize the chromosome only at very low frequency (ca. 10^{-8} per cell). However, by various genetic and recombinant DNA techniques variants of these plasmids have been constructed in which chromosome mobilization is enhanced by a factor of 10^3–10^5.

Other Plasmids

Several plasmids render fairly innocuous plasmid-free cells pathogenic. For example, the Ent plasmids of *E. coli* synthesize enterotoxins that are responsible for travelers' diarrhea. A plasmid called Hly (for hemolysis) has been found in *E. coli* strains isolated from pigs. Whereas the hemolysin destroys red blood cells in blood samples, the Hly plasmid does not seem to cause any pathogenicity. Certain plasmids residing in the human pathogen, *Staphylococcus aureus*, enhance pathogenicity. A penicillinase (penicillin-destroying) *S. aureus* plasmid has been carefully studied.

Many species of *Pseudomonas* can utilize several hundred organic compounds as carbon sources—in particular, such toxic substances as camphor, toluene, and octane. This metabolic ability resides in a set of plasmids collectively known as **degradation plasmids.** Each plasmid provides one or more metabolic pathways to degrade these compounds. Since many enzymes are needed, the plasmids are fairly large (molecular weight of $50-100 \times 10^6$). These plasmids enable bacteria to degrade many *synthetic* compounds and hence are making an important contribution in the removal of environmental pollutants. For example, some strains can degrade persistent chlorinated hydrocarbons, herbicides, and various detergents. Many laboratories are attempting to use genetic engineering to construct "super plasmids," which can be used for pollution control and chemical syntheses.

Plasmids have also been isolated that confer resistance to toxic metal ions. Such plasmids, whose origin is unknown, are found in environments containing these ions, such as the sludge produced by industrial reprocessing of photographic film (resistance to the Ag^+ ion). In only one case has the biochemistry of resistance been elucidated: resistance to Hg^{2+} ions results from a plasmid-encoded reductase that converts Hg^{2+} to metallic mercury, which is sufficiently volatile that it evaporates away.

PROBLEMS

1. Is there any relation between plasmid size and copy number?

2. What is a relaxation complex?

3. When DNA is isolated from cells, nicks are often produced by various nucleases. How does the nicking produced by a relaxation complex differ from the nicking of chromosomal DNA?

4. What mode of replication is used by plasmids for their reproduction?

5. Are there any circumstances in which plasmids undergo rolling circle replication?

6. During transfer, does replication occur in the donor, the recipient, or both?

7. What early enzymatic step is required in transfer replication but is not required for typical θ replication?

8. Are sex pili present on the donor or recipient cell?

9. What elements encode genes for pili synthesis?

10. What two roles are played by pili in bacterial conjugation?

11. What are the *tra* genes of F?

12. Define the terms conjugative, mobilizable, and self-transmissible.

13. State which of the terms in question 12 describe F and ColE1.

14. If a plasmid is mobilizable but nonconjugative, what function does it lack?

15. In a hybrid plasmid carrying two replication origins, one from a high-copy-number plasmid and the other from a low-copy-number plasmid, which origin will be used and what will the copy number be?

16. If two plasmids cannot be maintained in a single cell, what property is common to the plasmids?

17. What feature of F, not present in ColE1, enables F, but not ColE1, to integrate into the host chromosome?

18. If ColE1 could be altered to contain insertion sequences homologous to sequences in the chromosome, such that it could integrate in the chromosome, would Hfr-like cells arise?

19. What are the two components of R factors and which carries the genes for replication and transfer?

20. Could you have a plasmid with no genes whatsoever?

21. Lac$^+$ and Lac$^-$ cells form purple and white colonies, respectively, on EMB-lactose agar. A culture consisting of cells whose genotype is F′*lac*$^+$/*lac*$^-$ is mutagenized and plated on EMB-lactose agar at 30°C. Several purple colonies whose color is a little less intense than the others are studied further. These give white colonies at 42°C and light purple colonies at 30°C. Bacteria obtained from white (42°C) colonies remain white when replated at 30°C. What type of mutation has been acquired?

22. If, in a particular cell type, rifampicin were to inhibit DNA transfer, what would you conclude about the transfer mechanism?

23. What are the roles of DNA synthesis in the donor and the recipient?

24. A Lac$^-$ bacterial strain has a *dnaA*(Ts) mutation, which prevents colony formation at 42°C. An F′*lac*$^+$ plasmid is introduced into the strain by conjugation. The culture is grown for many generations at 30°C, and then 10^6 cells are plated at 42°C. A few colonies arise and these are capable of growth at both 30°C and 42°C.

   (a) Are these colonies Lac$^+$ or Lac$^-$?
   (b) Do the cells still carry the *dnaA*(Ts) mutation?
   (c) What feature of the cell has changed that enables them to grow at 42°C?
   (d) Can the cells grow at 42°C in the presence of acridine orange?

25. An $F'(Ts)lac^+$ plasmid has a temperature-sensitive mutation in its replication system.

    (a) What is the phenotype of an $F'(Ts)lac^+/lac^-$ cell at 42°C?

    (b) An $F'(Ts)lac^+/lac^-gal^+$ strain is grown for many generations and then plated at 42°C. Some Lac$^+$ colonies form at 42°C. How have these formed?

    (c) Some of the Lac$^+$ colonies in (b) are Gal$^-$. How have these formed?

26. Plasmids contained in Rec$^+$ cells frequently dimerize. A dimer can then be isolated and transferred by the $CaCl_2$ technique to a Rec$^-$ cell in which it will be stable and not revert to the monomer. If a monomer plasmid has a copy number of 10, what will the copy number of the dimer be?

27. On EMB-lactose agar Lac$^+$ and Lac$^-$ colonies are purple and white, respectively. If several thousand $F'lac^+/lac^-$ cells are plated, a few sectored colonies appear. This may have a purple half and a white half or a wedge of white in a predominately purple colony. Sectored colonies that are predominately white are not found. Explain the cause of sectoring.

28. An $F'lac^+/str\text{-}s$ is mated with a $lac^-str\text{-}r$ female that also carries a $dnaG(Ts)$ mutation. Mating is at a nonpermissive temperature. After 30 minutes of mating, streptomycin and an inducer of the lac operon are added. Will any β-galactosidase be made in the culture?

Chapter 12

Transposable Elements

A long-standing unspoken assumption about gene organization and chromosome structure is that each gene has a definite and unvarying location in a particular chromosome. This assumption has been strengthened over the years by the generation of genetic maps by both genetic and physical techniques. However, hints of gene rearrangements have come forth repeatedly over the past few decades. For example, gene inversion and gene relocation occurs at low frequency in both bacteria and in eukaryotes. In the 1940s in a genetic analysis of the mottling of the kernels of maize Barbara McClintock discovered regulatory elements that moved from one site to another and thereby affected gene expression. Her work was followed 30 years later by the recognition that bacteria contain mobile segments of DNA that relocate at low frequency. The frequency with which the elements move in both bacteria and eukaryotes is fairly low and depends on the particular element (10^{-7}–10^{-2} per generation), but nonetheless the concept of a static genome has gradually been replaced by the current view of the "genome in flux."

Because a chromosome is a continuous DNA molecule, movement of mobile elements is a DNA-exchange process, a kind of recombination. However, it differs from the more commonly observed recombination systems in which exchange occurs between homologous DNA sequences, for homology plays no part in the process. In bacteria, the evidence is clear, for homologous recombination always depends on the *recA*-gene product or its equivalent

(Chapters 9 and 14), whereas, in contrast, movement of these elements occurs at the same frequency in *recA*⁻ and *recA*⁺ cells.

The movement discussed in this chapter is called **transposition,** and the mobile segments are called **transposable elements,** or, in bacterial systems **transposons.**

TERMINOLOGY

Many bacterial transposons contain easily recognizable genes, which may or may not exist elsewhere in the genome. Antibiotic-resistance genes are common, and transposons carrying such genes are the ones most frequently studied because their presence is made evident by simple plating tests. Most antibiotic-resistance transposons were originally designated by the abbreviation Tn followed by a number (e.g., Tn5), which distinguished the different elements. By agreement, all newly discovered bacterial transposons are to be designated in this way even if no recognizable gene is present. When it is necessary to refer to the genes carried on a transposon, these are indicated by standard genotypic designations, for example, Tn1(*amp-r*), in which *amp-r* indicates that the transposon carries the genetic locus for resistance to ampicillin. The transposons first discovered did not contain any known host genes, and for historical reasons, they were called **insertion sequences** or **IS elements** and designated IS1, IS2, and so forth. Transposons have occasionally been designated in nonstandard ways—for example, γδ, an element contained in the F plasmid.

A transposon is frequently located within a particular gene; this creates a mutation in that gene, which is usually given a number. The following notation is used to designate mutation 87 in the *lac*Z gene produced by transposon number 3:

$$lacZ87::Tn3$$

The antibiotic-resistance genes present in transposons are usually quite different from antibiotic-resistance genes that arise by simple mutation of bacteria lacking transposons. In fact, the origin of the transposon genes is unknown. The antibiotic-resistance genes of most R plasmids are carried by transposons present in the plasmid DNA.

INSERTION SEQUENCES

The first transposable elements that were discovered in bacteria (in 1967) were the IS elements, and how they were discovered is of some interest. It is now known that they are just one type of transposon but nonetheless some of the essential characteristics of transposons and the various means of detecting transposition can be easily seen by examining this special class.

A collection of *gal⁻* and *lac⁻* mutations with unusual properties were found in *E. coli*. The mutations had the following features:

1. They were highly **polar mutations** in that each mapped in the first gene of an operon but proteins of the downstream genes were not synthesized. For example, the mutation might be in the *lacZ* gene, yet complementation will not occur with an F'*lacZ⁺lacY⁻* plasmid because of lack of *lacY* function both in the plasmid and in the chromosome. In each case the polarity resulted from the presence of a transcription-termination sequence.

2. These mutations could not be reverted by base-analogue or frame-shift mutagens, so the mutations could not be base substitutions or single-base additions or deletions. (Less polar mutations had been known since 1960, but as these were chain-termination mutations that caused decreased transcription of downstream genes, they were usually revertible by base analogues.)

3. If a plasmid was transferred to a cell containing one of these polar mutations, similar polar mutations occasionally appeared in genes carried on the plasmid. The locations of the chromosomal and plasmid polar mutations could be quite different, for example, in the *gal* operon of a chromosome and a *lac* operon on the plasmid. An example of such an event was the following: F'*lac⁺* was transferred by mating to a *gal⁻*(polar)*lacZ⁻* cell to form a *gal⁻*(polar)*lacZ⁻*/F'*lac⁺* cell. Lac⁻ mutants arose, which exhibited the same high-polarity phenotype. The mutation was located on the F', as demonstrated by the fact that when the F' was transferred to an F⁻ cell, it could not confer the Lac⁺ phenotype on a *lacY⁻* cell.

4. Physical studies of several plasmids containing such polar mutations in different genes provided the crucial evidence regarding the nature of the mutations. These experiments showed that in each case the plasmid was larger than the original plasmid, because *a segment of DNA had been inserted in the operon*. (We will see shortly that these segments contained a transcription stop signal.) It was hypothesized that these inserted segments were mobile elements that moved from one region of a DNA molecule chromosome to another site.

When various phages infected cells carrying these polar mutations, rare phage mutants could be isolated that contained polar mutations. Electron microscopic examination of the DNA of phages carrying polar mutations was especially informative. Hybridization of the DNA from a phage bearing the mutation with the DNA of a similar phage lacking a polar mutation showed a loop (Figure 12-1), which indicated that the mutation is an insertion. The length of the insertion was the length of the loop. In other experiments wild-type phage DNA was hybridized with DNA obtained from various mutant phage, each carrying a polar insertion (derived from a single bacterial strain) in different phage genes. In these experiments the insertion had a constant

Figure 12-1 A hybridization test showing that an IS element is a segment of inserted DNA. After renaturation half of the molecules have re-formed the normal and mutant phage DNA molecules (only the heteroduplex has been drawn). Mutant refers to a phage carrying a polar mutation.

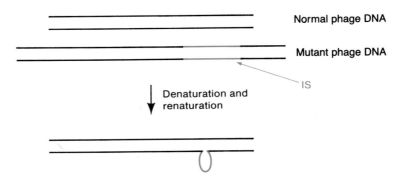

Normal phage DNA

Mutant phage DNA

IS

Denaturation and renaturation

size and a location that corresponded to the mutant site. Furthermore, if different phage species infected the same bacterial strain, the loop of the DNA from one phage mutant carrying a polar insertion could hybridize with the insertion present in another phage species. These observations together indicated that a mutant bacterium donated the *same* sequence to all of the phage that acquired a polar mutation.

A large number of phage mutants containing genes from numerous regions of the *E. coli* chromosome into which an IS element had been inserted have been compared to plasmids containing insertion sequences. This analysis demonstrated the existence of several different IS elements. Many of these elements have been isolated from mutant phage DNA and their base sequences have been determined. A number of significant features of these elements have become apparent from the sequence analyses:

1. Those elements that produce polar mutations contain transcription-stop signals (except for IS1) and also chain-termination mutations in all possible reading frames. These properties are sufficient to account for the polar effects.
2. The termini of each IS element have inverted-repeat sequences consisting of 16–41 base pairs, the number depending on the element (Figure 12-2).
3. At many sites of insertion (for example, in the *E. coli gal* operon) the sequences are inserted in either the left-to-right or right-to-left ori-

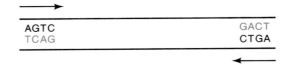

```
AGTC                                    GACT
TCAG                                    CTGA
```

Figure 12-2 An example of a terminal inverted repeat in a DNA molecule. The arrows indicate the inverted base sequences. Note that the sequences AGTC and CTGA are *not* in the same strand. In a so-called direct repeat, the sequence in the upper strand would be AGTC . . . AGTC.

Table 12-1 Properties of some *E. coli* insertion elements

Element	Number of copies and location*	Number of base pairs
IS1	5–8 in chromosome	768
IS2	5 in chromosome; 1 in F	1327
IS3	5 in chromosome, 2 in F	ca. 1400
IS4	1 or 2 in chromosome	1428
IS5	Unknown	1195
γδ	1 or more in chromosome; 1 in F	5700

entation. This is expected because of the symmetry of the inverted-repeat configuration of the termini.

4. Different IS elements contain different numbers of bases. Radioactive copies of the elements have been prepared and hybridized with denatured *E. coli* DNA and with denatured F plasmid DNA to determine the number of copies of the elements in the chromosome and in F. The results are summarized in Table 12-1.

5. The elements contain at least two apparent coding sequences initiated by an AUG and terminated with an in-phase stop codon. By using *in vitro* protein-synthesizing systems with isolated IS DNA as template, proteins have been made. Some of these proteins have also been isolated from cells containing the element. Study of a large number of bacterial transposons indicates that each encodes at least two proteins.

DETECTION OF TRANSPOSITION IN BACTERIA

The IS elements just described are a special class of bacterial transposon. In this section we examine the more general features of transposons and how these elements are observed.

In contrast with IS elements, which are normally detected by polarity effects, the more common class of bacterial transposon contains recognizable genes—for instance, genes for antibiotic resistance—that are often not present elsewhere in the genome. These genes provide the means for detecting transposition. An example is shown in Figure 12-3.

In this experiment an R plasmid having an *amp-r* (ampicillin-resistance) transposon is transferred by conjugation to a *recA⁻* recipient bacterium that carries a transposon having the *kan-r* (kanamycin-resistance) gene. On continued growth in the presence of both antibiotics, transposition occasionally occurs and yields an R plasmid carrying both drug-resistance markers and hence both transposons. The presence of this two-transposon plasmid is

Figure 12-3 An experiment to demonstrate transposition. A donor bacterium (left) containing an *amp-r* R plasmid transfers R to a recipient cell (center) containing a *kan-r* transposon. In subsequent generations the *kan-r* transposon is transposed to the R plasmid. The presence of the second transposon on R is demonstrated by simultaneous transfer of the *amp-r* and *kan-r* genes. The *amp-r* gene is shown as a transposon, as are most drug-resistant determinants on plasmids, but this fact is not essential to the experimental demonstration of transposition.

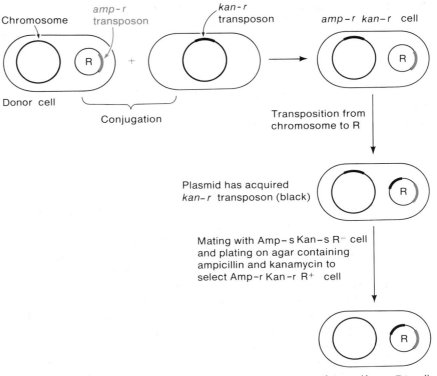

detected by transferring the plasmid by conjugation to a Kan-s Amp-s bacterium and plating on agar containing both antibiotics. As a further test, the plasmid containing the two markers could be hybridized to the original plasmid molecule and examined by electron microscopy; loops such as those in Figure 12-1 would be seen, indicating the presence of an inserted DNA sequence.

When transposition occurs, most transposons can be inserted at a very large number of positions. The most direct evidence comes from several types of experiments in which mutants formed by insertion of a particular transposon are isolated. For example, consider an Amp-s *E. coli* culture that, by the $CaCl_2$ transformation technique (Chapter 10), has taken up a DNA fragment carrying an *amp-r* transposon. The fragment will, in general, lack the replication origin and hence will be incapable of replication. Thus, Amp-r cells can arise only by transposition of the *amp-r* element to the chromosome. Plating the cells on nutrient agar containing ampicillin enables one to isolate colonies derived from cells in which transposition occurred. If a sufficiently large number of Amp-r bacterial colonies are examined and tested (by replica plating) for both nutritional requirements and ability to utilize particular sugars as a carbon source, it is observed with most transposons that a colony can be found bearing a mutation in almost any gene that is tested. This indicates that insertion sites for transposition are scattered throughout the *E. coli* chromosome.

Although there are certainly many sites of insertion in *E. coli*, and even in several phages, it is clear that the locations are not randomly distributed; certain positions seem to be excluded and others ("hotspots") are used repeatedly. This can be seen by examining the insertion sites in very small regions whose base sequences are known. Such studies show that there is a rather broad range of insertion specificities from one transposon to the next. At one extreme is IS4; 20 independent *galT*-gene insertion mutants were isolated and all insertions were at exactly the *same* position in the gene. At the other extreme is Mu, which can insert almost anywhere. Most transposons exhibit insertion specificities between these extremes. For example, the insertion sites of transposon Tn9 in the final 160 base pairs of the *lacZ* gene have been determined by base-sequencing; of 28 independent insertions, 16 different positions have been noted but 5 of these are represented by multiple occurrences of insertion. The transposon Tn10 is even more specific; for example, 22 different sites have been noted in the *his* operon of *Salmonella typhimurium* but 40 percent of all known transposition events have occurred at a single nucleotide position. Little is known about the cause of hot spots for insertion.

Figure 12-4 shows another way that transposition can be detected. In this case, a temperate phage acquires an *amp-r* gene from a *recA⁻* cell containing an *amp-r* R plasmid, in which the *amp-r* gene is carried on a transposon, and transfers the transposon to another cell. A *recA⁻* cell (chosen for the experiment to eliminate all possibility of homologous recombination) is infected with the phage, and in a small number of progeny phage (perhaps 1 in 10^7) the transposon carrying the *amp-r* gene has been picked up. The resulting phage population is used to infect an Amp-s culture of bacteria at a ratio of about 1 phage per 10 bacteria in a growth medium containing ampicillin. Uninfected bacteria are killed. More important, cells infected with a phage lacking the *amp-r* gene will not yield progeny phage because of premature lysis induced by the drug; however, if the phage carries the *amp-r* gene, the bacteria survive and progeny *amp-r* phage are produced. Thus, in the presence of ampicillin, only the λ carrying the *amp-r* transposon can grow; the lysate therefore consists of a homogeneous population of λ*amp-r* phage.

To demonstrate that the *amp-r* gene is part of a transposon, the λ*amp-r* phage are used to infect a *recA⁻ amp-s*(λ) lysogen. Since the lysogen is immune to infection by λ, the phage are unable to develop, and the bacteria survive. The infected cells are then plated on agar containing ampicillin. The *amp-r* phage enables the cells to grow, but, since the phage cannot replicate (because of the presence of the phage λ repressor), in each division cycle one daughter cell does not inherit a λ*amp-r* DNA molecule, and a bacterial colony cannot form. However, in a few bacteria, transposition does occur, and the bacterial chromosome acquires the *amp-r* transposon, and all daughter cells do inherit

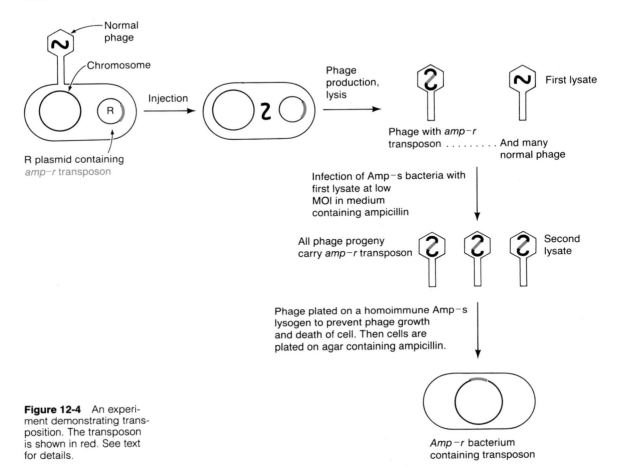

Figure 12-4 An experiment demonstrating transposition. The transposon is shown in red. See text for details.

an *amp-r* gene. Thus, formation of a colony indicates that transposition has occurred. Note that the resulting *amp-r* bacteria could be reinfected again by a normal phage to yield a second population of phage, a few of which would contain the transposon.

TYPES OF BACTERIAL TRANSPOSONS

Because of particular related features, bacterial transposons have been divided into several distinct classes. One is the IS group, which we have already discussed. Three more are described below.

1. *Composite transposons*. A variety of transposons carrying antibiotic-resistance genes are flanked by two identical or nearly identical copies of an

Inverted repeat

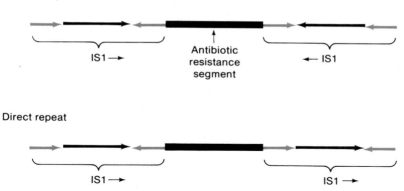

Direct repeat

IS element. These transposons are called composite type-I transposons. The IS elements in these composite units can be in an inverted or direct repeat configuration (Figure 12-5). Since the two ends of an IS element are themselves inverted repeats, the relative orientation (inverted or direct) of the flanking IS elements of a composite transposon does not alter its terminal sequences—that is, they remain the same in the inverted-repeat array. The ability to transpose is determined by the terminal IS elements. Three frequently studied composite transposons are Tn5, Tn9, and Tn10. Some properties of these and several other transposons are shown in Table 12-2.

Most of the composite type-I transposons were first detected as genetic elements that could be transposed from one plasmid to another or to a phage. When the DNA of some of these plasmids was denatured and renatured, single-stranded circles containing a stem-and-loop structure were observed by electron microscopy. Such a structure is indicative of an inverted repeat, as shown in Figure 12-6.

The composite transposons have two interesting features: (1) Some of the terminal IS elements are able to transpose by themselves. (2) For many

Table 12-2 Properties of selected composite type I transposons of *E. coli*

Element	Genes carried*	Size in base pairs	Terminal IS element and size in base pairs	Relative directions of terminal IS elements
Tn5	*kan*	5400	IS50 (1500)	Inverted
Tn9	*cam*	2638	IS1 (768)	Direct
Tn10	*tet*	9300	IS10 (1400)	Inverted
Tn204	*cam, fus*	2457	IS1 (768)	Direct
Tn903	*kan*	3100	IS903 (1050)	Inverted
Tn1681	*ent*	2088	IS1 (768)	Inverted

cam, chloramphenicol; *ent*, enterotoxin; *fus*, fusidic acid; *kan*, kanamycin; *tet*, tetracycline.

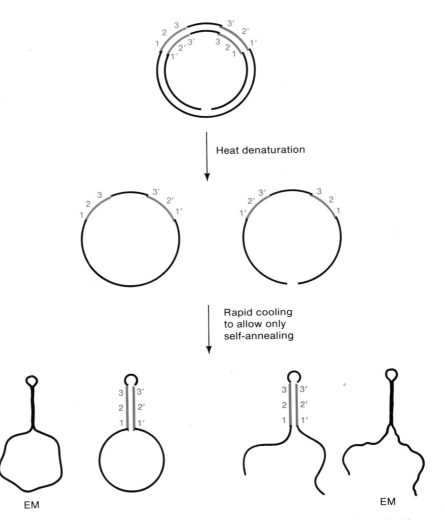

Figure 12-6 Formation of a stem-and-loop structure by denaturation and intramolecular annealing of a plasmid carrying an inverted repeat. The configurations labeled EM indicate the appearance of such a molecule in an electron microscope; light and heavy lines represent single- and double-stranded regions, respectively.

composite transposons, if the transposon resides in a small plasmid, it behaves like two different transposons. This is because, within the frame of reference of the terminal IS elements, the DNA they flank and the DNA that surrounds them are interchangeable. For example, consider a composite transposon having the structure IS-A-IS, where A is the central sequence. If the transposon inserts in a plasmid whose sequence we call B, the structure of the transposed plasmid will be

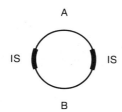

Because the plasmid is circular, both A and B are flanked by the IS elements and transposition of either IS-A-IS or IS-B-IS can occur.

2. *The Tn3 transposon family.* The Tn3 family of transposons consists of quite large elements (about 5000 bp). The prototype, Tn3, was first detected by the appearance of an inverted-repeat element in plasmid R1. Each transposon carries three genes, one encoding β-lactamase (which confers resistance to ampicillin) and two others needed for transposition (which will be discussed later). All Tn3-like transposons contain short (38-bp) inverted repeats and none are flanked by IS-like elements.

3. *The transposable phages.* Two related phages, Mu and D108, include transposition as an essential part of their life cycles. In fact, replication of the phage DNA occurs during transposition. Mu will be described in greater detail later in the chapter.

TRANSPOSITION

The end result of the transposition process is the insertion of a transposon between two nucleotide pairs in a recipient DNA molecule. Base sequence analysis of many transposons and their insertion sites reveals that there is no sequence homology between the transposon and the insertion site, as expected from the lack of a requirement for the *E. coli recA* system. However, the analysis of sequences of the regions in which the transposon is joined to the recipient DNA has demonstrated several features of transposition, which we document in this section.

Duplication of a Target Sequence at an Insertion Site

Insertion of a transposon always is accompanied by the duplication of a short nucleotide sequence (3–12 bp) in the recipient DNA molecule, called the **target sequence,** and the inserted transposon is sandwiched between the repeated nucleotides. This arrangement is shown in Figure 12-7. We repeat, emphatically, that *only one copy of the target sequence is present in the recipient DNA prior to insertion of the transposon and it is not present in the transposon itself.*

Figure 12-7 Insertion of
a transposon generates a
duplication of the target
sequence. The transposon
is never present as a free
DNA segment but comes
from another region of the
genome, as shown in
Figure 12-10.

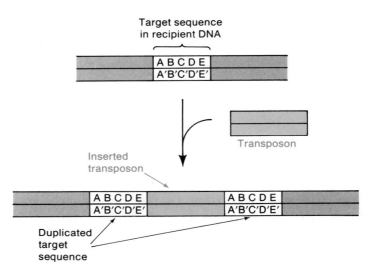

Figure 12-7 Insertion of a transposon generates a duplication of the target sequence. The transposon is never present as a free DNA segment but comes from another region of the genome, as shown in Figure 12-10.

The length of the target sequence varies from one transposon to the next but *is the same for all insertions of a particular transposon*. For example, an inserted IS1 is always flanked by a 9-bp sequence, whereas another transposon is always flanked by a duplication of 5 bp. Note that *for a particular transposon, the target sequence is different for each insertion site; only the length of the duplicated sequence is constant*.

The following basic mechanism, whose details are unknown and which is shown only in outline (Figure 12-8), is believed to be responsible for the duplication of the target sequence.

An enzyme, probably encoded in the transposon, acts on a target sequence in the recipient DNA by making two single-strand breaks, one in each strand, at the ends of the target sequence. The transposon DNA is then attached to the free ends generated by the nicks, such that one strand of the transposon is joined to one strand of the target sequence and the other strand of the transposon is joined to the other strand at the opposite end of the target sequence. This joining leaves two gaps, one across from each strand of the target sequence. Each gap is then filled in and the nicks are sealed; thus, the two strands of the recipient DNA are again continuous and two copies of the target sequence flank the transposon. Note that no homology is utilized in joining the terminal sequence of the transposon to the recipient DNA. Later in this chapter a more detailed proposal for generating the duplicated target sequence will be described.

Since every transposon that is integrated is flanked by a duplicated target sequence and since this is true even of those transposons maintained at fixed sites (for example, those in the F plasmid), it is reasonable to inquire whether this duplication is also necessary for transposition from a donor site. This is definitely not the case since DNA sequences have been constructed

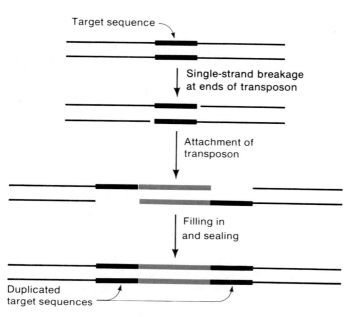

Figure 12-8 A schematic diagram indicating how target sequences might be duplicated. The transposon is never present as a free DNA segment but comes from another region of the genome, as shown in Figure 12-10.

(using recombinant DNA techniques) in which a transposon is not flanked by a directly repeating sequence, and these transposons are still able to transpose.

The Structure of Transposons and the Nature of the Transposon–Target-Sequence Joint

In studying numerous transposons base-sequencing analysis has been carried out on segments of DNA of known base sequence containing a particular transposon. Two important facts have emerged from these analyses:

1. A transposon has well-defined ends. This is shown by the observation that the transposon sequences adjacent to the two target sequences are identical for every insertion of a particular transposon (Figure 12-9). However, each kind of transposon has a specific sequence, so the transposition process consists of joining the ends of the transposon to the ends produced by nicking at opposite sides of the target sequence.

2. The two termini of the transposon consist of a single sequence in an inverted repeat array (one exception is known). This is also shown schematically in Figure 12-9. Note that for transposon I the sequence at the left end of the upper strand and the right end of the lower strand is 1234, reading each time in the $5' \rightarrow 3'$ direction. Depending on the transposon, the number of bases in the inverted repeat is 8 to 38.

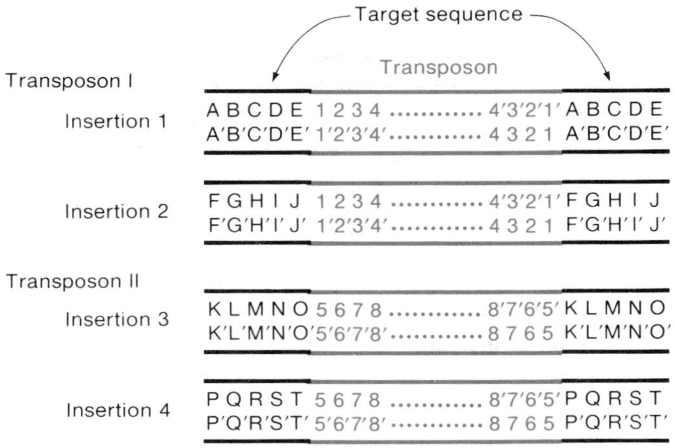

Target sequence

Transposon

Transposon I

Insertion 1

A B C D E 1 2 3 4 .... 4′3′2′1′ A B C D E
A′B′C′D′E′ 1′2′3′4′............ 4 3 2 1 A′B′C′D′E′

Insertion 2

F G H I J 1 2 3 4 .... 4′3′2′1′ F G H I J
F′G′H′I′ J′ 1′2′3′4′............ 4 3 2 1 F′G′H′I′ J′

Transposon II

Insertion 3

K L M N O 5 6 7 8 .... 8′7′6′5′ K L M N O
K′L′M′N′O′5′6′7′8′............ 8 7 6 5 K′L′M′N′O′

Insertion 4

P Q R S T 5 6 7 8 .... 8′7′6′5′ P Q R S T
P′Q′R′S′T′ 5′6′7′8′............ 8 7 6 5 P′Q′R′S′T′

Figure 12-9   Diagram showing that two insertions by the same transposon have the same sequences at the junction of the transposon and the target sequence. That is, 1234 is at the left terminus of all insertions by transposon I. For transposon II the sequence is always 5678. Each number and each letter represent a base; the prime denotes a complementary base. The identifications are merely symbolic; 5678 designates a sequence different from 1234 and *not* one that follows 1234 in numerical order.

A great deal of genetic evidence indicates that the termini of a transposon are essential for transposition. For example, site-specific mutagenesis (Chapter 20) has been employed to generate transposons with single base-pair changes, small deletions, and small insertions in the inverted repeats. These alterations invariably prevent transposition.

Replication of Transposons in the Course of Transposition Between Two Plasmids

Consider a cell possessing two plasmids, A and B, one of which contains a transposon. At a frequency of about 10^{-7} events per generation, the phenomenon shown in Figure 12-10 occurs. That is, after many generations of growth, cells are produced in which plasmid B (now called B′) also possesses the transposon originally in A. An important observation is that with many transposons *the transposon is contained in both plasmids*. Thus, for these

Figure 12-10 Transposition of a transposon from plasmid A to a plasmid B.

Bacterium

Many generations of growth

A   B   A   +   B′

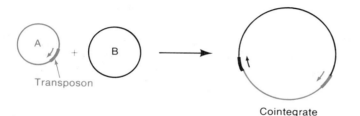

Figure 12-11 Formation of a cointegrate by transposon-mediated fusion of two plasmids A and B. Two copies of the transposon are generated in the process. The arrows denoted an arbitrary direction to show that both copies of the transposon in the cointegrate are in direct repeat.

transposons the original transposon sequence is duplicated in the transposition process—transposition of these elements does not consist of excision of the element from one site and insertion of the same DNA fragment at a second site. Thus, *transposition of these transposons is a replicative process.* For some transposons arguments have been presented that replication does not play a role in transposition. This situation is widely debated for certain transposons (e.g., Tn5), for which some research groups argue for replication and others for nonreplicative transposition. The situation should be resolved in the coming years. In this chapter we will primarily consider transposons for which there is agreement that transposition includes a replicative step.

A related phenomenon, which is important in understanding transposition, has also been observed in cells containing two different plasmids. At a frequency also of about 10^{-7} events per cell generation, the two plasmids fuse to form a single plasmid called a **cointegrate**. (This process is sometimes also called **replicon fusion.**) This hybrid plasmid is not simply a fusion of the two plasmids, because it contains *two* (not just one) copies of the transposon. Both copies are in the same orientation (that is, in direct repeat) and are located precisely at the junction between the donor and recipient plasmid sequences (Figure 12-11).

The fact that the cointegrate has two copies of the transposon, whereas only one copy was present before cointegration occurred, again indicates that the transposon has replicated. The arguments, mentioned above, that some elements transpose via a nonreplicative process are based, to a great extent, on the inability to detect a cointegrate containing these elements.

The Cointegrate as an Intermediate in Transposition of Tn3

Certain experimental results with transposons in the Tn3 family (which quite clearly includes replication in the transposition process) suggest that formation of a cointegrate is an intermediate step in the transposition of these elements.

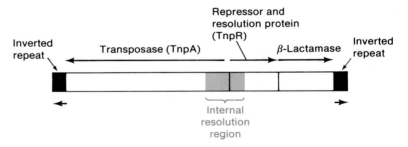

Figure 12-12 The physical map of transposon Tn3. There is a total of 4957 base pairs; each inverted repeat contains 38 base pairs. The three genes are indicated above the arrows, which show the direction of transcription. The internal resolution region *res* is discussed in the text.

The structure of Tn3 is shown in Figure 12-12. Tn3 has a pair of terminal inverted repeats, like all transposons, and these flank three genes. The three gene products have been isolated; from their size and from the base sequence of Tn3, we know that the genes include all of the bases between the inverted repeats. The leftmost gene encodes a large protein (1015 amino acids), denoted TnpA, called a **transposase;** it is responsible for formation of cointegrates. The rightmost gene encodes β-lactamase, an enzyme that inactivates ampicillin. The central gene encodes a small protein (185 amino acids) called a **resolvase** and symbolized TnpR; it has two functions—repression of the synthesis of TnpA, and catalysis of a site-specific exchange that resolves cointegrates in a second step of the transposition process. Near the boundary of the genes encoding TnpA and TnpR is a sequence of DNA called the **internal resolution site** or *res;* it includes the base sequence at which TnpR acts.

Genetic experiments have yielded the following information. All *tnpA⁻* mutants are unable either to transpose or to form cointegrates. Also, neither *tnpR⁻* mutations nor deletions of the internal resolution site interfere with the ability to form cointegrates, but they completely prevent transposition. That is, in a cell containing a plasmid with a *tnpR⁻* copy of Tn3 and a second plasmid lacking Tn3, the process shown in Figure 12-11 can occur, but the process shown in Figure 12-10 does not occur. These observations suggest that Tn3-mediated transposition proceeds by a two-step mechanism—first, TnpA induces formation of a cointegrate and then TnpR catalyzes a site-specific exchange at the internal resolution site. A schematic diagram of this sequence of events is shown in Figure 12-13.

Many transposons outside of the Tn3 family do not carry a TnpR-like site-specific recombination system but are still able to transpose. These systems may use a homology-dependent exchange system since, once a cointegrate has formed, there are two identical sequences—namely, copies of the transposon—so any homology-dependent process could carry out the exchange shown in Figure 12-13. However, since transposition of these elements occurs in *recA⁻* cells, the homology-dependent recombinase would

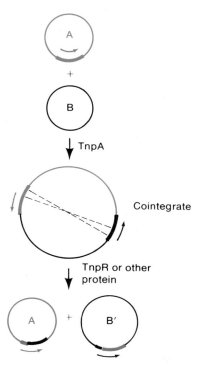

Figure 12-13 A model for transposition utilizing a cointegrate intermediate.

have to be encoded in the transposon. That a homology-dependent process could resolve the cointegrate is supported by the fact that a mutant of Tn3 lacking TnpR or the internal resolution site can carry out the transposition process of Figure 12-10 (that is, reduction of a cointegrate to individual plasmids) in a $recA^+$ host cell.

A more detailed model for transposition is presented in the next section for the reader interested in current speculation. The section may be omitted in most courses.

A Model for Replicative Transposition

The mechanism of replicative transposition is not known, but current proposals abound. The facts that must be accounted for are the following: (1) the retention of the transposon by the donor at the initial site of the transposon; that is, that the process is replicative, (2) the existence of cointegrates, (3) the generation of a short repeated sequence of target DNA on each side of the newly integrated element, and (4) the requirement by some transposons for an internal resolution site or, in others, for a homology-dependent exchange.

Many models for transposition have been proposed; some are simply subtle variations of one another and others are quite different. Two features are common to all models, namely:

1. Two single strands are joined together (ligated) without base pairing to determine the position of joining.
2. A replication fork is created by the ligation event.

The reader should note these features in the model to be described.

The most widely favored model (designed for Tn3) is shown schematically in Figure 12-14. In this model the cointegrate is an essential intermediate.

We consider a donor plasmid containing a transposon and a recipient plasmid. (Neither the donor nor the recipient need to be plasmids or even circular. However, the figure is drawn that way because it is the arrangement most often studied.) In the recipient there is a short base sequence, the target sequence, at which integration of the transposon is to occur. No assumption is made concerning how the transposon and the target sequence are to be linked. The following steps, numbered as in Figure 12-14, have been proposed:

    I. A pair of single-strand breaks is made (by the transposase?) in the target sequence. These breaks are staggered; thus, cleavage of the target sequence yields two complementary single strands.

    II. A pair of single-strand breaks is made in complementary strands at opposite ends of the transposon; this generates two free ends. Each 3' free end is attached by a single strand to a protruding strand (which carries a 5' group) of the staggered cut in the target site; this generates two replication forks.

    III. Replication begins by synthesis of a strand complementary to the protruding end of the target sequence. When replication is completed, a cointegrate is formed containing two copies of the transposon.

    IV. Finally double-strand exchange occurs between the two copies of the transposon. In Tn3, at least, this exchange probably takes place in the internal resolution site. The result of the exchange is to separate the cointegrate into the donor and recipient units, each of which now has a copy of the transposon. In the recipient the transposon is flanked by a direct repeat of the target site; the length of this repeat is the number of base pairs between the staggered cuts.

EXCISION OF TRANSPOSONS

Transposons are sometimes excised from their insertion sites. Whether the process is related to transposition is not known. Excision can be recognized

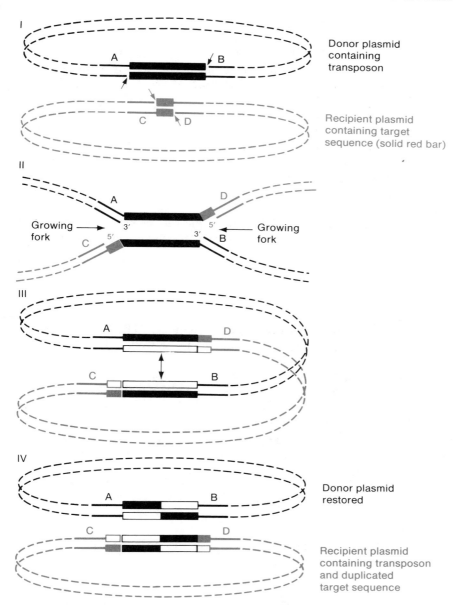

Figure 12-14 A model for transposition proposed by J. Shapiro. The model is drawn for transposition from one plasmid to another, as this is a common arrangement that is studied. (I) Nicks are made at sites in the DNA molecule indicated by arrows. (II) Strand separation between nicks and joining of nonhomologous strands. Two replication forks result. (III) DNA synthesis (hollow lines) forms this structure. A site-specific strand exchange occurs between the internal resolution sites (arrows), or possibly a homology-dependent strand exchange occurs anywhere in the transposon. (IV) Plasmids separate. Capital letters denote base sequences.

in a variety of ways. An example with Tn5, which carries the *kan-r* gene, inserted in the *lacZ* gene, is the following. In one study 19 insertions of Tn5 into *lacZ* were detected by the production of Lac⁻ Kan-r cells. These insertion strains were stable, and mapping experiments showed that the *lac⁻* and *kan-r* mutations were tightly linked. Cultures derived from each of the 19 mutants were prepared and tested for Lac⁺ reversion. These arose at a frequency of 10^{-6} per generation, and 99 percent of the revertants were Kan-s. Thus, associated with recovery of the Lac⁺ phenotype was complete loss of Tn5 from the cell. Interestingly, 1 percent of the Lac⁺ cells were also Kan-r, but the *kan-r* gene was excised from the *lac* site and relocated at a new site. A few percent of these Lac⁺ Kan-r revertants were auxotrophic for various nutrients, as would be expected if re-insertion occurred at random sites.

Excision occurs in *recA⁻* cells, which indicates that excision is not simply the result of homologous exchange between flanking target sequences.

GENETIC PHENOMENA MEDIATED BY TRANSPOSONS IN BACTERIA

Transposons mediate a variety of genetic phenomena such as gene rearrangement and plasmid-chromosome integration. These will be described in this section.

When a transposon is present in a chromosome or plasmid, the frequency of nearby deletions is increased about 100–1000-fold. When such a deletion occurs, it is often found that (1) the transposon is still present, (2) the deletion originates at the end of the element, and (3) the transposon is bounded by only one copy of the target sequence (that is, there is no sequence in direct repetition surrounding the transposon). These features are shown in Figure 12-15.

The frequency of inversions is also enhanced in the vicinity of transposons. In this process, beginning with a sequence

$$\ldots \text{transposon-}a\ b\ c\ d\ e\ f \ldots$$

in which *a–f* are genes adjacent to the transposon, an event occurs that leads to a sequence such as

$$\ldots \text{transposon-}[5]\text{-}b\ a\text{-transposon-}[5]\text{-}c\ d\ e\ f \ldots$$

in which [5] denotes an arbitrary 5-bp target sequence. Note that *a b* is inverted, and that a new target sequence and the transposon are duplicated and surround the inverted sequence. The molecular mechanisms for both deletion and inversion generation are not known but can be explained by the model of Figure 12-14, as shown in Figure 12-16.

In this figure the array shown in Figure 12-14 is repeated—that is, segments A and B are separated by a transposon and segments C and D surround a target sequence, as shown in panels I and 1 of the two figures.

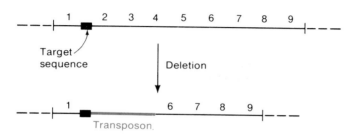

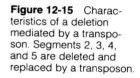

Panel 3 shows the array surrounding the two copies of the transposon after steps I through III of Figure 12-14 are completed. If each of the segments in panel 1 were circles, the two segments in panel 3 would be fused into one circle, as explained in the preceding section. However, if both segments are part of a single DNA molecule (for example, a chromosome), the transposition process would yield the two molecules shown in panel 5. Note that only one molecule in this panel contains the replication origin and that this molecule has also lost the segment containing B and C; that is, this molecule is deleted for B and C. The smaller circle containing B and C does not have a copy of the replication origin and therefore cannot replicate. Thus, if this entire process occurred in a bacterium, the smaller circle would not multiply as cell division occurred and progeny bacteria would contain only the deleted chromosome. Note that the segment deleted is immediately adjacent to one side of the transposon and includes all material between the transposon and the target sequence.

The lower part of Figure 12-16 shows how genetic inversions could be produced. In this case the initial gene array is exactly that which yielded the deletion but the target sequence has an orientation opposite to that of the transposon (that is, the strands ligated in panel I of Figure 12-14 are reversed); this is equivalent to insertion of the transposon in the opposite orientation. The transposition process then gives rise to the arrays shown in panel 3'. Note that the orientations of A and C are inverted with respect to the starting array, as are the sequences C and D. Thus, if the initial segments come from a single DNA molecule, the array shown in panel 6 results. In this case, no DNA is lost (that is, there is no deletion), but the segment between one side of the transposon and the target sequence is inverted with respect to the other segment—that is, there is a **genetic inversion.**

Although most of the transposon-mediated inversions are formed in this way—that is, without the resolution step—it is possible that others are formed by a mechanism using resolution or perhaps homologous *recA*-mediated recombination. A simple mechanism by which this might occur is shown in Figure 12-17. Here transposition occurs at a target sequence (indicated by a square in the left circle of the figure) in the molecule containing the transposon. Insertion can occur in either a direct-repeat orientation (I in the figure) or an inverted repeat (II in the figure). With a transposon capable of

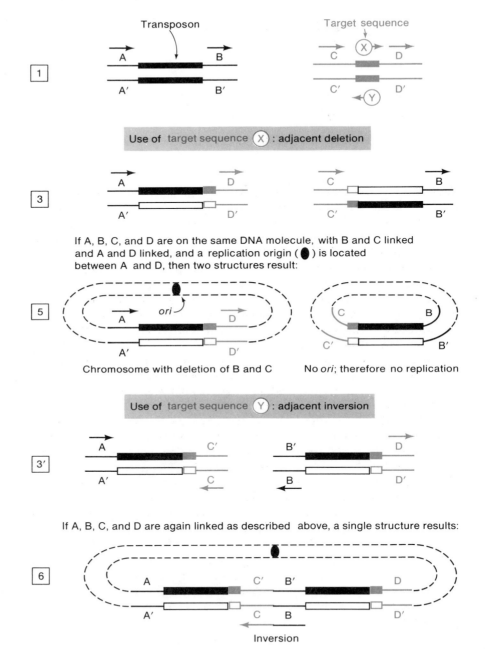

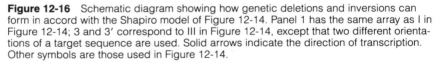

Figure 12-16 Schematic diagram showing how genetic deletions and inversions can form in accord with the Shapiro model of Figure 12-14. Panel 1 has the same array as I in Figure 12-14; 3 and 3′ correspond to III in Figure 12-14, except that two different orientations of a target sequence are used. Solid arrows indicate the direction of transcription. Other symbols are those used in Figure 12-14.

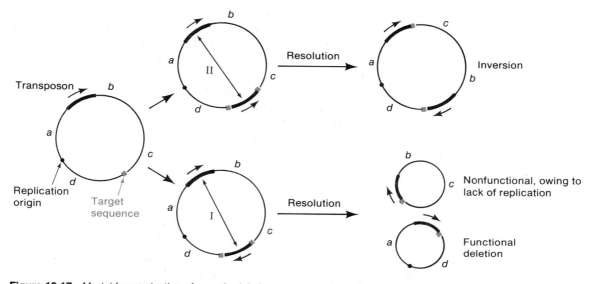

Figure 12-17 Model for production of genetic deletions and inversions. The transposon DNA is inserted into the target sequence in orientation I or II. The circle could be a plasmid or the chromosome. Resolution by a site-specific strand exchange at sites indicated by double-headed arrows or by exchange in homologous sequences yields a deletion from I and an inversion from II.

resolution a site-specific exchange could occur within the two transposons; alternatively, a homology-dependent exchange mediated either by a transposon gene or by the RecA protein, if the cell is RecA$^+$, may occur. If the transposons are in a direct-repeat array, the exchange would generate two circles. As in the model shown in Figure 12-16, since the chromosome or a plasmid generally contains a single replication origin, only the element possessing the origin can persist; this element has lost a DNA segment and thus is a deletion mutant. If the transposons are in the inverted-repeat array, the result is instead an inverted sequence, as shown in the right portion of the figure. Note that the end results of the schemes shown in Figures 12-16 and 12-17 are the same.

The models in the figures just shown are not necessarily the correct explanation for the phenomena since, as mentioned earlier, the model in Figure 12-14 is only one of many that have been suggested for replicative transposition. Other models are also able to explain deletions and inversions.

E. coli PHAGE MU

E. coli phage Mu is a temperate phage capable of both lytic and lysogenic growth. Its name, Mu, comes from *mu*tator inasmuch as about 3 percent of all Mu lysogens are mutant for some easily recognizable gene. The mutations arise because Mu inserts its DNA at randomly distributed sites in the *E. coli* chromosome and often these sites are within *E. coli* genes or regulatory elements, which are inactivated when a Mu prophage divides the gene or element. Another peculiarity of Mu is that insertion of DNA into the host chromosome is an obligatory stage in the lytic life cycle. Furthermore, insertion always results in a duplication of a target sequence; this and other observations make it clear that Mu is a giant transposon that has acquired phage functions that enable it to be packaged in a phage coat, lyse its host, and thereby leave the cell.

Mu DNA

Mu DNA is linear, consisting of about 38,000 base pairs. However, in a population of Mu phage particles each DNA molecule is different. This is made evident when a sample of Mu DNA is denatured and then renatured. Electron microscopy of the renatured double helices, which invariably contain two single strands derived from different double-stranded molecules, yields the two structures shown in Figure 12-18. All single-stranded regions consist of sequences that are noncomplementary. Further experiments have indicated the following:

1. The noncomplementary sequences at the termini are bacterial DNA segments, which differ from one DNA molecule to the next.
2. If a segment of Mu DNA in the α region is increased in length (by an insertion of foreign DNA), the nonhomologous region at the left (c) end retains the wild-type length whereas the segment at the right (S) end becomes shorter.
3. Renatured molecules may have an unpaired region called the **G loop.** The two strands of the G loop, which are phage DNA, are identical but one is reversed in direction with respect to the other. That is, a

Figure 12-18 The major structures of Mu DNA seen after denaturation and renaturation. The symbols α, β, G, and S refer to distinguishable regions of the heteroduplexes.

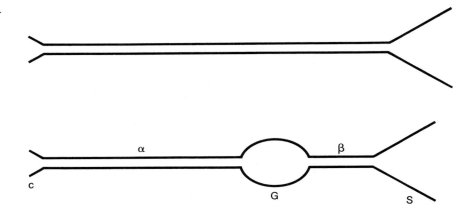

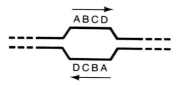

Figure 12-19 The structure of a G loop (Figure 12-18) of Mu DNA showing inversion of the G sequence denoted by ABCD.

G loop has the structure shown in Figure 12-19. For unknown reasons, occasionally in the phage life cycle the G segment becomes inverted by genetic recombination. The normal and inverted configurations are called *flip* and *flop* and annealing of a strand containing *flip* with a complementary strand containing *flop* yields a G loop. The G loop is not relevant to the transposition phenomenon and will not be discussed further.

Replication and Maturation of Mu DNA

Points 1 and 2 in the preceding section are explained by the fact that transposition is obligatory in Mu DNA replication. Shortly after infection, the incoming Mu DNA becomes inserted at a random position in the *E. coli* chromosome. The terminal bacterial sequences are not inserted: only the Mu segment is inserted, as is usually the case in transposition. Note that this transposition process is different from those seen earlier in the chapter in that the transposition occurs from a linear DNA molecule (the incoming phage DNA) to a circular molecule (a plasmid or the chromosome). Theories abound for the mechanism of this transposition, but each theory must at the present time be considered speculative. Replication of Mu DNA proceeds by repeated transposition to various sites in the chromosome, which explains the association of Mu DNA with various pieces of bacterial DNA described in point 1. Genetic evidence indicates that replication and insertion are coupled: Mu mutants unable to replicate their DNA also cannot insert, and mutants unable to insert are unable to replicate.

Numerous physical experiments show that insertion of progeny Mu DNA at various sites occurs throughout the lytic cycle. However, this was first indicated by several genetic experiments, one of which was the following. F'*lac*-containing cells were infected with a Mu mutant that was unable to kill the cells. Shortly after infection the cells were mated with appropriate females, and F'*lac* transfer was scored. The transferred F'*lac* often contained integrated Mu DNA, and the insertion site was not fixed. Use of various F' plasmids indicated that Mu could insert in various parts of the DNA.

Mu DNA is never found free in the cell. About halfway through the infectious cycle, circular DNA molecules are found that contain Mu plus

Figure 12-20 Diagram of inserted Mu DNA (red) showing how the cutting process generates host sequences at the termini.

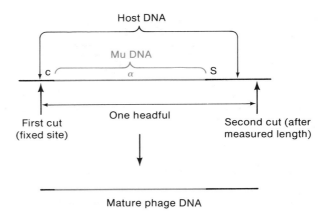

bacterial sequences. The role of these molecules in the life cycle is unknown, but the existence of variable bacterial segments supports the view that Mu inserts throughout the chromosome. As mentioned earlier, the DNA in the phage particles always contains terminal bacterial DNA. These sequences arise from the packaging process. Packaging of the DNA occurs in the following way (Figure 12-20): The phage-DNA maturation system recognizes the c end of the integrated Mu DNA and makes a cut in the host DNA about 100 bases to the left of the phage DNA. Packaging begins at this point and the phage head is filled. The second cut, at the S end, is determined by the amount of DNA that can fit in the phage head. One headful exceeds the length of the phage DNA sequences, which explains point 2. In Chapter 15 we will see that phage T4 also uses a headful packaging mechanism.

Enumeration of the location of inserted Mu in Mu lysogens indicates that the number of possible insertion positions is huge. If a culture is lysogenized by Mu and Mu-induced mutations are scored, it is found that the frequency of insertion in a particular gene is roughly proportional to the size of the gene. That is, if gene a is twice as large as gene b, there will, on the average, be twice as many a^- mutants as b^- mutants. However, collections of insertions into a single gene show that all sites are not equally probable. For example, among 75 insertions in the *lac*Z gene only 50 insertion sites were seen, and they tended to be in one region of the gene. Since the gene contains 3063 base pairs, one would expect that if insertion occurred at random, no insertion sites would be duplicated. No particular base sequences were seen at the insertion sites.

TRANSPOSONS AND EVOLUTION

Apart from conferring antibiotic resistance, transposons do not seem, at first glance, to be of any value to a cell. This point may be made even more strongly for the IS elements. Thus, it is reasonable to ask why transposons have been maintained over evolutionary time. Generally, a gene that serves no purpose to a cell will ultimately be mutated until it is unrecognizable, nonfunctional, and often deleted. For some time, it has been thought that transposons are an example of what has been termed "selfish" DNA—that is, a DNA segment that has persisted only because it has evolved to maintain itself. For example, a cell possessing a copy of a useless IS element will keep the transposon if transposition, which duplicates the element, occurs at a higher rate than mutation or loss of the element. This argument may indeed explain the persistence of many transposons. However, recent experiments have suggested that whereas a transposon does not provide a selective advantage to an individual cell, its presence will in the long run confer an advantage to the cell population. The argument is based on reasoning that since evolution proceeds by accumulating mutations a process that accelerates mutation may speed up evolution.

In Chapter 4 continuous growth of bacteria in a chemostat was described. Chemostat cultures are useful in studying mutation rates and evolution, because both constant conditions and cell multiplication can be maintained indefinitely. Several laboratories have compared the multiplication rate of a cell that contains a transposon (Tn5 and Tn10 have been studied) with an isogenic cell that lacks the transposon. The generation times of these cultures determined in the standard way are not measurably different in cultures grown in the standard way. However, in a chemostat it is possible to prepare an initial mixture consisting of, for example, 10^5 wild-type *E. coli* and one cell possessing Tn10. After prolonged growth (hundreds of generations) the ratio of transposon-containing cells to wild-type cells can be determined. Study of numerous chemostat cultures showed that some cultures had maintained the original 10^{-5} ratio but others consisted mainly of cells containing Tn10. No antibiotic was present, so cells containing Tn10 did not have any selective advantage for this reason. Cultures in which the Tn10-containing cells had "won" were examined for transposition to determine whether a transposition event accounted for the more rapid growth. The number of copies of the *tet-r* gene present in Tn10 had not increased, but the number of copies of the IS10R terminus of Tn10 had usually increased by one. (Tn10 is a composite transposon with two very slightly different forms of IS10—IS10L and IS10R—at the left and right termini, respectively.) No increase in either Tn10 or IS10R was observed in cultures in which the wild-type cells had won. Furthermore, biochemical analysis showed that the IS10R insertion in the Tn10-containing cells was localized in a fairly small region of the bacterial chromosome. Thus, this insertion was associated with a slightly higher growth rate.

The interpretation of this and related experiments is that transposition produced beneficial mutations—that is, cells containing the transposons evolved faster. A similar effect has also been observed with cells containing mutator genes (Chapter 10), that is, genes that confer on the cell an enhanced mutation rate.

It is clear that a significantly enhanced mutation rate produced by a transposon could also be deleterious. In bacteria, since transposition is mainly a replicative process, the number of elements would seem to be capable of increasing indefinitely. This would be harmful to the host cell and hence to the transposon, since ultimately the host genome would be damaged beyond its ability to carry out necessary cellular functions. Thus, it seems likely that transposition would be limited in some way, and indeed the existence of an as-yet-unknown regulatory system has been inferred. Evidence for such a system comes from experiments in which either plasmid transfer, phage infection, or $CaCl_2$ transformation is used to place a bacterial transposon in a cell that has no copies of the element. Transposition begins within several generations and occurs repeatedly. Soon the rate of transposition decreases and ultimately the cell acquires a stable number of copies of the transposon; the particular number is characteristic of each element. The mechanism of this inhibition of transposition is obscure.

The number of copies of a particular transposon in a bacterium is usually less than ten. However, in the higher eukaryotes, in which up to 30 percent of the DNA may consist of transposable elements, the number of copies of an individual element may be in the hundreds of thousands.

PROBLEMS

1. What two components are characteristic of transposable elements?

2. If you were determining the base sequence of a long DNA segment, what features of the sequence would make you confident that a transposable element was present?

3. A particular transposable element is observed at five different locations. Will the flanking target sequences be the same at each site?

4. A transposable element is observed at several different sites in different bacterial strains. Each element is flanked by a pair of target sequences. What characteristic is common to these target sequences?

5. What is a typical length of a target sequence?

6. What is a composite transposon?

7. Does a composite transposon necessarily have terminal inverted repeats?

8. Are all drug-resistance genes located on transposons?

9. Why does the presence of an IS element in a particular genetic region often cause polarity?

10. What is the evidence that transposition in bacteria includes a replicative step?

11. When transposition occurs in prokaryotes, two base sequences are duplicated. What are they?

12. When transposition occurs, is there any site selectivity with respect to the new site?

13. What term is used to describe a plasmid that results from transposon-mediated fusion of two plasmids, one of which contained a transposon? How many copies of the transposon are present in the fused molecule?

14. In the Tn3 system what enzyme is responsible for formation of a cointegrate?

15. In the Tn3 system what enzyme breaks down a cointegrate into two independent replicons? How many copies of the transposable element exist after action of this enzyme?

16. Name five genetic phenomena mediated by transposable elements.

17. In homologous recombination two DNA molecules break and rejoin. If the event is physically reciprocal, the amount of DNA present is the same as that present before recombination; if the event is nonreciprocal, the total amount of DNA present decreases. What can be said about the relative amounts of DNA before and after transposition?

18. Transposition is a type of nonhomologous exchange, suggesting that matching of sequences is not part of the process. However, there seems to be some sequence effect in that transposition occurs to some sites more than to others. The mechanism of this effect is unknown. Suggest what type(s) of things might account for this effect.

19. Explain how insertion of a transposon that lacks a transcription-termination signal can cause genetic polarity.

20. You have isolated a segment of DNA having about 500 base pairs. Several genetic experiments have suggested but have not proved that the segment contains a transposon. What physical or chemical experiment would you do to test the hypothesis?

21. An Amp-r plasmid whose replication is temperature-sensitive is introduced into an Amp-s cell by the $CaCl_2$ transformation method. After growth for several generations 10^7 cells are plated on agar containing ampicillin at 42°C. Fifty colonies form.

(a) Give three mechanisms which could explain the presence of these colonies.
(b) Which mechanisms would not occur in a RecA⁻ cell?

22. You have isolated two different insertion mutations in phage λ. How would you determine whether the two insertions are the same transposon?

23. A particular IS element has inserted into different positions in two λ phages. In one case the insertion is at position 55; in the other, at position 75 (both measured on a scale of 100 from the left end of the phage. The IS element has a length equal to 1 percent of the size of λ DNA. One of these phage variants shows polar effects on downstream genes; the other does not.

(a) Explain the difference in the polar effects.
(b) How could your explanation best be tested by electron microscopy?

24. Two bacterial genes (*chk* and *lat*) are very near one another. When chk^- mutants are isolated (they arise spontaneously at a frequency of about 10^{-6} mutants per cell per generation), about 1% of these mutants are also lat^-. This is a surprisingly high frequency for spontaneous formation of double mutants.

(a) Could you guess what might be the cause of this phenomenon?

(b) Would a measurement of this frequency in a $RecA^-$ cell be informative? Explain.

(c) Would studying the effect of chemical mutagens be informative? Explain.

25. It is sometimes desirable to remove a transposon from a bacterium. This is possible because excision of transposons occurs spontaneously, though at a very low frequency and by an unknown mechanism. If a transposon were inserted in the *lac* gene, one would only have to obtain a Lac^+ revertant colony to obtain cells that had lost the transposon. In some genes, revertants are not detected in such a straightforward way; however, the following technique is successful for transposons carrying some antibiotic-resistance markers. Consider a culture of cells in which a transposon carrying the *tet-r* gene is present in gene *x* rendering the cells X^-. Tetracycline is added to a growing culture and a few minutes later penicillin is added. Penicillin only kills growing cells. The Tet-s cells are inhibited by the tetracycline. Tet-r cells grow in the presence of tetracycline and hence are killed by the penicillin. After many hours of growth the cells are plated on agar lacking tetracycline. At a low but easily detected frequency colonies form whose cells lack a functional *tet-r* gene. Describe three mechanisms for generating these Tet-s cells and indicate which result in transposon loss.

26. When Mu infects a bacterium, DNA is inserted into the bacterial chromosome. Is all of the DNA inserted?

27. If a single burst experiment were done with Mu, would all progeny in a particular burst have the same bacterial sequences?

28. If Mu DNA is denatured and renatured and examined by electron microscopy, double-stranded molecules with unpaired ends result. Would you ever expect to see molecules without frayed ends?

29. Since Mu obligatorily inserts into the bacterial chromosome during its replication cycle and since insertion frequently occurs in a bacterial gene, thereby inactivating the gene, should one conclude that most infections do not lead to phage production?

Chapter 13

Bacterial Transformation

Bacterial transformation is a process in which a recipient cell acquires genes from free DNA molecules in the surrounding medium. In the laboratory, transfer is accomplished by isolating DNA from donor cells and then adding this DNA to a suspension of recipient cells; however, transformation also occurs in nature. At one time transformation was one of the more important mapping techniques in microbial genetics. Classical transformation is not often performed today, but for certain organisms it remains the only means of genetic mapping. A variant of the classical technique, in which plasmid or phage DNA is transferred to a recipient cell, is exceedingly important in genetic engineering (Chapter 20).

THE DISCOVERY OF TRANSFORMATION

The discovery of bacterial transformation was one of the most important events in biology, for it provided the foundation for modern microbial and molecular genetics. This discovery led to the demonstration that DNA is the genetic material.

Bacterial pneumonia in mammals is caused by certain strains of bacteria of the genus *Pneumococcus* (properly, *Streptococcus pneumoniae*). Cells of these strains are surrounded by a polysaccharide capsule that protects the bacterium from the immune system of the infected animal and thereby enables the bacterium to cause disease. When a pathogenic (disease-causing) *Pneumococcus* is grown on a nutrient agar, the enveloping capsule gives the bacterial colony a glistening, smooth (S) appearance. Some mutant strains of *Pneumococcus* lack the enzyme activity required to synthesize the capsular polysaccharide, and these bacteria form colonies that have a rough (R) surface. R strains do not cause pneumonia, because without their capsules the cells are inactivated by the host immune system.

Both R and S strains of *Pneumococcus* breed true, except for rare mutations in R strains that give rise to the S phenotype and rare mutations in S strains that give rise to the R phenotype. Since simple mutations can cause conversion between the forms, it was recognized quite early that the S and R phenotypes are determined by different allelic forms of a simple genetic unit.

In 1928 a seminal observation was made by Fred Griffith: mice injected with either living R pneumococci or with heat-killed S cells remained healthy, but mice injected with a *mixture* containing a small number of R bacteria and a large number of heat-killed S cells died of pneumonia. Bacteria isolated from blood samples of the dead mice produced *pure* S cultures having a capsule typical of the heat-killed S cells (Figure 13-1). Evidently, dead S cells could in some way provide the living R bacteria with the ability to withstand the immunological system of the mouse, multiply, and cause pneumonia. Furthermore, cultures derived from these changed or **transformed** bacteria retained the ability to cause pneumonia.

Figure 13-1 Production of disease-producing type S *Pneumococcus* by infecting a mouse with a mixture of living type R cells, which alone do not cause disease, and heat-killed type S cells.

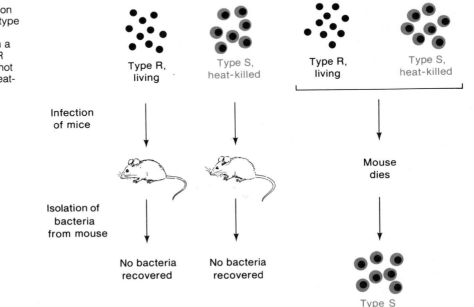

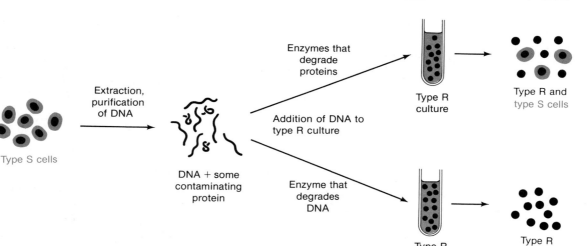

Figure 13-2 A diagram of the experiment that demonstrated that DNA is the active material in bacterial transformation.

Later experiments showed that living S cells can also be produced (at low frequency) when R cells are grown in a liquid medium containing heat-killed S cells. Furthermore, extracts of S cells, from which intact cells and large pieces of cellular debris (mostly cell walls and membranes) had been removed by filtration, were as effective as intact S cells in causing specific transformation of R cells to the S phenotype. In 1944 Oswald Avery, Colin MacLeod, and Maclyn McCarty carried out a critical experiment that led to our current understanding of the phenomenon. They purified DNA from S cells and found that addition of minute amounts of this DNA to growing cultures of R cells consistently resulted in production of smooth colonies whose cells contained type-S capsular polysaccharide. Their DNA samples contained traces of protein and RNA, which complicated the interpretation of the data somewhat. However, the transforming activity was not altered by treatment with enzymes that degrade proteins or by treatment with RNase, but was completely destroyed by DNases (Figure 13-2). These experiments, and several more sophisticated variants done over the following few years, demonstrated that the substance responsible for genetic transformation was the DNA of the donor cells and, hence, that DNA is the genetic material.

DETECTION OF TRANSFORMATION

Genotypic transformation is detected by standard plating tests. For example, purified DNA obtained from an erythromycin-resistant (*ery-r*) culture of

Pneumococcus is mixed with cells from an *ery-s* culture. After a period of incubation plus time for phenotypic expression of drug resistance the cells are plated on agar containing erythromycin. Formation of Ery-r colonies above the mutation frequency for Ery-s to Ery-r (about 10^{-8}) indicates that transformation has occurred. For *Pneumococcus* the maximum frequency of transformation for most antibiotic-resistance markers is 0.1–1 percent.

COMPETENCE

Transformation begins with uptake of a DNA fragment (the bacterial chromosome is usually broken extensively during isolation) from the surrounding medium by a recipient cell and terminates with recombinational exchange of part of the donor DNA with the homologous segment of the recipient chromosome. Probably, most bacterial species are capable of the recombination step, but the ability of most bacteria to take up DNA efficiently is limited. Even in a species capable of transformation, DNA can penetrate only a very small fraction of the cells in a growing population. However, incubation of cells of these species *under certain conditions* yields a population of cells in which the uptake of DNA is greatly enhanced, by a factor of 10^4–10^6. A culture of such cells is said to be **competent.** The conditions required to produce competence and the fraction of the cells that are competent vary from species to species.

A great deal of effort has gone into understanding the mechanism of competence, but still the phenomenon is far from clear. Competence appears to result from changes in the cell wall of the bacteria and probably is associated with the synthesis of cell-wall material at a particular stage of the life cycle of the bacterium. In the course of developing competence, receptors of some kind are either formed or activated on the cell wall; these receptors are responsible for initial binding of the DNA. The number of active receptors varies widely from one organism to the next—for example, about 80 for *Pneumococcus*, 50 for *Bacillus subtilis*, and 4 for *Hemophilus influenzae*. Competence usually arises at a specific stage of growth of a culture, typically late log phase, and is highly dependent on the growth medium and the degree of aeration of the culture. The particular medium and growth conditions that produce maximum competence vary from one bacterial species to the next. With *Pneumococcus* competence is generated just as the cells are entering stationary phase, but the competent state persists for only a few minutes (Figure 13-3). Competence initially arises in only a small fraction of the cells. Associated with early development of competence is the excretion by these cells of one or more proteins called **competence factors.** These proteins convert the remainder of the cells in the population to the competent state, probably by some direct action on the receptors. Purified competence factors can also induce competence in most cultures in a way that does not depend strongly on the growth phase of the culture, which suggests

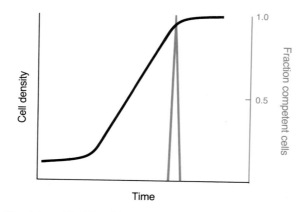

Figure 13-3 Time interval (red) in which cells of *Pneumococcus* are competent.

that the medium, temperature, degree of aeration, and so on, needed to produce a competent culture probably serve only to trigger synthesis or release of competence factors by a small number of cells. Competence factors are species-specific and induce competence only in very closely related species. The mechanism of action of competence factors is unknown.

The fraction of a culture that becomes competent and the duration of the competent state are also species-dependent. For example, conditions have been found that produce 100 percent competence in *Pneumococcus*, but competence persists for only a few minutes. In contrast, with *B. subtilis* no more than 20 percent of the cells become competent, but the state lasts for several hours. Possibly the inability of most bacterial species to be transformed is only a reflection of lack of knowledge of the treatments necessary to induce competence.

The ability of a cell to take up DNA can also be elicited in a wide variety of bacterial strains by the $CaCl_2$ transformation technique described in Chapter 11. However, the state of the cells produced by this treatment is quite different from that of naturally competent cells. Whereas almost any DNA molecule, of a wide range of sizes and either fragmented or intact, can be taken up by a cell treated with $CaCl_2$, stable genotypic change usually only occurs with intact replication systems such as plasmids and phage DNA molecules. We will see shortly that the mechanism of DNA uptake is quite different in competent cells versus $CaCl_2$-treated cells.

DNA UPTAKE

DNA uptake by competent cells has been examined in two ways. In one type of experiment cells are exposed to radioactive DNA and then to a DNase capable of degrading DNA completely. These experiments indicate that there are at least two stages of interaction of DNA with competent cells. The first is a brief reversible-binding stage in which the DNA can be washed from the cells and in which the DNA is also completely sensitive to DNase degradation. In the second, or irreversible, stage DNA cannot be washed from the cells and it is resistant to DNase. Irreversible interaction is an energy-requiring process that is associated with passage of the DNA through the cell membrane (to be discussed shortly).

Although stable genotypic transformation is highly species-specific, DNA uptake is normally not. This might be expected, since all DNA molecules probably look the same to the outside of a bacterium. Since uptake of foreign DNA cannot be detected by genetic transformation, a competition assay has been used to show that a particular DNA molecule can be taken up by a competent cell. In one such experiment different samples of penicillin-sensitive (Pen-s) *Pneumococcus* were mixed with a fixed amount of DNA isolated from a Pen-r culture and with various amounts of calf thymus DNA. Increasing amounts of calf thymus DNA reduced the number of Pen-r transformants (Figure 13-4). The foreign DNA bound to the receptor sites and competitively inhibited uptake of *Pneumococcus* DNA. An exception to the rule of non-specificity of DNA uptake is found in *Hemophilus influenzae*, which only takes up *Hemophilus* DNA. Apparently, particular sequences of about ten base pairs, which are rare in other DNAs, are needed for the initial binding to the cell surface. The mechanism of transformation of *Hemophilus* differs from that in most other bacteria in other ways also.

Denaturation of a DNA sample capable of transformation eliminates all transforming activity. Also, addition of competing but denatured calf thymus DNA does not inhibit transformation of *Pneumococcus* by *Pneumococcus* DNA (Figure 13-4), so presumably single-stranded DNA does not bind to

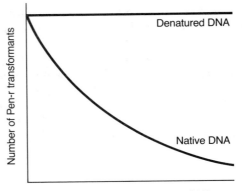

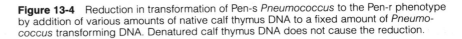

Figure 13-4 Reduction in transformation of Pen-s *Pneumococcus* to the Pen-r phenotype by addition of various amounts of native calf thymus DNA to a fixed amount of *Pneumococcus* transforming DNA. Denatured calf thymus DNA does not cause the reduction.

the receptors of competent cells. Studies with denatured radioactive *Pneumococcus* DNA confirm this conclusion. Proof that it is only the denatured state that is responsible for the lack of uptake comes from renaturation experiments. If the denatured DNA is renatured, both transforming activity and the ability to compete with normal transforming DNA are restored.

A simple genetic test shows that the state of DNA is altered during uptake. A Str-s culture of *Pneumococcus* is exposed to DNA from a Str-r strain. After 15 minutes the cells are washed free of excess unincorporated DNA, and divided into two portions. After a period of time sufficient to allow expression of the *str* gene, one part is plated on medium containing streptomycin, and a significant fraction of the cells prove to be transformed to the Str-r phenotype. DNA is isolated from the remainder of the 15-minute sample and tested for its ability to transform a fresh culture of competent Str-s cells, and indeed transformation to streptomycin-resistance occurs. Thus, transforming DNA is present in the 15-minute sample. In a related experiment cells are plated and DNA is isolated only 2 minutes after initial exposure to the DNA rather than after 15 minutes. In this case, plating yields the same number of Str-r colonies, indicating that washing the cells at this time (2 minutes) does not affect the transformation process. However, the DNA isolated from the cells shows no transforming activity. In summary, active transforming DNA is adsorbed to and enters competent cells; at first (2 minutes) it is in a non-transforming state, after which (before 15 minutes) it is converted to a transforming state. The period during which no transforming DNA can be isolated from potentially transformed cells (about the first 10 minutes) is called **eclipse.** Physical studies with radioactive transforming *Pneumococcus* DNA have shown what is happening. Associated with irreversible binding and DNA uptake is degradation of one strand of the transforming DNA by a nuclease contained in a multiprotein assembly of the cell membrane (Figure 13-5). Similar results have been obtained with *B. subtilis*.

A great many physical experiments indicate that only a single strand of

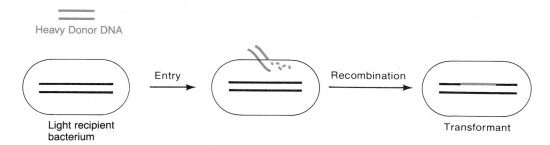

Figure 13-5 An experiment showing that in *Pneumococcus* one strand of transforming DNA is destroyed and the complementary strand is inserted. This conclusion is based on the finding that transformants contain hybrid (heavy-light) segments but no fully heavy segments.

DNA enters a competent *Pneumococcus* cell. However, none of these experiments show whether the partial degradation is essential for DNA uptake. As discussed in Chapter 1, a common approach to answer such a question is to attempt to isolate a degradation-deficient mutant and determine whether it can still take up DNA. Such a mutant has been isolated on a color-indicator plate, using a principle quite different from that used to identify sugar-utilization mutants—namely, a color change that occurs when the dye methyl green binds to DNA. Colonies formed on agar containing both methyl green and free DNA are surrounded by a ring whose color differs from that of the bulk of the agar. This color difference results from the excretion of DNases by aging cells in mature colonies. The DNase degrades the DNA in the agar, causing the methyl green to changes color locally to that of DNA-free dye. When a mutagenized *Pneumococcus* culture was plated on DNA-methyl green agar, some colonies were found that lacked the colored ring. These mutants were termed Noz$^-$. Biochemical analysis of Noz$^-$ cells showed that they lack a particular membrane nuclease, they do not degrade transforming DNA to the single-stranded form, and transforming DNA is not taken up by the cell. The properties of these mutants strongly suggest that degradation of one of the DNA strands is essential for DNA uptake in *Pneumococcus*.

With *Hemophilus influenzae* an eclipse period does not occur and degradation does not accompany DNA uptake. In fact, many studies indicate that transformation in *Hemophilus* proceeds by a mechanism that is quite different from transformation in other bacteria; the mechanism is, however, not known in any detail.

MOLECULAR MECHANISM OF TRANSFORMATION

In attempting to understand transformation in *Pneumococcus* two possible mechanisms were originally considered: (1) the incoming donor allele replaces the allele of the recipient, that is, DNA substitution occurs, or (2) the DNA fragment containing the donor allele is added to the genome of the recipient, by insertion adjacent to the recipient allele, thereby forming a gene duplication (partial diploid). That the first mechanism is correct was first demonstrated by a study of reciprocal transformation. First, an x^- strain was transformed with DNA from a strain carrying a dominant x^+ allele. Then, the newly formed x^+ strain was exposed to DNA from an x^- strain, and it was found that x^- transformants arose at the same frequency as did x^+ transformants in the first experiment. If DNA addition occurred, the initial transformant would have had the genotype x^+x^-. If DNA from this transformant were used as donor DNA, a transforming DNA fragment would almost always carry both the + and − alleles, so x^- cells would be exceedingly rare in the second transformation. Thus, it was clear that DNA addition does not occur in transformation.

Another type of experiment using two very closely linked markers made the point even more convincingly. In this experiment the recipient bacterium

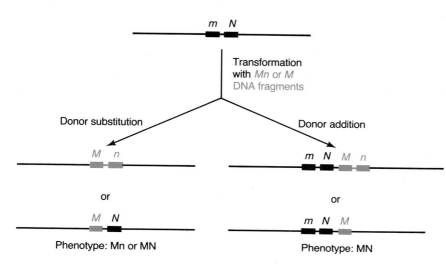

Figure 13-6 An experiment that shows that transformation occurs by donor substitution rather than donor addition. If the donor is *Mn* and the recipient is *mN*, transformants selected for the M phenotype are either *Mn* or *MN*. The addition model predicts that only *MN* will be recovered.

(recipient 1) contained two closely linked loci *m* and *n* for antibiotic sensitivity, and the donor was resistant to both, having the genotype *MN*. Three genotypes, *Mn*, *nM*, and *MN* were found with roughly the same frequency, which shows that *M* and *N* are linked (i.e., often carried on the same DNA molecule—discussed in a later section). In a second experiment donor DNA with the genotype *Mn* was used to transform an *mN* recipient, and the *M* marker was selected. If the mechanism of transformation were donor addition, the genotypes of *M* recipients would be *MmN* and *MmNn*, both of which would have the N phenotype (Figure 13-6). However, colonies with the MN and Mn phenotypes were equally frequent, which agrees with the allelic substitution idea. That is, the transformation event that resulted in incorporation of the *M* marker was often associated with disappearance of the *N* marker of the recipient and replacement by *n*, which was carried on the same DNA molecule.

As stated earlier, in transformation with *Pneumococcus* (the best-understood case) uptake of the transforming DNA into the cell is accompanied by digestion of one of the input strands. Consequently, only a single strand is available for interaction with the recipient DNA. The single strand is, in an unknown way, protected from nuclease attack—possibly it is coated with the same protein (SSB protein) used to maintain single-stranded regions in a replication fork (Chapter 8). Physical experiments in which high-density-labeled (heavy) DNA is used to transform a bacterium whose DNA has low density (light DNA) show that a single strand of the donor DNA, or a portion thereof, is linearly inserted into the recipient DNA (Figure 13-5). The currently accepted model for this process is shown in Figure 13-7. Aided by a bacterial protein, probably equivalent to the *E. coli* RecA protein, which facilitates DNA pairing in recombination (Chapter 14), the incoming single-

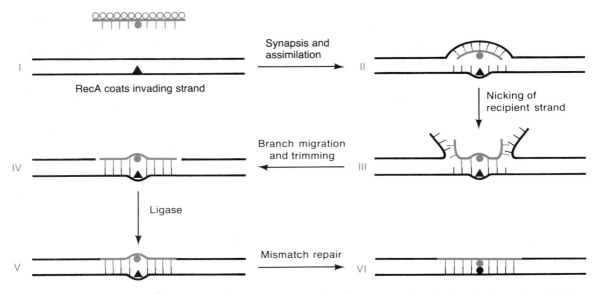

Figure 13-7 Bacterial transformation. The accepted mechanism for integration of single-stranded donor DNA.

stranded fragment causes local unwinding and invades the DNA of the recipient, presumably from the 5′ end. By an unknown mechanism the displaced single strand is cut. Unwinding of the recipient DNA continues at the end of the assimilated DNA, which allows the fraction of the invading strand that is base-paired to the recipient strand to increase. (This process is called **branch migration,** since portions of the unpaired strands move.) Trimming nucleases act simultaneously to remove free ends, which may be in donor or recipient DNA, and ultimately DNA ligase seals the nicks. The result is a heteroduplex region containing a mismatched base pair. The outcome of the process—that is, whether the donor marker is or is not recovered in the progeny cells—depends on whether mismatch repair occurs and, if so, whether the unpaired base in the donor or recipient strand is removed. For some markers, known as high-efficiency markers, either the mismatch is always corrected to match the donor genotype or there is little or no repair. In the latter case, following replication of the chromosome and cell division one cell has the donor genotype and the other has the recipient genotype. Because the plating conditions used to detect recombinants are usually chosen to allow growth only of recombinants (for example, an antibiotic is added or a necessary amino acid is absent), the colony that forms consists exclusively of recombinant bacteria. For the low-efficiency markers, mismatch repair usually removes the mismatched base from the donor and the cell retains the recipient genotype. Proof that this is the correct explanation for the two types of markers comes from the isolation of *Pneumococcus* mutants, called Hex⁻, that have a higher spontaneous mutation frequency. These mutants seem to have normal editing activity and methylate DNA at the normal rate and hence are thought to be defective in mismatch repair. When Hex$^-$ mutants are transformed, all markers appear to be transformed as high-efficiency markers, probably because one of the two daughter cells following cell division always has the donor genotype.

MAPPING BY TRANSFORMATION

Transformation is the only technique available for gene mapping in some species. The technique is different from, but related to, that described in Chapter 1. For example, genetic mapping is usually accomplished by measuring recombination frequencies between markers. However, in transformation several effects make frequency measurement an unreliable procedure. The main problem is that the probability of recovery of a donor marker in a recipient depends on the molecular weight of the donor DNA fragment and, more significantly, on the marker itself. In the previous section we saw the distinction between a low-efficiency and a high-efficiency marker. Since this distinction is based on the probability of mismatch repair in a particular direction, which is probably determined by local base sequences (among other things), different alleles of the same gene are incorporated at different frequencies. Furthermore, reciprocal transformations do not yield the same transformation frequencies; that is, transformation of a donor A allele to a recipient a strain will not, in general, occur at the same frequency as transformation of a donor a allele to an A strain, because of differences in mismatch repair. Clearly, these effects cause the recombination frequency between a donor marker and a nearby resident marker to depend on features other than the distance between the markers, so recombination frequency cannot be used to generate a meaningful map.

Mapping by transformation experiments is based instead on the principle that *two markers will transform together if they are near enough to be carried on the same DNA fragment*. When DNA is isolated from a donor bacterium, each chromosome is invariably broken into several hundred small fragments. With highly competent recipient cells and excess DNA, transformation of most genes occurs at a frequency of about one cell per 10^3. If two genes a and b are so widely separated in the donor chromosome that they are always carried on two different DNA fragments, then at the same total DNA concentration the probability of simultaneous transformation (**cotransformation**) of an $a^- b^-$ recipient to wild type is roughly the square of that number, or one $a^+ b^+$ transformant per 10^6 recipient cells. However, if the two genes are so near one another that they are often present on a single DNA fragment, the frequency of cotransformation will be nearly the same as the frequency of single-gene transformation, or nearly one wild-type transformant per 10^3 recipients. Thus, *cotransformation at a frequency substantially greater than the product of the two single-gene transformations implies physical proximity of two genes*. Studies of the ability of various pairs of

Figure 13-8 Cotrans-
formation of linked mark-
ers. Markers *a* and *b* are
near enough to each other
that they will often be
present on the same frag-
ment, as are markers *b*
and *c*. Markers *a* and *c*
are not. The gene order
must therefore be *a b c*.

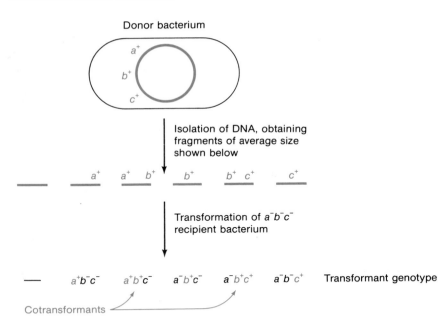

Note: No $a^+b^+c^+$ or $a^+b^-c^+$ cotransformants

genes to be cotransformed yields gene order. For example, if genes *a* and *b*,
and genes *b* and *c*, can be cotransformed, but genes *a* and *c* either cannot,
or cotransform at exceedingly low frequency, then the gene order must be
a b c (Figure 13-8).

The use of cotransformation frequencies in mapping is not always
straightforward. For example, when all or most cells in a population are not
competent, the existence of a high transformation frequency is not sufficient
to conclude that two markers are linked. This can be seen in the following
example. Consider two markers *A* and *B*, each of which transforms at a
frequency of one transformant per 100 recipient cells (frequency $= 10^{-2}$).
If *A* and *B* are unlinked, the probability of cotransduction should be the
product of the individual frequencies, or one cotransformant per 10^4 cells.
However, suppose in a particular experiment only 5 percent of the population
were competent. In that case, the single transformants would arise not at a
frequency of 10^{-2} but actually at 1 per 5 competent cells, or a real frequency
of 0.2. In the absence of linkage the expected number of double transformants would then be $(0.2)(0.2) = 0.04$, or 1 transformant for every 500 cells
in the entire population. That is, the expected number of doubles arising at
random would be only 1/5 the number of singles. If indeed doubles were
found at a frequency of $1/500 = 0.002$, which is high compared to the value
of 10^{-4} that might be expected if all cells were competent, one would be

tempted to conclude that the doubles arose by cotransformation. The fact that the number of cotransformants is 1/5 the number of single transformants would probably not make one suspicious, because DNA molecules are fragmented at random and not all molecules would be expected to carry both markers. Thus, unless competence is very high, some test is needed to assess whether a particular transformation value does indeed represent linkage. The most quantitative test available is a dilution test. For a population of cells the overall transformation frequency of a single marker is clearly a function of DNA concentration: as fewer DNA molecules are added to a culture, there will be fewer transformants. Except for extremely high DNA concentrations at which the cells are saturated with DNA, the transformation frequency decreases as a direct linear function of DNA concentration. Thus, if we are studying transformation of the markers A and B, the rate of decrease of the frequency of transformation of A, B, or linked AB with decreasing DNA concentration will be the same (Figure 13-9). However, if there are no DNA molecules that contain both markers, the production of AB transformants depends on the cell interacting with both A-containing and B-containing molecules. In this case, a tenfold decrease in DNA concentration will lower the probability of the double interaction by a factor of 10^2, so the curve relating frequency and DNA will be steeper.

When carrying out transformational mapping, a tool unique to transformation is available—namely, the size of isolated DNA fragments can be controlled, so the probability of cotransformation of two genes can be related to the average molecular weight of the donor DNA. Such a relation enables a physical map to be constructed from transformation data. For example, by

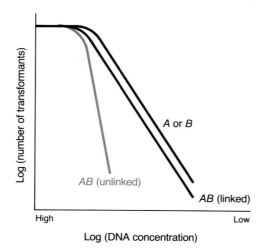

Figure 13-9   A dilution test for cotransformation. If A and B are linked, the cotransformation frequency falls off with decreasing concentration at the same rate as the individual genes. If they are unlinked, the decrease is much greater.

measuring the cotransformation frequency as a function of the molecular weight of the transforming DNA, the physical separation between the genes can be determined roughly. The frequency of transformation of any gene ultimately decreases, since for small fragments the trimming seen in Figure 13-7 can destroy the gene in the donor DNA. However, this decrease does not occur until the average molecular weight is less than about 5×10^5. A reduction in cotransformation frequency with decreasing molecular weight of the donor DNA occurring at high average molecular weights is a reliable sign that the markers do in fact cotransform.

For many bacterial species transformation yields only a minimal genetic map that provides little information about gene organization. The problem is that many bacteria—for example, *Pneumococcus* and *Hemophilus*—only grow in complex media, so the only genetic markers that are available are resistance to antibiotics and a few traits concerning polysaccharide metabolism. The situation has been quite different with *B. subtilis*, which grows well in minimal medium. Furthermore, nutritional mutants are easily obtained, and there are numerous phages, which also allows the use of phage-resistance markers. Several hundred markers are known for *B. subtilis*. However, even with *B. subtilis* a problem with using cotransformation in mapping arises. The problem follows from the fact that the bacterial chromosome always breaks during isolation and the size of the donor DNA fragment limits the size of the region that can be examined. Thus, cotransformational mapping invariably yields distinct linkage groups. It is possible with *B. subtilis* to transform with DNA having a mean molecular weight of about 25×10^6. Such a molecule can contain up to about 30 genes, but only some of these will yield genetic markers (for example, most mutations may not be simply detectable by plating tests). Such a fragment is about 1 percent the size of the *B. subtilis* genome, so each fragment can yield information about only a few genes in a small region of the genome. Of course, fragmentation of the chromosome during DNA isolation occurs at random, so the region of the chromosome contained in one fragment will overlap the region contained in many other fragments. If available genetic markers were distributed uniformly over the chromosome, this overlap would result in a map that is a single linkage group. However, if there were regions of the chromosome, no more than 1 or 2 percent of the chromosome in length, that are devoid of markers, these regions would break the sequence of overlaps and thereby fragment the map. Indeed, cotransformational mapping of the *B. subtilis* chromosome yields about five distinct linkage groups. However, a classical physical experiment provided the missing information for ordering these groups.

Consider a circular DNA molecule whose replication begins at a unique site and which replicates unidirectionally. Genetic markers are replicated in order $A, B, C, \ldots, Z$ from the replication origin to the replication terminus. After replication has begun, there will be twice as many A markers in the

DNA than Z markers. As replication proceeds, the number of copies of each marker will double successively in the order in which they are replicated. Thus, in a synchronized culture in which all DNA molecules are in the same stage of replication, the successive times at which the number of copies of each marker doubles will tell the temporal sequence of replication of each marker. If DNA could be isolated from such a culture at various times and used as donor DNA in a transformation experiment, a simple measurement of transformation frequencies would yield the gene order (because transformation frequency is related to DNA concentration). For a variety of reasons, the primary one being that it is difficult to measure transformation frequencies with sufficient accuracy for the experiment, this technique is inadequate. However, the following variation on the experiment works. *B. subtilis* is a spore-forming bacterium. **Spores** are a stable inactive form of bacteria that form when certain bacterial species are subjected to extreme stress, such as starvation for nutrients. In forming a spore, replication of the bacterial chromosome is completed and not reinitiated. Spores do not divide and are stable for years. When spores are exposed transiently to high temperature and placed in a good growth medium, they germinate—that is, they are converted to a normal bacterial form and grow and multiply. Initiation of DNA replication occurs synchronously, cells divide at nearly the same time, and the cells remain in fairly good synchrony for about two generations. If spores are germinated in a growth medium containing D_2O (heavy water), the daughter DNA strands that are produced will contain deuterium (D) instead of hydrogen and will thus have a higher density than the parental strands. Thus, semiconservative replication can be followed by isolating DNA at various times and sedimenting the DNA to equilibrium in CsCl solutions (that is, the classical technique of Meselson and Stahl, described in Chapter 8). As each region of the DNA replicates, fragments obtained from that region will be shifted from the normal low density to the density of hybrid deuterium-hydrogen DNA. With this in mind the following experiment was done. Spores were germinated in heavy medium as just described, and at various times DNA was isolated and centrifuged in CsCl. The hybrid (replicated) and light (unreplicated) fractions were isolated from each time sample and tested by transformation for the presence of each of a large number of markers (Figure 13-10). Replication of a marker was detected as a switch from finding the marker in the light DNA to finding it in the hybrid DNA. The switching time for each marker indicates when that marker replicates and the sequence of times provides a physical map that locates each marker along the chromosome. Such a physical map can be combined with the numerous linkage maps derived from independent cotransformation experiments to join the various linkage groups into a single map.

Unfortunately there is a major complication with this analysis, namely, that *B. subtilis*, like *E. coli*, replicates its DNA bidirectionally (Chapter 8). Thus, transfer of two markers from the light to the hybrid region at nearby

Figure 13-10 An experiment showing the temporal order of replication of individual genes. (a) A unidirectionally replicating DNA molecule, showing genes *a–k*, that has replicated for a portion of the bacterial life cycle in medium containing D$_2$O (heavy medium). (b) The results of CsCl density-gradient centrifugation of DNA extracted from cells at three different times after the density shift. The individual letters show the genes carried on transforming DNA from two regions of the density gradient. Note that, as always, DNA is fragmented during isolation from bacteria.

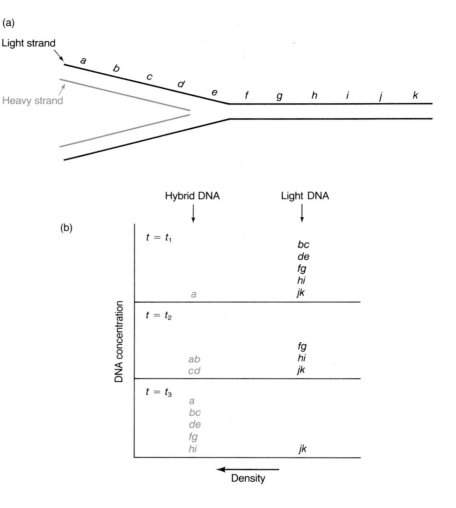

times does not mean that they are nearby in the chromosome; they could be duplicated by replication forks moving in opposite directions, which would mean that the markers are widely separated. Thus, a marker *A* that replicated after one-fourth of a generation could be located in a 100-unit circular map at either 25 or 75, and marker *B* that replicated slightly later (say, 3 units later) could be at either 28 or 72. However, if one of the linkage segments includes both *A* and *B*, then they are both either at positions 25 and 28 or at positions 72 and 75. By examining enough markers, the mapping puzzle can be solved and indeed an excellent genetic map of *B. subtilis* has been obtained. This map has been supplemented by ordering genes by the generalized transduction procedures described in Chapter 18.

TRANSFORMATION MISCELLANY

Transformation can be used for purposes other than mapping. One is to determine whether certain chemical agents acts directly on DNA. When a bacterium or an animal is exposed to a particular chemical and mutagenesis occurs, one cannot be sure, just from the production of mutants, that the chemical acts directly on DNA. Alternatively, a metabolite of the chemical could be the active substance, or the chemical could induce a mutagenic pathway of some kind. However, if pure DNA is exposed to the chemical and then used in a transformation experiment, the production of mutations would indicate a direct effect on DNA. Such an experiment would normally be done in a cotransformation experiment. Consider two closely linked genes, for example, *str* and *lac*. If DNA from a donor strain with the genotype *lac⁻ str-r* was used to transform a *lac⁺ str-s* recipient and *str-r* was the selected marker, some of the *str-r* cells would also be *lac⁻*. However, if the donor strain were *lac⁺*, the *lac⁻* marker could not appear in the recipient. Thus, if DNA from a *lac⁺ str-r* donor was treated with a known mutagen and the mutagen acted directly on DNA, some *lac⁻* cells would be found among the *str-r* transformants. Thus, the existence of *lac⁻ str-r* transformants would indicate that the mutagen is directly active on DNA.

Transformation is also useful in identifying DNA. For example, in an experiment in which DNA isolated from a cell infected with a phage is fractionated by zonal sedimentation or electrophoresis, it is often necessary to locate the bacterial and phage DNA among various fractions. With *B. subtilis* testing each fraction for transformation of bacterial markers can be used to identify bacterial DNA.

Transformation is usually carried out in the laboratory with purified DNA. However, the question has frequently been raised whether transformation occurs in nature. Given the fact that evolution has proceeded in part by mutation and in part by recombination, it seems likely that transformation could have evolved in organisms that lack other mechanisms for exchanging DNA. What is necessary for transformation to occur in nature is the release of DNA from cells and the generation of competence. Many Gram-positive bacteria, for example, *B. subtilis*, lyse and release their DNA when cells age or are subjected to starvation or various deleterious environments. Thus, there is a ready supply of DNA in the environment. More informative is the fact that transformation has been observed when genetically marked cultures of *B. subtilis* are mixed. Presumably, some DNA is released in the cultures and a small fraction of the cells are always competent, leading to occasional transformation. Furthermore, transformation has been observed when different marked bacteria are placed in the gut of a mouse. Transformation is nature is certainly an infrequent event. However, given the time allotted for evolutionary change, it has probably been one of the important forces in microbial evolution.

PROBLEMS

1. What experimental result showed that transformation consists of a permanent genetic change?

2. What argument was used by Avery and co-workers to rule out protein or RNA as the genetic material?

3. Following publication of the transformation experiments of Avery, MacLeod, and McCarty, opponents of the DNA = gene theory, who believed that genes were made of proteins, argued that the transformation was caused by proteins that were contaminating the DNA sample.

    (a) If transformation was indeed carried out by protein rather than DNA molecules and if the DNA preparation used contained at most 0.02 percent protein, how many protein molecules (each consisting of about 300 amino acids) would have been present in 1 milliliter of a DNA solution at a concentration of 10^{-7} mg/ml?

    (b) If protein was the active agent in transformation, would the number calculated in part (a) account for the fact that in a typical transformation experiment 1000 transformants result from 0.0001 g of *Pneumococcus* DNA?

4. Suppose you wish to prove to a skeptic that transformation is mediated by DNA and not by protein. You do not have available pure enzymes that degrade DNA or protein. However, you have worked hard and have shown (i) that 50 different genetic traits can be transformed, and (ii) that transformation is very inefficient and highly dependent on DNA concentration. The only piece of nonmicrobiological laboratory equipment you possess is an ultracentrifuge and with it you are able both to measure the molecular weight of DNA and fractionate DNA according to molecular weight. Design a simple experiment to prove that the transforming principle is DNA. Hint: Think about linkage.

5. A critic of the interpretation of the transformation experiment might say that protein is the genetic substance and the protein can penetrate the cell only when the protein is bound to DNA. Thus, the loss of transforming activity that accompanies boiling of transforming DNA might be a result of dissociation of protein from the DNA. Assume that you have current knowledge about the denaturation of proteins and of DNA, about the effect of low and high pH on the chemical and physical properties of DNA and protein, and about the ionic strength dependence of the binding of protein to DNA. Design an experiment to prove that the critic is incorrect.

6. Molecular biology was in a formative stage when Avery and his colleagues performed the transformation experiment. Many kinds of instruments and techniques are now available that would have simplified their investigation. Suggest a way that CsCl density gradient centrifugation might have been used to determine whether DNA or protein is the genetic material.

7. A Lac⁻ *Pneumococcus* is treated with DNA from a Lac⁺ strain and plated on a nonselective color-indicator medium in which Lac⁺ and Lac⁻ cells produce red and white colonies, respectively. What types of colonies will form on the medium?

8. Transformation sometimes occurs in nature by spontaneous lysis of a small number of cells and uptake of the released DNA by naturally competent cells. You are examining a number of bacterial species to see which ones can engage in intercellular exchange. A pair of strains of a bacteria that grow in Camembert cheese and have different genetic markers are mixed together. After a period of time the bacteria are plated on selective media and recombinant colonies form. You have reason to believe that transformation is occurring but your colleague thinks that it is a new example of bacterial conjugation (like an $F^+ \times F^-$ mating). What simple experiment might you do to distinguish these two alternatives?

9. A transformation experiment is carried out using donor DNA that is $A^+B^+C^+$ and a recipient that is $A^-B^-C^-$. A^+ transformants are selected and tested further. Of these, 64% are B^+ and none are C^+. Also, B^+ are selected, and 8% are also C^+. What is the gene order?

10. In a transformation experiment using DNA at a concentration of 10 μg/ml, 3.5×10^6 A^+ transformants, of which 3×10^5 were also B^+, were detected. In another experiment the DNA was instead at 1 μg/ml; the results were: A^+, 2.3×10^5; A^+B^+, $<10^3$. What can be said about the linkage of genes A and B?

11. Two different *B. subtilis* strains, carrying different nearby *lac* mutations, *lacZ1* and *lacZ2*, were transformed by a single DNA sample from a Lac$^+$ cell. Lac$^+$ recombinants were detected by plating on minimal medium containing lactose as the sole carbon source. With the *lacZ1* strain, the number of Lac$^+$ transformants was 5×10^4 per ml; for the *lacZ2* strain the value was 2×10^3. Explain the difference between these values.

12. A competent culture of *Pneumococcus* is mixed with [^{14}C]DNA having a radioactivity of 10^4 counts per minute (cpm). After 5 minutes the cells are centrifuged. No acid-precipitable radioactivity can be found in the supernatant fluid. (Recall that only nucleic acid polymers are acid-precipitable; nucleotides are not.) How much radioactivity do you expect to find associated with the cells?

13. A transformation experiment is done in which the donor strain is $A^+B^-C^+$ and the recipient is $A^-B^+C^-$. In one part of the experiment A^+ cells are selected, and 250 are tested further. The associated unselected markers and the number of each genotype are: B^-C^+, 12; B^-C^-, 3; B^+C^+, 100; B^+C^-, 135. In the second part of the experiment C^+ transformants are selected and 250 of these are also tested. The unselected genotypes and their numbers are: A^+B^-, 17; A^+B^+, 85; A^-B^-, 13; A^-B^+, 142. What is the gene order of the three genes?

Chapter 14

Bacterial Conjugation

In Chapter 11 plasmid transfer between donor and recipient cells was described. It was also mentioned briefly that some plasmids can integrate into the host chromosome; when they do so, their ability to transfer is retained, and in this way the chromosome itself becomes a giant transferable element. The best-known example of chromosome transfer is the F-mediated process in *E. coli*. Properties of chromosome transfer and the recombination events that occur in the recipient cell following transfer are the topics of this chapter.

INSERTION OF F INTO THE *E. coli* CHROMOSOME

In addition to the variety of genes carried by the F plasmid that mediate transfer of F to an F^- cell and regulate plasmid replication, F also contains three transposons, namely, IS1, IS2, and $\gamma\delta$ (Figure 11-8). These elements do not participate in any of the systems responsible for replication, maintenance, or transfer, but they provide a mechanism for integration of F into the chromosome. A cell with an integrated F is called an **Hfr** cell. Hfr is an acronym for *high frequency of recombination*, which refers to the fact that chromosomal genes are transferred from an Hfr cell to an F^- cell with considerably higher frequency than from an F^+ cell.

Integration is a reciprocal DNA exchange much like that occurring when phage λ lysogenizes a bacterium. This is known from base-sequencing studies, which indicate that an integrated F sequence is always flanked by two copies (in direct repeat) of one of the IS elements found in the F plasmid.

There are two ways by which Hfr cells might arise in a cell containing F:

1. *Homologous recombination between two identical IS elements, one in the chromosome and the other in F.* If the recombinational event is physically reciprocal, this mechanism would yield flanking IS elements and might utilize the bacterial Rec system (described later in this chapter), a transposon gene product, or conceivably some product of a gene in F (Figure 14-1).

2. *Replicon fusion.* An Hfr cell could arise by formation of a cointegrate mediated by an IS element in F, and by a duplication of a target sequence in the chromosome.

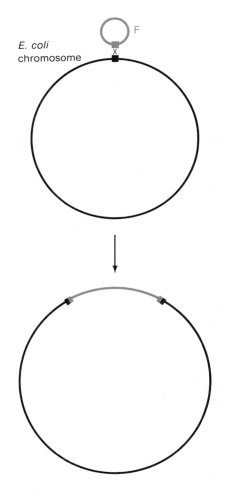

Figure 14-1   Integration of F by a reciprocal exchange between an IS element in F and a homologous sequence in the bacterial chromosome.

Mechanism 1 requires a copy of the transposon in the chromosome, but mechanism 2 does not. Note that the consequences of these two mechanisms are the same—namely, F flanked by two copies of the IS element in direct repeat. It is likely that Hfr strains arise by both mechanisms for the reasons that follow.

Integration of F into the *E. coli* chromosome has the following characteristics that differ from λ prophage formation:

1. The exchange site in F is not unique; whereas exchange occurs predominantly in the IS3 element located near 94.5 on the physical map, exchange also occurs in the other IS3 element and in γδ, as shown in Figure 14-2.

2. There are many sites in the chromosome at which integration can occur, and more than 20 *major* sites are known (Figure 14-3). The

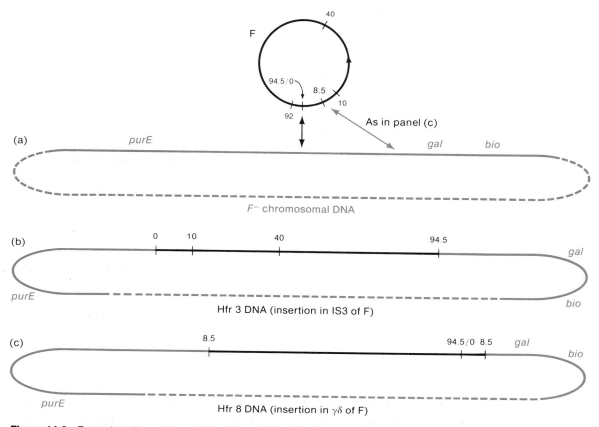

Figure 14-2 Formation of two different Hfr strains. (a) The *F⁻* chromosome and F. The exchange in both molecules occurs at the points indicated by the black double-headed arrow. (b) The result of the exchange shown in panel (a). (c) The result of the exchange indicated in panel (a) by the double-headed red arrow.

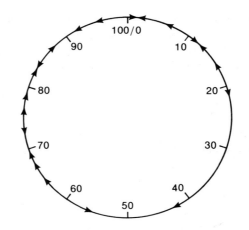

Figure 14-3 *E. coli* genetic map showing twenty different origins for Hfr transfer. The arrows show the direction of transfer.

affinity of F for each site is not the same in that the frequency of formation of a particular Hfr strain varies from one site to the next. Base-sequencing studies have shown that at each of these sites there is a copy of IS2, IS3, or γδ in the chromosome, and the affinity represents the probability of exchange with these elements. Integration at the major sites requires a functional RecA product.

3. F can integrate in both clockwise and counterclockwise orientations, which probably depends on the orientation of the IS element in the chromosome. We will see shortly that an inserted F mediates chromosome transfer in an ordered fashion, just as chromosomal genes are transferred on a F′ plasmid. Since F transfer is initiated by looped rolling circle replication from the transfer origin *oriT* and transfer occurs in a single direction, some Hfr strains transfer the chromosome in one direction and others transfer genes in reverse order. In fact, there are a few instances in which F can integrate in either orientation at a single site (note the sites at positions 12 and 83 in Figure 14-3). That is, if F integrates between genes *p* and *q*, one orientation would cause the transfer order to be *q r s . . . z a . . . n o p*, and the other orientation would yield the order *p o n . . . a z . . . s r q*.

Hfr TRANSFER

When F becomes integrated into the bacterial chromosome, the chromosome remains a single, circular DNA molecule and F behaves as if it were part of this chromosome, increasing the size of the chromosome. F is now able to transfer the entire chromosome from these Hfr cells to an *F⁻* culture.

The stages of transfer are much like those by which F is transferred to F^- cells—namely, pairing of donor and recipient, rolling circle replication in the donor, and conversion of the transferred single-stranded DNA to double-stranded DNA by replication in the recipient—but the transferred DNA does not circularize and is not capable of further replication in the recipient (the reason is given in item 2 below). Furthermore, the replication and associated transfer of the Hfr DNA (called transfer replication), which is controlled by F, is initiated in the Hfr chromosome at the same point in F at which replication and transfer begin within an F plasmid (*oriT*). Thus, a portion of F is the first DNA transferred, chromosomal genes are transferred next, and the remaining part of F is the last DNA to enter the recipient.

Several differences between F transfer and Hfr transfer are notable.

1. It takes 100 minutes for an entire bacterial chromosome to be transferred, in contrast with about 2 minutes for transfer of F. The difference in time is a result of the relative sizes of F and the chromosome. (A common "preparatory" period of a few minutes precedes transfer in both.)

2. During transfer of Hfr DNA to a recipient cell, the mating pair usually breaks apart before the entire chromosome is transferred, owing to Brownian motion (movement resulting from bombardment by solvent molecules). On the average, several hundred genes are transferred before the cells separate.

3. In a mating between Hfr and F^- cells the F^- recipient usually remains F^-, because cell separation almost always occurs before the final segment of F is transferred. The recipient does not gain the ability to engage in subsequent transfer unless it has received all of the F DNA from its donor—that is, unless chromosome transfer has been fully completed.

4. In Hfr transfer, although the transferred DNA fragment does not circularize and cannot replicate, one or more of its regions is frequently exchanged with the chromosome of the recipient, thereby generating recombinants in the F^- cell. For example, in a mating between an Hfr *leu*$^+$ culture and an F^- *leu*$^-$ culture, F^- *leu*$^+$ cells arise. Note that the genotype of the donor is unchanged, because the transferred DNA strand is replaced by concurrent replication in the donor.

The overall transfer process is shown schematically in Figure 14-4.

The presence of the new chromosomal fragment in the female sets in motion a recombination system that causes genetic exchanges to occur, and a recombinant F^- cell often results. Thus, in a mating between an Hfr *leu*$^+$ culture and an F^- *leu*$^-$ culture, F^- *leu*$^+$ recombinants arise (Figure 14-5). As in F and F′ matings (Chapter 11), in order to recognize the recombinants, some means is needed to distinguish a male from a female cell. As before, this is done by using cells that have genetic differences that can be recognized

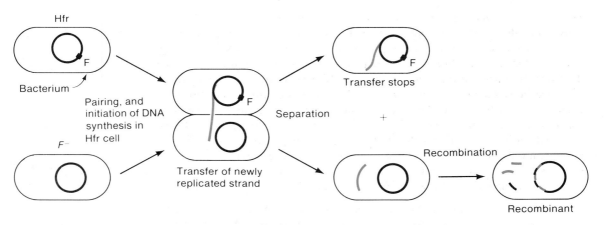

Figure 14-4 A diagram showing mating between an Hfr male bacterium and an *F⁻* female bacterium. The two bacteria form a pair. Then, under the influence of a unit called F, DNA replication begins in the male, adjacent to F, and a replica of F is transferred to the female. By random motion the mating cells break apart. At a later time a portion of the transferred DNA exchanges with the corresponding piece in the female. The intact female chromosome then replicates; the fragments do not replicate and are ultimately lost in the course of cell division. Although not shown, only a single strand is transferred and this is converted to double-stranded DNA in the recipient.

by growth of a colony on agar. Thus, in a mating Hfr *leu⁺ str-s* × *F⁻ leu⁻ str-r,* plating on minimal medium lacking leucine and containing strepto-mycin selects against both the original *F⁻* (Leu⁻) cells and the Hfr (Str-s) cells, and allows Leu⁺ Str-r recombinants to form colonies. Recall that the transferred allele that is selected by means of the medium composition (*leu⁺* in this case) is called a selected marker, and the allele used to prevent growth of the male (*str-r* in this case) is called the counterselective marker.

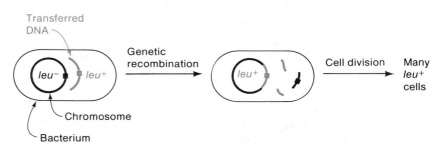

Figure 14-5 Conversion of a *leu⁻* cell to a *leu⁺* cell by incorporation of a segment of the transferred DNA. The fragments remaining after recombination do not replicate and are gradually diluted as the recombinant cell divides. This drawing is very schematic and does not attempt to distinguish between various proposed mechanisms of exchange; for example, possibly only a single strand of transferred DNA is inserted.

Interrupted Mating and Time-of-Entry Mapping

Shortly after Hfr and F^- cultures are mixed, transfer of the chromosome from the Hfr begins. Transfer is not synchronous but is initiated over a period of several minutes. Each type of Hfr transfers genes in an order determined by the site of insertion and the orientation of F. If there is a single transfer origin, the order of transfer will reflect the gene order in the chromosome, and the transfer order can be used to generate a gene map. A variety of experiments indicate that transfer of the Hfr chromosome proceeds at a constant rate (evidence will be given shortly). Thus, the *times* at which particular genetic loci enter a female are simply related to the positions of these loci on the chromosome, and a map can be obtained from the time of entry of each gene.

Certain features of the transfer process in Hfr cells and information concerning the arrangement of bacterial genes in Hfr and F^- cells have been obtained by mechanically interrupting transfer during mating. The time at which a particular gene is transferred can be determined by purposely breaking the mating cells apart at various times and noting the earliest time at which breakage no longer prevents recombinants from appearing. (Violent agitation of the suspension of mating cells in a kitchen blender is a standard method.) This procedure is called the **interrupted-mating technique.** When this is done with Hfr $\times$ F^- matings, it is observed that the number of recombinants of any particular genotype increases with the time during which the cells are in contact. The reason for the increase is slow transfer of the Hfr DNA. This phenomenon is illustrated in Table 14-1.

Table 14-1 Data showing the production of Leu$^+$ Str-r recombinants in a cross between Hfr *leu$^+$ str-s* and F^-*leu$^-$ str-r* cells when mating is interrupted at various times

Minutes after mating	Number of Leu$^+$ Str-r recombinants per 100 Hfr cells
0	0
3	0
6	6
9	15
12	24
15	33
18	42
21	43
24	43
27	43

Note: Extrapolation of the data to a value of zero recombinants indicates a time of entry of 4 min.

Greater insight into the transfer process can be obtained by observing the results of a mating with several genetic markers. Consider the mating

$$\text{Hfr } a^+ b^+ c^+ d^+ e^+ str\text{-}s \times F^- a^- b^- c^- d^- e^- str\text{-}r$$

Again, at various times after mixing the cells, samples are taken, agitated violently, and then plated on media containing streptomycin and one of four different combinations of the five substances A through E—for example, B, C, D, and E, but no A, or A, C, D, and E, but no B. Colonies that form on the medium lacking A are $a^+ str\text{-}r$, those growing without B are $b^+ str\text{-}r$, and so forth. All of these data can be plotted on a single graph to give a set of curves, as shown in Figure 14-6(a). Four features of this set of curves are notable:

1. The number of recombinants in each curve increases with time of mating.

Figure 14-6 Construction of the circular map of *E. coli*. (a) Time-of-entry curves for a particular Hfr strain. (b), (c), (d), (e) Four linear maps obtained from four different Hfr types. (f) The circular map derived from the data in panels (a)–(e).

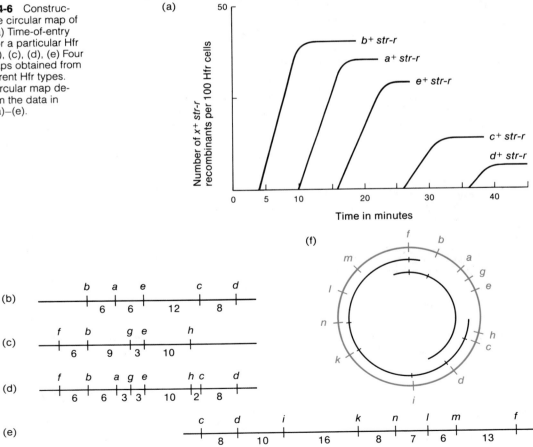

2. For each type of recombinant there is a time before which no recombinants are detected.
3. Each curve has a linear region that can be extrapolated back to the time axis, defining a **time of entry** for each locus a^+, b^+, . . . , e^+.
4. The number of recombinants of each type reaches a maximum, the **plateau value,** the value of which decreases with successive times of entry.

The explanation for the time-of-entry phenomenon is the following. Transfer begins at a particular point in the Hfr chromosome, which we now know is the transfer origin of F. Genes are transferred in linear order to the recipient. The time of entry of a gene is the time at which that gene first enters a recipient in the population. All donor cells do not start transferring DNA at the same time, so the number of recombinants increases with time; separation of a mating pair prevents further transfer and limits the number of recombinants seen at a particular time. At a time much later than the time of entry, the transferred DNA undergoes genetic recombination in the recipient, forming a recombinant cell.

The times of entry of the genes used in the mating just described can be placed on a map, as shown in Figure 14-6(b). Mating with a second recipient whose genotype is $b^- e^- f^- g^- h^-$ str-r can then be used to locate the three genes f, g, and h. Data for the second recipient will yield a map such as that in Figure 14-6(c). Since genes b and e are common to both maps, the two maps can be combined to form a more complete map, as shown in Figure 14-6(d).

Studies with other Hfr cell lines (panel (e)) are also informative. It is usually found that different Hfr strains (for example, those of panels (c) and (e)) yield different sets of curves, distinguishable by their origins and directions of transfer, indicating that F can integrate at numerous sites in the chromosome and in two different orientations. Combining the maps obtained for different Hfr strains yields a composite map that is *circular*, as illustrated in Figure 14-6(f). The circularity of the map is a result of the circularity of the *E. coli* chromosome in F^- cells and the multiple points of integration of the F plasmid; if F could integrate at only one site, the map would be linear. Notice, however, that the map circularity does not indicate whether the Hfr chromosome is circular; this has been demonstrated mainly by physical experiments, such as the autoradiogram of Figure 8-3.

At the present time, roughly 1000 genes have been mapped in *E. coli*—about half of all of the genes. Actually, time-of-entry mapping is only accurate to about one minute of the map and each minute contains about 20 genes. Thus, precise mapping is accomplished by transduction, as will be described in Chapter 18. An abbreviated *E. coli* map, showing a small number of genes, is reproduced in Figure 14-7. To give an idea of the detail with which *E. coli* genes have been mapped, a photograph of several regions of an encyclopedic *E. coli* map is shown in Figure 14-8. Furthermore, dozens of mutations have been mapped within certain genes, such as *lacZ* and *trpE*.

Figure 14-7 A 1980 map of *E. coli* showing the positions of 52 genes that are well-mapped and commonly used in conjugation experiments. The units of the map are minutes. The map is roughly 100 minutes long, which represents the time required to transfer an entire chromosome at 37°C. The threonine (*thr*) locus has, for historical reasons, been chosen as 0. The light regions of the circle are segments in which very few markers had been isolated by 1980. However, recent studies show that these are not gene-free regions. For details of segments of the map, see Figure 14-8.

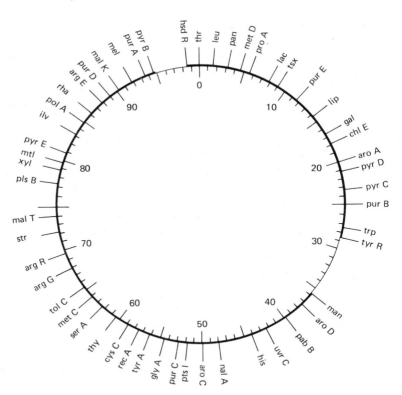

Figure 14-8 Detailed segments of the 1980 *E. coli* map, in which about 1000 genes have been accurately mapped by P1 transduction (Chapter 18). The large numbers represent minutes. This is part of a reference map compiled by B.J. Bachman and K.B. Low (*Microbiol. Rev.*, 44, 1, 1980).

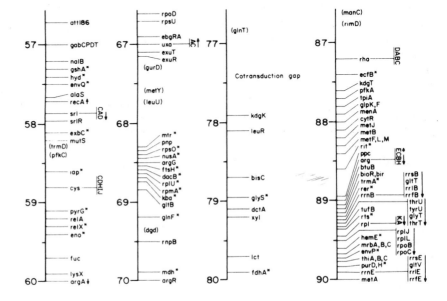

The Rate of Chromosome Transfer

F can integrate at numerous sites in the chromosome, generating Hfr cells with different origins of transfer. Each Hfr strain can be used to obtain maps, as we have just seen. An important feature of these maps is that the time interval between the entry of different genes is independent of either the Hfr cell line or the F^- strain. For example, the times of entry of the *thr*, *lac*, and *trp* genes with HfrH are 5, 13, and 32.5 minutes, yielding the intervals: *thr–lac*, 8 minutes, and *lac–trp*, 19.5 minutes. With HfrB7, which transfers in the opposite order, the times of entry for *trp*, *lac*, and *thr*, are 3, 22.5, and 30.5 minutes, respectively. These values again yield the distances: *trp–lac*, 19.5, and *thr–lac*, 8 minutes. In fact, it is just experiments of this sort that led to the suggestion that DNA transfer occurs at a constant rate. Physical experiments confirmed this point. In these experiments Hfr *tsx-r* cells, whose DNA was ^{14}C-labeled, were mated with F^- *tsx-s* cells. At various times after mixing the cells, a great excess of phage T6 was added, which caused lysis of the sensitive female cells within about one minute. The amount of [^{14}C]DNA released was measured and found to increase linearly with time after mixing.

Linkage Mapping

Short regions of *E. coli* can be mapped by analysis of recombination frequencies. This technique was first used to map the individual genes of the *lac* operon.

Consider an Hfr that transfers in the order *pro lac purE*, in which a *purE*⁻ mutation introduces a nutritional requirement that can be satisfied by adenine. The gene order of the *lacZ* and *lacY* genes with respect to *purE* can be determined by comparing the results of the crosses Hfr *purE*⁺ *lacZ*⁺ *lacY*⁻ *str-s* × *F*⁻ *purE*⁻ *lacZ*⁻ *lacY*⁺ *str-r* (cross I) with the reciprocal cross (II) Hfr *purE*⁺ *lacZ*⁻ *lacY*⁺ *str-s* × *F*⁻ *purE*⁻ *lacZ*⁺ *lacY*⁻ *str-r* (Figure 14-9). In each case Pur⁺ Str-r recombinants are selected by plating on minimal agar lacking adenine and containing streptomycin, and these recombinants are tested for the Lac⁺ phenotype by replica plating onto lactose color-indicator plates. Note that if the order is *purE lacY lacZ*, the number of *pur*⁺ *lacY*⁺ *lacZ*⁺ *str-r* bacteria is smaller in cross I than in cross II, since more exchanges are required in cross I. With order II *pur*⁺ *lacY*⁺ *lacZ*⁺ *str-r* types predominate in cross I. Experimental data show that the order is *purE lacZ lacY*.

Note that this technique of looking at relative recombination values yields the gene order but not the map distances, which must be obtained from the actual values of recombinant frequencies. This can be done in a straightforward way once the gene order has been determined, using either the recombination frequencies in the three-factor cross or frequencies obtained from separate two-factor crosses. Consider three mutations *lacZ1*, *lacZ2*, and

Figure 14-9 Determination of the relative positions of the *lacZ* (denoted *z*) and *lacY* (denoted *y*) genes with respect to the *purE* gene, by comparing the results of reciprocal crosses.

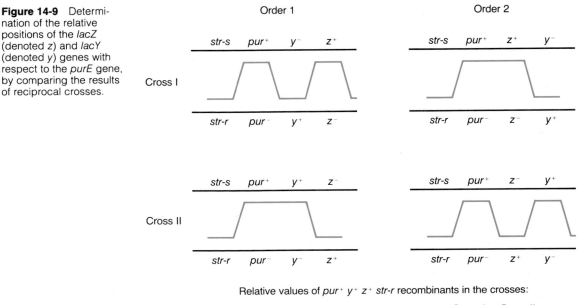

Relative values of *pur⁺ y⁺ z⁺ str-r* recombinants in the crosses:

Cross II > Cross I   Cross I > Cross II

lacZ3, which have the gene order with respect to *purE* of *purE lacZ1 lacZ2 lacZ3*. Two crosses are performed:

1. Hfr *pur⁺ lacZ1⁺ lacZ2⁻ str-s* × F⁻ *purE⁻ lacZ1⁻ lacZ2⁺ str-r*
2. Hfr *pur⁺ lacZ2⁺ lacZ3⁻ str-s* × F⁻ *purE⁻ lacZ2⁻ lacZ3⁺ str-r*

and Pur⁺ Str-r colonies are selected. These are further tested for the Lac⁺ phenotype. For the first cross, the fraction (number of Pur⁺ Lac⁺ Str-r)/(number of Pur⁺ Str-r) is a measure of the recombination frequency between *lacZ1* and *lacZ2*. The same proportion for the second cross measures the recombination frequency between *lacZ2* and *lacZ3*. Assume that the values are 0.12 and 0.08, respectively; then, the genetic map is *lacZ1*–0.12–*lacZ2*–0.08–*lacZ3*. Carrying out this procedure for numerous *lac* mutations—that is, determining the gene order from reciprocal crosses and evaluating genetic distances from the recombination frequencies—can yield a fairly complete genetic map.

Mapping by this procedure is limited by the reversion frequencies of individual mutations and the number of recombinant colonies that can conveniently be tested by replica plating. The smallest frequency measured in an early study of the *lac* operon was 3×10^{-5}.

It should be noted that a linkage map generated in this way is in terms of map units, whereas a time-of-entry map has units of minutes. Relating map units to time units is not a straightforward process, and in fact the *E. coli* map is most often given in minutes. In Chapter 18 we will see how transduction can yield a fine-structure map in minutes.

Mapping of Recessive Markers:
The Method of Unselected Markers

Consider a met^- (methionine-requiring) marker in an F^- cell. Determining the time of entry of this marker is straightforward since one can simply perform a time-of-entry experiment with an Hfr that is Met^+. Plating the mating mixture on agar lacking methionine enables Met^+ recombinants to grow. However, suppose the met^- mutation was originally isolated in an Hfr strain. In that case, mapping is less direct because *inability* to grow on a plate lacking methionine is the trait brought to the recipient by the Hfr, and transfer of that trait would be recognized only by a decrease in the number of colonies with time. Furthermore, the mutation might be located such that only a few percent of the cells receive the marker; in that case, entry of the marker would decrease the number of recipients by only a few percent, which would be unrecognizable against the statistical fluctuation in number of colonies on different plates corresponding to the same time.

The met^- mutation can easily be mapped by linkage analysis, as will be seen in the following example. Consider the mating Hfr $met^- leu^+ lac^+ str$-s $\times$ $met^+ leu^- lac^- str$-r. Interrupted mating will be carried out and samples will be plated on two different media: (I) lactose color-indicator agar, which contains all amino acids and (II) medium containing methionine but no leucine. Both types of media contain streptomycin to select against the donor cells. The plates select for the following recombinants: (I) $lac^+ str$-r and (II) $leu^+ str$-r. These recombinants are tested for the met^- mutation by picking colonies from plates I and II and touching them to plates containing leucine but no methionine. Data are shown in Table 14-2. Note that at first all leu^+ colonies are met^+, but at later times $leu^+ met^-$ recombinants have formed.

Table 14-2 Data showing mapping of a met^- mutation as an unselected marker in the cross Hfr $met^- leu^+ lac^+ str$-s $\times$ $F^- met^+ leu^- lac^- str -$

Minutes after mixing	Leu$^+$Str-r recombinants		Lac$^+$Str-r recombinants	
	Number	% Met$^-$	Number	% Met$^-$
0	0	—	0	—
3	0	—	0	—
6	7	0	0	—
9	14	36	0	—
12	23	48	3	100
15	32	60	6	100
18	41	70	14	100
21	42	79	20	100
24	42	81	26	100
27	42	81	27	100

Note: The times of entries from these values are: *leu*, 4 min; *met*, 7 min; *lac*, 11 min.

This result shows clearly that *leu* enters the recipient before the *met⁻* mutation, because the Hfr marker is present before the marker in the recipient. Also, all *lac⁺* recombinants are *met⁻*, indicating that the *met⁻* marker was transferred before *lac*. The gene order is clearly *leu met lac* with this Hfr strain. Note that by plotting the fraction of *leu⁺* recombinants that are *met⁻* as a function of time, one obtains the time of entry of the *met⁻* mutation. In this experiment the *met⁻* marker is not used in the initial selection; it is said to be an **unselected marker,** and this mapping procedure is occasionally referred to as the method of unselected markers.

Some markers can only be mapped as unselected markers. An example is a phage-resistance marker. Consider a cross between an Hfr that is resistant to phage T6 (Tsx-r) and a recipient that is Tsx-s—for example, a cross Hfr *lac⁺ tsx-r* × *lac⁻ tsx-s*, in which *str-r* is present in the recipient to prevent growth of the Hfr. The Tsx-r and Tsx-s phenotypes can be tested by plating on a medium that has been previously spread with about 10^8 T6 phage: Tsx-r cells can form colonies on such a medium, and Tsx-s cells cannot. If the mating mixture were plated directly on plates containing streptomycin and T6, no colonies would form because (1) the Hfr cells will be inhibited by the streptomycin, (2) the Tsx-s recipients would be lysed by the T6, and (3) cells that have received the *tsx-r* allele would still possess normal (Tsx-s) T6 receptors on their cell walls and hence would also be killed. Note that the reciprocal cross with a Tsx-s Hfr and a Tsx-r recipient would also not work because all recipients would grow except for the few that had acquired the dominant Tsx-s allele; hence, recombination could only be detected by a small decrease in the total number of cells. However, the *tsx* gene can be mapped easily if the allele is treated as an unselected marker. That is, *lac⁺ str-r* recombinant colonies are selected and then tested for phage sensitivity. As in the mapping of the *met* marker, one would observe that many *lac⁺* cells would be Tsx-s if *lac* entered the recipient before *tsx*.

Chromosome Transfer by F^+ Cultures

F was originally detected by virtue of its ability to mediate transfer of chromosomal genes to a recipient cell. For the most part, this transfer is a result of the presence of rare Hfr cells in a predominately F^+ culture. This was first demonstrated by a fluctuation test of the type used to show the nature of spontaneous mutations (Chapter 10, Figure 10-1, Table 10-1). In this test the ability to transfer a particular gene (*thr*) by 50 small cultures was compared to the same ability of 50 aliquots of a large culture. It was found that the number of recombinants produced by mating each of the 50 aliquots of the large culture with an appropriate recipient culture ranged from 10 to 23. The mean value was 16 and the variance was 13. In contrast, with the 50 individual cultures, the number of recombinants ranged from 1 to 116; the mean was 15 but the variance was 351. As seen in Chapter 10, the large variance implies that Hfr cells arose at various times (as "mutations" of F^+ cells) in the small cultures, producing clones of Hfr cells in each culture. If, on the contrary, every F^+ cell was capable of a transient plasmid-chromosome association that could lead to the ability to transfer, Hfr "jackpot" cultures would not have been observed and the variance among the 50 small cultures would have been no greater than the variance of the 50 aliquots of the large culture.

Isolation of Hfr Strains

The previous section shows that a typical F^+ culture will contain rare Hfr cells. These cells can be isolated from the culture by a series of replica platings in the following way. About 10^7 F^+ str-s cells are spread on a nutrient agar surface and allowed to grow until a confluent layer of cells forms (this is the master plate). This layer will contain a few microscopic clones derived from individual Hfr cells. A minimal agar plate containing streptomycin is also covered with a thin (nonconfluent) layer of a Str-r auxotrophic strain— for example, Leu$^-$. A velvet pad is touched to the surface of the master plate and then replicated to the minimal plate. Since DNA transfer is not immediately inhibited by streptomycin, the few Hfr cells transferred to the surface of the minimal plate will engage in conjugation with the recipient cells. Neither the Hfr cells nor the recipients can produce any visible growth on the plate, because they are Str-s and Leu$^-$, respectively. However, Leu$^+$ Str-r recombinants will grow and produce small colonies. The positions of these colonies indicate regions on the original F^+ plate where there are microclones of Hfr cells. These regions consist of some Hfr cells but will be contaminated with a great excess of F^+ cells. The regions are scraped from the plate, diluted, and spread on a new nutrient agar plate, one plate for each region, yielding new master plates. These plates are considerably enriched for Hfr cells compared to the original F^+ plate. Each of these plates is replicated onto minimal-streptomycin plates, as before, and regions in which recombinants arise are again noted. Cells are then scraped from the appropriate region of the confluent donor plates, spread on a new set of nutrient agar plates, and retested. After several cycles of replica plating the ratio of Hfr cells to F^+ cells becomes high enough that the master plate can be seeded with a few hundred cells so that individual Hfr colonies can be found.

Chromosome Transfer Mediated by F' Plasmids

Chromosome transfer by F'-containing strains occurs at a considerably greater frequency than by F^+ strains—by a factor of about 10^5. This phenomenon is called **chromosome mobilization.** Insight into the mechanism comes from the observation that efficient transfer requires an active *recA* gene in the F' strain. (We will see shortly that, in an Hfr cross, recombination normally requires an active *recA* gene in the recipient. However, transfer from an Hfr *recA*$^-$ occurs at the same frequency as from an Hfr *recA*$^+$.) Chromosome mobilization is a result of reciprocal recombination between the chromosomal segment of the F' and the homologous region of the chromosome itself. The result is insertion of the plasmid into the chromosome (much like that seen in Chapter 5 for insertion of phage λ) effectively generating an Hfr cell. This cell differs from a typical Hfr in that the chromosomal genes of the F' are present in the recombinant donor in two copies, one that is transferred immediately after pair formation and the other transferred as the final markers (the reader should draw the recombinational event to confirm this point). Further evidence for the recombination model comes from the fact that if the chromosome carries a large deletion that includes all of the chromosomal genes of the F', the plasmid cannot cause chromosome mobilization and the transfer of chromosomal markers occurs at the same low frequency as from an F^+ cell.

Mechanism of Chromosomal Transfer

Figure 11-7 of Chapter 11 showed the mechanism of transfer of a plasmid. The essential features were initiation of rolling-circle replication in the donor, transfer of a single strand through a conjugation bridge, and conversion to double-stranded DNA in the recipient. This same mechanism applies to Hfr transfer (Figure 14-4), except that the transferred DNA does not circularize. Interestingly, if DNA synthesis in the F^- cell is prevented by raising the temperature of a temperature-sensitive *dnaE* (polymerase III) mutant, transfer begins but continues for only a few minutes. For unknown reasons, conversion of the transferred single-stranded DNA to double-stranded DNA seems to be necessary for continual transfer. Possibly, a topological problem exists.

RECOMBINATION IN RECIPIENT CELLS

The final stage of bacterial conjugation is the incorporation of a transferred DNA fragment into the chromosome of the recipient to generate a recombinant cell. In this section we examine several features of this process.

Necessity for a Double Exchange

At a time characteristic of each genetic marker, a time-of-entry curve shows a rise in the number of recombinants (Figure 14-6). This rise is followed by a plateau region, whose value depends on gene position. To some extent, the value of the plateau is a function of the probability of gene transfer, that is, whether the mating pair has broken apart before transfer has occurred. However, it is also a measure of the probability of recombination.

When a DNA fragment enters a recipient cell, it cannot be stably maintained through subsequent cell divisions, because the fragment is unable to replicate. Two factors prevent replication: (1) the fragment will in general lack a replication origin, and (2) in bacteria, except for a few phages, only circular DNA can replicate. Also, a linear fragment will gradually be degraded by scavenging nucleases. The situation is quite different when a plasmid is transferred, since the plasmid is an intact self-replicating unit. Thus, maintenance of the genes in the fragment requires DNA exchange, exactly as was seen in Chapter 13 for DNA transformation. Because the transferred DNA is linear and the chromosome is circular, two exchanges are necessary, which flank the genes that are incorporated (Figure 14-5). Multiple exchanges can also occur, but the number of exchanges must always be an even number. The mechanism by which the exchange occurs, which has been investigated for more than 20 years, remains unknown. Physical evidence suggests that only a single strand is integrated, as in bacterial transformation, and that the mechanism is more complex than that used in transformation. An unusual feature is a requirement for the presence of specific base sequences (*chi* sequences, roughly 5000 base pairs apart) in both the donor and recipient DNA.

The frequency of exchange is quite high in *E. coli*. This is made clear by examining crosses in which the first 10 percent of the transferred chromosome contains several markers. In such crosses multiple exchanges are frequent, and recombination frequencies between markers 5 minutes apart are sufficiently high that the markers appear to be unlinked. Thus, the linkage mapping described earlier for the *lac* genes cannot be applied to markers separated by more than a few minutes. In fact, only time-of-entry experiments give a clear picture of the overall structure of the *E. coli* genome.

Anomalous Plateau Values

The value of a plateau region reflects both the efficiency of transfer of a marker and the probability of recombination. In general, the transfer efficiency is the major factor in determining plateau values, so the value continually decreases with time of entry, as seen in Figure 14-6. However, two types of markers do not follow this pattern: those that enter very early and those near the counterselective marker.

Since two recombination events are needed for integration of a marker, the recombination frequency will be affected by the distance of the marker from the ends of the fragment. Thus, if transfer is interrupted shortly after a marker enters a recipient, the probability of recombination will be very low, because little space will be available for DNA exchange. However, in

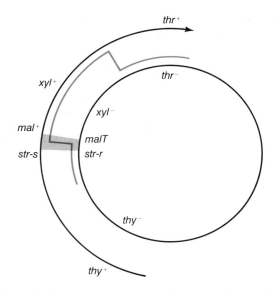

Figure 14-10   Recombination near a counterselective marker. The red line indicates a possible Mal⁺ Str-r recombinant. The first exchange can be anywhere clockwise from the *mal* marker, but the second exchange *must* be in the shaded region. A later marker, such as *thy,* may have a higher plateau region; even though fewer *thy⁺* markers are transferred, the possible exchange regions are fairly large. Recall that an exchange occurs, on the average, once in every 5-minute segment.

general there will be a significant length of DNA between a marker and the end of the fragment since transfer will continue after the marker enters the cell. Markers near the transfer origin present a special case, since the distance between a very early marker and the beginning of the early segment of F may be quite small. This is reflected in an anomalously low value of the plateau frequency for very early markers.

Figure 14-10 shows that later markers may also have anomalously low plateau values. This will occur whenever a marker is quite near the counterselective marker. For example, if the Hfr is Str-s and the F^- is Str-r, and Str-r is the counterselection, a donor marker within a minute or so of the *str* gene (*malT,* in the figure) will have a low probability of being recovered in a recombinant because an exchange must occur between the donor marker and the *str* gene. Otherwise the recipient would be Str-s and would not form a colony on the selective plate.

Efficiency of Transfer from an Hfr

Transfer of an F′ occurs with essentially 100 percent efficiency in that each donor cell can transfer a plasmid copy within 20 minutes after mixing donor and recipient cultures. This is made evident in that if, in a mating, the ratio of F'*lac*/*str-s* cells to $F^- lac^- str$-*r* cells is 1:10, roughly 10 percent of the F^- cells will acquire F'*lac* in 20 minutes. However, in Hfr crosses, plateau values for early markers range from 20 to 50 recombinants per 100 Hfr cells, the maximum depending on the donor-recipient pair. One would, of course, not expect plateau values ever to reach 100 percent for several reasons: (1) the presence of a transferred DNA fragment does not mean that DNA exchange must occur, (2) recombination events within the fragment do not necessarily lead to integration of a particular genetic marker contained in the fragment, and (3) recombination may lead to base-pair mismatches that could be corrected to the donor allele, the recipient allele, or randomly (recall high- and low-efficiency markers in bacterial transformation, Chapter 13). Nonetheless, it might be expected that some information about transfer and recombination could come from knowledge of the efficiency of transfer.

Transfer efficiency can be measured genetically by mating an Hfr strain that is lysogenic for phage λ with a recipient that is not lysogenic. Recall from Chapter 5 that lysogens are immune to infection by a phage that is the same as the prophage. This phenomenon, which will be discussed in greater detail in Chapter 17, is a result of the synthesis of a repressor protein by the prophage. That is, if λ infects a λ lysogen, the incoming phage DNA is repressed by the λ repressor, which prevents both transcription and replication of the infecting phage DNA. However, when λ infects a nonlysogenic cell, no repressor is present and a lytic cycle normally occurs. Mating of an Hfr cell that is lysogenic for λ with a nonlysogenic recipient is like infecting a nonlysogen with λ. Early in the mating, the transferred Hfr DNA behaves in a normal way, stimulating recombination in the recipient. However, once a λ prophage enters a nonlysogenic recipient, which does not contain any λ repressor, transcription of the prophage begins and a lytic cycle ensues. Thus, a nonlysogenic recipient receiving a prophage will lyse and hence can never make a colony. This phenomenon is called **zygotic induction.**

Zygotic induction is made evident in two ways. In one test an Hfr *str-s* lysogen is mixed with nonlysogenic *str-r* recipients and without separating paired cells, the mixture is plated on a lawn of a *str-r* indicator cells in soft agar containing streptomycin. Cells receiving a prophage behave like a phage-infected cell (they are infective centers) and form plaques in the lawn of indicator bacteria. This test can be used with any phage for which an indicator strain is available. The second procedure applies to any lysogenic Hfr strain even if the prophage is not derived from a known phage or has not been previously detected. This procedure is based on the aberration produced by zygotic induction on a set of time-of-entry curves. Consider an Hfr strain whose genes *a*, *b*, *c*, *d*, *e*, and *f* have times of entry 5, 10, 15, 20, 25, and 30 minutes, respectively, and which is mated with a nonlysogenic recipient. The mating would result in six standard time-of-entry curves. However, if this strain were made lysogenic for λ and the prophage attachment site were between genes *b* and *c*, two changes in the set of curves would be observed: (1) no recombinants containing any of the genes *c*, *d*, *e*, or *f* would be formed, and (2) the number of recombinants containing gene b would be markedly reduced. Observation 1 is a result of the fact that genes c, d, e, and f enter the recipient after the λ prophage, so all recipient cells that acquired these genes have already received λ and hence are destined to die. Observation 2 can be explained by the fact that many of the cells that have received gene b will also have received the λ prophage several minutes later and hence are killed. Thus, the main evidence for zygotic induction is the absence of time-of-entry curves for genes that enter after a particular time. It was by this observation that zygotic induction was discovered.

The principal importance of zygotic induction is that it is a measure of DNA transfer, because it is independent of subsequent genetic exchange processes. That is, any cell that receives the prophage becomes an infective center, so the number of infective centers indicates the number of cells to which DNA has been transferred. If the prophage is transferred within about 15 minutes after mixing the cultures, the number of infective centers equals the number of Hfr cells. Thus, transfer, or at least initiation of transfer, is 100 percent efficient. Comparison of this value with the number of recombinants for fairly early markers shows that certainly half, if not all, of the recipients engage in genetic recombination following receipt of a DNA fragment.

Rec⁻ Mutants

In 1965 a search began among a collection of mutagenized F^- cells for mutants that could not engage in recombination with an Hfr culture. Three genes were identified and labeled $recA$, $recB$, and $recC$. These mutants were not defective in any stage of conjugation other than recombination, since F′ plasmids transferred at the usual rate and efficiency to each of the mutants. The phenotype of these mutants with respect to conjugation was that recombinants were not found for any early markers when mated with an Hfr. The recombination frequency was decreased by a factor of 10^6 or more with a $recA^-$ female and by about 10^3 with either $recB^-$ or $recC^-$ cells. The residual recombination observed with $recB^-$ and $recC^-$ mutants is probably a result of alternative functions provide by several other genes, namely, $sbcA$, $sbcB$, and $recF$. The rec genes were mapped roughly by noting that a few recombinants appeared for late markers. The explanation was that once a particular rec gene entered the recipient from the Hfr, recombinants could form. Recombination frequencies were, however, quite low, since some degradation of transferred DNA occurred with the rec^- cells; degradation was most extensive with $recA^-$ cells. Precise mapping of these genes was accomplished by transduction, a procedure to be described in Chapter 18.

Shortly after the isolation of $recB^-$ and $recC^-$ mutants, it was found that the products of the two genes were subunits of a dimeric protein, now known as the RecBC protein. The RecA gene product is a multifunctional protein whose function in recombination is to bring two DNA molecules together. The RecA and RecBC proteins will be described in more detail shortly.

Isolation of F′ Plasmids

An Hfr cell is produced when F integrates stably into the chromosome, as we have already stated. At a very low frequency (10^{-7} per generation), F can also be excised. Excision is often imprecise in that the excised circular DNA frequently contains genes that were adjacent to F in the chromosome (Figure 14-11); that is, the excised DNA is an F′ plasmid.

F′ plasmids can be isolated from Hfr cultures by two straightforward techniques. One procedure is based on the fact that in an Hfr mating the F

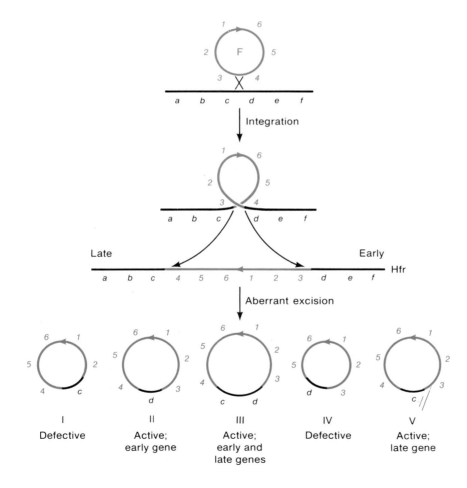

Figure 14-11 Formation of various F′ plasmids by aberrant excision from a particular Hfr strain. Plasmids I and IV have lost F genes and hence are defective. If the plasmids are replication-defective, they cannot be maintained and hence will not be detected. The usual means of detection of F′ plasmids is by the presence of genes, normally transferred late by an Hfr cell, at a time sufficiently early that the genes could not have been transferred by the Hfr cell. Thus, I and IV are normally not detected because the defects in these plasmids are defects in transfer; similarly, a type II plasmid will not be found because it only contains early genes.

segment is transferred in two stages. If cells are separated any time before 100 minutes after mixing donor and recipient cells, no recipient will receive a complete copy of F. Furthermore, if mating is interrupted at 30 minutes, no markers that enter after 30 minutes will appear in recombinants. However, when an F' forms by a rare aberrant excision, the plasmid contains a complete copy of F and the chromosomal segment that would normally be transferred last or first. Thus, in a brief mating a recipient cell that has become recombinant for a terminal marker will in general carry an F'. As a test, one need only prepare a culture of the recombinant and perform a mating with a suitable recipient; if the late marker is again transferred shortly after mating, the recombinant carries an F'. This procedure has a problem in that if a small number of mating pairs are not separated by the technique used to interrupt mating, transfer can continue on the plate and a late marker will appear to have been transferred early. However, these recombinants will not show up as F'-containing cells in a second mating and can be discarded. Nonetheless, the formation of an F' is such a rare event that the number of apparent F' plasmids recovered by inefficient interruption can be comparable to or greater than the number of cells containing an F', which requires testing a large number of recombinants. A variation of the procedure avoids this problem—namely, mating with a $recA^-$ cell. In such a mating, recombination is very rare and one can virtually count on the fact that any colony that is recombinant for a late marker consists of cells containing an F'. This technique also allows isolation of an F' plasmid containing an early marker.

Note in Figure 14-11 that a given Hfr strain can produce several F' plasmids, because the positions of imprecise excision and hence the extent of the bacterial segment can vary. Use of Hfr strains with different transfer origins allows the isolation of F' plasmids that include any region of the *E. coli* chromosome. In fact, F' plasmids have been isolated that cover the entire *E. coli* genetic map.

PROPERTIES OF SYSTEMS LACKING RECOMBINATION PROTEINS

Mutant cells lacking the ability to recombine have an additional phenotype—namely, they are exceedingly sensitive to damage to their DNA. For example, both $recA^-$ and $recBC^-$ bacteria are killed by much smaller doses of ultraviolet light than are rec^+ cells; the data for a $recA^-$ mutant was shown in Figure 9-10 in Chapter 9. The main reason for this sensitivity is that in both $recA^-$ and $recBC^-$ cells a major repair pathway—recombination repair (Chapter 9)—is absent.

The DNA of $recA^-$ cells is unstable—that is, it is continually degraded and resynthesized. The degradation is most striking following irradiation with ultraviolet light; then, more than half of the bacterial DNA is enzymatically degraded to short oligonucleotides. Since there is no degradation in a *recArecB* or *recArecC* double mutant, it is inferred that the product of these genes (a nuclease to be described shortly) is responsible for the degradation. This rampant degradation of DNA is called the **reckless** phenotype. Since the RecBC nuclease is responsible for DNA degradation in *recA*⁻ cells, it would seem that either its activity or its synthesis is regulated by the RecA product. However, biochemical studies of cell extracts of various mutants have yielded ambiguous results concerning the activity and amount of the RecBC product. Hence, this question remains open.

Another feature of *recA*⁻ cells is their slow growth compared to *rec*⁺ strains; this is indicative of other defects, most of which are unknown. Rec⁺ *E. coli* divide roughly every 25 minutes at 37°C in rich growth media. In contrast, a typical *recA*⁻ mutant may have a doubling time of 40–60 minutes. Part of the slowness of growth is probably due to reckless DNA degradation. The growth defect is more extreme for *recBC*⁻ cells, which typically divide every 100–120 minutes. These growth defects are probably the result of many processes that are either inoperative or functioning inefficiently in *rec*⁻ cells, or possibly to newly turned-on aberrant processes. For unknown reasons *recBC*⁻ mutants occasionally grow and divide without DNA replication and a daughter cell completely lacking DNA results. Of course, such cells cannot grow further.

THE RecA PROTEIN AND ITS FUNCTION

Pairing of DNA molecules is essential to all modes of homologous recombination. An understanding of how it occurs began with the purification of the RecA protein in 1976, 11 years after the first *recA*⁻ mutant was isolated.

The RecA protein has two principal biochemical activities: (1) it binds to single-stranded DNA, and (2) it is an enzyme that cleaves certain proteins. Its DNA-binding activity is the feature that is relevant to recombination (the other property is regulatory.) When acting as a DNA-binding protein, the RecA protein mediates nonspecific pairing of DNA molecules and homology-dependent **strand invasion**. Figure 14-12 shows several RecA-mediated DNA-DNA interactions that have been carried out with purified RecA protein and DNA molecules at room temperature. The structures shown are very stable and are held together by AT and GC base pairs between complementary base sequences. Study of the three interactions shown in the figure (and others that are more complex) and of pairs of DNA molecules that will not interact has shown that stable pairing is dependent on two things:

1. One molecule must be single stranded or have a single-stranded region.
2. Either one of the molecules must have a free end.

The first requirement can be (and often probably is) provided by supercoiling of the participant lacking a free end. For example, the *E. coli* chromosome is supercoiled.

Figure 14-12 Three interactions mediated by the RecA protein; sc indicates that the circle is supercoiled.

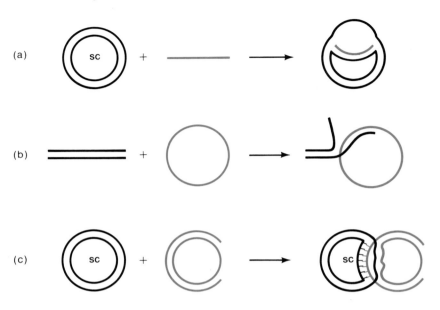

The RecA-mediated interactions shown in Figure 14-12 are the end result of a sequence of three steps: presynaptic binding of RecA protein to single-stranded DNA, synapsis, and postsynaptic strand exchange (Figure 14-13). These stages were first elucidated in a study of the RecA-mediated pairing of a double-stranded circle and a linearized form of one of the complementary strands. Details of these steps follow:

1. *Polymerization of RecA protein on single-stranded DNA.* If single-stranded DNA is mixed with RecA protein, a nucleoprotein filament forms in which the single-stranded DNA is coated with RecA protein.
2. *Synapsis.* In the presence of ATP the nucleoprotein forms a complex with the double-stranded circle. The initial interaction is *not* between homologous regions. After the initial sequence-independent interaction (termed **conjunction**), the two DNA molecules move relative to one another until homologous sequences come into contact; this is called **homologous alignment.** When homologous sequences are aligned, the two strands are not yet intertwined.
3. *Post-synaptic strand exchange.* Homologously aligned, but not intertwined, strands are bound together only weakly, and the structure is quite unstable. Once a homologously aligned region forms at the end of the single strand, RecA actively promotes the displacement of a strand from the double-stranded molecule and assimilation of the new strand. The mechanism is not yet understood, but it is clear that RecA acts as a helicase, unwinding the double-stranded DNA in advance of the forming heteroduplex.

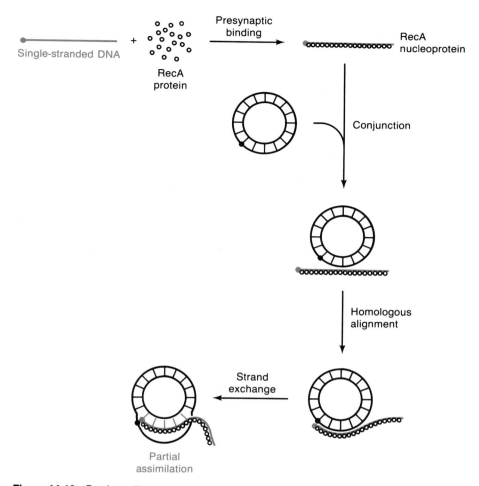

Figure 14-13   RecA-mediated pairing that leads to assimilation of a complementary strand.

Each of the interactions shown in Figure 14-12 (the three single-strand and double-strand interactions) can be explained by the multistep RecA-mediated process just described. However, this does not seem to have any obvious bearing on the homologous pairing of two double-stranded DNA molecules, which occurs in phage-phage recombination, transduction, and meiotic recombination in eukaryotes. Since it seems clear that both single-stranded DNA and a free end are required, almost all models of recombination include an early step in which one DNA strand is nicked and, in a variety of ways, unwound from the nick. No model of recombination between two double-stranded DNA molecules has yet been proved; hence we leave a presentation of such models to references at the end of this book.

The role of RecA protein in pairing of single-stranded DNA to a double-stranded molecule is consistent with physical experiments (mentioned earlier quite briefly) that indicate that only one strand of the transferred DNA combines with the chromosome of the recipient.

THE RecBC PROTEIN

In bacterial conjugation and phage transduction (Chapter 18) in *E. coli* a dimeric protein encoded by the genes *recB* and *recC* is needed in addition to the RecA protein. The RecBC protein is a multifunctional protein with several nuclease activities. Under certain experimental conditions the nuclease activities can be inhibited and the enzyme instead exhibits the ability to unwind double-stranded DNA. The unwinding that occurs when the nuclease activity is inhibited is unusual, as shown in Figure 14-14. The enzyme binds to one end of the DNA molecule and unwinds the DNA. It moves along the strand but remains in contact with two regions of that strand, allowing the end of the strand to pass through the enzyme, thus forming the single-stranded loop shown in panel (b). As the length of the single-stranded tail

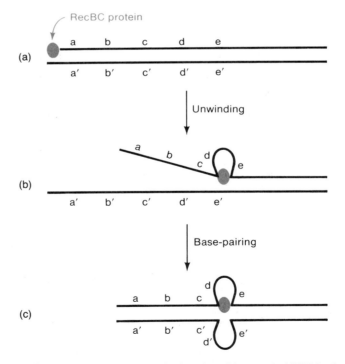

Figure 14-14   The proposed mode of unwinding of double-stranded DNA by the RecBC protein.

increases, homologous base-pairing occurs with the other strand to form the doubly-looped structure shown in panel (c). If the nuclease activity is not inhibited, nicking occurs in the single strands several thousand nucleotides part. The RecBC protein also forms a complex with DNA polymerase I but the properties of this complex are not known. The numerous enzymatic activities of the RecBC enzyme have been incorporated into many models for homologous recombination but its precise role has not been demonstrated experimentally for any particular stage of recombination.

One experimental result seems to implicate the nuclease activities as the function of the RecBC protein in recombination. *RecBC* ⁻ mutants show only a thousandfold reduction in recombination frequency compared to the almost complete block in *recA* ⁻ mutants. The residual recombination results from products of three other genes—*sbcA*, *sbcB*, and *recF*—which can be seen by the fact that multiple mutants, such as *recBC* ⁻ *recF* ⁻, are considerably more defective in recombination. Whereas the role of these genes in recombination is unknown and probably complex, the gene products have been shown to be a variety of nucleases.

ROLE OF THE *rec* GENES IN PHAGE REPLICATION

The RecBC nuclease seems to have deleterious effects on phages such as λ that engage in rolling circle replication in one stage of their life cycle (Chapter 5). An early λ gene called *gam* is responsible for synthesis of a protein that binds directly to the RecBC protein and inhibits its nuclease activity. Rolling circle replication does not occur in *gam* ⁻ mutants, which causes a significant decrease in burst size. However, progeny phage are still made, but by an alternative replicative pathway. Interestingly, this pathway includes a recombinational step mediated by a λ gene called *red*, which is responsible for most genetic recombination in λ. Thus, *red* ⁻ *gam* ⁻ phage are fairly defective in that neither the recombination pathway nor the rolling circle pathway is active. However, a few phage are still made because the RecA system provides an alternative pathway. Thus, λ*red* ⁻ *gam* ⁻ produces no progeny when infecting a *recA* ⁻ cell. Phage production is restored in a *recA* ⁻ *recBC* ⁻ host, since the Gam product is no longer needed to eliminate activity of the RecBC protein, and rolling circle replication is available. Humorously, the inability of λ *red* ⁻ *gam* ⁻ to grow in a *recA* ⁻ host is called the "feckless" or Fec⁻ phenotype. This phenomenon is analyzed in greater detail in Chapter 16.

CHROMOSOME TRANSFER IN BACTERIA OTHER THAN *E. coli*

F can be transferred to several enterobacteria other than *E. coli*. For example, F has been placed in *Salmonella typhimurium*, Hfr strains have been obtained, and more than 430 genes have been mapped. The map is almost identical to that of *E. coli* except for a large segment, about 20 minutes long, that is inverted. Hfr strains of *Citrobacter freundii* and *Erwinia chrysanthemi*, a plant pathogen, have also been obtained. F is not stably maintained in all enterobacteria, nor can it be integrated in all species in which the plasmid is stable. Thus, because of the value of genetic maps in facilitating genetic manipulations, geneticists have sought other mating systems.

In Chapter 11 broad host range plasmids, which can be maintained in many bacterial genera, were described. Two of these plasmids, RP4 and R68.45, have been particularly valuable and have been used to generate circular maps of *Acinetobacter* and *Rhizobium* species. However, these plasmids are somewhat difficult to use because they transfer at a frequency of about 10^{-8} per cell and do not form stable Hfr cell lines. However, genetically modified R'-like forms of the plasmids have been developed that transfer with greater frequency and from a single origin. These modified plasmids share a homologous region between plasmid and chromosome, which enables F'-like RecA-dependent chromosome mobilization to occur. Three methods have been used to create the necessary homology:

1. Recombinant DNA techniques (Chapter 20) have been used to create an RP4 variant that contains a segment of chromosome in the plasmid.
2. Mu is inserted into the plasmid and the plasmid is transferred into a cell whose chromosome also contains Mu inserted in a gene (the Mu insertion is made evident by a Mu-mediated mutation in the gene).
3. A transposon other than Mu is inserted into both plasmid and chromosome.

For unknown reason, chromosome mobilization using these R' plasmids does not occur in all bacterial species, even though sequences are shared with the chromosome. For example, an RP4-Mu plasmid behaves like an F' and mobilizes the chromosomes of *E. coli* and *Klebsiella pneumoniae*, but not of *Rhizobium leguminosarum* or *Proteus mirabilis*, even though the plasmid can be transferred to and from these bacteria.

So far, all bacteria discussed in this chapter have been Gram-negative. Conjugative plasmids have also been found in three genera of Gram-positive bacteria—*Streptococcus*, *Streptomyces*, and *Clostridium*. Some plasmids in *Streptococcus faecalis* and *Streptomyces coelicolor* transfer as efficiently as F in *E. coli*, but in most cases, chromosome transfer is entirely of the F^+ type (that is, at a frequency of about 10^{-7} per donor). However, plasmid SCP1 of *S. coelicolor* integrates and produces an efficient Hfr system, and accurate mapping has been done with this bacterium. The mechanism of conjugation in Gram-positive bacteria is totally different from that of *E. coli*. Contact is not made via pili. Instead, diffusible proteins, called pheromones, released by donor cells, mediate clumping of large numbers of donor and recipient cells, and transfer occurs in these clumps in an unknown way.

PROBLEMS

1. Does the F plasmid make a cell a donor or a recipient?

2. How does an Hfr cell differ from an F^+ cell?

3. Is there only one or several possible sites of integration of F in the chromosome?

4. What is the function of a counterselective marker in an Hfr $\times$ F^- mating?

5. Distinguish F^+ and Hfr transfer with respect to the amount of genetic material (DNA) transferred and the intactness of the transferred unit.

6. In a time-of-entry experiment, recombination frequency is determined for a particular gene at various times. How are these frequencies used to determine the time of entry of that gene?

7. In a set of time-of-entry curves all curves do not have a plateau at the same recombination value and the plateau value decreases with increasing time of entry. What is the major reason for this phenomenon?

8. Genes p, f, and q have times of entry of 7, 11, and 19 minutes, respectively. What is the gene order and what are the map distances, in time units, with respect to the transfer origin?

9. How are F' plasmids produced?

10. An Hfr strain with genotype $met^-his^+leu^+trp^+$ and which transfers the met gene very late is mated with a $leu^-met^+trp^-his^-$(Ts) female. The his^-(Ts) mutation introduces a requirement for histidine at 42°C. After mating for several hours the mixture is diluted and plated on media having four different supplements. The plates are incubated at 42°C. The supplements in the plates and the number of colonies per plate are the following: His + Trp, 250; His + Leu, 50; Leu + Trp, 500; His, 10.

   (a) What is the purpose of the met^- mutation in the Hfr strain in this experiment?
   (b) Which genes entered first, second, and third?
   (c) You now know the relative order of these three markers, but you know nothing about their exact location on the chromosome. What type of experiment would tell you where these markers are located on the chromosome?

11. An Hfr donor whose genotype is $a^+b^+c^+str\text{-}s$ mates with a female whose genotype is $a^-b^-c^-str\text{-}r$; the order of transfer is a b c. None of these genes are transferred early, the distance between a and b is the same as the distance between b and c, and none of the markers are near str. Recombinants are selected as usual by plating on agar lacking particular nutrients and containing streptomycin. Which of the following are true (several answers are)? Explain.

   (a) $a^+str\text{-}r$ colonies > $c^+str\text{-}r$ colonies.
   (b) $b^+str\text{-}r$ colonies < $c^+str\text{-}r$ colonies.
   (c) $a^+b^+str\text{-}r$ colonies < $b^+str\text{-}r$ colonies.
   (d) $a^+b^+str\text{-}r$ colonies = $b^+str\text{-}r$ colonies.
   (e) Most $a^+c^+str\text{-}r$ colonies will also be b^+.
   (f) Most $b^-c^+str\text{-}r$ colonies will also be a^-.
   (g) $a^+b^+c^-str\text{-}r$ colonies < $a^+b^-c^-str\text{-}r$ colonies.

12. Suppose you collect a large number of galactose-requiring (Gal$^-$) bacterial mutants and identify three closely linked genes (designated galA, galB, and galC) by complementation and rough mapping studies. You wish to order these genes and learn something about the genetic structure of the galactose region of DNA. To order the genes, you mate an Hfr having the genotype $bio^+ gal^+ str\text{-}s$ with various F^- strains having the genotype $gal^- bio^- str\text{-}r$. The Hfr transfers the bio locus later than the gal locus. You select $bio^+ str\text{-}r$ recombinants and measure the fraction of these that have the genotype gal^+. For each of the mutations the following fractions are observed: galA$^-$, 0.65; galB$^-$, 0.72; galC$^-$, 0.84. What is the gene order relative to the bio locus?

13. The order of four genes in an Hfr strain is $a\ b\ c\ d$. In a cross between an Hfr donor that has genotype $a^+ b^+ c^+ d^+ x^- str\text{-}s$ and a female that has genotype $a^- b^- c^- d^- x^+ str\text{-}r$, 90% of the $d^+ str\text{-}r$ recombinants are x^-, and 100% of the $c^+ d^+ str\text{-}r$ recombinants are x^-. The times of entry of a, b, c and d are 5, 10, 15, and 20 minutes; the str gene enters at 55 minutes. Where is x located?

14. An Hfr donor of genotype $a^+ b^+ c^+ d^+ str\text{-}s$ is mated with an F^- recipient having genotype $a^- b^- c^- d^- str\text{-}r$. Genes a, b, c, and d are spaced equally. A time-of-entry experiment is carried out and the data shown in the table below are obtained. What are the times of entry for each gene? Explain the low recombination frequency in the plateau region for $d^+ str\text{-}r$ recombinants.

Time of mating, in min	Number of recombinants of indicated genotype per 100 Hfr			
	$a^+ str\text{-}r$	$b^+ str\text{-}r$	$c^+ str\text{-}r$	$d^+ str\text{-}r$
0	0.01	0.006	0.008	0.0001
10	5	0.1	0.01	0.0004
15	50	3	0.1	0.001
20	100	35	2	0.001
25	105	80	20	0.1
30	110	82	43	0.2
40	105	80	40	0.3
50	105	80	40	0.4
60	105	81	42	0.4
70	103	80	41	0.4

15. Suppose you have isolated two independent arginine-requiring (Arg$^-$) mutant strains from a parent E. coli strain which already requires methionine (Met$^-$) and is resistant to streptomycin. You mate the two mutants (1 and 2) with an Hfr strain whose genotype is $arg^+ met^+ str\text{-}s$. Using the interrupted mating technique you obtain the time-of entry curves with the following characteristics. With mutant 1 the $arg^+ str\text{-}r$ recombinant curve extrapolates to 4 minutes, and the time of entry of met is 6 minutes. With mutant 2 the time of entry of met is again 6 minutes, but the data for the $arg^+ str\text{-}r$ recombinants extrapolate to 20 minutes. Explain the difference observed in the two matings.

16. After a brief mating between an Hfr whose genotype is $pro^+ pur^+ lac^+$ and a female whose genotype is $F^- pro^- pur^- lac^-$ str-r, many $lac^+ pur^+ pro^-$ str-r recombinants are found. A few $pro^+ lac^- pur^-$ str-r recombinants also arise, and all of these are Hfr donors. Explain this result and state the location of F in the Hfr. (The three genes *pro*, *lac*, and *pur* are very near one another.)

17. List two biochemical activities of the RecA protein.

18. What steps in the RecA-mediated synapsis process could occur between a double-stranded circle and a completely nonhomologous single strand?

19. In RecA-mediated synapsis has base-pairing occurred in the stage of homologous alignment?

20. Name two properties of RecA that are important for recombination and state the stage at which each is important.

21. An Hfr strain transfers genes in alphabetical order. Would you expect to obtain $F'y$ plasmids lacking gene z?

22. An Hfr cell transfers genes in the order $ghi \ldots def$. Which types of F' plasmids could be derived from this strain?

23. An Hfr strain transfers genes in alphabetical order. When using tetracycline sensitivity as a counterselective marker, the number of h^+ tet-r colonies is 1000-fold lower than h^+ str-r colonies found when using streptomycin sensitivity as a counterselective marker. Explain the difference.

24. An Hfr strain transfers genes in order *abc*. In an Hfr $a^+ b^+ c^+$ str-s $\times$ $F^- a^- b^- c^-$ str-r mating, will all b^+ str-r recombinants have received the a^+ marker, and will all b^+ str-r recombinants also be a^+?

25. An Hfr transfers genes in alphabetical order. A variant strain V13, known to be lysogenic for a phage XP1, is found that transfers only to gene e. That is, genes past e never seem to be transferred (at least, *f*-containing recombinants are never formed), and the frequency of transfer of genes a–d is less than for the normal Hfr strain. A female strain S132 has the property that transfer from V13 to S132 is normal; that is, time-of-entry and plateau values are the same as between a normal Hfr strain and a normal female strain. Suggest an explanation.

Chapter 15

Phage Genetics I:
Phage T4

Bacteriophages have been exceedingly important in the development of molecular genetics. Studies of these organisms, in particular, *E. coli* phages T4 and λ, have led to the discovery of many basic phenomena concerning replication, transcription, regulation, and recombination. Numerous examples have been seen throughout this book. In this chapter we will be concerned primarily with recombination processes, mapping, genome organization, and genetic procedures used to understand features of the life cycles of particular phages. The chapter begins with some general features of phages, supplementing the material presented in Chapter 5. However, phage T4 will comprise most of the chapter. For the biology of phages having single-stranded DNA or RNA, the reader should consult the references at the end of the book. It is recommended that the reader review Chapter 5 before studying this chapter.

PHAGE MUTANTS

Phage mutants have been important in research in phage biology. In the 1950s and early 1960s most available phage mutants were plaque-morphology mutants, in which the appearance of a plaque differs from that of the wild-type phage. A common variation is plaque size. For example, wild-type T4 produces a fairly small plaque, about 1 mm in diameter. The size results from a poorly understood phenomenon called **lysis inhibition.** That is, when the multiplicity of infection is high, as is the case when a plaque approaches maturity, lysis is delayed or inhibited. Thus, late in the development of a wild-type T4 plaque, growth of uninfected bacteria catches up with phage multiplication, and nutrient is exhausted, which terminates enlargement of the plaque. T4 *rII* mutants do not undergo lysis inhibition and produce plaques several millimeters in diameter (Figure 15-1). Other T4 mutations lead to very turbid plaques (*tu* mutants), very small plaques (mutations that reduce the enzymatic activity of lysozyme), and enable plaques to form on media containing inhibitory acridine compounds (*ac* mutants). Another important type of mutation is the *h* mutation, which enables a plaque to form on a cell that is resistant to wild-type phage; these mutations alter the adsorption site and allow adsorption of T4 both to wild-type *E. coli* B and to the mutant B/4 (Chapter 5). T4 and T4*h* can be distinguished by plating on a mixture of B and B/4; wild-type phage grows only only B and hence forms a very turbid plaque (owing to unimpeded growth of B/4), whereas T4*h* multiplies in and hence lyses both bacteria and yields a normal plaque. Phage λ plaque-morphology mutants of the type just described also exist, as well as the clear-plaque mutations that prevent lysogenization (Figure 5-13, Chapter 5).

A great deal of genetic research has been possible with these mutants but because plaque-morphology mutations are limited to a few genes, a fairly complete description of the phage genome was not possible with these alone. A breakthrough in the evolution of phage genetics came with the discovery of conditional lethal mutations. These were primarily chain-termination mutations that allow growth and hence plaque formation only on bacterial strains containing a chain-termination suppressor (Chapter 10). That is, the

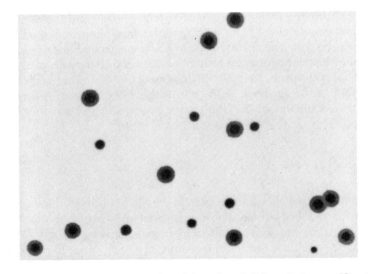

Figure 15-1   A portion of a plate showing *r*+ (small) and *rII* (large) plaques. [Courtesy of A. H. Doermann.]

mutants fail to grow on wild-type (*sup⁻* or *sup0*) bacteria, but make normal or nearly normal plaques on *sup⁺* bacteria. These mutations were isolated by brute-force screening techniques. A phage population was heavily mutagenized, plated on a *sup⁺* host, and then the plaques were replica-plated onto a *sup⁻* lawn. Hundreds of thousands of plaques were tested, which led to the isolation of thousands of mutants. In the early literature such T4 mutants were labeled *am* and the λ mutants were denoted *su;* however, once the mutations were sorted into individual genes by complementation tests, these designations were replaced by gene names and mutation numbers, such as λ*O29*. The existence of so many chain-termination mutations provided material for rather complete mapping of the phage genomes, and in fact for a few phages (e.g., λ and φX174) all phage genes have been identified.

The chain-termination mutations just described were isolated by their preventing phage mutants from plating on wild-type bacteria; hence, they are located only in genes whose activity is essential to the phage. The concept of essential and nonessential activity requires some clarification. A gene is usually considered to be nonessential if a mutation in that gene does not prevent plaque formation. If plaque formation is prevented, the gene is essential. These definitions must be interpreted carefully because, whereas a typical burst size for an infected bacterium is 50–100, a burst size of 4 is usually sufficient to form a plaque. Indeed, mutations in many so-called nonessential genes reduce the burst size markedly, though not enough to prevent plaque formation. An example is the *red* gene of phage λ, mutations of which eliminate one of two major and alternative pathways for phage production (discussed in Chapter 16).

A gene may be nonessential for three reasons: (1) An identical gene may be present in the bacterium, or a phage gene may have the same function as a gene present in a bacterium. This duplication may be of some value to the phage, because it might increase the burst size by providing a higher concentration of an essential enzyme; alternatively, in nature there may be hosts that lack the gene, and the gene will then be essential for growth. (2) The gene does not duplicate a bacterial gene but in some way increases either the rate of phage production or the burst size. In both cases (1) and (2) the gene confers an evolutionary advantage. (3) The gene is not needed in the laboratory but it enables the phage to cope with special situations met in nature. A fourth possibility—that the gene is always unnecessary—is less likely because a truly useless gene would in general lack the survival advantage needed for the phage to retain it in the face of continued and long-term spontaneous mutagenesis.

GENETIC RECOMBINATION IN PHAGES

Genetic recombination in phages was discovered by Alfred Hershey and Raquel Rotman in the late 1940s when an *E. coli* culture was infected with two different genetically marked T2 phages (T2 is closely related to T4)—

rh^+ and r^+h—each at a multiplicity of infection of about five. Roughly 98 percent of the progeny had the parental genotypes, but 2 percent consisted of about equal numbers of rh and r^+h^+ recombinants. On replating, the recombinants bred true, showing that the genotypes were stable. Furthermore, if *E. coli* was mixedly infected with the recombinants, genetic recombination occurred, again producing 98 percent rh and r^+h^+ and 2 percent rh^+ and r^+h (the original parents). This discovery was a milestone in phage genetics, rendering the T2 and T4 phages suitable objects for genetic analysis. In this section we examine several features of the recombination process in phages T4.

Effect of the Parental Ratio on Recombination Frequencies

Phage crosses differ from a typical cross in a eukaryotic cell in that a cell must be infected with several copies of the parental phages in order for recombination to occur. Furthermore, the multiplicity of infection must be high enough that all bacteria receive both parents. However, because individual phages will be distributed among host cells according to the Poisson distribution (Chapter 5), all bacteria will not receive the same number of both parents. This is an important consideration because the recombination frequency depends on the parental ratio.

In order to obtain reproducible recombination frequencies in different experiments the parental ratio (among other things) must be kept constant. Ideally one would also like to obtain maximum recombination values. A simple calculation shows the optimal ratio. Consider an infection by two phages for which the proportion of each phage is p and $1 - p$. If phage DNA molecules pair at random (which is certainly the case), the fraction of pairings between identical parents, which lead to no recombination, is $p^2 + (1 - p)^2$, in which each term is the contribution of one of the parents. *Recombination results only from pairings between different parents*, and the fraction of such pairings is $2p(1 - p) = 2p - 2p^2$. To determine the value of p that maximizes the value of this expression, we set the derivative of this expression, $2 - 4p$, equal to zero and solve for p. This shows that $p = \frac{1}{2}$ gives a maximum: recombination frequency is maximal when both parents are present in equal numbers. The Poisson distribution does not affect this conclusion since with an MOI = 5 of each parents, there will be as many cells infected with 4 of one parent and 1 of the other, as 4 of the second parent and 1 of the first.

Reciprocity in Genetic Recombination

In the cross described in the introduction to this section, both recombinants were produced in equivalent numbers, a situation that is common in phage crosses. When this occurs, recombination is said to be **reciprocal.** In eukaryotic systems recombinant gametes are always produced in reciprocal pairs. In an attempt to determine whether this is the case for phage systems single burst analysis (Chapter 5) was carried out. In these experiments cells infected by the rh^+ and r^+h parents were diluted shortly after infection and placed in individual culture tubes such that, on the average, one infected cell was present in every ten tubes. After an incubation period sufficient for lysis to occur each tube was plated, and the genetic composition of the plaques was determined for each plate. The observation was surprising: in individual bursts recombinants did not occur in equal numbers. Some bursts contained parental phages plus only one of the recombinants, and others contained greatly unequal numbers of the two recombinants (of course, most bursts contained no recombinants). A clue to what was happening came from the observations that (1) the parental types also occurred in rather unequal numbers, and (2) summing all of the genotypes from *all* bursts yielded equal numbers of both recombinants. These observations indicated that some statistical process was occurring. Later, biochemical studies of infected cells showed that only about half of the progeny phage DNA was packaged into phage heads and that DNA molecules were selected at random for packaging. Thus, the apparent nonreciprocity could be a result of fluctuations caused by random selection of molecules from the pool of newly synthesized DNA. Some years later, rather sophisticated statistical analyses of the composition of hundreds of single bursts were carried out to determine whether the existence of one recombinant type correlated with the presence of the other recombinant in the same burst, which would be expected if recombination was truly reciprocal. Indeed this proved to be the case, and recombination with phage T4 and with λ is believed to be a physically reciprocal process.

The original observation of apparent lack of reciprocity raises a general question of whether reciprocal recombinants should be expected as the outcome of a biochemical exchange process that is itself reciprocal. There is an enormous difference between being genetically nonreciprocal and being physically nonreciprocal. This distinction is exemplified in Figure 15-2. In panel (a), an exchange occurs between markers *a* and *b*. No DNA is lost; the exchange is physically reciprocal and both recombinant types are found, and thus, the exchange is also genetically reciprocal. In panel (b), breaks occur on both sides of marker *a;* the products in the first row are physically reciprocal. However, a^+/a^- heteroduplexes are generated in the overlap region (second row). This is significant because physical experiments indicate that overlaps are very large. If there were no mismatch repair, each DNA molecule would replicate and two parental genotypes and the two recombinant genotypes would result; the exchange would appear to be reciprocal. If mismatch repair were to occur prior to replication and the negative allele were converted to the positive allele in both products of the first row of (b), then the products would be one parental molecule and one recombinant molecule, as shown in the third row; genetically, the exchange would be nonreciprocal.

(a) Genetically and physically reciprocal exchange

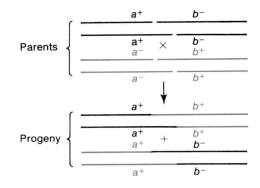

(b) Physically reciprocal but genetically nonreciprocal exchange

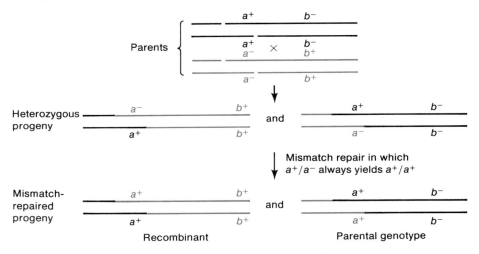

Figure 15-2 The mean of reciprocal and nonreciprocal exchanges.

Actually, there are many ways by which a physically reciprocal exchange can show genetic nonreciprocity; however, there are no simple models that can generate genetic reciprocity from an exchange that is physically nonreciprocal. The purpose of determining whether an exchange is genetically reciprocal is to put constraints on hypotheses about the physical event.

Recombination by Breakage and Rejoining of DNA Molecules

In 1931 studies with both *Drosophila* and maize demonstrated that recombination was associated with physical exchange of chromosomes. These experiments utilized mutants whose chromosomes carried physically recognizable features (knobs) that were either associated with or very near the mutant site. Recombinant organisms were selected and cells were examined microscopically. It was found that the chromosomes of recombinants possessed both morphological features, suggesting that the individual chromosomes had broken and reassembled during gamete formation, a physically reciprocal process. This idea persisted until the single-burst experiments described in the preceding section were done. In the time interval between the original observation and the ultimate explanation of the lack of apparent reciprocity resulting from statistical sampling, other suggestions were made for the mechanism of recombination. The discovery in the 1950s that all genetic information in phages resides in DNA placed the various models on a DNA level. There is no need here to go through the various models for physically nonreciprocal recombination, now known to be incorrect, for they were laid to rest in 1962 by experiments with $E.\ coli$ phage λ. These experiments utilized the density-labeling technique (Chapter 2) to determine the contribution of parental phage DNA molecules to recombinant progeny and showed that in forming recombinants parental DNA molecules break and rejoin (the earliest theory). The original experiments, performed by Matthew Meselson and Jean Weigle, are somewhat complex to interpret (because of the simultaneous activity of several recombination systems), so a variant that is more easily understood will be described.

Two types of λ phage were prepared: A^+R^-, whose DNA contained a heavy isotope in both strands, and A^-R^+, whose DNA contained a light isotope in both strands. (A and R are genetic markers at the termini of the λ map.) In addition, the DNA of both phages carried mutations to eliminate all recombination except that determined by the bacterial recombination genes; this enabled one to examine the mechanism of the bacterial recombination system specifically. Mutations in the phage and in the bacteria also prevented the initiation of normal DNA synthesis in order that density changes caused by physical exchange of parental DNA were not obscured by shifts caused by replication. Bacteria were infected with both phages, and the infection was allowed to proceed until lysis occurred. Progeny phage, which contained only parental material, were centrifuged to equilibrium in a CsCl solution and the density distribution of the genotypes of all phage particles was determined. Figure 15-3(a) shows that recombinant phage (A^+R^+ and A^-R^-) were found to range in density from fully heavy to fully light. That is, each recombinant particle contained material from both parents, indicating that physical exchange of material had occurred in forming the recombinants. When a central marker (in the c gene) was included (panel (b)), the density distribution showed that the c^+ marker from the A^+ parent appeared in A^+R^+ recombinants only if breakage and reunion occurred to the right of the marker. More complex experiments, in which DNA replication was permitted and in which other recombination systems were active, also gave evidence for breakage and reunion.

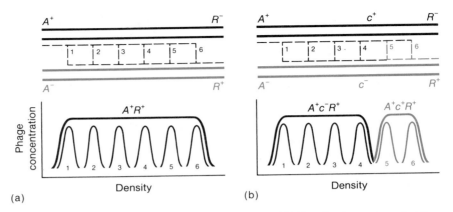

Figure 15-3 An experiment demonstrating breakage of DNA molecules and reunion to form recombinants. In the upper part of each panel, heavy lines represent high-density (black) and low-density (red) *E. coli* phage λ carrying the indicated markers. After crossing the phage, using conditions that prevent DNA replication, as described in the text, progeny phage are centrifuged to equilibrium in CsCl. The thin lines in the distribution curve in the lower part of each panel show the expected density distribution for each of the numbered exchanges shown in the upper part of the panel. The heavy curves are the expected and observed distributions for all recombinant progeny taken together. (a) The uniform distribution obtained with two markers. (b) When three markers are used, the density distribution shows that A^+R^+ phage are also c^+ only if breakage and reunion occurs to the right of the *c* marker, as expected.

Detection of Deletions by Recombination

Genetic recombination in phage is used primarily as a mapping technique, and we will examine this in some detail throughout this chapter. However, measurement of recombination frequencies can also be used to detect a large deletion, a procedure we will wish to make use of later.

Consider two markers *a* and *b* that recombine with a frequency of 25 percent. A mutation *d* is found that from three-factor crosses maps between these markers. In two-factor crosses the recombination frequencies for the intervals *a–d* and *b–d* do not sum to 25 percent, which is not unusual for such a large interval (there are numerous explanations). One possibility is that *d* is a large deletion. If so, the recombination frequency between *a* and *b* if both parental phages contain *d* will be severely depressed because there is less material between *a* and *b*. Such an observation has often led to the discovery of large deletions.

GENETIC MAPPING OF PHAGE T4

The earliest genetic mapping experiments were carried out with *E. coli* phages T4 and λ. The results of a typical cross, between T4 *r48* (large plaque) and *tu42* (plaque with a light turbid halo), namely,

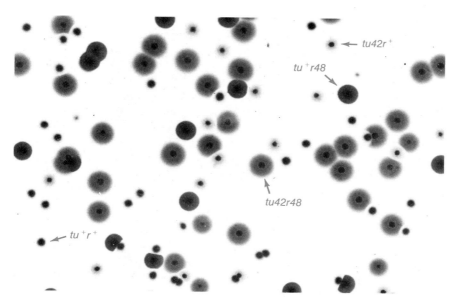

tu42r⁺

tu⁺r48

tu42r48

tu⁺r⁺

Figure 15-4 Progeny of a cross between *E. coli* T4 phage *tu⁺r⁺* (a plaque is labeled at the lower left) and *tu42r48* (center). Two types of recombinant plaques are found; representative plaques are labeled. [Courtesy of A. H. Doermann.]

$$tu42r48(\text{turbid, large}) \times tu^+r^+ \text{ (clear, small)}$$

is shown in Figure 15-4. Four plaque types appear—the parental types and the $tu42r^+$ (turbid, small) and tu^+r42 (clear, large recombinants)—as indicated. Note that both parental and recombinant types are easily identifiable. Combining the data from numerous crosses, in which recombination frequency is defined, as usual, as

$$(\text{Number of recombinant phage/total number of phage}) \times 100$$

generates a typical linkage map. Frequently different bacteria are used to detect recombinants, and all recombinant types are not seen. For example, in a cross between two chain-termination mutants, one usually plates the lysate on an sup^+ host to score the total number of progeny phage and on an $sup0$ host to count the number of wild-type recombinants. The doubly mutant recombinants are not seen. However, since reciprocity in bulk lysates is the rule, the total number of recombinants is assumed to be twice the number of wild-type recombinants counted.

In this section we examine several features of the T4 map and show how these led to an understanding of unexpected aspects of genome organization and phage production.

The Circular Map of T4

T4 contains quite a large DNA molecule (173,000 base pairs). In the phage life cycle many progeny DNA molecules are made rather quickly, so *in toto*

T4 DNA molecules engage in several (about five) rounds of mating. This has the effect that the recombination frequency per unit length of DNA is fairly high, and in fact the T4 map has a length of about 1500 map units. This high frequency of recombination means that markers do not have to be very far apart (only a few percent of the genome length) before they appear to assort randomly. Thus, genetic mapping in which genetic distances are determined can only be carried out with markers that are quite near one another. In the early days of T4 mapping, when the number of available markers was quite limited, the genetic map appeared to consist of several linkage groups, which suggested that T4 phage might contain as many as seven different DNA molecules. In 1959 physical experiments made it clear that T4 has only one DNA molecule, so a search began for additional mutants that would enable the different linkage groups to be joined together in a single map. Figure 15-5 shows one of the early unitary maps.

The map shown in Figure 15-5 showed no unusual features until attempts were made to confirm map positions by three-factor crosses. Recall from Chapter 1 that a three-factor cross gives the gene order unambiguously because of the six possible recombinant classes, the two rarest classes can be assumed to be a result of double exchanges. When the cross $r67h42ac^+ \times r^+h^+ac41$ (see the figure for positions of these genes) was carried out, the rarest recombinant classes were observed to be $r67h^+ac^+$ and $r^+h42ac41$, which indicates the map order $r67–h42–ac41$. This order conflicted with the map shown in the figure, which says that $r67$ is closer to ac than to h. Other three-factor crosses also placed the ends of the linear map adjacent to one another. The resolution of this paradox was the proposal that the ends of the map were closed on themselves, forming the circular map shown at the bottom of the figure.

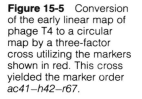

Figure 15-5 Conversion of the early linear map of phage T4 to a circular map by a three-factor cross utilizing the markers shown in red. This cross yielded the marker order *ac41–h42–r67*.

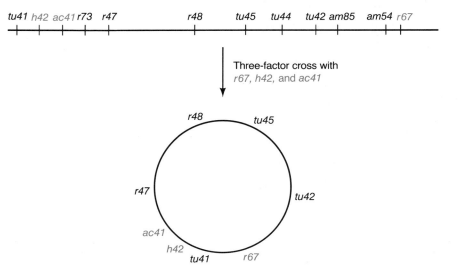

A more complete map of T4, in which gene function is identified, will be shown later in the chapter.

Possible Explanations for the T4 Circular Map

The most obvious explanation for a circular genetic map is that the DNA itself is circular. Whereas today circularity could be ascertained simply by electron microscopy, this technique was not available at the time that the circular map was discovered. A simple measurement of the viscosity of a T4 DNA sample exposed to a nuclease provided the necessary information. The viscosity of a solution of a macromolecule is affected primarily by the shape of the molecule at constant molecular weight: a long thin molecule yields solutions with a higher viscosity than a short molecule. Indeed, solutions of DNA have very high viscosities because, for a given molecular weight, DNA molecules are much longer than most macromolecules. The crucial experiment consisted of placing a sample of T4 DNA molecules in a viscometer with a small amount of a DNase and measuring the viscosity of the solution repeatedly. The viscosity of a linear DNA sample decreases continually as strand breakage causes the molecules to become shorter. However, for a circular DNA molecule the first break will linearize the circle, extending the molecule, and hence cause the viscosity to rise; subsequent breaks will then decrease the viscosity. Thus, if T4 DNA were linear, the viscosity of DNase-treated T4 DNA should decrease continually with time; if it were circular, the viscosity should first rise and then decrease. Accurate viscosity measurements showed that there was no rise in viscosity and hence that the DNA is linear.

Given a linear DNA, a simple explanation of the circular map would be that the DNA assumes a circular form after infection. However, an alternative hypothesis was also considered: (1) Individual DNA molecules are terminally repetitive (redundant) in the sense that the gene order is *abc . . . xyzabc;* (2) a population of T4 DNA molecules is cyclically permuted in the sense that individual molecules have the gene orders *bcd . . . yzabcd, cde . . . zabcde, def . . . abcdef,* etc. This somewhat bizarre hypothesis was based on an unusual property of T4 heterozygotes described in the next section.

Phage Heterozygotes

A heterozygote contains two alleles of a gene—in most experiments, wild-type and mutant. In diploid organisms this arrangement arises when one member of a homologous pair of chromosomes contains the wild-type allele and the other homologue contains the mutant allele. Furthermore, heterozygous diploids breed true. A phage contains only a single chromosome, so heterozygotes, if they were to exist, would have another structure. Phage

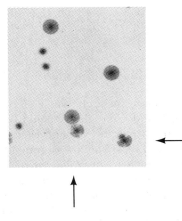

Figure 15-6 Results of a mixed infection with r^+ (small plaques) and rII (large plaques) phage showing mottled plaques (arrows).

T4 heterozygotes were first observed in mixed infections with r^+ (small plaques) phage and rII (large plaques) phage. A small fraction of the progeny produce unusual plaques termed mottled (Figure 15-6). If phage are isolated from these plaques and replated on fresh bacteria, half of the resulting plaques are r^+ and half are rII. Henceforth, both of these phage breed true. A simple explanation for heterozygotes of this type is that the phage that produce mottled plaques are **overlap heterozygotes.** That is, in the course of breakage and rejoining of two parental DNA molecules, the breaks are situated such that the overlapping heteroduplex region consists of one strand with the r^+ allele and the complementary strand with the rII allele. Thus, the mottled plaque would not contain heterozygous particles but would consist of the progeny of an original heterozygote that produced both parental types when it replicated. Such heterozygotes would not be unexpected and their existence was confirmed in two experiments, one with T4 and one with λ. (1) If *E. coli* is infected with both T4r^+ and T4rII in medium containing a partial inhibitor of DNA replication, the fraction of progeny that are r^+/rII heterozygotes increases—that is, the fraction of the plaques that are mottled increases. A reduced number of rounds of replication would decrease the number of overlap heterozygotes that would normally be converted to two parental types by DNA replication. (2) In the experiment shown in Figure 15-3, in which two λ phage were crossed, cI^+/cI^- heterozygotes (which produce a plaque with clear and turbid regions) were found but only at a density that would correspond to breakage in the central part of the λ DNA, where the cI gene is located.

Another feature of T4 heterozygotes indicated that a second type of heterozygote must also be present. Numerous deletions in T4 have been isolated. These deletions have little influence on most genetic processes other than bringing flanking genes nearer (see earlier section). However, the

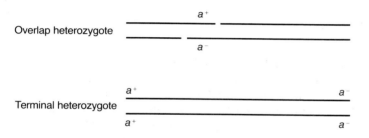

Figure 15-7 Two types of heterozygotes in T4. The overlap heterozygote is an immediate product of recombination, such as that in panel (b) of Figure 15-2. The a^+ and a^- alleles segregate when the DNA replicates. The terminal heterozygote occurs when the two alleles are present in the terminally redundant region. These alleles do not separate when the DNA replicates.

deletions also increased the length of the heterozygous region. If phages bearing several closely linked markers are used in a mixed infection, a small fraction of progeny are found to be heterozygous for more than one marker. One explanation is that two markers can be present in an overlap region, which is certainly true. However, if the phages also both carry a distant deletion (so distant that the deletion cannot possibly affect local recombination in any way), the number of heterozygotes for the two markers increases. Furthermore, when the deletion is present, heterozygotes containing more than two markers can be found that are undetectable if the deletion was absent. It is not obvious how a distant deletion could affect the length of an overlap heterozygote. Furthermore, the number of these double heterozygotes is not increased by inhibiting DNA replication. Thus, it seems clear that T4 possesses two types of heterozygotes—those that increase in number when DNA replication is inhibited and those that increase in length when a deletion is present.

With great insight George Streisinger and Frank Stahl suggested that the second type is a **terminal heterozygote**. That is, they proposed that T4 DNA is **terminally redundant** so that some heterozygotes, for example, r^+/rII, could have one allele at one terminus of the molecule and the other allele at the other terminus (Figure 15-7). That is, the gene order of the molecule would be $qr^+st \ldots abcd \ldots nopqrIIs$. They then explained the effect of a deletion by assuming that *the amount of DNA contained in a phage head is fixed*. Thus, if the deletion eliminated a gene or part of a gene, the heterozygous segment could be included in a longer terminally redundant region. Such a heterozygote would not produce a mottled plaque, since progeny would also be terminal redundancy heterozygotes; to generate mottling they proposed a replicating and packaging scheme by which a circularly permuted set of molecules would be generated. In a circularly permuted set the markers in a terminal redundancy heterozygote will separate from one another, because the terminal redundancy of most unit-sized molecules will not include these markers. This scheme is described in the following section.

Packaging and the Production of Cyclically Permuted, Terminally Redundant DNA Molecules

The mechanism of packaging of T4 DNA in a phage head is not yet elucidated, though the process can be carried out *in vitro*. It is known that replication of T4 DNA generates enormously long molecules (Figure 15-8) called **concatemers** and that packaging proceeds by cutting individual phage-DNA units from these concatemers. The cuts are not made in unique base sequences in the DNA because, if they were, T4 DNA could not be cyclically permuted. Instead, the cuts are made at positions that are determined by the amount of DNA that can fit in a head. Presumably a free end of the DNA molecule enters the head and this continues until there is no more room; then the concatemer is cut. This is known as the **headful mechanism** and it explains how both terminal redundancy and cyclic permutation arise (Figure 15-9). The essential point is that *the DNA content of a T4 particle is greater than the genome length*. Thus, when cutting a headful from a concatemeric molecule, the final segment of DNA that is packaged is a duplicate of the DNA that is packaged first—that is, the packaged DNA is terminally redundant. The first segment of the second DNA molecule that is packaged is not the same as the first segment of the first phage. Furthermore, since the second phage must also be terminally redundant, a third phage-DNA molecule must begin with still another segment. Thus, the collection of DNA molecules in the phage produced by a single infected bacterium is a cyclically permuted set.

Figure 15-8 An electron micrograph of the replicating complex of T4 DNA. Note how it resembles the structure of *E. coli* DNA. [Courtesy of Joel Huberman.]

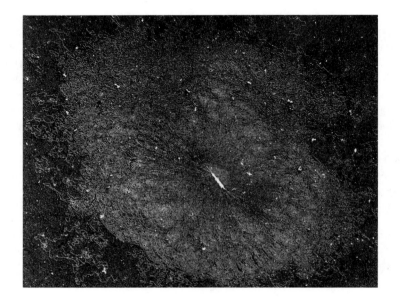

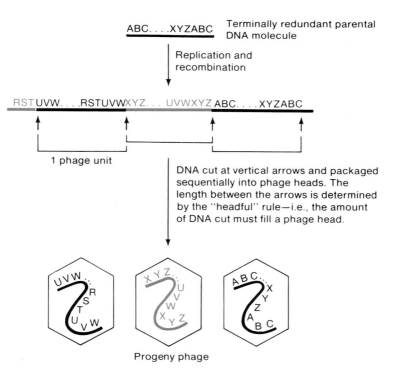

ABC....XYZABC   Terminally redundant parental
DNA molecule

Replication and
recombination

RST UVW....RSTUVWXYZ...UVWXYZ ABC....XYZABC

1 phage unit

DNA cut at vertical arrows and packaged
sequentially into phage heads. The
length between the arrows is determined
by the "headful" rule—i.e., the amount
of DNA cut must fill a phage head.

Progeny phage

Figure 15-9   Origin of cyclically permuted T4 DNA molecules. Alternate units are shown in different colors for clarity only.

Proof of the explanations just given for the circular map came from physical experiments that directly demonstrated terminal redundancy of individual DNA molecules and cyclic permutation of the DNA population. Terminal redundancy can be shown by treating DNA with a DNase that removes bases only from the two 5′ ends of the DNA (Figure 15-10). If the amount removed is greater than the terminally redundant segment, which from genetic arguments is about one percent of the genome length, each molecule will be terminated by complementary single-stranded regions. For example, one terminus will have the sequence 3′-*abc* and the other the sequence *a′b′c′*-3′, in which a prime denotes a complementary base. Hence, if such treated DNA is exposed to renaturing conditions, the complementary termini will renature and a circular molecule will form. Such circular molecules were seen by electron microscopy, confirming the prediction of terminal redundancy. The length of the terminally redundant region was also determined by examining molecules in which the circle contained a small double-stranded segment flanked by two single-stranded regions. These arise in the following way. If the terminal single-stranded segments were 3′-*abcdef* and *xyza′b′c′*-3′, a double-stranded region consisting of renatured *abc* and *a′b′c′* will be flanked by single-stranded *xyz* and *def*. The length of the double-stranded segment is the length of the terminally redundant region.

Figure 15-10 A terminally redundant molecule and its identification by means of exonucleolytic digestion and circularization. A nonredundant DNA molecule cannot be circularized in this way.

(a) Terminally redundant DNA

5' ——————————————————————————————————————— 3'
   A B C D E F G   W X Y Z A B C
 3' A'B'C'D'E'F'G'   W'X'Y'Z'A'B'C' 5'

(b) After digestion with a 3' exonuclease

5' ——
   A B C D E F G   W X
                F'G'   W'X'Y'Z'A'B'C' 5'

(c) After circularization of the molecule in (b)

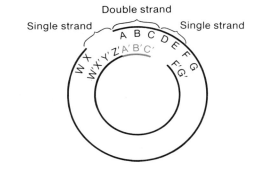

Cyclic permutation was demonstrated by denaturing a T4 DNA sample and then allowing renaturation to occur. Since renaturation is a random process, pairing of unlike but complementary strands will occur and circular molecules bearing single-stranded termini will result (Figure 15-11). Electron micrographs of renatured DNA showed the predicted branched circles, and thereby demonstrated that T4 DNA is cyclically permuted.

Further confirmation of the headful model came from studies of T4 chain-termination mutants with altered head proteins. When infecting *sup0* cells these mutants produced aberrant heads that were sometimes larger and sometimes smaller (depending on the particular mutation) than the wild-type head. It was found that independent of head size the aberrant heads were always filled to capacity. Some of the giant phage heads had as much as six times the normal amount of DNA.

Genetic Maps of Other Phages

Genetic maps have been obtained for many phages. Circularity is not a universal feature of these maps. For example, the map for phage T7 is linear, which corresponds to the linear DNA molecule contained in the phage. The DNA is terminally redundant, but not cyclically permuted. T7 DNA also

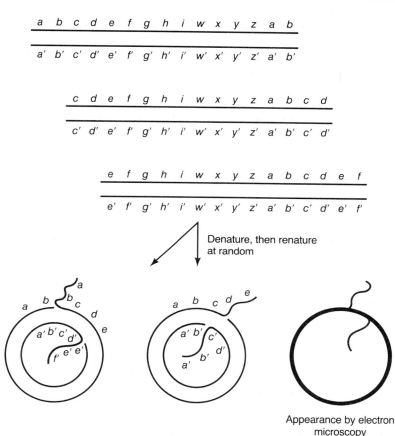

Figure 15-11 An electron microscopic test of cyclic permutation. A DNA sample is denatured until strands separate completely. Renaturation of strands with different termini produces double-stranded circular molecules (heavy line) with short single-stranded branches (thin lines).

remains linear throughout the life cycle of the phage. In contrast is phage λ, which has a circular map (described later), though its DNA molecule is linear. However, λ circularizes shortly after infection. *Salmonella typhimurium* phage P22 has a circular map, and, like T4, its DNA is terminally redundant and cyclically permuted. Of the hundreds of phages whose DNA has been examined, the following types of linear DNA have been observed: terminally redundant and cyclically permuted, terminally redundant but not permuted, and neither redundant nor permuted. No DNA has ever been observed that is permuted but not terminally redundant. Furthermore, both linear and circular molecules have been isolated from various phages. Most DNA molecules, if they are linear, circularize after infection; T7 and closely related *E. coli* phages are rare exceptions.

FINE-STRUCTURE MAPPING OF THE T4 *rII* LOCUS

Until the mid-1940s there were no examples of recombination between mutations within a gene. Such exchanges were not observable because of the lack
of a system to detect low recombination frequencies. The first indication of
intragenic recombination came from a study of the *lozenge* locus of *Drosophila*. However, the significance of the observation was unclear since little
was known about the substructure of the gene or, on a molecular basis, what
a gene is. The situation changed dramatically in the mid-1950s—first, with
the recognition that genes are segments of DNA and, second, with the work
of Seymour Benzer, who carried out extraordinarily detailed mapping of the
T4 *rII* locus. In his experiments about 2400 independent mutations were
mapped at 308 sites, with the goal of learning something about the internal
organization of a gene.

Benzer's work exploited specific features of the T4 *rII* locus. Figure 15-
1 showed the plaques formed by wild-type (T4r^+) phage and an *rII* mutant.
The T4 r^+ plaques are small with fuzzy edges, whereas the *rII* mutants
produce larger plaques with sharp edges. Because recombination occurs with
high frequency in T4, recombination between *rII* markers could be detected
by the appearance of small r^+ plaques—for example, certain *rII* markers
recombine with frequencies as high as 8 percent. T4 *rII* mutants are easily
isolated, as large-plaque mutants that arise at a frequency of about 1 mutant
per 10^5 wild-type phage. The mutation frequency can be enhanced considerably by treatment with various mutagens, so 20 plates with 500 plaques
per plate will yield several mutants. Large-plaque mutants fall into three
classes, *rI*, *rII*, and *rIII*, which map in different regions of the genome. The
rII mutants were of special interest for fine-structure mapping, since they
have another property that Benzer realized would enable him to detect
exceedingly low frequencies of recombination between different *rII* mutants—
that is, *rII* mutants are conditional lethals that fail to form plaques on *E. coli*
strain K12 that is lysogenic for phage λ. The inability to plate on K12(λ) is a
result of the activity of a λ gene, called *rex*, which is transcribed in a lysogen.
How the *rex* gene causes this inhibition has eluded geneticists now for about
30 years, but understanding the mechanism underlying the plating deficiency
is unnecessary to the use of the system. The significant features of the *rII*-
K12(λ) system are the following:

1. The biochemical block to *rII* growth in K12(λ) is so complete that
   the rare plaques that do form are always revertants of some kind.
   The reversion frequency for many *rII* mutants is sufficiently low that
   recombination frequencies as low as 0.00001 percent could be detected.
   Actually, the recombination frequencies in two-factor crosses are never
   that low, but this sensitivity makes possible multifactor crosses within
   the locus.

2. *E. coli* strain B supports the growth of both *rII* and r^+ phages, and it is on this strain that the distinction in plaque morphology is easily made.

3. Crosses can be performed between different *rII* mutants by infecting *E. coli* B. The number of recombinants can be measured simply by plating the progeny both on *E. coli* B (to detect all phage) and on K12(λ), on which only the r^+ recombinants grow. Since phage recombination is reciprocal, a double recombinant forms for every r^+ recombinant; thus, the recombination frequency is twice the number of plaques formed on K12(λ) divided by the number of plaques formed on strain B.

4. Deletions in the *rII* locus are found at a reasonable frequency and can be identified as mutations for which no plaques appear on plates containing *E. coli* K12(λ) and 10^8 mutant phage particles. In contrast, a typical *rII* point mutant will yield a few, or perhaps as many as 100, plaques.

5. Point mutants (that is, revertable mutants) fall into two complementation groups, *rIIA* and *rIIB*. In a typical complementation test K12(λ) bacteria are infected with two *rII* mutants. Infection by only one phage mutant yields no progeny. However, mixed infection with some pairs of mutants produces progeny, which consist predominately of the two infecting phage types. More will be said about complementation shortly.

In the preliminary study of the *rII* system about 50 *rII* mutants were sorted into complementation groups, and mapped in the conventional way by performing two-factor crosses. However, the formidable goal of mapping the entire collection of about 2400 mutants could not be achieved by the direct method of carrying out two- or three-factor crosses, because the number of required crosses would be more than a half million. One of the important technical contributions of Benzer's study was his introduction of procedures that vastly reduced the required number of crosses. One procedure was to subdivide the *rII* region into several segments and locate each mutation roughly within that segment. The technique, which was first exploited in mapping *Drosophila,* was **deletion mapping.** This procedure is based on a simple principle: a phage carrying a deletion cannot recombine with another phage having a mutation located in the region covered by the deletion, to form wild-type recombinants, because a wild-type allele is not present in either phage.

To locate mutations by deletion mapping, it was necessary to determine, at least, approximately, the location and size of the deletion. This was accomplished first by crossing the deletions against one another. If two phage with overlapping *rII* deletions are crossed, r^+ recombinants cannot form because certain regions of the gene will be missing in both phage. However, if the

Figure 15-12 Mapping of deletions by the overlap method. The red boxes represent three deletions. In cross 1 (I × II) and cross 2 (I × III), no wild-type recombinants are produced (−), because the deletions overlap. Wild-type recombinants are produced—(+) in cross 3 (II × III)—because the deletions do not overlap. Indication of overlap by the lack of appearance of wild-type recombinants yields an unambiguous order of the deletions.

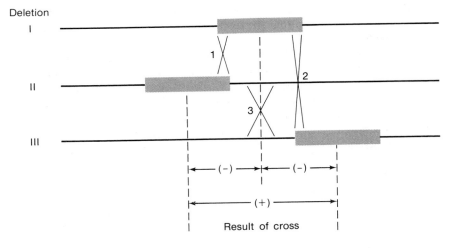

deletions do not overlap, wild-type recombinants will form. Thus, by cross-ing many pairs of deletion mutants, deletions could be ordered; the principle is shown in Figure 15-12. Of many *rII* deletions studied, seven large dele-tions were found that spanned the region in which all of the 50-or-so mapped *rII* point mutations were located. This yielded the relative order of the dele-tions. Then, crossing a number of individual mutations against the deletions provided more detailed information about the limits of the deletions.

By using crosses between deletions and point mutations Benzer was able to localize each of the 2400 mutations into one of seven major regions. He then used a set of smaller deletions, roughly four corresponding to each major deletion, to localize each mutation more precisely. In this way, the approximate location of each mutation could be determined by 11 crosses (seven with major deletions and four with small deletions), so with about 25,000 crosses all mutations could be localized. To perform such a huge number of crosses, though much reduced from the more-than-500,000 needed for mapping without deletions, would be a formidable task if each cross had to be done in the standard way—namely, mixed infection of a growing bac-terial culture, incubation until lysis occurred, and plating the lysate on both *E. coli* B and K12(λ). To reduce the labor, a special technique—a **spot test**—was developed by which 10–20 crosses could be carried out on a single plate. Furthermore, in a spot test, no phage mutant had to be handled more than once, and plaque counting was unnecessary. These spot tests were performed in the following way. A plate was prepared in which the soft agar contained about 10^8 K12(λ) cells, 10^6 B cells, and 10^7 phage carrying a dele-tion. If this plate were incubated, roughly 90 percent of the deletion phage would adsorb to the K12(λ), where they could not multiply. The remainder would infect the *E. coli* B and produce progeny, most of which would then adsorb to K12(λ). The K12(λ) always outnumber the B bacteria, so the small amount of lysis of the B cells could not produce a visible plaque. Thus, these plates would produce a uniform lawn, appearing as if no phage had been present in the agar. In performing a spot test a drop containing about 10^7 particles of an *rII* mutant was placed in a small spot on the surface of the agar. After incubation of the plate three types of spots were found:

1. *Complete clearing of the spot*. This observation is the result of complementation. If the mutant and the deletion are in different complementation groups, some of the K12(λ) infected with both phages will, by complementation, produce a burst containing both phage types. These will continue to mixedly infect K12(λ) and ultimately lyse all bacteria in the region.
2. *A few plaques in the spot*. If both the deletion and the mutation are in the same complementation group, massive phage production, such as that in 1, is not possible. However, if the deletion does not include the mutant site, recombination can occur in the B cells, producing a few r^+ phage that can then form a plaque on the excess K12(λ).
3. *No plaques*. If the deletion includes the mutant site, neither complementation nor recombination can occur and no phage are produced.

Thus, the spot test enabled Benzer to determine whether a mutant overlapped a particular deletion, and if it did not, whether it was in the *rIIA* or *rIIB* complementation group. Usually about 20 different mutants can be spotted on a single plate.

Another spot test was used to identify deletions and at the same time to localize the deletions. This is exemplified in Figure 15-13. In this case two deletions are taken as standards: *r164* is in complementation group A, and *r196* is in group B. A mutant to be tested is mixed with *E. coli* B and K12(λ), as above, such that 10 percent of the bacteria are infected. The left panel of

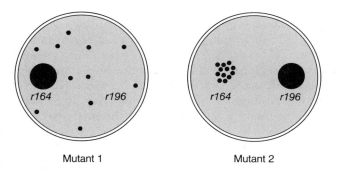

Mutant 1   Mutant 2

Figure 15-13 A spot test used to classify *rII* mutations. Dark regions indicate plaques (small spots) or cleared regions (large spot). Scattered plaques in the background indicate that the tested phage is a point mutation; absence of plaques implies a deletion. A large clear spot represents complementation. Small plaques in the spot result from recombination. See text for details.

the figure, in which mutant 1 is tested, shows a background of plaques throughout the plate, indicating that this mutant is a point mutant (the plaques represent revertants). In addition, mutant 1 complements *r164* (complete clearing) and fails to recombine with *r196* (no plaques). Thus, mutant 1 is located in group B and must be within the *r196* deletion. The right panel shows that mutant 2 is a deletion, because there is no background of plaques. Furthermore, mutant 2 recombines but does not complement *r164* (a few plaques) and complements *r196* (complete clearing). Thus, mutant 2 is a deletion in region A and does not overlap *r164*.

Once each mutation had been localized in a single region (determined by deletion mapping) Benzer used individual two-factor crosses for precise determination of recombination frequencies and hence derived a detailed map of the *rII* locus. The following conclusions could be drawn from these data:

1. All of the 2400 mutants mapped into only 308 sites. The distribution was not random, for some sites are much more mutable than others. These sites were called hot spots (see also Chapter 10, Figure 10-6).

2. All mutations in complementation group A mapped at one end of the map, and all in group B were at the other end of the map. This observation was the first real proof that a complementation group represents a contiguous region of DNA, likely to be that corresponding to a single polypeptide chain. Benzer introduced the term **cistron** to represent a region of the DNA that encoded a single polypeptide chain. This term was designed to replace the word gene, but has never really done so. However, it remains in the term polycistronic mRNA (Chapter 6):

3. The smallest recombination frequency observed between nearby markers was 0.02 percent. Since the total genetic map of T4 is about 1500 map units long, this represents $0.02/1500 = 1.33 \times 10^{-5}$ of the map. If one assumes that recombination occurs with the same frequency in all regions of the DNA, from the size of the T4 DNA (1.73×10^5 nucleotide pairs), the smallest region in which recombination could occur is 2.3 nucleotide pairs. Since the assumption of uniform recombination at the nucleotide level is probably incorrect, and certainly mutations would not be recovered at all possible nucleotide positions, it seemed likely that recombination could occur between any pair of nucleotide pairs. In later years physical measurements showed that the *rIIA* and *rIIB* genes consist of about 1800 and 850 nucleotide pairs, respectively. These numbers enabled map distances to be correlated roughly with physical distances. These data suggested that genetic exchange could probably occur between mutations in adjacent nucleotide pairs, a fact that was proved later by direct sequencing of recombining mutant forms of the *E. coli trpE* protein.

The fine-structure map of the *rII* region has been widely used in the last 25 years. For example, it was first shown that the genetic code is a nonoverlapping triplet code by analysis of the effects of certain types of single-nucleotide insertion or deletion mutations in the *rIIB* cistron. Furthermore, the mechanism of action of numerous mutagens was worked out by studying the pattern of induction and reversion of mutations at particular sites in the *rII* region. The heterozygotes discussed earlier in this chapter that led to an understanding both of the arrangement of genes in T4 DNA and the mechanism of packaging of DNA in the phage head were *rII* heterozygotes. Finally, detailed studies of the mechanisms of recombination have been carried out by studying exchanges between *rII* mutations separated by various distances. The sensitivity of the *rII* system is great enough to detect multiple exchanges between as many as ten markers; these experiments have yielded information about the clustering of exchanges and what may happen in or near a region of genetic exchange.

FEATURES OF THE T4 LIFE CYCLE

The T4 life cycle is not atypical but has certain features that are worthy of note. In brief, the timing of the life cycle is outlined below, with times in minutes at 37°C.

$t = 0$ Phage adsorbs to bacterial cell wall. Injection of phage DNA probably occurs within seconds of adsorption.

$t = 1$ Synthesis of host DNA, RNA, and protein is totally turned off.

$t = 2$ Synthesis of first mRNA begins.

$t = 3$ Degradation of bacterial DNA begins.

$t = 5$ Phage DNA synthesis is initiated.

$t = 9$ Synthesis of "late" mRNA begins.

$t = 12$ Completed heads and tails appear.

$t = 15$ First complete phage particle appears.

$t = 22$ Lysis of bacteria; release of about 300 progeny phage.

The main feature to be noticed at this point is the orderly sequence of events. Figure 15-14 illustrates these events. In the sections that follow, a few of these stages are described further.

Taking Over the Cell

Shortly after infection, several events occur that enable the phage to turn off many bacterial functions necessary for continued bacterial growth. For example, the host RNA polymerase is modified in such a way that host promoters are recognized poorly. Second, by an unknown mechanism host macromolecular synthesis is turned off. Finally, the first phage mRNA encodes

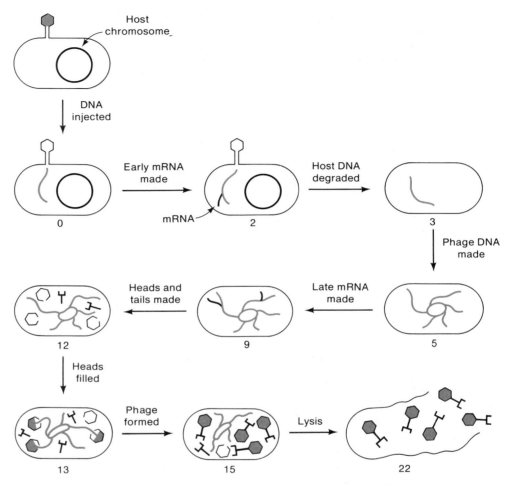

Figure 15-14   A schematic diagram of the life cycle of phage T4. The numbers represent time after injection in minutes at 37°C. For clarity, mRNA is drawn only at the time at which its synthesis begins.

DNases that rapidly degrade host DNA to nucleotides (this is not a common feature with phages).

Part of the takeover process is a series of events that allow transcription of T4 to occur in an orderly way. The complete pattern of early transcription of T4 DNA is very complex. However, the basic pattern, which has also been observed in several other phages, is the following. Transcription of early mRNA starts at a single class of promoter by means of *E. coli* RNA polymerase (because that is the only polymerase in the cell). Then, the polymerase is modified first by the addition of a small molecule, ADP-ribose, and later by the addition of phage-specified proteins, with the result that it no longer recognizes host promoters but goes on instead to initiate transcription of later species of mRNA at phage promoters. Successive modifications continue and provide one means of temporal control of the synthesis of many species of T4 mRNA.

Replication of T4 DNA

T4 DNA differs from typical DNA molecules in that it contains no cytosine (C). Instead a modified base, 5-hydroxymethylcytosine (HMC), pairs with guanine. Furthermore, the hydroxymethyl group has various glucose-like sugars covalently linked to it, so T4 DNA can be considered to be cloaked in sugar chains (Figure 15-15). The presence of HMC and its glucosylation introduces particular requirements on the phage life cycle.

Five aspects of T4 DNA replication are especially interesting: (1) the source of nucleotides; (2) the synthesis of HMC; (3) the prevention of incorporation of cytosine; (4) glucosylation of T4 DNA; and (5) the enzymology of replication. The first four are discussed below. The enzymology is quite complex.

1. *Source of T4 DNA nucleotides—degradation of host DNA*. An early event in the T4 life cycle is the degradation of host DNA to deoxynucleoside monophosphates (dNMP). The responsible enzymes, which are active *only on cytosine-containing DNA*, cleave the host DNA to double-stranded fragments, which are then degraded to dNMP by a phage-encoded exonuclease. These mononucleotides are then built up to dATP, dTTP, dGTP, and dCTP by the usual *E. coli* enzymes, and this provides sufficient dNTP to synthesize 30 T4 DNA molecules. DNA precursors are synthesized *de novo* but, to ensure an abundant supply of dNTP, five phage-encoded enzymes, which are virtually identical in activity to the *E. coli* enzymes, are also synthesized.

2. *Synthesis of 5-hydroxymethylcytosine (HMC)*. *E. coli* does not possess enzymes for forming HMC; therefore, this is accomplished by two phage enzymes, which convert dCMP to dHDP. The *E. coli* enzyme, nucleoside phosphate kinase, which forms all triphosphates in *E. coli*, then converts dHDP to dHTP, the immediate precursor of the HMC in the DNA.

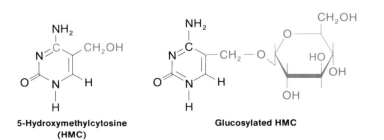

5-Hydroxymethylcytosine (HMC)

Glucosylated HMC

Figure 15-15
Nonglucosylated and glucosylated 5-hydroxymethylcytosine. If the CH_2OH (red) in HMC were replaced by hydrogen, the molecule would be cytosine.

3. *Prevention of incorporation of cytosine into T4 DNA*. T4 DNA polymerase cannot distinguish dCTP from dHTP, both of which can hydrogen-bond to guanine. It is essential that no cytosine be incorporated into daughter T4 DNA strands, because such cytosine-containing DNA would be a substrate for the T4 nucleases that degrade host DNA. *E. coli* itself has no use for enzymes that prevent incorporation of C into DNA; thus, the phage must encode these enzymes. For C to become part of daughter DNA molecules, dCMP must be converted to dCDP and then to dCTP. A phage enzyme, called dCTPase, degrades both dCDP and dCTP to dCMP.

Another phage enzyme, dCMP deaminase, converts dCMP to dUMP, which then acquires a methyl group and becomes dTMP. This enzyme duplicates the activity of a similar *E. coli* enzyme (and hence is a product of one of the nonessential genes) but has an interesting economic function. The base compositions of *E. coli* DNA and T4 DNA are 50 percent $(A+T)$ and 66 percent $(A+T)$, respectively. In *E. coli*, the ratio of dTTP and dCTP is about 1:1 in keeping with the T:C ratio in DNA. The bacterial and phage dCMP deaminases, acting together, increase the amount of dTMP with respect to dCMP, so the ratio of dTTP to dHTP is 2:1, as is the T:HMC ratio in T4 DNA.

Occasionally some C appears in progeny phage DNA. Presumably, the T4 nucleases degrade this DNA shortly after synthesis; since the C is in only one strand of each daughter double helix, the DNA that is removed can be replaced by normal repair synthesis.

4. *Glucosylation of T4 DNA*. The presence of HMC in T4 DNA creates another problem for the phage because *E. coli* possesses an endonuclease that attacks certain sequences of nucleotides containing HMC (the normal role of this enzyme in *E. coli* is not known). To avoid this damage the HMC residues in T4 DNA are glucosylated. This is accomplished by two phage enzymes that successively add two glucoses to HMC that is already in DNA. Thus, glucosylation is a postreplicative modification. The *E. coli* endonuclease is inactive against glucosylated DNA; therefore, glucosylation is a protective device. A simple genetic experiment shows that this is the only essential function of glucosylation. A T4 αgt^- mutant cannot carry out glucosylation, so its newly synthesized DNA is destroyed by the *E. coli* HMC nuclease. However, if an *E. coli* mutant $(rglB^-)$ that lacks this nuclease is used as a host bacterium, T4 αgt^- mutants grow normally even though nonglucosylated DNA is produced.

Production of T4 Phage Particles

Production of complete phage particles can be separated into two parts—assembly of heads, tails, and other structures, and packaging of DNA of a sufficient length to provide a little more than one set of genes in the phage head. Packaging has been discussed in an earlier section.

Assembly of T4 (and many other phages) has been studied by two techniques, both of which require a large collection of phage mutants unable to produce finished particles. In one method, different cultures of cells, each infected with a particular mutant, are lysed and examined by electron microscopy. This procedure shows that heads are made in the absence of tail synthesis and that tails are made by a mutant unable to synthesize heads. Thus, head and tail assembly are independent processes. Electron microscopic examination of cell extracts infected with various head, tail, and tail fiber mutants has indicated the presence of partially assembled structures. Study of these structures and the components they contain has given information about the order of assembly of the various gene products. The second technique is a type of complementation assay. If two extracts of infected cells, one lacking heads and the other lacking tails, are mixed, phage particles form *in vitro* (Figure 15-16). The "headless" extract can also be fractionated and a component can be isolated that allows a "tailless" extract to make tails. In this way, a protein in the tail assembly pathway can be isolated and identified.

These studies have shown that there are two types of components—**structural proteins** and **morphogenetic enzymes.** Some of the structural components assemble spontaneously to form phage structures, whereas others do so exceedingly slowly and hence need the help of enzymes. Genetic analysis shows that a few host-encoded factors are also needed for head assembly; that is, *E. coli* mutants have been found that do not support the multiplication of T4. Infection of these mutants yields phage with aberrant heads. An abridged diagram of the assembly pathway is shown in Figure 15-17.

T4 GENE ORGANIZATION

To date, 135 T4 genes have been identified. These genes account for about 90 percent of the DNA; thus, 15 to 20 genes remain to be found. T4 genes can be divided into two classes—82 metabolic genes and 53 particle-assembly genes. Of the 82 metabolic genes only 22 genes—namely, those involved in DNA synthesis, transcription, and lysis—are essential. The remaining 60 metabolic genes duplicate bacterial genes; particles in which these genes are mutated will grow, though occasionally they will have a smaller burst size. Of the 53 assembly genes, 34 code for structural proteins and 19 code for the synthesis of enzymes and protein factors that are required catalytically for assembly. Thus, 17 percent of the DNA of phage T4 encodes essential metabolic functions, 39 percent is necessary for phage assembly, and 44 percent serves nonessential metabolic functions.

A genetic map of some of these genes is shown in Figure 15-18. A notable feature of the map is that genes having related functions are often adjacent

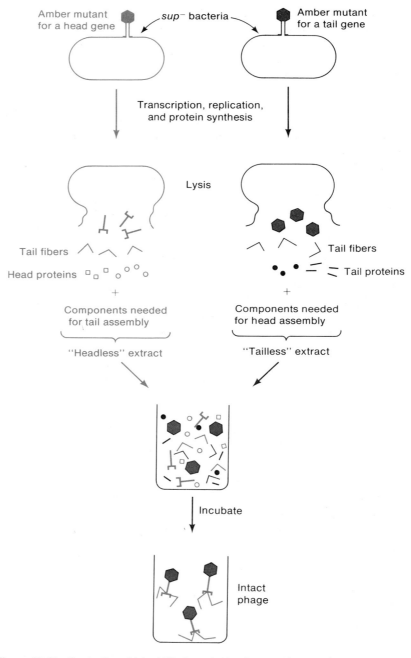

Figure 15-16   Production of intact T4 phage by *in vitro* complementation.

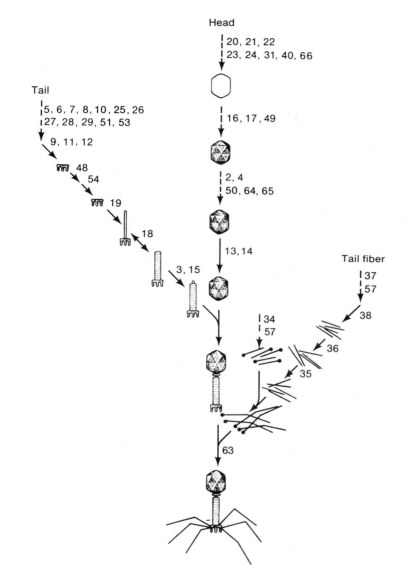

Head
| 20, 21, 22
| 23, 24, 31, 40, 66

Tail
| 5, 6, 7, 8, 10, 25, 26
| 27, 28, 29, 51, 53
9, 11, 12

48
54
19
18

| 16, 17, 49

| 2, 4
| 50, 64, 65

13, 14

3, 15

Tail fiber
| 37
| 57

38

| 34
| 57

36

35

63

Figure 15-17
Morphogenetic pathway of
T4 phage. The numbers
designate T4 genes.
[Courtesy of William
Wood.]

and transcribed as part of polycistronic mRNA molecules. This is an efficient arrangement, allowing the synthesis of functionally related proteins to occur at nearly the same time and minimizing the number of regulatory elements required. However, not all functionally related genes are part of single transcription units and some transcription units contain functionally discrete genes. The tendency to cluster related genes is common in many phage

Figure 15-18 Genetic map of phage T4 showing some, but not all, genes. The clustering of genes with related functions should be noted; although the tail-baseplate and tail-fiber genes form large clusters, other tail genes are distributed throughout the map. The solid and open regions indicate the locations of essential and nonessential genes respectively. Control refers to genes needed to initiate various modes of transcription. The inner red arrows indicate the direction and origins (but not the lengths) of various transcripts.

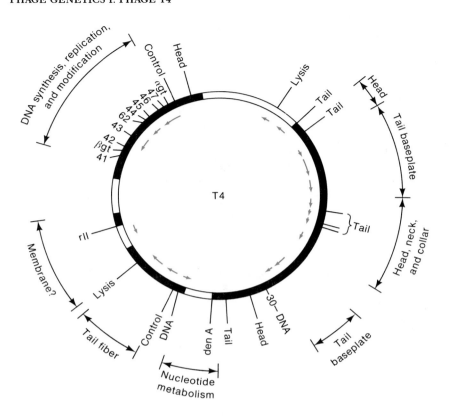

systems and occurs almost without exception with *E. coli* phage λ, as will be seen in Chapter 16.

PROBLEMS

1. If an *E. coli* culture is simultaneously infected by phages T4 and T7, each at MOI = 5, only T4 phage will be produced. From what you know about T4 biology, propose a simple explanation.

2. What is the function of T4 dCTPase (deoxycytidine triphosphatase) in a T4 infection?

3. Describe the course of T4 phage DNA synthesis following infection of *E. coli* with a T4 mutant which cannot synthesize (1) cytidine hydroxymethylase or (2) α-glucosyl-transferase.

4. Can one determine whether the early genes or the late genes of phage T4 are injected first into a host bacterium? Explain your answer.

5. What is the basic principle used by T4 in regulating the transcription sequence of the mRNA molecules made before DNA replication begins?

6. If a T4 phage has a large deletion which of the following will be true? (1) The activity of one or more proteins will be greatly altered or altogether missing. (2) The phage DNA will be smaller. (3) The terminal redundancy will be larger. (4) The phage DNA will be the same size. (5) Cyclic permutation will be eliminated.

7. Suppose you have a phage whose linear DNA is synthesized by the rolling circle mode and is packaged by "the headful rule" (that is, DNA is added to a head of fixed size until no more DNA can fit). However, the DNA is normally neither terminally redundant nor cyclically permuted. You find a mutant strain of this phage, the DNA of which has a deletion in a nonessential gene. This phage is used to infect a bacterium, and many phage are produced. The DNA is isolated and is treated with an exonuclease, which removes a few bases from the 5'-P end. The treated DNA is then exposed to conditions which could circularize T4 DNA (if it were also pretreated with the exonuclease). This DNA is examined by electron microscopy. Will circles be found?

8. Most double-stranded DNA phages have several classes of mRNA that can be divided into two major groups, early mRNA and late mRNA. The genes carried on the early mRNA species vary from one phage to the next. Nonetheless, there are certain genes that are usually on early transcripts and some which are invariably on late transcripts. What are these genes?

9. What is the appearance of r^+ and rII phage when plated on a 1:1 mixture of E. coli B and K12(λ)?

10. In a cross between $rIIA254$ and $rIIB82$, 26 plaques are found on a K12(λ) plate and 482 on a B plate, using phage at the same dilution. What is the recombination frequency between $rIIA254$ and $rIIB82$?

11. Four rII deletions, A–D, are crossed against one another. Of the six crosses the following yielded recombinants: (1) $A \times B$, (2) $A \times C$, and (3) $C \times D$. The remaining three—(4) $A \times D$, (5) $B \times C$, and (6) $B \times D$—yielded no recombinants.

   (a) What is the relation between these deletions?
   (b) An rII mutation recombines with A, B, and C, but not D. Where is this mutation located?

12. With most phages, of which T4 is certainly an example, all have life cycles in which a large fraction is devoted to late transcription. Why is the duration of time allotted to late transcription greater than the time for early transcription?

13. Three T4 rII deletions, A, B, and D have the following properties: $B \times D$ yields r^+ recombinants, but $A \times B$ and $A \times D$ do not. What is the map order of the deletions?

14. Consider the deletions in Problem 13. A mutant yields r^+ recombinants with B and D, but not with A. Locate the mutant with respect to the deletions.

15. A phage that produces large plaques maps in the rII region. If 10^7 particles are plated on K12(λ) cells, no plaques result. What do you know about the mutation?

16. A plate is prepared with 10^7 particles of a known $rIIA$ mutant, 10^8 K12(λ) cells, and 10^6 B cells. Two mutants are tested in spot tests. The background is plaque-free. Mutant 1 gives a totally clear spot, and mutant 2 yields a few plaques. What do you know about the mutants?

17. Mutations *rIIA1* and *rIIB2* complement. Also, both mutations can recombine with the deletion *rIIX4* to yield r^+ recombinants. However, *rIIX4* fails to complement with either *rIIA1* or *rIIB2*. Explain.

18. A phage that uses the headful-packaging rule is subjected to the 5′-exonuclease test for terminal redundancy. It is found that the double-stranded segment that is flanked by single strands has a length of 0.6 μm. The entire native phage DNA molecule is 57 μm. A phage mutant is also tested, and the length of the double-stranded segment is 1.1 μm. What does this observation tell you about the mutation?

19. The genome of a phage that uses the headful packaging rule is represented as *ABCDEF . . . XYZAB*. It is terminally redundant and cyclically permuted, and individual molecules can be represented as *ABCDE . . . XYZAB, EFGHI . . . BCDEF*, and *JKLMN . . . GHIJK*. A deletion that removes genes *C* and *D* is isolated, and a phage lysate is prepared by infecting cells with this deletion. Which of the following sets of sequences should be observed among the phage progeny? (1) *ABEFG . . . XYZAB, EFGHI . . . ZABEF, YZABE . . . VWXYZ;* (2) *ABEFG . . . ZABEF, BEFGH . . . ABEFG, ZABEF . . . YZABE*.

20. Consider a phage with the strange property that productive infection occurs only if at least two phage adsorb to the bacterium. If you have 10^8 bacteria and add to this culture 3×10^8 phage, how many bacteria will be productively infected?

21. A new protein X appears in infected cells. Describe various experiments that you might perform to prove that the gene coding for X is phage-encoded and is not encoded in host DNA.

22. How would you show whether a phage-encoded gene product is required throughout the infectious cycle or only at a unique time?

Chapter 16

Phage Genetics II:
Phage λ

T he *E. coli* phage λ has two alternate life cycles—the lytic and the lyso-genic. In this chapter we consider only the former, which differs sig-nificantly from that seen for T4 in the preceding chapter; the lysogenic cycle is described in detail in Chapter 17.

λ DNA AND ITS GENE ORGANIZATION

λ contains a linear double-stranded DNA molecule, consisting of 48,514 base pairs of known sequence. At each end of the DNA molecule the 5′ terminus extends 12 bases beyond the 3′-terminal nucleotide. The base sequences of these single-stranded terminal regions, which are known as **cohesive ends,** are complementary to one another (Figure 16-1). Thus, by forming base pairs between the cohesive ends, the linear λ DNA molecule can circularize, yielding a circle with two single-strand breaks. Immediately after injection, λ DNA circularizes in this way, and *E. coli* DNA ligase converts the molecule to a covalent circle.

Forty-six λ genes have been identified; of these, 14 are nonessential to the lytic cycle but only 7 are nonessential to both the lytic and lysogenic cycles. Most λ proteins have been either identified by gel electrophoresis or purified. All regulatory sites, promoters, and termination sites are known.

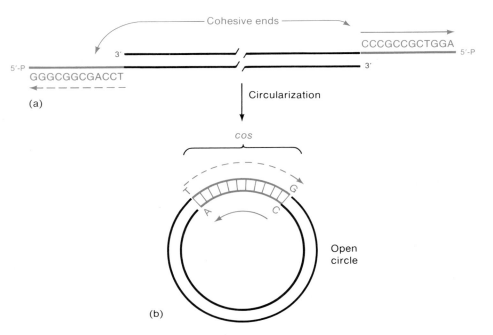

Figure 16-1   (a) A diagram of a λ DNA molecule showing the complementary single-stranded ends (cohesive ends). Note that 10 of the 12 bases are G or C. (b) Circularization by means of base-pairing between the cohesive ends. The double-stranded region that is formed is designated *cos*.

The genetic map of λ is shown in Figure 16-2. The map is circular, owing to the circularization of the linear DNA after injection. A striking feature of the map is the clustering of genes according to function (considerably more than for phage T4). For example, the head, tail, replication, and recombination genes form four distinct clusters. Even the genes needed for head-tail attachment lie between the head and tail genes. Many λ proteins—for example, regulatory proteins and those responsible for DNA synthesis—act at particular sites in the DNA. In general these proteins are situated adjacent to their sites of action (when there is a single site). For instance, the origin of DNA replication lies within the coding sequence for gene *O*, which encodes a protein for initiation of DNA replication, and the gene that generates the cohesive ends is located adjacent to one of the ends.

OUTLINE OF THE LIFE CYCLE OF λ

The schedule of the lytic cycle of λ is fairly complex, probably because certain genes are used in both the lytic and lysogenic cycles. The life cycle in outline is the following (time in minutes at 37°C):

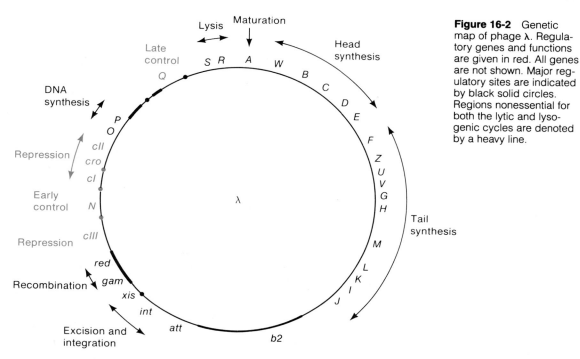

Figure 16-2 Genetic map of phage λ. Regulatory genes and functions are given in red. All genes are not shown. Major regulatory sites are indicated by black solid circles. Regions nonessential for both the lytic and lysogenic cycles are denoted by a heavy line.

$t = 0$	Phage adsorbs and DNA is injected.
$t = 3$	First ("pre-early") mRNA is synthesized.
$t = 5$	Two classes of early mRNA are synthesized.
$t = 6$	DNA replication begins.
$t = 9$	Synthesis of late mRNA begins.
$t = 10$	Structural proteins begin to be made.
$t = 22$	First phage particle is completed.
$t = 45$	Lysis and release of progeny phage.

Note that the cycle is 45 minutes long rather than the 22–25 minutes we saw for T4. The life cycles of most phages vary between 22 and 60 minutes at 37°C.

THE TRANSCRIPTION SEQUENCE OF λ

With phage T4, and many other phages as well, timing of synthesis of the various mRNA molecules is accomplished primarily by mechanisms that determine the availability of promoters—namely, the synthesis of a new RNA polymerase (phage T7), or the modification of the host polymerase (T4). In λ the host RNA polymerase is also modified but not for the purpose of

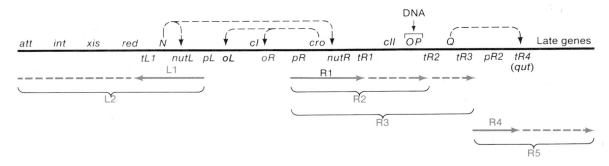

Figure 16-3 A genetic map of the regulatory genes of phage λ. Genes are listed above the line; sites are below the line. The mRNA molecules are red. The dashed black arrows indicate the sites of action of the N, Cro, and Q proteins.

recognition of phage promoters. Instead, *the modification enables the polymerase to ignore certain termination sites*. Figure 16-3 shows a detailed version of a genetic map of λ, which includes the three regulatory genes *cro*, *N*, and *Q*; three promoters *pL*, *pR*, and *pR2*; the DNA-replication genes *O* and *P*; and five termination sites *tL1*, *tR1*, *tR2*, *tR3*, and *tR4*. Seven mRNA molecules are also shown; the L and R series are transcribed leftward and rightward, respectively, from complementary DNA strands. The λ map is often drawn as a linear map, matching the linear DNA molecule in the phage head. In the standard orientation, gene *A* is at the left end and gene *R* is at the right end.

An essential feature of the life cycle of T4 is rapid killing of the host and degradation of the host DNA. λ differs in this respect because in the lysogenic cycle the host must survive. Instead, even in the lytic cycle, λ multiplies while the host cell continues its normal function. Since inactivation of host functions is not required, the transcription pattern that might be expected is the following. First, the *O* and *P* genes, whose products are necessary for DNA synthesis, would be transcribed. This would be followed by transcription of the genes encoding the structural proteins, and finally the packaging system and the lytic proteins would be made. This is basically what occurs, with the modification that, both before and after transcription of genes *O* and *P*, other small transcripts are formed that encode the regulatory proteins responsible for turning transcription on and off at the appropriate times.

A detailed analysis of the sequence of transcriptional events is beyond the scope of this book and can be found in references at the end of the book. However, certain features can be seen by examining Figure 16-3 and referring to Table 16-1. λ has two early promoters, *pL* and *pR*, from which synthesis of the RNA molecules L1 and L2 are initiated. Transcription initially terminates at the sites *tL1* and *tR1*. L1 encodes only the gene-*N* protein, which is a major positive regulatory element that controls when certain regions of the DNA are transcribed. Once synthesized, the N protein binds to the

Table 16-1   Some sites and gene products in phage λ

Site or gene product	Description
Site	
oL, oR	Left and right operators
pL, pR	Left and right promoters
tL (1,2)	Termination sites for leftward transcription
tR (1,2,3,4,5)	Termination sites for rightward transcription
Gene product	
Cro	Protein inhibitor of transcription from pL and pR
cII	Protein delaying late transcription (plays other roles in the lysogenic pathway)
N	Antitermination protein acting at tL1, tR1, and tR2
O, P	Proteins required for DNA replication
Q	Antitermination protein acting at tR4
L	Messenger RNA synthesized in leftward direction
R	Messenger RNA synthesized in rightward direction
R4	Constitutively synthesized mRNA

sites *nutL* and *nutR*, which are to the left and right, respectively, of the promoters. RNA polymerase, moving along the DNA, picks up the N protein, which enables the polymerase, by an unknown mechanism, to ignore the termination sites *tL1* and *tR1* and form the longer transcripts L2 and R2. N protein is called an **antiterminator**.

Once antitermination sets in, rightward transcription allows synthesis of the products of the DNA-replication genes *O* and *P*. The leftward transcript contains the *red* locus. This encodes two genes needed for genetic recombination, which plays an important role late in the life cycle. Since the Red proteins and the O and P proteins are catalytic, they do not have to be made continuously. Thus, when sufficient Cro protein (encoded in R1) is made, a repressor activity of Cro turns off synthesis of all leftward mRNA by binding to the leftward operator *oL* . Rightward transcription terminates at *tR3*, which is a termination site for RNA polymerase even when it is modified by N protein. During this early-transcription period rightward transcription provides enough mRNA that the concentrations of the O and P proteins reach values sufficient for efficient DNA replication. Somewhat later than the time when the Cro protein inhibits transcription from *pL*, the concentration of Cro increases to the point that it also binds to the rightward operator *oR* to block rightward mRNA synthesis. This ensures that wasteful synthesis of O and P proteins does not occur and also prevents a kind of aberrant and deleterious DNA synthesis that results from excessive amounts of these proteins. During the early-transcription phase the gene-*cII* protein, encoded in the same transcript containing *O* and *P,* builds up in concentration sufficient to act at a late promoter to delay late mRNA synthesis. Once Cro has turned off rightward transcription, cII protein is no longer made and the inhibition of late mRNA synthesis is relieved. However, a positive regulator is needed to turn on late mRNA synthesis, and that is the gene-Q protein.

Since the time of infection, the tiny transcript R4 has been synthesized continually from $pR2$, terminating at $tR4$. This transcript does not encode any known genes but is a leader for the late mRNA. Once R3 has been made, the product of gene Q is made. This protein is responsible for turning on synthesis of the late mRNA molecule that encodes the structural and assembly proteins, the maturation system, and the lysis enzymes. The Q protein is also an antiterminator. It binds to a sequence called *qut* and is picked up by RNA polymerase, thereby enabling RNA polymerase to ignore $tR4$. Thus, R4 is then extended to form transcript R5, the late mRNA, which encodes the head, tail, assembly, and lysis proteins.

Let us now review the essential features of this highly efficient (and perhaps mind-boggling) regulatory system.

1. A λ-specific RNA polymerase is not made; the *E. coli* RNA polymerase is used throughout the life cycle (as is true of T4) and is modified by accessory proteins to alter its specificity toward various DNA base sequences. However, at no time is its activity with respect to promoters modified; instead, its ability to terminate transcription at certain termination sites is altered.

2. Inhibition of transcription occurs, but only as a result of the repressor activity of Cro on the promoters pL and pR. Wasted synthesis is avoided by this repressing activity.

3. All structural components are encoded in a single giant mRNA molecule, which is translated sequentially. Thus, synthesis of the complete set of components takes many minutes, thereby delaying synthesis of intact heads and of a functional phage maturation system until the DNA replication system has provided many copies of λ DNA.

Note the series of delays—times to transcribe particular early regions, and times to synthesize regulatory proteins. The result of these delays is that about 30 copies of λ DNA form before the maturation system is established and 50–100 completed phage particles form before the onset of lysis.

GENETIC EVIDENCE FOR SOME FEATURES OF TRANSCRIPTION

Most of the information about transcription in λ was derived from experiments in which newly synthesized RNA was labeled at various times after infection, and this labeled RNA was hybridized to various segments of λ DNA. In each case, the λ genes present in each segment of the DNA were known, so it was possible to determine the temporal sequence of transcription of all regions of the λ genome. Regulatory elements were identified by the fact that mutations in certain genes prevented synthesis of particular species of mRNA. For example, Q^- mutants were unable to synthesize the late species of mRNA, and N^- mutants failed to synthesize all mRNA, other than L1 and R1. However, a great deal of information came from strictly genetic experiments, a few of which will be described.

In the discussion that follows, several mutants will be described and we will make statements about their inability to grow. One might reasonably ask how samples of these mutants are ever obtained. The answer is that, as is invariably the case with mutations in essential genes, the mutations are conditional. With λ almost all mutations in use are suppressor-sensitive mutations, which enable the phage to be grown on suppressor-containing cells. Experiments in which the phage are to behave as mutants are done with strains lacking known suppressors (*sup0* strains).

By several techniques it is possible to construct bacterial strains that contain defined segments of λ DNA. For example, a $recA^-$ strain was constructed that contained an integrated segment starting just to the right of gene Q and extending through gene J. (Strains with a portion of an integrated phage genome are occasionally called **cryptics**.) This strain (strain II) was used in complementation tests by infecting the bacterium with various λ mutants and determining whether progeny phage could be produced (Table 16-2). The bacterium was also $recA^-$, so no phage could be produced by acquisition of a wild-type gene from the bacterium by recombination. If strain II was infected with λQ^-, no phage were produced because neither the phage nor the bacterium contained the Q gene. However, if the strain carried an integrated gene Q (strain I), still no phage were produced. This observation

Table 16-2 Result of infection of a cryptic bacterium, carrying a λ segment, with various λ mutants

Strain	λ genes in cell*	Genotype of infecting λ	Phage production	Modern explanation[†]
I	$Q\bullet SRA...D...J$	Q^-R^+	No	Q in bacterium not tran-
		Q^+R^-	Yes	scribed. Transcription occurs if
		Q^+D^-	Yes	Q in the infecting phage anti-
		Q^+J^-	Yes	terminates R4
II	$\bullet SRA...D...J$	Q^-R^+	No	Transcription occurs if Q in the
		Q^+R^-	Yes	infecting phage antiterminates
		Q^+D^-	Yes	R4
		Q^+J^-	Yes	
III	$RA...D...J$	Q^-R^+	No	No R4
		Q^+R^-	No	
		Q^+D^-	No	
		Q^+J^-	No	

*The heavy raised dot represents the promoter for R4.

[†]These explanations could not be given at the time these experiments were done because neither the antitermination effect of N nor the existence of R4 RNA were known.

suggested that Q was not transcribed in strain I; presumably its promoter was not present. We now know, of course, that the Q promoter is $pR2$. If the bacterium that lacked Q (strain II) was infected with a λR^- mutant (defective in lysis enzymes), the cells lysed and progeny R^- phage were produced. Apparently the infecting phage supplied the Q function. However, if the phage DNA contained in the bacterial chromosome began just to the right of gene S (strain II) and R^- was the infecting phage, no phage were produced. In fact, no phage were produced with phages mutant in any gene between R and J. These observations showed that transcription of genes downstream from Q required some element between Q and R. We now understand that this is the promoter for the small RNA, R4, which is a leader for the late mRNA.

Some of the properties of gene N were also uncovered by genetic experiments. λN^- is unable to grow on an $sup0$ host, primarily because the Q product is not synthesized and hence no late mRNA is made. The *red* gene *exo*, which is downstream from N and encoded in L2, synthesizes an easily assayed exonuclease. N^- mutants do not make this exonuclease since L1 is never extended to form L2. However, a small deletion mutation was isolated between N and *exo;* when coupled with an N^- mutation, this deletion enables the N^- phage to make the exonuclease. This experiment, plus the observation that L1 is extended to form L2 shortly after infection, provided the first evidence for an N-sensitive termination sequence downstream from $N;$ that is, the deletion removes this site. The N-sensitive termination site between genes P and Q was also identified by a genetic test. λ N^- mutants were plated on $sup0$ bacteria to seek revertants. Some of these revertants mapped in N and restored N-gene activity, as might be expected. However, one revertant was a mutation that mapped between P and Q. This mutation, called *nin5* (for N-*in*dependence) was a substitution of phage DNA by bacterial DNA; this substitution replaces the transcription-termination site and allows synthesis of Q protein without the antitermination activity of N protein.

In the preceding section we mentioned the acquisition of N protein by RNA polymerase at a *nut* site to form an RNA polymerase that can ignore certain termination sequences. The interaction with N protein and RNA polymerase was also first recognized by a genetic observation. In an attempt to seek bacterial functions needed for λ phage production, bacteria were mutagenized and mutants on which λ would not form plaques were sought. Many such mutants were found. Some of these were unable to adsorb λ, which was detectable by the ability of a λh mutant to plate on these mutants, and these were not studied further. The remainder were called *gro* mutants. One of these mutants, termed *groN*, had the unusual property that several (but not all) N^- mutants could grow on the bacteria (Table 16-3). Growth of N^- mutants required that the cell also harbor a suppressor. However, since N^+ mutants would *not* grow, simple suppression of a phage mutation was not the explanation for the phenomenon. Since a suppressor was needed, it was clear that growth required a complete N protein (not a prematurely

Table 16-3 Growth of λ *N* mutants on *E. coli groN* mutants

Bacterial genotype	Growth*	
	λ *N*[-][†]	λ *N*[+]
gro[+] *sup0*	−	+
gro[+] *sup0*[+]	+	−
groN sup0	−	−
groN sup[+]	+	−

* + = plaque formation; − = no plaques.
*† + occurs only for certain *N*[-] mutants.*

terminated fragment). The following hypothesis, now known to be correct, was made: normal N protein interacts with the GroN product, and this interaction is essential for λ to multiply. A mutant GroN protein could not bind to wild-type N protein, but certain (but not all) suppressed mutant N proteins could bind to the mutant GroN protein. The likely molecular explanation was that the *groN* mutation altered the binding site in the GroN protein, and some suppressed *N*[-] genes produced an N protein whose binding site for GroN was also modified such that interaction could occur. Mapping of the *groN* gene located the gene within a cluster of mutations in a gene encoding one of the subunits of RNA polymerase. Other nearby RNA polymerase mutations were tested, and some of these had the GroN phenotype. Ultimately N protein was purified, and, as might be expected, one of the first experiments done was to test binding to RNA polymerase. Indeed, N protein and RNA polymerase do form a complex.

λ DNA REPLICATION AND PHAGE PRODUCTION

Most of the steps involved in λ DNA replication and the required proteins are known. This information has come from a variety of genetic and physical experiments and recently from studies with an *in vitro* replication system. Some of the observations are described in this section. We will also see that replication and conversion of DNA to a form that can be packaged are coupled processes.

Genetic Experiments Providing Information about λ DNA Replication

In the initial major search for λ mutants two classes of mutations that prevent λ DNA replication were found. These were mapped in the two adjacent genes, *O* and *P*, as already discussed. Further analysis provided evidence for a requirement for bacterial proteins—for example, λ could not grow on *E. coli* carrying mutations in known DNA replication genes, such as those encoding DNA polymerase III and the DnaJ protein. Some bacterial mutants were found in another way. For example, some of the Gro mutants mentioned in the preceding section proved to be a type called GroP; these supported the growth of λ carrying certain suppressed *P⁻* mutations. Similar experiments showed that the *groP* gene was identical to the *dnaB* gene, whose product is essential for many stages of *E. coli* DNA synthesis.

Further information about the genetics of λ DNA replication came from a study of genes *O* and *P*, whose gene products are essential for DNA replication. In a search for revertants of *O⁻* mutations most of the revertants mapped in gene *O*, as expected; however, some mapped in gene *P*. Furthermore, only revertants of certain *O⁻* mutations mapped in *P*. Similarly, one revertant of a particular *P⁻* mutation mapped in gene *O*. These experiments suggested that the O and P proteins interact. No GroO bacterial mutants were ever found, so O protein probably does not interact with any bacterial protein. Hence, the genetic results suggest that O, P, and DnaB proteins form a complex with contact points between O and P proteins and between P and DnaB proteins.

In Chapter 8 it was pointed out that replication always begins at a unique site called an origin. A genetic experiment first located the λ origin *ori*. A λ mutant was isolated that produced a very small plaque; it was called *ti12* for tiny. Infection by this mutant yielded only a few phage per cell, rather than a normal burst of about 50–100. In a mixed infection with wild-type λ, *ti12* appeared not to grow at all. For example, in a mixed infection with λ*ti12* and a genetically marked *ti⁺* phage, the infected cells produced about 50 *ti⁺* progeny phage per cell and no (less than 0.01) detectable *ti12* phage. This and several other experiments showed that *ti12* is unable to compete with another phage. Mapping of the mutation placed it adjacent to known gene-*O* mutations. The common clustering of λ genes according to function suggested that *ti12* affects DNA replication. However, it is clearly a site, since it could not be complemented. Physical experiments using density-labeled *ti12* phage showed that it cannot initiate replication in a mixed infection. Thus, it was concluded that *ti12* is a mutation in *ori*. It is not an absolute-defective because, if so, it would never have been isolated (it would not be able to replicate at all). Rather, it is a leaky mutation that fails to compete effectively for the replication-initiation system when a wild-type *ori* is present.

DNA Replication and Maturation: Coupled Processes

Following synthesis of the *O* and *P* products, replication of circular λ DNA begins. There are two modes of λ DNA replication—θ and rolling circle replication (Figure 16-4; see also Chapter 8). The θ mode increases the num-

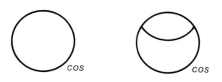

Figure 16-4 Three species of λ DNA present at the time maturation begins. The region containing the joined complementary single-stranded termini (of the linear DNA molecule present in the phage head) is called *cos* (for *co*hesive *s*ite).

ber of templates for transcription and further replication; the rolling circle mode provides the DNA for phage progeny. As the life cycle proceeds, θ replication stops and the rolling circle becomes the predominant form.

Little is known about the control of initiation of rolling circle replication. However, for our purposes at this point the most important feature is the relation between replication and maturation of the DNA, since the DNA-cutting mechanism of λ differs from that used by T4 or Mu.

The DNA found in a λ phage particle is linear and has single-stranded termini. These ends are joined when a circle forms, and the double-stranded region so formed is called a *cos* site (for *co*hesive site). Thus, every monomeric λ circle contains one *cos* site; however, a multimeric branch of a rolling circle contains many *cos* sites.

Since the ends of the DNA molecule in the phage particle are always the single-stranded termini, there must be a mechanism for cleaving a *cos* site to generate these termini. This is accomplished by a sequence-specific cutting system called the **terminase** or **Ter system,** and the DNA-cutting is often called Ter-cutting.

Figure 16-4 shows the three major species of intracellular λ DNA— circles, θ molecules, and rolling circles. Early replication is by the θ mode, but in time there is a gradual cessation of this mode and a transition to rolling circle replication. By the time heads and tails have been synthesized and the Ter system is active, rolling circles predominate. Note that since a rolling circle has two classes of *cos* sites—the one in the circle and those in the linear branch—some mechanism must exist for preventing cleavage of the one in the circle, because if it were broken, replication would cease. This difficulty is circumvented by a site requirement of the Ter system: *efficient cleavage of a single* cos *site does not occur; if there are two* cos *sites and both are present on a single segment of DNA, cutting can occur.* A λ DNA molecule can be cut from a linear branch by cleavage of two neighboring *cos* sites, but a pair of cuts in which one *cos* site is in the branch and the other is in the circle cannot be made. However, Ter cutting does not require that the DNA molecule be linear, inasmuch as a single λ unit can be cut from a dimeric

Figure 16-5 Two rules of packaging. In one mode (black) each λ unit is packaged. In the more limited mode (red) alternate units are packaged. The more economical black mode is used by λ.

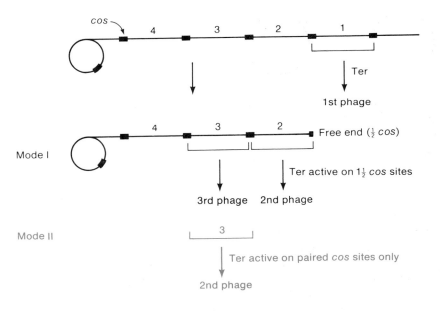

circle that has been formed by genetic recombination. Physical experiments show that a single cut in a monomeric circle does not occur efficiently.

The "two-*cos*-sites" rule explains how the first λ DNA unit is cut from a concatemeric branch of a rolling circle. However, this rule would not allow excision of the second (adjacent) λ unit, because this unit would be flanked by only one *cos* site and a free cohesive end. Hence, only half of the DNA would be usable because only alternate segments of DNA would be packageable (Figure 16-5). The solution to this apparent lack of economy is that a free cohesive end and an adjacent *cos* site are also sufficient for DNA cutting to occur, and allow sequential packaging. Thus, the Ter-cutting rule may be restated as follows: *Ter-cutting requires two* cos *sites or one* cos *site and a free cohesive end on a single DNA molecule.* An experiment depicted in Figure 16-6 shows that the free end must be the end near the A gene (Figure 16-2). (The A and R genes are not significant. We have used these genes to name the two ends of λ DNA to avoid possible ambiguity in the terms left and right.) In this experiment three types of λ DNA molecules were prepared by *in vitro* cohesive-end joining of either two intact DNA molecules (I) or one intact molecule plus one fragment containing either the gene-A cohesive end (II) or the gene-R cohesive end (III). (The fragments were prepared by breaking intact molecules near the center and then separating the fragments bearing a particular end from one another.) The molecules also contained a mutation in the P gene, so DNA replication was not possible. Three cultures of E. coli were separately transfected (by the $CaCl_2$ transformation technique) with these fragments. There was no DNA replication in the infection but transcription occurred that resulted in synthesis of heads, tails, and the ele-

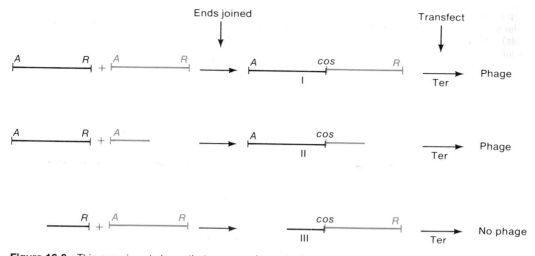

Figure 16-6 This experiment shows that one *cos* site and a free *A*-gene cohesive end are sufficient for activity of the Ter system. The cohesive ends are depicted by a vertical line. All molecules contained a mutation in the λ *P* gene to prevent DNA replication. Transfection refers to infection of a competent bacterium with phage λ DNA.

ments of the Ter system. Phage would be produced if the Ter system could cut λ units from these hybrid molecules. As shown in the figure, phage were obtained when cells were transfected with the dimer or with the molecule containing a free *A* end, but no phage were obtained when the fragment containing only a free *R* end was used. This and other experiments show that the packaging of λ DNA from the concatemeric branch of the rolling circle is polarized and proceeds from the *A* end to the *R* end.

Cutting at the *cos* sites and packaging of λ DNA are somehow coupled. In fact, the Ter system is virtually inactive unless the Ter proteins are components of an empty λ head. Thus, when a bacterium is infected with a λ mutant unable to cause formation of an intact phage head (for example, an E^- mutant, which fails to make the major head protein), the linear branch of a rolling circle of DNA is not cleaved. With such a mutant, dimers, which contain two *cos* sites, are also not cleaved.

The Ter system was first identified by genetic analysis of tandem dilysogens (cells with two adjacent prophages). Properties of dilysogens will be described in Chapter 17; at this point we will just assume that they exist. Figure 16-7 shows a dilysogen that was constructed. The prophage on the left had the genotype A^+R^- and that on the right was A^-R^+. Both prophage were *int⁻*, which means that they lacked the ability to be excised from the bacterial DNA by the normal route. In Chapter 5 it was pointed out that lysogens can be induced to produce phage; the details of the process were not given. Indeed when the dilysogen shown in the figure was induced,

Figure 16-7 Phage produced by *int⁻* tandem dilysogens by Ter cutting at *cos* sites (dots). Both prophages are *int⁻*.

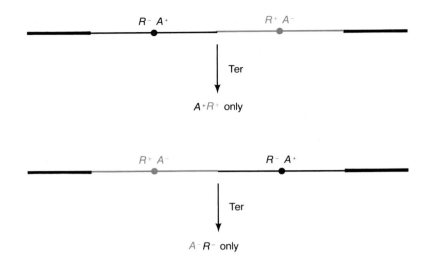

phage were produced; however, all phage were A^+R^+. When the prophage order was reversed, that is, an A^-R^+ prophage on the left and an A^+R^- prophage on the right, all phage produced were A^-R^-. Note that the phage produced always possessed the A allele to the right of the left *cos* site and the R allele to the left of the right *cos* site. These experiments demonstrated the existence of a genetic system that makes a cut between genes A and R, that is, in the *cos* sites. The use of this system in further genetic analysis of lysogens will be found in Chapter 17.

Particle Assembly

Assembly of a completed λ phage particle is accomplished by both phage and bacterial genes. As in the study of T4 assembly, the pathway has been elucidated primarily by examining lysates of cells infected with various λ mutants. The basic observation is that there are four classes of phage mutations: those that (1) prevent the head formation, (2) eliminate functional tails, (3) allow synthesis of heads and tails but not intact phage, and (4) prevent filling of the head with DNA. Isolation of *gro* mutants of *E. coli*, in which wild-type λ fails to produce heads, identified a role of a bacterial protein. The observation that phage production is restored to an infected *gro⁻* cell by the presence of certain mutations in the λ gene *E* (which encodes the major head protein) indicates that the E protein and the bacterial Gro protein (GroE) interact. Complementation of *groE* mutations shows that there are two bacterial genes, *groES* and *groEL*. Furthermore, certain *groES* mutations are reverted by mutations in the *groEL* gene, indicating that the GroEL

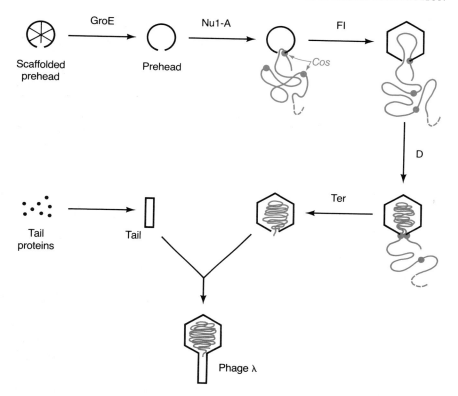

GroE

Scaffolded prehead

Nu1-A

Prehead

Cos

FI

D

Tail proteins

Tail

Ter

Phage λ

Figure 16-8   Diagram showing the pathway for assembly of phage λ. Most steps are fairly well understood, except for the condensation of DNA into the phage head, the mechanism of which remains obscure.

and GroES proteins interact, a fact that has been confirmed with purified proteins.

Figure 16-8 shows various stages of assembly that have been worked out, with details omitted. The process begins with an aggregation of many copies of four head proteins to form a scaffolded prehead, that is, a sphere with an internal supporting structure. This type of structure is the first assembly product in many phage systems. The GroES-GroEL complex then binds to the scaffold prehead, and the scaffolding is removed by the activity of bacterial proteases. (With some phages the scaffolding falls away spontaneously and can be reused.) Two phage-gene products, Nu1 and A, which also comprise the Ter system, catalyze interaction of a short base sequence adjacent to one *cos* site with a point on the head (the neck) that will later become the head-tail attachment region. λ DNA then folds into the phage head; after a small amount of DNA enters, a conformational change in the E protein occurs, which causes an expansion of the spherical particle and formation of an icosahedron. The λ FI protein plays some role in this process. A small number of λ D-protein molecules are inserted into the head; at the same time, in an unknown way, λ DNA fills the head. Head-filling terminates when the next *cos* site is reached, at which point the Ter complex makes appropriate pairs of staggered cuts, thereby forming the single-stranded cohesive termini and releasing the unpackaged DNA from the filled head. A few bases at the end of one of the single-stranded termini of the λ DNA project from the head. In the meantime, tails have been assembled from several tail proteins. The tail is terminated by a head-tail connector protein, and the completed tail is bound to the head via both the short piece of single-stranded DNA and the λ head proteins in the neck. The free single-stranded terminus of the released DNA is available for binding to the neck of a second prehead, in accord with the observations mentioned in an earlier section.

RECOMBINATION IN THE λ LIFE CYCLE

λ can engage in genetic recombination in a $recA^-$ cell; hence, it is obvious that λ encodes its own recombination system. A search for recombination genes began in the usual way—by seeking recombination-deficient mutants. These mutants were found by the following genetic test. A $recA^-$ bacterial strain was prepared that contained a small segment of λ DNA, that including gene J, which encodes the tail-fiber protein. The J gene carried a mutation of the host-range (h) type that enables a phage to plate on a λ-resistant (Lam-r) bacterium. We denote this strain $recA^-$(λcryp h), using λcryp to indicate that only a part of the phage DNA is present. This bacterium was infected with wild-type λ, and some phage were produced that carried the h marker from the bacterium and hence could plate on Lam-r cells. Since the cell was $recA^-$, the h marker must have been acquired by an exchange mediated by the λ recombination system. This observation was also made with a spot test. Soft agar was seeded with a 1:1 mixture of the $recA^-$(λcryp h) cells and Lam-r cells. A drop containing λ was spotted on the agar and the plate was incubated. A completely clear spot developed because the wild-type λ lysed the cryptic strain and the h recombinants (produced by infecting cryptic cells) lysed the Lam-r cells. When mutagenized λ was used in this spot test, occasional turbid spots were found. These were formed by phage mutants that could not acquire the h marker and hence could only lyse one of the indicator cells (the cryptic strain). The mutations possessed by these phage were termed red, for recombination deficiency. Proof that they were indeed unable to carry out genetic recombination came from combining a red^- mutation with one genetic marker in one phage and another marker in a second phage and showing that recombination could not occur between the markers. For example, in a mixed infection with $λred^-A^-$ and $λred^-G^-$, no A^+G^+ recombinants formed. Complementation tests between various red^- mutants showed that there are two red genes, now called bet (β in the older literature) and exo. The exo gene encodes an exonuclease; the function of the bet gene is unknown. Many physical and biochemical experiments have been carried out with purified Bet and Exo proteins, but to date the mechanism of λ-mediated recombination has not been elucidated.

An additional phenotype of Red⁻ mutants is their reduced burst size—roughly 20 percent that of a Red⁺ phage. An understanding of this phenomenon has come from physical studies in which intracellular DNA from infected cells was examined by electron microscopy. It was stated earlier that as the λ life cycle proceeds, θ replication halts and rolling circle replication predominates. However, by the middle part of the life cycle only a few rolling circles exist, which generate long concatemers from which λ DNA molecules are cut; in addition, there is a rather large amount of circular λ DNA. Because these have only a single *cos* site, they cannot be packaged and would be wasted were it not for the activity of the Red system: Red-mediated recombination converts most of these monomeric circles to dimeric circles, which have two *cos* sites and from which a single λ DNA molecule can be cut and packaged. Thus, the Red system serves as a scavenger, avoiding wastage of the many circular DNA molecules that are generated early in the life cycle to serve as transcription templates.

In Chapter 14 it was pointed out that rolling circle replication is inhibited by the *E. coli* RecBC protein, and that λ has a gene, *gam*, whose product inactivates the RecBC enzyme. A *gam⁻* mutant, like a *red⁻* mutant, has a decreased burst size because the rolling circle pathway to the production of packageable λ DNA is absent (Table 16-4). A *red⁻gam⁻* phage is even more defective, since both pathways—the rolling circle one and the recombinant dimer one—are absent. However, a few phage are still formed from dimers produced by the *E. coli* Rec system. As might be expected, λ*red⁻gam⁻* cannot grow at all on a *recA⁻* cell, since no pathway whatsoever exists for

Table 16-4 Plaque formation and maturation systems active when various λ mutants infect bacterial *rec* mutants

Phage mutant	Bacterial genotype	Plaque formation	Maturation system active	
			Rolling circle	Dimer formation
red⁺gam⁺	*recA⁺*	+	+	+
red⁺gam⁺	*recA⁻*	+	+	+
red⁻gam⁺	*recA⁺*	+ (small)	+	−
red⁻gam⁺	*recA⁻*	+ (small)	+	−
red⁺gam⁻	*recA⁺*	+ (small)	−	+
red⁺gam⁻	*recA⁻*	+ (small)	−	+
red⁻gam⁻	*recA⁺*	+ (very small)	−	Few (by Rec)
red⁻gam⁻	*recA⁻*	−	−	−
red⁻gam⁻	*recA⁻recBC⁻*	+ (small)	+	−
red⁻gam⁻chi	*recA⁺*	+	−	+ (by Rec)
red⁻gam⁻chi	*recA⁻*	−	−	−

the production of DNA molecules with two *cos* sites. Instead, θ replication continues and large numbers of nonmaturable λ circles are produced.

An interesting event occurs when a *red⁻ gam⁻* infects a RecA⁺ cell. As just mentioned, the infection is quite defective, with a burst size of about four. Consequently, plaques are exceedingly small. However, with continual growth of this double mutant in a Rec⁺ culture, phage that produce large plaques begin to accumulate. These mutant phage have acquired a third mutation, called *chi;* this mutation generates a DNA sequence that stimulates recombination mediated by the bacterial Rec system. Because of the increased Rec-mediated recombination, these *red⁻ gam⁻ chi* mutants produce a normal burst size, which accounts both for the large plaque size and their accumulation with continued infection (that is, they quickly outgrow the phage that lack the *chi* sequence). *Chi* mutants have attracted considerable attention among researchers who study the mechanism of Rec-mediated recombination, since roughly one *chi* sequence exists in *E. coli* for every 5000 base pairs.

MAINTENANCE OF A SET OF NONESSENTIAL GENES

When a phage is sedimented to equilibrium in CsCl solutions, it comes to rest at a density that is determined by the ratio of DNA (density 1.7 g/ml) and protein (density 1.3). For λ, which is 50 percent DNA by weight, the density is roughly 1.5. However, if a centrifuge tube containing λ phage at equilibrium in CsCl is fractionated by collecting successive drops, which are formed from successive layers of the tube, and each drop is plated, about 0.001 percent of the phage are present in a band at a slightly lower density, corresponding to a phage with 13 percent less DNA. Genetic analysis of these phage shows that they are deletions, extending from, but not including, the *J* gene to the attachment site *att*. (This deletion is shown in the lower part of the circular map in Figure 16-2.) The deleted region is called *b2* (*b* for buoyancy). Gel electrophoresis of the proteins synthesized by wild-type λ and the *b2* mutants indicates that seven proteins are missing in the deletion mutant. Clearly, these proteins are not essential to λ, for if they were, the phage could not form plaques.

With phage T4, nonessential genes normally duplicate bacterial genes, and like the λ *red* gene, which is involved in one pathway to DNA maturation, mutations in some nonessential genes reduce the burst size. However, the *b2* mutants have an apparently normal burst size. In fact, it has not been possible to detect any defect in the growth of a *b2* mutant. One could imagine that the *b2* genes might have a more subtle effect: for example, their absence might produce such a small decrease in the burst size that only on an evolutionary scale would *b⁺* (wild-type) λ outgrow λ*b2*. To test this, λ*b⁺* and λ*b2* were mixed in equal numbers and allowed to infect a bacterial culture.

The phage progeny that were produced after several cycles of multiplication were taken, and another culture was infected with them. This process was continued through many cycles, and finally the ratio of b^+ and $b2$ phage was measured (a simple plating test that is sensitive to the amount of DNA in a λ phage distinguishes the phage types). The expectation that there would be many more b^+ phage than $b2$ phage was not realized. In fact, the $b2$ phage outgrew the wild-type phage. A straightforward calculation shows that the difference in the multiplication rate is caused by the fact that it takes a slightly shorter time to replicate the deletion phage (because it has less DNA) and hence in one cycle of infection, a few more $b2$ phage are made than wild-type phage. This phenomenon raises the following question: If λb^+ is at a slight disadvantage compared to $\lambda b2$, how has the $b2$ segment managed to survive evolutionary time? It seems likely that any deletion in a nonessential region would lead to ultimate replacement of the wild-type phage by the deletion phage.

This phenomenon, which is representative of numerous observations in the microbial world, is probably explained by the fact that laboratory conditions are never identical to natural conditions. In the laboratory, excellent growth media, controlled temperature and pH, and other idealized conditions are provided to growing microorganisms, for the purpose of obtaining reproducible data. However, in nature conditions are far from ideal. Thus, we may safely assume that the $b2$ genes do have value in nature. Confirmation of this hypothesis comes from the fact that in the laboratory repeated propagation of λ samples, without occasional purification of a phage stock by preparing a new stock from a plaque, results in gradual accumulation of $b2$ mutants. In contrast, λ has been isolated from the environment on various occasions and in several parts of the world (Europe, North America) and only λb^+ has been isolated. This observation eliminates the possibility that most λ in nature are $b2$ deletions and that it was just by chance that λ research began with a b^+ strain.

PROBLEMS

1. Following injection, what structural transitions does phage λ undergo, prior to transcription?

2. What two modes of DNA replication are used by λ?

3. What rules, which are different, determine the cutting sites in concatameric T4 and λ DNA?

4. What enzymes are needed to form a supercoiled λ DNA molecule after infection? State them in order of activity.

5. Which transcripts of phage λ are made by an N^- mutant?

6. What functions are lacking in a λ Q^- mutant?

7. There exist certain λ particles that carry the genes for synthesizing tryptophan. The locus of the *trp*-gene insertion is illustrated below. The tryptophan (*trp*)

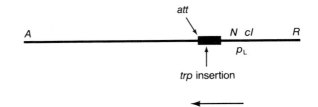

operon is transcribed in the direction indicated by the horizontal arrow, that is, in the same direction as the phage mRNA that goes leftward from *pL*. There are several classes of aberrant phage that contain *trp* DNA (*trp*-transducing particles, Chapter 18) that differ according to the size of the DNA molecule replaced by *E. coli trp* DNA. In each class, the size of the inserted bacterial DNA is roughly the same size as the DNA which is absent, the bacterial DNA contains the tryptophan synthetase gene, the insertion begins at *att* and moves to the right, and no essential phage genes are missing. These particles have the property that when they infect a bacterium lacking tryptophan synthetase, this enzyme is made.

(a) If the phage carries a point mutation in the λ *N* gene, no tryptophan synthetase is made. Why not? What part of the *trp* operon must be missing for this to be the case?

(b) Consider a *trp*-containing particle whose piece of bacterial DNA is so large that part of the *N* gene is replaced by bacterial DNA. Will any tryptophan enzymes be made? Explain briefly.

8. How many λ particles can be packaged from a dimer circle? A trimer circle?

9. If λ P^- infects *E. coli* at a low MOI, no phage are produced because there is no DNA replication. If MOI = 10 and either the bacterial or phage recombination systems are active, about one-third of the infected cells release one phage. Explain.

10. The genes *J* and *E* of phage λ encode a tail protein and the major head protein, respectively. The E^- and J^- mutations are chain-termination mutations and the bacteria used in the following experiments do not contain a suppressor. If *E. coli* is infected with an E^- phage, no viable phage are produced but lysis occurs. Such a lysate is called an E^- lysate. A similar infection with λ J^- yields a J^- lysate, which also contains no viable phage. If E^- and J^- lysates are mixed, viable phage form in numbers reaching about 50 phage per original cell. To study this phenomenon the following procedure is carried out. The E^- lysate is mixed with *E. coli* which has an amber suppressor, and the bacteria are then removed by centrifugation. The supernatant is devoid of activity; that is, complementation with the J^- lysate is abolished. However, the sediment, which contains all of the added cells, is active, in the sense that if the J^- lysate is added, the cells develop phage and lyse. What is happening in this experiment, and what is the genotype of the phage produced by the infective centers?

11. Suppose a λ phage has infected *E. coli* and is actively replicating its DNA. No DNA has been packaged, since late mRNA has not yet been made. If a T4 phage is then added, will the λ phage still be made?

12. A large number of N^- phage are plated on a lawn of *sup0* bacteria. A few plaques arise, most of which are reversions in gene *N*. However, one of the revertants, a small plaque-former, carries a point mutation that maps between genes *P* and *Q*. What do you think this mutation is? Why do you think the plaques are small?

13. λ DNA has a density in CsCl of 1.708 g/ml. Most proteins have a density of about 1.300. λ phage are 50 percent DNA by weight and hence have a density of 1.500. A phage mutant carrying a deletion is found whose density in CsCl is 1.492. What can be said about the size of the deletion?

14. The *E. coli* gene *dnaA* is needed to initiate a round of replication of the bacterial chromosome. Using a temperature-sensitive *dnaA* mutant (mutant above 40°C), suggest a simple plating test to determine whether the *dnaA*-gene product is used by λ. Do not suggest plating on the mutant bacteria, because that will not work (why not?).

15. A mutant that makes a very tiny plaque on both *sup0* and sup^+ bacteria is isolated. It makes normal-sized plaques on both *groN sup0* and $groN\ sup^+$ bacteria. What kind of mutant is it?

16. A λ mutant fails to form a plaque on *sup0*, *supI*, and *supII* bacteria but produces normal plaques on *supIII* cells. Explain this phenomenon.

17. In a study of the size of the linear branch of the rolling circle in λ-infected cells, it is found that the branch increases in length until about 125 minutes after infection and then remains at a fairly constant length. What do you expect about the length of the branch if the cells are infected with a λ E^- mutant?

18. A lawn is prepared with a 1:1 mixture of *sup0* cells and a cryptic strain carrying genes *Q* through *J*. A spot test is carried out with both λR^- and $\lambda R^- red^-$, in which R^- is a suppressor-sensitive mutation. What do the spot tests look like after incubation of the plates?

19. λN^- mutants are unable to produce progeny phage when infecting *sup0* bacteria. Furthermore, the infected cells do not die. In fact, in such an infected culture and in the progeny of the infected cells, a large fraction of the cells contain λ N^- DNA in plasmid form. Explain this phenomenon.

Chapter 17

Phage Genetics III:
Lysogeny

Infection of *E. coli* by phage T4 and by λ in its lytic cycle results in lysis of the host cell and release of many progeny phage. In Chapter 5 an alternative life cycle, the lysogenic pathway, was described, in which phage are not produced. Instead, a copy of phage DNA becomes inserted into the host chromosome, and the host cell survives and divides indefinitely. The host cell containing integrated phage DNA (a prophage) is called a lysogen. The prophage retains the potential to multiply: many bacterial generations later, if environmental conditions are right, the lysogenic state can be terminated and a lytic cycle started anew from the prophage. When this occurs, the host cell is killed, the prophage is excised from its chromosomal site, and progeny phage are formed and released. Except for the fact that the infection begins with an internal phage DNA molecule, this lytic pathway is no different from a lytic cycle initiated by infection with a phage particle. In this chapter we examine lysogeny in some detail, referring primarily to λ. In particular, we will emphasize the role that genetic analysis has played in understanding the lysogenic state and how it is achieved.

IMMUNITY AND REPRESSION

In Chapter 5 the appearance of plaques of temperate phage was described. Virulent phages, such as T4, form clear plaques because all bacteria in the region of the developing plaque are killed and lysed. However, temperate phages such as λ form a plaque with a turbid center (Figure 5-13). The turbidity is caused by the growth in the plaque of lysogenic bacteria, which cannot be infected by the phage in the plaque. A simple plating test demonstrates this resistance. A sterile wire is stabbed into the turbid center of a λ plaque and then streaked across a nutrient agar plate. After incubation of the plate, colonies are selected and inoculated into liquid medium. The resulting cultures are used for plating λ. However, no plaques are produced. Genetic tests (one will be described shortly) demonstrate that these cells are not simply phage-resistant bacterial mutants unable to adsorb phage. The cause of this resistance will be examined in this section in some detail.

Causes of Immunity

The resistance of a λ lysogen to infection by λ is called **immunity.** The cause of the phenomenon is the following. Phage λ contains a repressor-operator system (Figure 17-1). The repressor gene is called *cI;* the repressor protein binds to two operators, *oL* and *oR,* which are adjacent to and control two promoters *pL* and *pR.* The letters L and R mean leftward and rightward and refer to the direction of synthesis of two early mRNA molecules, when the genetic map is drawn as a linear map in a standard orientation (see Chapter 16, Figures 16-2 and 16-3).

In a lysogen the cI repressor is synthesized continuously and in sufficient quantity that both operators, *oL* and *oR,* contain bound repressor molecules; thus, *pL* and *pR* are unavailable to RNA polymerase, and transcription from the two early promoters is prevented. This block is sufficient to keep the prophage in an "off" state, enabling a lysogenic cell to grow indefinitely. When a normal (nonlysogenic) cell is infected by λ, the two operators of the incoming λ DNA molecule are unoccupied, since phage repressor has not yet been made; thus, transcription from *pL* and *pR* will occur. However, when a phage infects a lysogen, excess repressor molecules already present in the cytoplasm of the cell bind rapidly to the two operators in the infecting DNA molecule before an RNA polymerase can bind to *pL* and *pR.* This operator-binding prevents the phage from proceeding into lytic development. This inhibition is referred to as **resistance to homoimmune superinfection.** The superinfecting DNA forms a covalent circle—phage gene prod-

Figure 17-1 The repressor-operator system of λ showing the two early mRNA molecules. Symbols: *cI*, repressor gene; *p*, promoter; *o*, operator; *L*, left; *R*, right.

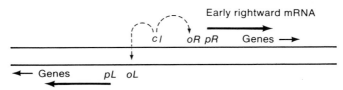

ucts are not needed for circularization—but it cannot replicate, as it lacks λ-specific replication-initiation proteins. The bacterium is unaffected by the presence of this DNA molecule and grows and divides normally, so the superinfecting DNA is progressively diluted out by bacterial multiplication.

Phage with mutations in the immunity system have been isolated. The two most important types are cI^- and *vir* mutants:

1. A cI^- mutant does not make a functional repressor and hence can engage only in a lytic cycle. Thus, a cI^- mutant makes a clear plaque. These mutants will be discussed in more detail shortly.

2. A *vir* (**virulent**) mutant carries mutations in both *oL* and *oR*. It also makes a clear plaque because it cannot establish repression; furthermore, it can superinfect and grow in a lysogen because it is insensitive to the repressor already present in the cell.

There are many temperate phages of *E. coli* other than λ. Two of these, which are closely related to λ (in that they have extensive regions of homology), are phages 21 and 434. Each of these phages has its own immune system—that is, its own repressor and repressor-specific operators. Thus, the 434 repressor cannot bind to a λ operator, and a λ repressor cannot bind to a 434 operator. Such a pair of phages is said to be **heteroimmune** with respect to one another. A temperate phage can form a plaque on a heteroimmune lysogen because the repressor made in the lysogen does not bind to the operator of the superinfecting phage. This is summarized in Table 17-1. The immunity region of the DNA, which includes the *cI* gene, operators, and promoters (and other elements to be described shortly), is given the genotypic symbol *imm*—specifically, *imm*λ, *imm21*, and *imm434*. Interesting hybrid phages, which have been quite useful in research, have been created by crossing two heteroimmune phages and selecting a recombinant containing the immunity region of one phage and the remaining genes of the other heteroimmune phage. A prominent example is a λ-434 hybrid, which is genotypically designated λ*imm434* (also, occasionally, λ*434hy*); this phage is shown in Figure 17-2.

Table 17-1 Ability of different phages to form plaques on homoimmune and heteroimmune lysogens

Superinfecting phage	Lysogen		
	B(λ)	B(434)	B(21)
λ	−	+	+
434	+	−	+
21	+	+	−

Note: The notation B(P) denotes a bacterium B lysogenic for phage P . + = forms a plaque; − = does not form a plaque.

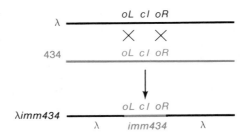

Figure 17-2 Formation of λ*imm434*.

Clear-Plaque Mutants of λ

Several types of λ mutants that produce clear plaques have been observed. The majority fall into three complementation groups, termed *cI*, *cII*, and *cIII*. Since turbidity results from immune cells growing in a plaque, these mutants must be defective in the immunity system. If a culture is infected with any one of these mutants, using conditions that would produce a high frequency of lysogenization with wild-type λ, most of the infected cells are killed. Rare survivors of *cI*⁻ and *cII*⁻ infections (10^{-6}) are invariably λ-resistant; that is, a *cI* mutant always kills a cell to which it adsorbs, if no other repressor is present. However, some cells survive infections with *cIII* mutants (10^{-3}) and are lysogens. Another difference between these mutants can be seen in a mixed infection with any of the mutants and a wild-type λ, or between pairs of mutants, as shown in Table 17-2. The results indicate that in no case can a *cI* mutant ever produce a stable lysogen, but *cII* and *cIII* mutants can be helped to lysogenize by a wild-type λ or by a phage with the corresponding allele. These results have led to the conclusion that the *cI*-gene product is needed to *maintain* lysogeny, whereas the cII and cIII proteins are needed only to *establish* lysogeny. Thus, in a mixed infection the

Table 17-2 Complementation of various clear mutants of λ with respect to lysogenization

Phages	Lysogenization	Prophage in lysogen
cI⁻	No	—
cII⁻	No	—
cIII⁻	Rare	*cIII*⁻
c⁺ + *cI*⁻	Yes	*c*⁺ only
c⁺ + *cII*⁻	Yes	*c*⁺ or *cII*⁻
c⁺ + *cIII*⁻	Yes	*c*⁺ or *cIII*⁻
cI⁻ + *cII*⁻	Yes	*cII*⁻
cI⁻ + *cIII*⁻	Yes	*cIII*⁻
cII⁻ + *cIII*⁻	Yes	*cII*⁻ or *cIII*⁻

cII and cIII proteins can be supplied by a c^+ phage, and either the wild-type or the mutant phage can become a prophage with equal probability. However, whereas λc^+ can prevent the cI^- phage from killing the cell, a single cI^- prophage cannot be maintained. Biochemical experiments have shown that the cI gene encodes the repressor and the cII-gene product is needed to activate transcription of the cI gene. The cIII protein is needed to stabilize the labile cII protein. In a small fraction of cells infected with $\lambda cIII^-$ the labile cII protein survives long enough to establish repression, which accounts for the fact that rare survivors of $cIII^-$ infections are lysogens. The biochemistry of this phenomenon will be discussed in the next section.

The *vir* mutant described above makes clear plaques. However, *vir* mutants, which require mutations in both operators, are not usually isolated as clear mutants on sensitive cells because the probability of finding a multiple mutant is quite low. These mutants are obtained by plating mutagenized λ on a lysogen and isolating phage from the exceedingly rare (10^{-8} or less) plaques that form.

Two Promoters for the Synthesis of the λ cI Repressor

At the outset of a lysogenic cycle, repressor molecules must be synthesized more rapidly than DNA replication increases the number of operators. If there were too many operators for the number of repressor molecules, transcription leading to phage development and cell death would occur. Thus, when the lysogenic pathway is to be followed, the repression system has an initial burst of repressor synthesis just after infection. In contrast, a lysogen contains only a single λ DNA molecule (the prophage), so less (but some) cI repressor is needed to maintain repression. Clearly, transcription of the cI gene must occur in a lysogen, but transcription need not be very strong.

In λ, high-level and low-level synthesis of the repressor is achieved by the existence of two promoters (Figure 17-3)—(1) the establishment promoter, *pre*, which functions in an infected cell and which requires activation by the cII protein, and (2) the maintenance promoter, *prm*, which functions in a lysogen and *is regulated by the repressor itself*. Thus, when lysogenization occurs (that is, when the phage has made enough cII protein to activate *pre*), there is a burst of synthesis of establishment mRNA and the repressor made from it activates the *prm* promoter. As the amount of repressor increases, transcription of the *cII* gene is repressed and no more cII protein is made. The cII protein is short lived, so soon its concentration is too low to maintain *pre* in an active state and synthesis of *pre* mRNA terminates. Synthesis of *prm* mRNA continues throughout successive generations, and translation of this mRNA maintains a concentration of the repressor sufficient to repress the prophage. As we have already stated, less repressor is needed in a lysogen; thus, for the sake of efficiency, less should be made when transcription

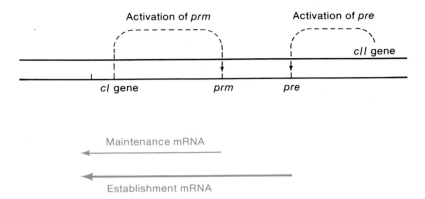

Figure 17-3   The *cI* gene of phage λ, and adjacent regions, showing the two promoters for synthesis of *cI*-encoded mRNA. The heavy black lines are the two DNA strands. The thin and heavy red arrows represent two mRNA molecules, both of which are transcribed from the lower DNA strand. The cII protein is translated from a mRNA molecule (not shown) transcribed rightward from the upper strand; this protein activates *pre*, yielding the establishment mRNA (heavy red arrow). The cI protein activates *prm*, yielding the maintenance mRNA.

is occurring from *prm*. Actually, in the fully "on" state the two promoters are equally active in binding RNA polymerase; however, less repressor is made from the transcript initiated at *prm* because the *pre* transcript has a strong ribosome binding site and the *prm* transcript binds only weakly to the ribosome. Thus, the amount of repressor made from the two mRNA molecules is determined by the efficiency of translation. A summary of the events in the sequence of repressor synthesis is given in Table 17-3.

We have just said that the repressor is needed to turn on transcription from *prm*. However, when the repressor concentration is very high, transcription from *prm* does not occur because the repressor also negatively regulates its own transcription. This regulatory mechanism enables λ to maintain a fairly constant repressor concentration, which is advantageous for two reasons: (1) In nature bacteria have continually varying growth rates

Table 17-3   An outline of the sequence of events in cI repressor synthesis in the lysogenic pathway

1. Infection of cell
2. Transcription of rightward mRNA from *pR*
3. Activation of *pre*
4. Transcription and translation of the cI repressor from *pre* (high-level synthesis)
5. Activation of *prm* by the cI repressor
6. Transcription and translation of the cI repressor from *prm*
7. Turning off of *pR* by the cI protein
8. Degradation of the cII protein, so that *pre* becomes inactive
9. Continued synthesis of the cI protein from *prm* (low-level synthesis)

owing to fluctuations in the supply of nutrients. Regulation ensures that the repressor concentration never diminishes so much that induction occurs spontaneously. (2) If unregulated, the repressor concentration might become so great that stimulated induction could not occur when needed.

One might ask why *prm* should be stimulated by the repressor. Since less repressor is made from *prm* than from *pre*, then if λ only had a lysogenic pathway, synthesis of *prm* mRNA could be constitutive. However, if a lytic cycle is also to be possible, then constitutive synthesis of *prm* mRNA should be avoided—such synthesis might allow premature repression to occur, and this would prevent initiation of the lytic cycle.

In order to understand the mechanisms of these regulatory circuits, knowledge of the structure of the operators is required; this is presented in the following section.

Structure of the Operator and Binding of the Repressor and the Cro Product

The operators *oL* and *oR* can be subdivided into six regions: *oL1*, *oL2*, *oL3*, *oR1*, *oR2*, and *oR3* (Figure 17-4). The base sequences of these regions are similar and each region is capable of binding cI repressor. The relative affinities of each region for the repressor are not the same and are

$$oL1 > oL2 > oL3 \quad \text{and} \quad oR1 > oR2 > oR3$$

Binding to *oL1* and *oL2* and to *oR1* and *oR2* is sequential and cooperative. Binding to *oR3* and *oL3* is not cooperative.

At low concentrations of repressor, *oL1* and *oR1* are occupied. Since the promoters *pL* and *pR* are adjacent to *oL1* and *oR1*, respectively, when repressor is bound to these sites, RNA polymerase has reduced access to both

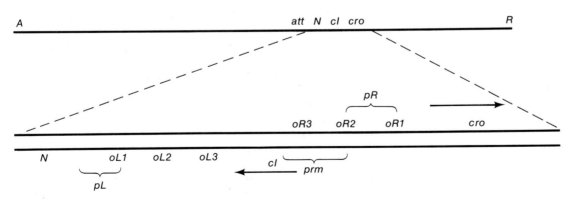

Figure 17-4 A close look at the operator-promoter region of λ. The arrows denote the direction of transcription of the *cro* and *cI* genes.

promoters. The block is even more complete when *oL2* and *oR2* are also filled, which is the state of the operators in a lysogen. The promoter *prm* is between *oR2* and *oR3*, so if the repressor concentration is very high, *prm* is blocked; this is the means by which the repressor negatively regulates its own synthesis. Furthermore, *prm* is not accessible to RNA polymerase unless *oR1* is occupied by repressor. Recent work has given some insight into the mechanism of *prm* activation. X-ray-diffraction analysis of the *cI*-operator complex indicates that repressor bound to *oR1* may provide a protein-protein binding interaction with RNA polymerase, which would significantly strengthen the binding of the polymerase to *prm*.

Thus, if a cell infected by a λ phage is destined for lysogeny, the following events occur:

1. Transcription from *pL* and *pR* begins.
2. The cII protein is translated from the *pR* transcript (Figure 16-3).
3. The cII protein enables RNA polymerase to transcribe from the *pre* promoter in order that repressor will be synthesized.
4. Repressor binds to *oL1* and *oR1*, turning off transcription and hence synthesis of the cII protein. The cII protein is unstable and soon there is no transcription from *pre*.
5. Occupation of *oR1* (item 4) causes *prm* to be activated, so the repressor continues to be synthesized from this promoter, though at a lower rate.
6. The cI repressor continues to accumulate, so *oL2* and *oR2* become occupied and repression of transcription from *pL* and *pR* becomes complete.
7. Transcription from *prm* continues unless the repressor concentration becomes so high that the protein also binds to *oR3*. Henceforth, the activity of *prm* is turned on and off to accommodate mild fluctuations in repressor concentration.

As we have described the system, there is no possibility for a lytic cycle. Entry into the lytic cycle requires the Cro protein, which is necessary for several features of the lytic cycle and to prevent synthesis of the cI repressor. The Cro protein is also a repressor that binds to the operators *oL* and *oR*. Its mode of action is based on its affinity for the subsites in the operators; *its affinity is opposite to that of the cI repressor*—namely,

$$oR3 > oR2 = oR1 \qquad \text{and} \qquad oL3 > oL2 > oL1$$

The sequence is the following:

1. The *cro* gene is transcribed from *pR* (Figure 17-4).
2. After transcription of the *cro* gene, the *cII* gene, whose product is required to activate the *pre* promoter, is transcribed.
3. Once the Cro protein is synthesized, it binds to *oR3*; hence, activation of *prm* is prevented.

4. The concentration of the Cro protein increases and *oR2* and *oR1* gradually become occupied. When these sites are filled, RNA polymerase loses access to *pR*, so synthesis of the Cro protein eventually stops.

Items 3 and 4 indicate that the Cro protein represses its own synthesis as well as repressing synthesis of the cII protein.

Recall that the lysogenic cycle requires leftward transcription from *pre* and *pL*, and the lytic cycle requires extensive rightward transcription. With this in mind, we can summarize the competitive roles of the cI and Cro proteins. As shown in the discussion of Figure 17-4, cI and Cro engage in a mutually exclusive competition for the operator sites and thereby control the transcription events that involve *oR* (except for the earliest synthesis of *N* mRNA, *oL* is primarily involved in excision, as we will soon see). Binding to *oR1* and *oR2* blocks *pR*, and binding to *oR3* inhibits transcription from *prm*. Cro and cI bind to these sites in a different order and hence constitute a bidirectional switch. When cI predominates, rightward transcription of *cro* from *pR* is off because *oR1* and *oR2* are occupied; leftward transcription of *cI* from *prm* is autoregulated by cI (on, when there is little cI, and only *oR1* and *oR2* are filled; off, when there is excess cI, and *oR3* is also occupied). When there is more Cro than cI, *oR3* is occupied first, so leftward transcription from *prm* is always off; rightward transcription from *pR* continues unless the amount of Cro is quite high, because binding of Cro to *oR1* is fairly weak. Thus, excess cI yields stable leftward transcription from *prm* and the lysogenic pathway ensues, and a relative excess of Cro produces rightward transcription and the lytic pathway. The question that remains is what determines whether cI or Cro are in excess at the point in the life cycle when the choice must be made. Later we will see that since the cII protein is the regulator of *cI* transcription, cII is the critical element.

LYSOGENIZATION AND PROPHAGE INSERTION

Lysogenization occurs when cells in late log phase, at which point nutrient depletion begins, are infected with λ at a high multiplicity of infection (Chapter 5). The result is establishment of immunity and insertion of a prophage. In this section we examine how lysogenization is detected and the genetic requirements for insertion.

Tests for Lysogenization

When a culture is infected with λ under conditions that lead to efficient lysogenization, most of the cells survive and, if plated, form colonies. In studies of lysogenization and in searches for mutations that introduce deficiencies in lysogenization it has been necessary to have tests that can indicate

Figure 17-5 A test for lysogenization. The vertical pink bars are regions containing either *imm*λ or *imm434* phages. The horizontal bands represent growth of three types of cells that have been streaked from left to right.

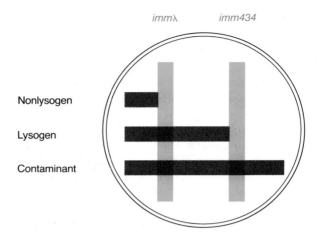

whether surviving cells are indeed stable lysogens. Most of these tests are based on the immunity to superinfection.

Figure 17-5 shows one test. A plate is prepared by streaking a drop of wild-type λ phage across the agar surface. (Soft agar ordinarily used in plating phage is not used in this test.) A second parallel streak is made with a heteroimmune phage such as λ*imm434*. A colony is then picked up with a sterile toothpick from a plate containing potential lysogens and drawn across the plate perpendicular to the phage streaks. The direction of streaking is such that the toothpick first passes through the line of λ and then through the line of *imm434* phage. The plate is then incubated for several hours, which allows the bacteria in the streak to form a visible band. If the colony is a nonlysogen, growth occurs only up to the first line of phage, because at this point the bacteria-laden toothpick picks up phage, which infect many of the cells on the toothpick. These phage multiply in the infected cells, and their progeny lyse all of the remaining bacteria. The pattern formed by a colony of a lysogen is quite different. In this case, the phage picked up from the *imm*λ line cannot infect the immune cells; thus, growth continues past this line of phage. However, when the toothpick reaches the *imm434* streak, λ immunity cannot protect the lysogen from the *imm434* phage, so cells are lysed downstream from that phage line. The reason for including the *imm434* phage in the test is to eliminate the possibility of scoring either a λ-resistant (adsorption-defective) mutant as a lysogen, or a contaminating bacterium (something that fell onto the plate from the air, for example). Both the *lam-r* cell and the contaminant will be resistant to λ; however, they will also be resistant to λ*imm43* and hence the bacteria will grow through both streaks. Thus, a lysogen is unambiguously identified by the pattern shown in the figure, namely, resistance to λ*imm*λ and lysis by λ*imm434*.

Prophage Integration

When λ DNA integrates, it is inserted at a preferred position in the *E. coli* chromosome. This site is between the *gal* operon and the *bio* (biotin) operon; it is called the λ attachment site and designated *att*. Integration at a unique site was discovered by various mapping experiments:

1. Zygotic induction in a cross between a lysogenic Hfr strain and a nonlysogenic female (Chapter 14) placed the prophage just after *gal* in the clockwise direction, when the *E. coli* map is drawn in a standard way.
2. Linkage analysis carried out by phage P1 transduction (to be discussed in Chapter 18) showed that in a lysogen λ genes are linked both to the *gal* and *bio* loci.
3. In Hfr crosses in which both the Hfr and the F^- cells were lysogenic and in measurements of linkage of *gal* and *bio* by transduction, it was found that the map distance between *gal* and *bio* is considerably greater in a lysogen than in a nonlysogen.

The most significant feature of the insertion mechanism was derived from the work of Allen Campbell, who obtained a λ prophage map by three-factor crosses with two λ genes and a *gal* marker. To do so, he used the specialized transducing phage λ*gal*, described in Chapter 18. Mapping experiments with these particles showed that the *gal* genes are contained within the central region of the λ map. He discovered that the prophage map is a permutation of the map obtained by standard phage crosses (Figure 17-6). This permutation was explained by a mechanism of integration shown earlier in Figure 5-12. The essence of the mechanism, which is frequently called the Campbell model, is circularization of λ DNA followed by physical breakage and rejoining of phage and host DNA—precisely, between the bacterial DNA attachment site and another attachment site in the phage DNA that is located near the center of the phage DNA molecule. A phage protein, **integrase** (its gene designation is *int*), recognizes the phage and bacterial attachment sites and catalyzes the physical exchange. The result is integration of the λ DNA

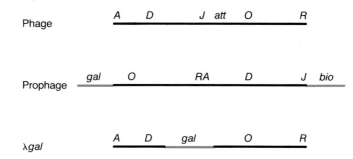

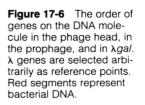

Figure 17-6 The order of genes on the DNA molecule in the phage head, in the prophage, and in λ*gal*. λ genes are selected arbitrarily as reference points. Red segments represent bacterial DNA.

molecule into the bacterial DNA. Because the phage exchange site is not at the point where the ends of the DNA join to form a circle, the phage map is turned inside out, with the ends of the phage map in the middle of the prophage map.

When Campbell's model was proposed, it was widely accepted as correct. However, it was clear that the experiments that led to it did not rigorously prove that the prophage was inserted linearly into the bacterial DNA and did not exclude such unusual possibilities as the prophage being a branch to the *E. coli* chromosome. (The reason for the uncertainty was that *bio* transducing particles had not yet been discovered, so he was unable to do a three-factor cross that included a *bio* marker.) However, a variety of experiments, both genetic and physical, proved the point. One significant genetic experiment made use of the fact that a segment of the *E. coli* chromosome was especially prone to the development of large deletions, namely, the region covering the *trp* (tryptophan-synthesizing) and *tonB* (sensitivity to phage T1) genes. A λ variant was prepared by crossing λ with another temperate phage called φ80. Hybrid phages were obtained (much like the λ-434 hybrids used in immunity studies) that carried the attachment site of φ80 and much of the remainder of λ. This hybrid phage, λ-φ80, inserted at a site adjacent to the *tonB* locus, yielding the gene order λ-φ80 *tonB trp* (Figure 17-7). A lysogen was prepared, and cells were plated on agar that had previously been spread with phage T1. Only T1-resistant colonies formed. These were tested for the ability to grow on minimal agar lacking tryptophan to obtain *tonB trp⁻* double mutants. These double mutants were found with fairly high frequency, suggesting that they were deletions that included both genes. To confirm that they were deletions, about 10^8 cells of each mutant were replated on agar lacking tryptophan. Mutants that showed no Trp⁺ revertants were considered to be deletions. These were then tested for phage production to obtain cells with deletions that penetrated the λ-φ80 prophage. These deletion strains were then infected with a set of λ mutants that were

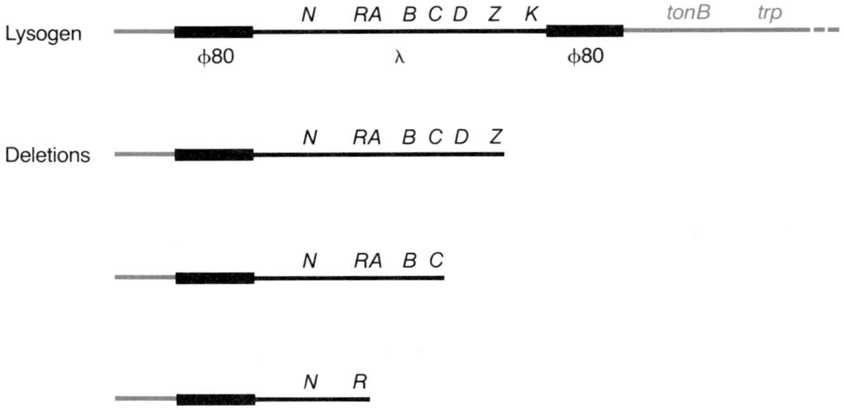

Figure 17-7 Various deletions entering the λ segment of a λ-φ80 hybrid prophage. The deletions remove the bacterial genes *tonB* and *trp*. The prophage is black, with the φ80 segment heavy. Bacterial DNA is red.

defective in one of several genes that spanned the prophage map (e.g., λA^-, D^-, Z^-, N^-, R^-). Production of a wild-type λ by recombination with prophage genes would indicate that a particular λ gene was present in the prophage. In this way the prophage genes that remained in each deletion strain were determined. These phage-prophage crosses yielded a map that showed the order in which genes could be deleted by a deletion extending into the prophage from *tonB*. If a prophage were a branch on the *E. coli* chromosome, one would expect that either all prophage genes would be present or all would be absent. If a prophage were linearly inserted into the chromosome, deletions should remove contiguous genes starting from the *tonB* side of the prophage. As the figure shows, the latter was observed.

A simple physical experiment confirmed that λ is linearly inserted into bacterial DNA. In this experiment a bacterial strain was used in which the λ attachment site was deleted (by virtue of a large *gal–bio* deletion). This strain also contained a plasmid that carried the attachment site F'*gal attλ bio*. This strain was lysogenized and several λ lysogens were obtained. To confirm that the prophage was in the plasmid the strains were mated with nonlysogenic F^- cells; zygotic induction indicated a plasmid location. Plasmid DNA was then isolated from both the lysogens and the original nonlysogen, and then tested by physical techniques for circularity and size. It was found that the F'(λ) plasmids were completely circular (without a branch), like the nonlysogenic plasmid, and furthermore that the molecular weight was one λ unit greater than that of the nonlysogenic plasmid. In later experiments heteroduplexes were prepared between F'*gal attλ bio* and F'*gal* (λ) *bio* and examined by electron microscopy (using the technique illustrated in Figures 2-9 and 2-10). Molecules consisting of a double-stranded circle the size of F'*gal attλ bio* with a single-stranded loop of λ length were seen, as expected of a plasmid with an inserted prophage.

Insertion-Defective Mutants

Stable lysogenization requires prophage insertion as well as repression, for without insertion λ-free cells will continually segregate out by cell division. In an effort to understand the insertion mechanism λ mutants were sought that were defective in insertion. The basis of the isolation procedure was the expectation that an insertion-defective phage would still be able to establish repression. That is, it was predicted that infection of cells by an insertion-defective mutant, using conditions that normally lead to a strong lysogenic response, would result in survival of the cells, but none of the cells would be lysogens. Thus, plaques formed by these mutants would still be turbid, but perhaps not quite as turbid as those produced by wild-type phage. The mutant hunt was carried out in the following way.

A λ sample was mutagenized and then plated. Many thousands of plaques were tested by stabbing a sterile needle first into the turbid center of a plaque and then onto the surface of a color-indicator (EMB) plate. The plate did not contain a carbohydrate as a carbon source, so cells on this medium should produce white colonies. Indeed, most of the colonies obtained from the plaques were white. However, a small fraction were small purple colonies with ragged edges. Since each plaque contains both phage and bacteria and since a lysogen is immune to superinfection, stable lysogenic cells obtained from the plaque will grow normally. However, if prophage integration had not occurred, the cells in the plaque would be a mixture of immune cells lacking an integrated prophage and sensitive cells that segregated out by division of an immune cell without an integrated prophage. Lysis of the sensitive cells would occur continually on the color-indicator plate because phage carried with the bacteria would infect these cells. This would cause the ragged edges and small size of the region of growth. Furthermore, phage-infected cells cause a decrease in the surrounding pH so the colony would be purple. Indeed some purple patches were found. Phage were then isolated from the plaques whose central cells produced purple growth, and phage lysates were prepared. These were tested for their ability to lysogenize in liquid medium. Streak tests of the sort depicted in Figure 17-5 showed that surviving colonies were lysed by both *imm*λ and *imm434* phages, indicating that none of the survivors were lysogens.

Two types of phage mutants are unable to lysogenize—those lacking an active integration enzyme (or enzymes) and those with a mutant *att*. These classes were distinguished by the following complementation test. A culture was lysogenized by a mixture of equal multiplicities of infection of the λ mutant and λ*imm434*, the latter being a "helper" phage to provide the missing function. Conditions were used such that repression of both phages would be established. Survivors were tested for the presence of both the *imm*λ and the *imm434* phages. Most of the lysogenization-deficient mutants were helped to lysogenize, so their mutation resided in a protein. A small fraction could not be helped, so their mutation presumably was in *att;* actually most of these proved to be deletions of all or part of the attachment site. The mutations that affected a protein were then tested against one another for complementation. All were found to be in a single complementation group, and hence in one gene. Mapping studies placed all mutations in a single region of the λ map, adjacent to *att*. The gene defined by these mutations was called *int*, for integrase.

Physical experiments confirmed the fact that *int⁻* mutants fail to insert. In these experiments cells in which the only *att*λ is in a plasmid (described above) were used. Such cells were infected with λ under conditions that lead to a lysogenic response, and the F′ circle was observed to disappear and be replaced by a circle that was larger by one λ unit. However, when the cells were infected by an *int⁻* mutant, the F′ was unaltered: it remained circular with constant size.

Several important questions were raised by the discovery of the *int* gene: (1) Does integrase act at the phage and bacterial attachment sites and is this an example of homologous recombination or site-specific recombination? (2) Does integrase also participate in prophage excision? (3) Can integrase alone catalyze excision? These questions are answered in the following two sections.

Int-Promoted Recombination

In Chapter 16 recombination in mixed infection by genetically marked λ was described, and it was pointed out that this recombination is catalyzed by the *red* genes. However, recombination between some markers can occur even if the participating phages are red^-. For example, with red^- parents no wild-type recombinants are observed in a cross between $\lambda A^- K^+$ and $\lambda A^+ K^-$, but in the cross $K^+ P^- \times K^- P^+$ wild-type recombinants arise with a frequency of about 8 percent. The difference between these crosses is the position of the markers with respect to *att*. A large number of such crosses shows that if all phages are red^-, recombination will occur whenever the two markers are on opposite sides of *att* (e.g., *J* and *P*), but not if they are on the same side of *att* (like *A* and *J*), as shown in Figure 17-8. If the parents are $red^- int^-$, no recombinants form with any pair of markers. (In all of the experiments described in this section, the bacteria used in the cross were $recA^-$ to eliminate any recombination that might occur via the bacterial recombination system.) The recombination observed between red^- phage is mediated by integrase and is called **Int-promoted recombination.** An additional experiment showed that recombination occurs in *att:* if either of the phages contained a point mutation in *att*, no recombinants arose.

Physical experiments in which density-labeled red^- phage were crossed (see Figure 15-3) showed the conservative nature of the exchange. That is, Int-promoted recombinants had a density commensurate with an exchange at *att* (Figure 17-9).

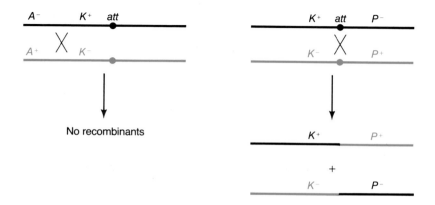

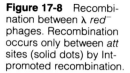

Figure 17-8 Recombination between λ *red⁻* phages. Recombination occurs only between *att* sites (solid dots) by Int-promoted recombination.

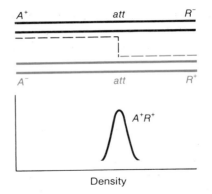

Figure 17-9 An experiment similar to that in Figure 15-3, which demonstrated breakage of DNA molecules and reunion to form recombinants. In the upper part, heavy lines represent high-density (black) and low-density (red) *E. coli* phage λ carrying the indicated markers. The phage are *red⁻* and the bacterium is *recA⁻*, so only the *int* gene is active in recombination. After mixed infection with the two phage, using conditions that prevent DNA replication, progeny phage are centrifuged to equilibrium in CsCl. In contrast with the data shown in Figure 15-3, in this experiment, a single narrow band is produced, because breakage and reunion occur only at *att*.

Note: For the following discussion and the material in the remainder of this chapter, the reader should refer to Figure 5-12 (Chapter 5) for the symbols used for the various *att* sites.

In Chapter 18 we will describe unusual λ particles that arise by aberrant excision. If excisional cuts are made to the left of the *gal* gene and hence of the left prophage attachment site (*BOP'*) and also left of the right prophage attachment site (*POB'*), a DNA molecule would result that carries *gal* and has the attachment site *BOP'*. Similar aberrant excision with cuts to the right of attachment sites could generate particles containing *bio* and *POB'* (see Figure 18-6 for a drawing of these excision events). These particles are called *gal*- and *bio*- transducing particles, respectively. In a mixed infection with λ*gal* and λ*bio* particles, both of which are *red⁻*, Int-promoted recombination can generate both a normal phage (lacking bacterial genes and with the phage attachment site *POP'*) and a phage (λ*gal bio*) carrying both the *gal* and *bio* genes and the *bacterial* attachment site *BOB'* (Figure 17-10).

These four phages—λ normal, λ *gal*, λ*bio*, and λ*gal bio*—each *red⁻* and carrying appropriate genetic markers, were used in pairwise crosses to determine the frequency of Int-promoted recombination between different attachment sites. The rationale behind the experiment was the following. If the phage and bacterial attachment sites are identical, the prophage attachment sites would also be the same as these, and all four phage types would possess the same attachment site. In that case, the frequency of Int-promoted recombination should be the same in all crosses. However, if *BOB'* and *POP'* are different, then possibly (but not necessarily) each cross would yield a

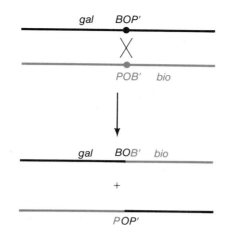

Figure 17-10   Int-promoted recombination between λ*gal* and λ*bio* phages (upper pair). The lower pair shows the two recombinants—λ*gal bio,* which carries *BOB',* and normal λ, which carries *POP'.*

unique recombination frequency. Table 17-4 shows the results of these crosses: the recombination frequencies are different. Thus, the phage and bacterial attachment sites must also be different. Later experiments showed that the sequences of *B, B', P,* and *P'* are quite different and the sequence of the *O* region (as in *BOB'*), often called the **core sequence,** is very rich in AT pairs (Figure 17-11).

Table 17-4   Recombination frequencies (percent) in Int-promoted crosses in which markers are on opposite sides of *att.**

	λ (POP')	λgal (BOP')	λbio (POB')	λgal bio (BOB')
λ	9.7	28	15	10.3
λgal	—	4.8	10	0.07
λbio	—	—	n.t.†	n.t.
λgal bio	—	—	—	n.t.

*All phages are *red⁻* and the host is *recA⁻*.

†n.t. = not tested, because neither phage carries an *int* gene.

```
- - -   GCTTTTTTATACTAA   - - -
- - -   CGAAAAAATATGATT   - - -
```

Figure 17-11   Base sequence of the core of the λ *att* region.

The Integrase Reaction

Integrase has been purified, and the integration reaction can be carried out *in vitro*. Integrase has DNA-binding activity, binding strongly to *POP′*, and is a kind of enzyme called a type I topoisomerase; that is, it can cause a strand break in one strand of a double helix, rotate one branch of the broken strand about the continuous strand, and then rejoin the ends. In addition to integrase, the reaction requires a host protein called IHF (integration host factor); this protein also binds to the attachment site, but its biochemical function is not yet known.

The mechanics of the integration reaction are fairly well known. First, two single-strand breaks are made in each attachment region in the complementary strands seven base pairs apart, as shown in Figure 17-12. Second, an exchange occurs in which two overlapping joints containing strands from two core sequences are formed, also shown in the figure. The geometry of the integrase reaction is shown in Figure 17-13. The easily melted, high-A + T core sequence in each attachment site opens and then complementary strands from different attachment sites pair to form a four-stranded segment (panel (b)). Pairs of single-strand breaks in homologous regions are made (panel (c)). Then the topoisomerase activity of integrase causes each double-stranded region to rotate (panel (c)) after which corresponding strands from the two core sequences are joined (panel (c)). This process (nick–rotate–join) is repeated on the other side of the base-paired region (panel (d)), and the two newly formed core sequences with overlapping joints separate (panel (e)).

Figure 17-12 The exchange that occurs in the integrase reaction showing the approximate positions of the strand breaks (arrows). The lowercase letters are bases in the flanking sequences *B*, *B′*, *P*, or *P′*; the capital letters are in the *O* region.

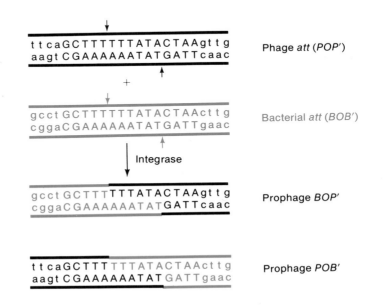

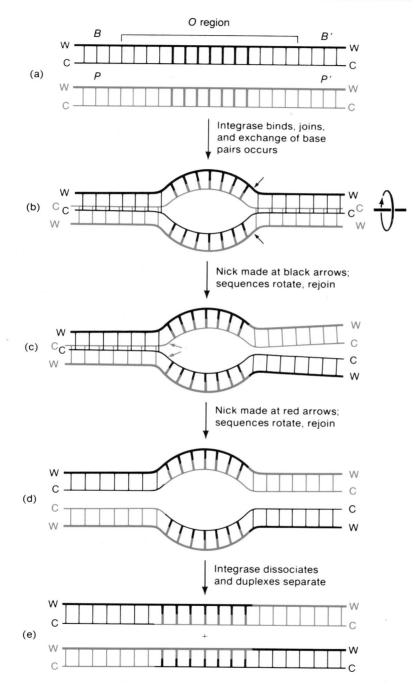

Figure 17-13 A model for the integrase reaction. The segments shown represent the 15-bp O region and small portions of the flanking B, B', P, and P' sequences. The heavy bars represent the only base pairs that are broken. The letters W and C are only for distinguishing the complementary strands. [Modified from a drawing provided by Howard Nash.]

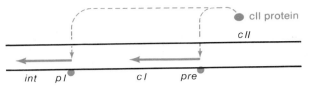

Figure 17-14 The role of the cII protein in stimulating synthesis of *cI* and *int* mRNA from *pre* and *pI* respectively. The mRNA molecules are drawn as red arrows that indicate the direction of RNA synthesis. The arrows are located nearer the DNA strand that is copied. Dashed arrows point to the sites of binding of the cII protein (red dots).

Coupling of Synthesis of Integrase and the cI Repressor

The synthesis of integrase is coupled to the synthesis of the cI repressor. This arrangement is efficient because integrase and the repressor are both needed in the lysogenic cycle and neither is needed in the lytic cycle. An outline of the regulatory pathway responsible for this coupling is shown in Figure 17-14. In the absence of a positive regulatory element (the product of the λ gene *cII*) the promoters for both the *int* and *cI* genes are unavailable to RNA polymerase. Shortly after infection, the cII protein is synthesized (see Chapter 16). If the concentration of the protein is high enough, the cII protein binds to sites near the promoters for the *cI* and *int* genes (designated *pre* and *pI*, respectively) and thereby renders them accessible to RNA polymerase. (In this respect the cII protein is similar to the cAMP-CRP complex in the *lac* operon.) RNA polymerase then transcribes both genes and the gene products are made. We will see later that the immediate cause that determines whether a lysogenic or lytic cycle occurs in an infection is the concentration of the cII protein.

PROPHAGE EXCISION

A lysogen is generally quite stable and can replicate nearly indefinitely without release of phage. However, if a lysogenic bacterium were to become damaged, it would be to the advantage of the phage to become derepressed and initiate a lytic cycle. This process, which is called **prophage induction,** includes a derepression step and excision of the prophage and is the topic of this section.

The *xis* Gene

In an earlier section it was pointed out that *int⁻* lysogens could be produced by mixed infection with an *int⁻* phage and a helper phage. These lysogens cannot be induced to produce phage, which indicates that integrase is also necessary for excision. This observation raises the question whether integrase alone carries out excision. A single consideration suggests that it does not.

If integrase alone can catalyze excision, we may ask why it does not excise the prophage shortly after integration occurs. Indeed, excision would seem to be the preferred reaction since, kinetically, two attachment sites in the same DNA molecule (that is, the host chromosome) ought to interact with each other more rapidly than two attachment sites in different DNA molecules (phage and bacterial DNA). A simple explanation is that the insertion reaction $BOB' \times POP'$ is efficient, but the excision reaction $BOP' \times POB'$ is not. However, true excision does occur with 100 percent efficiency (that is, all cells in a lysogenic population produce phage when induced), so another explanation is necessary. The initial part of the explanation is that another protein is required. The first indication for this protein was the discovery of a mutation that prevents excision but not integration. This mutation was isolated in the following way, making use of a valuable mutation in the cI gene.

Of the many mutations known in the cI gene one temperature-sensitive mutation $cI857$ has been of particular value. This mutation results in the formation of a cI repressor that is inactive at 42°C but active at 30°C; phage carrying $cI857$ make clear plaques at 42°C and turbid plaques at 30°C. The main value of this mutation in the studies of λ biology has been that transcription of the prophage in a lysogen can be initiated merely by raising the temperature of the growth medium to 42°C. Another useful feature of the mutation, which is quite unusual for a temperature-sensitive mutation, is that the temperature-induced alteration in protein structure is reversible; that is, if the temperature is lowered after heating, the cI protein renatures and repression is reestablished. This property gives rise to a phenomenon called **prophage curing.** If a bacterium lysogenic for λ$cI857$ is heated to 42°C and this high temperature is maintained, after 45 minutes (the length of the phage life cycle of an infecting λ) the cells lyse and phage are produced. However, if the cells are instead heated for 5 minutes at 42°C and then the temperature is dropped to 30°C, the cells survive. If these cells are allowed to grow for several generations and then plated, a large number of the colonies are nonlysogenic—they are cured of the prophage. Thus, 5 minutes at 42°C is sufficient to activate the excision system, but not sufficient to establish a stable lytic cycle. The explanation is that restoration to 30°C re-establishes repression but cannot turn off the excision process if the mRNA encoding the excision system has been made.

Curing served as the basis for isolation of mutations in the excision system. A culture of λ$cI857$ lysogens was mutagenized and then plated at 30°. After colonies formed, they were replica-plated onto a fresh plate and incubated for several generations at 30°C to initiate growth of the cells. The replica plates were then rapidly warmed to 42°C for 5 minutes and then cooled to 30°. After growth for a few generations to allow segregation of cured cells, the plates were returned to 42°C and allowed to grow until colonies were visible. A lysogen cannot produce colonies at 42°C because of the presence of the *cI857* mutation, which leads to induction and lysis. However, most of the cells transferred by replica plating formed visible colonies since the cured cells produced by the temperature cycling were able to grow at 42°. However, in a few cases transferred cells did not grow: these blank spots represented colonies on the master plate that did not segregate cured cells. These colonies were removed from the master plate and tested further. First, it was shown that heating these cells did not lead to phage production. However, infection of cultures derived from these mutants with wild-type λ*imm434*, which could provide all excision functions, and heating to 42°, allowed *imm*λ progeny phage to develop from the prophage. Phage with the *imm*λ character were selected from these lysates and tested for the ability to lysogenize. Two types of mutants were present. Those that could not lysogenize were *int* mutations, whose recovery would be expected since integrase is clearly involved in excision. However, some of these mutants lysogenized normally, but the lysogens could not be induced to produce progeny phage; that is, heating the lysogens resulted in cell death without phage production. The mutations carried by these phage were termed *xis*. Complementation tests showed them to be in a single gene, and mapping experiments located the *xis* gene immediately adjacent to the *int* gene, in accord with the usual observation in λ of clustering of genes with related functions.

The Int-promoted crosses described in Table 17-4 were repeated with an *xis*⁻ mutation in each phage. The results showed that the Xis product was needed only for the cross between the *gal* and *bio* phages, which contained the left and right prophage-attachment sites (*BOP′* and *POB′*), respectively. This observation is exactly what would be expected.

The *xis*-gene product, excisionase, has been purified. It forms a complex with integrase allowing the latter to recognize the prophage attachment sites *BOP′* and *POB′*; once bound to these sites, integrase can make cuts in the core sequence and re-form the *BOB′* and *POP′* sites. The physical role of the Xis protein has recently been elucidated in experiments that demonstrated a tight complex in which *POB′* is wrapped around the Xis protein; this complex is thought to contain the Int protein also. Formation of the complex is required before *POB′* and *BOP′* can interact. Thus, the Xis-Int complex reverses the integration reaction, causing excision of the prophage (Figure 5-12). Note that the reactions between all attachment sites can be written

$$BOB' + POP' \underset{\text{Int,Xis}}{\overset{\text{Int}}{\rightleftarrows}} BOP' + POB'$$

The Xis-Int complex fails to catalyze the rightward reaction, so when excess excisionase is present, excision is an irreversible reaction.

The product of excision of a prophage from the *E. coli* chromosome is an intact chromosome and a circular λ molecule, which is also the arrangement present in an infected cell immediately after infection.

Avoidance of Excision During Lysogenization

In the lysogenic pathway, before cI repressor is made, transcription from *pL* occurs, so both the *int* and *xis* genes are transcribed. If both integrase and excisionase were present, a prophage that had just been inserted into the chromosome would immediately be excised, which clearly must be avoided. A simple mechanism prevents this excision (Figure 17-15).

As part of the progress along the lysogenic pathway, the activity of the cII protein turns on synthesis of the cI repressor from the *pre* promoter and of integrase from the *pI* promoter. The cI repressor then acts on *pL* and turns off transcription of the *int* and *xis* genes from *pL*. The promoter *pI* is located *within* the *xis* gene downstream from the ribosomal binding site for synthesis of the *xis* product. Thus, this mRNA molecule provides integrase but not excisionase.

Prophage Induction

Earlier in this section it was explained that whereas lysogens are generally quite stable, conditions that could cause death of the cell lead to derepression

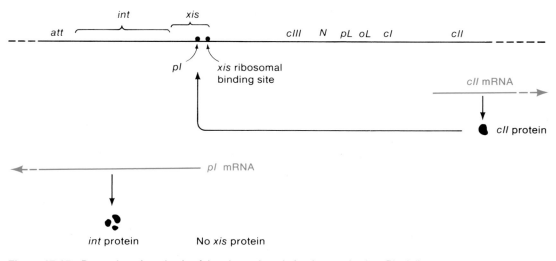

Figure 17-15 Prevention of synthesis of the *xis* product during lysogenization. Black lines are DNA molecules; red lines are mRNA molecules. The cII protein activates *pI*, the *int*-gene promoter, which is downstream from the ribosomal binding site of the Xis protein.

(prophage induction) and phage production. The usual signal for derepression is damage to the DNA. So far, all inducing agents, of which ultraviolet radiation has been studied most extensively, cause DNA damage that results in the activation of the SOS repair pathway (Chapter 9). Furthermore, $recA^-$ lysogens are not inducible by these agents, which is consistent with an SOS effect since the RecA protein is an essential component of the SOS response. Evidence that a $recA^-$ mutation only blocks the damage-to-derepression step comes from experiments in which $recA^-(\lambda cI857)$ lysogens are heated; with these strains derepression only requires thermal inactivation of the cI repressor and indeed phage are made equally well in $recA^-$ and $recA^+$ lysogens of $\lambda cI857$.

The role of the RecA protein in SOS repair is based on its ability to bind to damaged sites in DNA and inhibit the editing function of DNA polymerase III. However, the RecA protein has another function—namely, it is activated in an unknown way to become a rather specific protease. Following this activation, it cleaves the LexA protein, which is the repressor for all of the SOS genes, including $recA$ itself, and causes synthesis of the SOS gene products to fairly high levels (Chapter 14). The RecA protease also cleaves the λ cI repressor (and the repressors of many temperate phages), and this cleavage is responsible for induction. Following cleavage, derepression occurs, the early promoters pL and pR become available to RNA polymerase, and a lytic cycle ensues.

In the absence of a DNA-damaging agent derepression also occurs, but at quite a low rate—about 10^{-4} cells per generation. Whether it results from unknown DNA damage, random fluctuations in cI repressor concentration, or occasional activation of the RecA protease, is unknown. Possibly spontaneous derepression has evolved to ensure that some λ always exist in nature.

DECISION BETWEEN THE LYTIC AND LYSOGENIC CYCLES

The mechanism that determines the choice between the lytic and lysogenic cycles has not yet been fully worked out. However, it is clear that the cII protein plays a central role. This protein has three important functions—(1) it activates the *pre* promoter, (2) it activates *pI*, the *int* promoter, and (3) it delays synthesis of late mRNA (which encodes phage heads, phage tails, and the lytic system). For a lytic cycle to occur, either the synthesis or the activity of cII protein needs to be reduced to the point that *pre* is not activated. If cII abundance or activity initiates the decision between lysis or lysogeny, these features must be responsive to various signals. The critical property of the cII protein is its low stability: its half life is about two minutes, owing to various proteolytic activities in the cell. Another phage protein, critical to cII, is cIII, which increases the half-life of cII, in part by binding to cII. In a way that is unclear, cIII senses the ratio of infecting phage to bacteria, enhancing lysogeny when the ratio is high. The bacterial proteins HflA and HflB also affect the stability by some effect on the protease activities, and these proteins respond in some way to the physiological state of the cell. These points of control affect cII *activity*. Another control is cII *synthesis*, and the bacterial proteins HimA and HimD regulate the translation of the mRNA encoding cII. Interestingly, HimA and HimD form the dimeric protein, integration host factor, needed in the integrase reaction. The complex decision phenomenon requires more study before it will be understood.

POLYLYSOGENY

Earlier we described an experiment in which a plasmid containing an inserted prophage was isolated and its DNA examined. In this experiment most plasmids had increased in molecular weight by one λ unit, as stated; however, some plasmids were found to be two λ units larger, suggesting that the plasmid contained two prophages. This explanation was supported by another observation. When the culture was allowed to grow for many generations, the plasmid population became a mixture of those containing one λ unit and those containing two λ units. Furthermore, if the cell was *recA⁻*, the larger plasmid (two λ prophages) was stable. Thus, it was hypothesized that these strains initially had a plasmid with two λ prophages in tandem and, in time, intramolecular reciprocal recombination occurred between homologous regions in the prophages resulting in excision of one λ unit (Figure 17-16), a common observation when any DNA segment exists in tandem duplication.

In studying lysogens of *xis⁻* mutants another unusual observation was made. A culture was lysogenized with λcI857 *xis⁻*, and 100 lysogens were selected by the immunity streak test (Figure 17-7). These lysogens were then

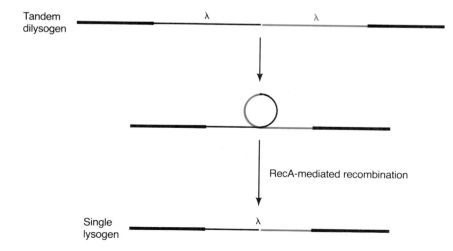

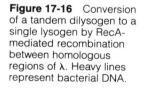

Figure 17-16 Conversion of a tandem dilysogen to a single lysogen by RecA-mediated recombination between homologous regions of λ. Heavy lines represent bacterial DNA.

stabbed into soft agar containing indicator bacteria, and the plate was incubated at 42°. If an *xis*⁺ phage had been used, all lysogens would produce clear regions at the stab sites, because the phage produced by the lysogen would lyse the indicator cells. However, with *xis*⁻ lysogens 60 percent of the cells did not produce clearing at the stab sites, and 40 percent did. This experiment was repeated with the bacterial strain described earlier in which insertion could only occur into a plasmid, and the plasmids were isolated from lysogens that produced phage and from those that did not. It was found that plasmids from phage-producing *xis*⁻ lysogens contained two λ prophages, whereas those from the nonproducers possessed only one prophage. The tandem dilysogens produced progeny phage by the Ter system described in Chapter 16. This system, which is responsible for cutting λ units from concatemers during packaging into the phage head, makes cuts in pairs of *cos* sites. In the central region of a tandem dilysogen a complete λ unit is always located between a pair of *cos* sites (Figure 17-17), which accounts for the phage production in these tandem *xis*⁻ dilysogens. The significance of this finding is that phage production by *xis*⁻ lysogens is a measure of the fraction of lysogens that are dilysogens. This assay was used to study the mechanism of formation of dilysogens. The unexpected finding was that when a culture is lysogenized, roughly 40 percent of all lysogens are tandem dilysogens. Actually, this is not too surprising since two straightforward models could account for dilysogens. Once a prophage is inserted, it is flanked by *BOP'* and *POB'*. Int-promoted phage crosses (Table 17-4) show that the exchanges *BOP'* × *POP'* and *POP'* × *POB'* occur efficiently. Thus, a second prophage can be inserted into one of the prophage attachment sites. This mechanism is called sequential insertion. In addition, two infecting phage can recombine via Int to form a circular dimer that contains two *POP'* sites; indeed such dimers account for about 10 percent of the DNA in the mid-period of an infection. Insertion of a dimer would yield a dilysogen. The sequential-insertion and dimer-insertion mechanisms have been tested in several experiments, and it has been found that both mechanisms occur with roughly equal frequency.

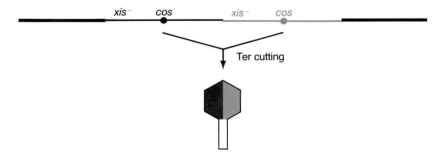

Figure 17-17 Ter cutting at prophage *cos* sites from an *xis*⁻ tandem dilysogen.

OTHER MODES OF LYSOGENY

Most temperate phages form lysogens in the way described for λ—namely, a prophage is inserted at a unique site in the host chromosome. Phages have been observed for which there are several chromosomal *att* sites, but this is rare. *E. coli* phage P1 is markedly different in that its prophage is not inserted into the chromosome. Following infection, P1 DNA circularizes and is repressed. In the lysogenic mode it remains as a free supercoiled DNA molecule, roughly one or two per cell. Once per bacterial life cycle the P1 DNA replicates, and somehow this replication is coupled to chromosomal replication (such replication does not occur with λ*int⁻*, because the *O* and *P* genes are fully repressed). When the bacterium divides, each daughter cell receives a P1 circle; how this orderly assortment is accomplished is unknown, but P1 possesses a Par function, like that which maintains plasmid number in cells containing low-copy-number plasmids (Chapter 11) The mechanism of prophage maintenance is not as foolproof in phage P1 as in temperate phages that insert their phage DNA into a chromosome; for example, in each round of cell division about 1 cell per 1000 fails to receive a P1 circle. It is not known whether this is due to occasional failure in replication or to imperfect segregation of plasmids into the daughter cells.

SOME PROPERTIES OF LYSOGENS

Lysogenization has been presented as a means of propagating a phage when a supply of sensitive bacteria has been exhausted. The ability to lysogenize also has value to a bacterium, for the bacterium survives the infection and is immune to homoimmune superinfection.

Lysogenization also has other effects on bacteria—that is, often a lysogen obtains properties not present in the original bacterium. This is called **lysogenic conversion.** One example is the resistance to phage T4 gene-*rII* mutants conferred on *E. coli* by a λ prophage. A λ gene, *rex*, also transcribed from the *prm* promoter, blocks an early stage of infection by these mutants. A second example, also already encountered, is immunity. A third example is found in *Corynebacterium diphtheriae*, the causative organism of diphtheria, which produces the disease only when the bacterium is lysogenic for the phage β. A site of insertion may also be within a bacterial gene so that prophage formation inactivates the gene. An example of a phage having this property is Mu, which was discussed in Chapter 12.

USE OF λ TO SCREEN FOR CARCINOGENS

Apart from their intrinsic interest, phages have served molecular biology principally as a means of understanding fundamental molecular processes.

Recently, λ phage has also proved to be a valuable tool for screening for carcinogens. In Chapter 10 a test (the Ames test) was described in which carcinogens are detected by their ability to induce reversions of certain His$^-$ mutations in the bacterium *Salmonella typhimurium*. A similar test (known as the **Devoret test**) utilizes *E. coli* strains lysogenic for phage λ. Lysogens are spontaneously induced at a fairly low frequency and in certain bacterial mutants the frequency is lower. However, many carcinogens cause DNA damage and are inducing agents. The induction can be detected quite simply. If one lysogenic cell is mixed with 10^8 bacteria and put on agar (exactly as might be done in detecting phage by plaque counting), the lysogen, which is not induced by this process, forms a microcolony (of at most 1000 cells) in the bacterial lawn. Occasionally, late in the development of the colony, a cell will be induced; this does not usually produce a plaque because depletion of nutrients in the agar limits cell growth, so the released phage cannot multiply (remember: phage multiply only in growing cells). However, if an inducing agent (in this case, a carcinogen) is in the agar, the original lysogenic cell will be induced to make phage at the time all of the bacteria in the agar begin to grow, so that a plaque will form. If 10^5 lysogenic cells are put in the agar, the effectiveness of the inducing agent can be measured by the fraction of cells yielding plaques. This test, which, like the Ames test, uses liver extracts in the agar to activate carcinogens, successfully detects those carcinogens already observed by the Ames test. The Devoret test is often more sensitive, though, requiring lower concentrations of carcinogens; thus, it is capable of indicating some weak carcinogens whose detection is beyond the sensitivity of the Ames test. Together these tests are being used to examine a very large number of industrial chemicals and food additives.

PROBLEMS

1. Which phage can plate on a λ lysogen: $\lambda imm\lambda$, $\lambda imm434$, $\lambda imm21$?

2. Which phage can plate on a lysogen of $\lambda imm434$: $\lambda imm\lambda$, $\lambda imm21$, $\lambda imm434$?

3. Two temperate phages are homoimmune. What phage elements must they have in common?

4. What is meant by a virulent mutant of a phage?

5. What plaque morphology is produced by a cI^- mutant and by a *vir* mutant?

6. What do you think is the difference, if any, between the appearance of a plaque of a *vir* phage on a homoimmune lysogen or a nonlysogen?

7. Phage 434 does not insert between the *gal* and *bio* genes, for its attachment site is at another position. At what locus in the chromosome would you expect $\lambda imm434$ to insert if only the promoter-operator region of phage 434 were present in the hybrid phage?

8. What symbol designates the bacterial, phage, and prophage *att* sites, and which is at the *gal* and at the *bio* end?

9. A phage has gene order *A B C att D E F*. What is the gene order in a prophage?

10. What two λ proteins are regulated by the cII protein?

11. What proteins are needed for integration and for excision?

12. What gene products activate the *pre* and *prm* promoters for cI synthesis?

13. What attachment sites are carried by *gal-* and *bio*-containing particles?

14. Mutations in what genes or sites would prevent lysogenization?

15. An *int⁻* mutant can integrate if the cell is coinfected with an *int⁺* phage, the latter "helper" phage provides the missing function. Could a phage with an operator mutation (e.g., *λvir*) be helped to integrate?

16. A λ*gal* phage can lysogenize a heteroimmune lysogen to form a tandem dilysogen. In theory, it could insert into the left (*gal*) or right (*bio*) prophage attachment sites. Genetic tests show that the gene order of the dilysogen is *gal BOP' R A gal BOP' R A POB' bio*, in which *A* and *R* are genes at the termini of the phage (given for reference only). Into which prophage *att* site did insertion occur?

17. You have isolated two independent mutant strains of the phage λ; both of these mutants form clear plaques. When either mutant alone infects *E. coli* at MOI = 10, no lysogenic survivors can be isolated. However, when *E. coli* is coinfected with 5 phage per cell of the first mutant plus 5 phage per cell of the second mutant, then 10 percent of the cells survive the infection, and almost all of these survivors are lysogenic. When the lysogenic survivors are induced, 99 of the plaques produced are clear.

   (a) What λ function or functions do you think are eliminated in these mutants?
   (b) What is the most likely explanation for the increased frequency of lysogenization which occurs during coinfection by the two mutants?

18. Which of the following λ mutants cannot lysogenize at normal frequency: N^-, P^-, Q^-?

19. Could an A^+R^+ recombinant be produced in a cross between $λA^-R^+bio$ and $λA^+R^-bio$, if both the Rec and Red systems are inactive? Explain. The *A* and *R* genes are at opposite ends of the λ DNA molecule.

20. If a lysogen is irradiated with UV light and the cells are plated with an indicator bacterium, plaques form, roughly one per irradiated cell. If the cell is *recA⁻*, no plaques are formed. If the cell contains a temperature-sensitive cI repressor, phage are produced with both *recA⁺* and *recA⁻* bacteria. Explain this observation.

21. Infection of a λ lysogen with 10 λ phage particles does not result in phage development. This is the immune response. If the multiplicity of infection is 30, phage development occurs. In a dilysogen (two λ prophages) 60 phage particles per cell are needed to initiate a successful infection. Explain.

22. The *cI857* temperature-sensitive repressor is reversibly denaturable in that if, after heating, a cell is returned to 34°C or less, the repressor renatures and becomes active.

   (a) At 40°C is the *cI* gene of a lysogen still transcribed?
   (b) If the temperature is returned to 34°C, will the *cI* gene be transcribed?

(c) If a growing λ*cI857* lysogen is heated for 10 minutes and then cooled, phage development occurs on continued incubation in growth medium. If the lysogen is instead cooled after 5 minutes, there is no phage development and the bacteria survive. Explain.

23. In an *E. coli* tandem dilysogen, the left and right prophages have the genotypes *cI857int⁻P⁻* and *cI857int⁻A⁻*, respectively. The repressor of each is inactive at 42°C and the prophages lack a functional integrase. The mutations *P⁻* and *A⁻* are, in this experiment, just genotypic markers located as follows: *A F J att N P R*. If the lysogen is heated to 42°C, what will be the genotype or genotypes of the phage progeny?

24. Suppose you have a λ mutant that makes a clear plaque on *E. coli* strain A and a turbid plaque on strain B. How can this difference be explained?

25. If a λ lysogen is induced with ultraviolet light, a lysate results in which about one clear-plaque mutant per 10^4 phage is present. If a culture of a lysogen is grown without any prior induction and the culture fluid is analyzed for free phage that are spontaneously released, the fraction that form clear plaques is always higher than 10^{-4}.

(a) Explain the higher frequency of clear-plaque mutants among the spontaneously released phage.
(b) Among the spontaneously released phage, in which of the genes *cI*, *cII*, or *cIII* do you suppose the mutations reside that yield a clear plaque?

26. A λ phage mutant called λ*b2* cannot lysogenize because it has a deletion starting just to the left of the *O* region of the attachment site; that is, its attachment site is Δ*OP′*, in which Δ represents a deletion. In a mixed infection with λ*bio* using conditions favoring the lysogenic response, dilysogens containing both λ*b2* and λ*bio* are found. Very few single lysogens (these can be ignored in your thinking) and no (*b2,b2*), or (*bio,bio*) dilysogens are found. How have the (*b2, bio*) dilysogens formed?

27. Replication of λ DNA can be detected by hybridization of labeled λ mRNA with all of the DNA isolated from an infected cell or from an induced lysogen. As replication proceeds and the number of copies of λ DNA increases, more λ mRNA can be hybridized. Interpret the following observations.

(a) If an *xis⁻* lysogen is induced, the amount of hybridizable λ mRNA increases with time.
(b) Also, using *bio* mRNA and *gal* mRNA (both of which are easily obtainable) the amount of these mRNAs that can hybridize also increases with time.
(c) If an *int⁺* lysogen is induced, the amount of hybridizable *bio* and *gal* mRNA (as well as λ mRNA) increases, though fewer of these hybridize than in part (b).

28. A λ*gal* phage can integrate into the left prophage attachment of a heteroimmune lysogen. However, integration is not observed into the right prophage attachment site. Why not?

29. Suppose you had a λ lysogen and you wanted to add a mutation that would prevent adsorption of λ. How might you go about it? (Think about what you could plate on the bacteria.)

Chapter 18

Phage Genetics IV: Transduction

Transduction is a phenomenon in which bacterial DNA is transferred from one bacterial cell to another *by a phage particle containing bacterial DNA*. Such a particle is called a **transducing particle.** Two types of transducing phage are known—generalized and specialized. Generalized transducing phage produce some particles that contain *only* DNA obtained from the host bacterium, rather than phage DNA, and the bacterial DNA fragment can be derived from *any* part of the bacterial chromosome. A **specialized** transducing phage produces particles containing both phage and bacterial genes linked in a single DNA molecule, and the bacterial genes are obtained from a *particular* region of the bacterial chromosome. In this chapter we examine *E. coli* phage P1, a well-studied generalized transducing phage, and specialized transducing particles formed by λ lysogens.

DNA TRANSFER BY MEANS OF TRANSDUCTION

During infection of *E. coli* by phage P1, or in induction of a P1 lysogen, a phage-encoded nuclease is made that causes fragmentation of bacterial DNA (a common event in the life of many phage species). In contrast with the degradation that occurs in infections by T4, endonucleolytic degradation of these fragments is quite slow, so the fragments remain fairly large throughout the P1 infectious cycle. Occasionally, a single fragment of bacterial DNA comparable in size to P1 DNA is packaged into a phage particle instead of P1 DNA. The positions of the nuclease cuts in the host chromosome are fairly random, so a transducing particle may contain a fragment derived from any region of the host DNA, and a sufficiently large population of P1 phage will contain particles, at least one of which carries each host gene. About one particle per 10^3 progeny is a transducing particle. On the average, any particular gene is present in roughly one transducing particle per 10^6 viable P1 phage. Actually all genes are not equally prevalent, since chromosome breakage is not truly random. When a transducing particle adsorbs to a bacterium, the DNA carried within the particle head is injected into the cell and becomes available for crossing over with the homologous region of the bacterial chromosome.

Let us now examine the events that follow infection of a bacterium by a generalized transducing particle obtained, for example, by growth on wild-type *E. coli* containing a *leu*⁺ gene. If such a *leu*⁺ particle adsorbs to a bacterium whose genotype is *leu*⁻ and injects its DNA into the bacterium, the cell will survive because the phage head contains only bacterial genes and no phage genes. A double-recombination event exchanging the *leu*⁺ allele carried by the phage for the *leu*⁻ allele carried by the host will convert the genotype of the host cell from *leu*⁻ to *leu*⁺. In such an experiment, typically about one *leu*⁻ cell in 10^6 will be converted to *leu*⁺ (Figure 18-1). Such frequencies are easily detected on selective growth medium—that is, as long as the *leu*⁻ allele does not have a high reversion frequency.

Figure 18-1 Transduction. Phage P1 infects a *leu*⁺ donor, yielding predominately viable P1 phage with an occasional one carrying bacterial DNA instead of phage DNA. If the phage population infects a bacterial culture, the viable phage will produce progeny phage and the transducing particle will yield a transductant. Notice that the recombination step requires two crossovers. For clarity, double-stranded DNA is drawn as a single line.

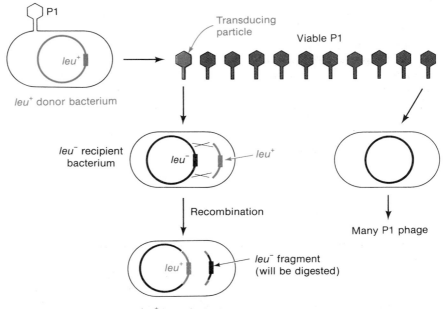

In certain circumstances—for example, in a Rec$^-$ cell or in a cell carrying a deletion—the bacterial DNA is not integrated. However, the gene present on the transducing fragment can be transcribed and produce a usable gene product, which allows limited growth of a mutant cell on a selective plate. For example, a cell with a *gal–bio* deletion can grow on a galactose-minimal plate if the cell contains a DNA fragment with the *gal* and *bio* genes. However, since the DNA fragment cannot replicate, *only one* of the daughter cells receives the fragment. The other cell will contain sufficient *gal* and *bio* enzymes in its cytoplasm to grow and divide a few times. The cell with the fragment divides again, producing one fragment-free daughter cell that can divide a few times and one daughter cell with the original fragment. The result of this repeated process, which is called **abortive transduction,** is that a clone does not develop by exponential multiplication of the initial cell (just a little faster than linear multiplication), so a visible colony rarely appears. However, by microscopy microcolonies of abortive transductants can be seen.

The molecular mechanism of transduction is not known in much detail. However, since integration occurs only in homologous regions, it clearly involves homologous recombination, probably similar to that which occurs in bacterial conjugation. As might be expected, transduction requires that the recipient cell be Rec$^+$; no transductants are observed with *recA*$^-$ or *recBC*$^-$ recipients. The donor cell can be either Rec$^+$ or Rec$^-$.

Transduction experiments are not performed with one bacterium and one transducing particle but with about 10^7–10^8 bacteria and a comparable number of phage, of which most are virulent (contain only phage genes). Because of the small fraction of the phage that are transducing particles and the low frequency of transduction, the number of productively infected cells on each plate is quite high. Thus, when plating the infected cells, it is necessary to prevent reinfection of transductants by phage progeny produced by the cells that have been infected with normal P1 phage. This is easily accomplished in several ways. A common procedure is to mix phage and bacteria, allow adsorption to occur and then to spread the infected cells on an agar coated with rabbit antiserum directed against P1 phage (obtained by infecting a rabbit with purified P1 phage). In this way, P1 progeny released from cells infected with normal phage will be inactivated soon after release from the cell and hence be incapable of subsequent infection. Another procedure involves including sodium citrate in the agar. P1 requires the Ca^{2+} ion for adsorption, and citrate binds this ion tightly; thus, in the presence of citrate, P1 cannot adsorb.

Another well-studied phage that produces generalized transducing particles is phage P22, which infects the bacterium *Salmonella typhimurium*. In fact, transduction was originally discovered with this phage. For technical reasons, physical studies are carried out more easily with P22 than with P1, and P22 provided the first experimental proof that generalized transducing particles contain only bacterial DNA. The experiment used the DNA density-labeling technique (Chapter 2) in the following way (Figure 18-2). Host bac-

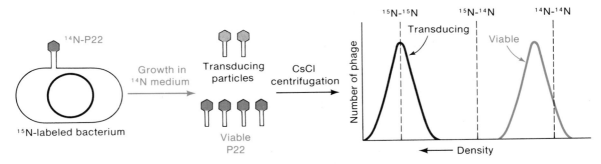

Figure 18-2 Demonstration that P22 transducing particles contain only bacterial DNA. Phage and bacterial DNA are shown in red and black, respectively.

teria were grown for many generations in growth medium containing ^{15}N and then infected with ^{14}N-labeled P22 in medium containing ^{14}N. The phage progeny were then centrifuged to equilibrium in concentrated CsCl, which separates the particles by density. The centrifuge tube was fractionated by puncturing the tube bottom and collecting successive drops, and each fraction was tested for the presence of both viable phage and transducing particles. The viable phage had the density of particles containing ^{14}N-labeled DNA (there was a small shift to higher density resulting from incorporation of some ^{14}N-labeled nucleotides obtained from degraded bacterial DNA), but all transducing particles were found at the density of particles having ^{15}N-labeled DNA, indicating that they contained little, if any, phage DNA.

COTRANSDUCTION AND LINKAGE

The small fragment of bacterial DNA contained in a transducing particle carries about 50 genes, so transduction provides a valuable tool for linkage analysis of short regions of the bacterial genome. Consider a population of P1 prepared from a *leu*$^+$*gal*$^+$*bio*$^+$ bacterium. This sample will contain particles able to transfer any of these alleles to another bacterium; that is, a *leu*$^+$ particle can transduce a *leu*$^-$ bacterium to *leu*$^+$, or a *gal*$^+$ particle can transduce a *gal*$^-$ bacterium to *gal*$^+$. Furthermore, if a *leu*$^-$*gal*$^-$ culture is infected with an excess of phage, both *leu*$^+$*gal*$^-$ and *leu*$^-$*gal*$^+$ bacteria will be produced. However, if the ratio of phage particles to bacteria is much less than 1, *leu*$^+$*gal*$^+$ colonies will not arise (Figure 18-3), because the *leu* and *gal* genes will be too far apart to be carried on the same DNA fragment, and few bacteria will be infected with both a *leu*$^+$ particle and a *gal*$^+$ particle.

The situation is quite different with a recipient bacterium whose genotype is *gal*$^-$*bio*$^-$, because the *gal* and *bio* genes are separated by only about 2.3×10^4 base pairs. A P1 phage can contain a DNA molecule with $8.5 \times$

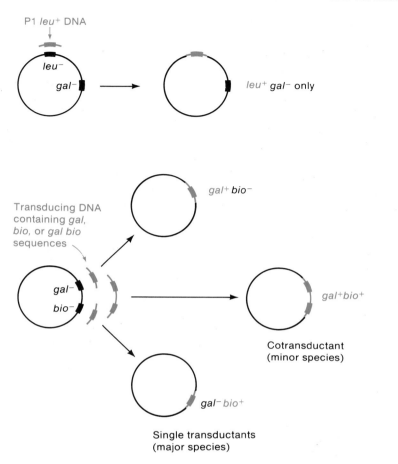

Figure 18-3 Demonstration of linkage of the *gal* and *bio* genes by cotransduction.

10^4 base pairs, so both sets of genes can be present in one DNA fragment carried in a transducing particle, namely, a *gal-bio* particle (Figure 18-3). However, all *gal*$^+$ transducing particles will not also be *bio*$^+$, and all *bio*$^+$ particles will not also be *gal*$^+$, because the nuclease cuts that produce the bacterial DNA fragments will sometimes be made between the *gal* and *bio* genes. The probability that both markers will be in a single particle and hence the probability of simultaneous transduction of both markers (**cotransduction**) depends on how close to each other the genes are. (Note that the presence of two markers in one particle does not mean that both will be transduced, for integration of the two markers must also occur; this will be discussed shortly.) Cotransduction of the *gal-bio* pair can be detected by plating infected cells on the appropriate growth medium. If *bio*$^+$ transductants are selected (by spreading the infected cells on a glucose-containing medium lacking biotin), both *gal*$^+$*bio*$^+$ and *gal*$^-$*bio*$^+$ colonies will be produced. If these colonies are tested for the *gal* marker by plating on galactose color-indicator plates, roughly half will be *gal⁺bio⁺* and half will be *gal⁻bio⁺*; similarly, if *gal⁺* transductants are selected (by growth on galactose-biotin medium and retested on a medium lacking biotin, half—the *gal⁺ bio⁺* ones—will be able to form colonies. Thus, not only will *gal⁺bio⁻* and *gal⁻bio⁺* transductants be formed, but also *gal⁺bio⁺* transductants will be produced at a frequency about half that for either single allele.

Cotransduction of *gal* and *bio* with λ genes was used to confirm that a λ prophage is linearly inserted into the chromosome. In one series of experiments, P1 was grown on a *gal⁺* lysogen and used to transduce *gal⁻* lysogens containing prophages mutant for various λ genes. Gal⁺ transductants were selected and the cotransduction frequencies for various λ genes were determined. These frequencies gave the order of the λ genes with respect to the *gal* locus. Similar experiments were done with a *bio⁺* lysogen transducing *bio⁻* lysogens containing various prophages. These two experiments confirmed the prophage gene order with respect to the *gal* and *bio* genes. In a second experiment P1 was grown on a *gal⁺bio⁺* lysogen and used to transduce a *gal⁻bio⁻* lysogen. In the second case, cotransduction of *gal* and *bio* did not occur, presumably because the λ prophage moved the genes too far apart to be included in a P1 particle.

Studies of cotransduction can be used to determine genetic distances between genetic markers and can yield a map. This procedure is examined further in the next section.

MAPPING BY COTRANSDUCTION

The use of cotransduction to determine gene order is based on the fact that *the frequency of cotransduction increases with decreasing distance between two markers*, precisely as for cotransformation, and the order of genes can be determined by the linkage procedure illustrated in Figure 13-8 (Chapter 13). When genes are very near, the method fails because of experimental uncertainty. As an example, consider three genes *a*, *b*, and *c*, for which the frequencies of cotransduction determined by two-factor crosses are: *a–b*, 90 percent; *a–c*, 33 percent; and *b–c*, 32 percent, each value having an error of about ±2 percent. The values indicate that *b* is closer to *a* than to *c* (90 > 32), but the order of the genes cannot be deduced from the data alone, because the difference between 32 and 33 percent is not significant. That is, although the order *a c b* can be excluded from consideration, the orders *a b c* and *b a c* are both consistent with the data.

Three-factor crosses can be used to determine the order of closely linked markers, such as the *a* and *b* genes just described, with respect to a third gene, or the order of mutant sites within a single gene. This method is based on the principle stated several times in this book that the probability of appearance of a genotype requiring four exchanges for its formation is much smaller than the probability of one formed by two exchanges. In this method, the results of reciprocal crosses—a pair of crosses in which the $+$ and $-$ markers are interchanged between the parents—are compared. These crosses are shown for three hypothetical genes in Figure 18-4. In the first cross, the donor (P1) genotype is $a^+ b^- c^+$ and the recipient genotype is $a^- b^+ c^-$. In the second cross, $+$ and $-$ alleles are interchanged, so the donor genotype is $a^- b^+ c^-$ and the recipient genotype is $a^+ b^- c^+$. In each cross, the $+ + +$ genotype is selected. The figure shows that for order I, production of $+ + +$ in cross 1 requires four exchanges (a quadruple crossover), whereas in cross 2, only two exchanges (a double crossover) are needed. Thus, for order I, the expected frequency from cross 1 is much lower than that from cross 2; in contrast, for the alternative order II the expected frequencies are roughly comparable for the two crosses, the differences depending only on the spacing between the markers selected. Let us assume that for cross 1 the frequency of $+ + +$ is observed to be much less than that of cross 2; then, the order must be $a \ b \ c$. Notice that we have not considered the order $a \ c \ b$, because the possibility of that order was eliminated by the two-factor cotransduction data described in the preceding paragraph. Actually the two-marker

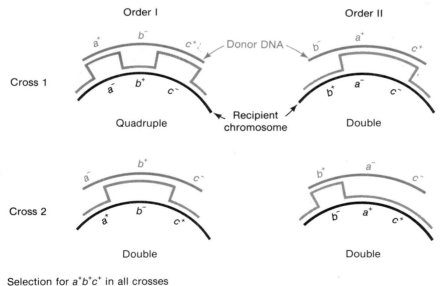

Selection for $a^+ b^+ c^+$ in all crosses

Expected results:

$$(a^+ b^+ c^+)_1 \ll (a^+ b^+ c^+)_2 \qquad\qquad (a^+ b^+ c^+)_1 \approx (a^+ b^+ c^+)_2$$

Figure 18-4 Use of reciprocal crosses to determine the order of markers *a* and *b*, relative to *c*, by transduction. The chromosome of the recipient is circular. The gray lines show the product of crossing over. The principle underlying the technique is that a double crossover has a greater probability of occurrence than a quadruple crossover.

cross does not really have to be done, since two-factor cotransduction data can obviously be obtained from the values for the three-factor cross, simply by ignoring one of the markers.

Sometimes it is inconvenient to prepare the strains needed for performing reciprocal crosses, and another analysis is possible. An example, in which actual data are shown, will illustrate this method.

Consider transduction from a donor strain whose genotype is $A^+B^-C^-$ to a recipient that is $A^-B^-C^+$. One phenotype (A^+) is scored by plating on medium lacking substance A, and 100 A^+ colonies are tested further for the B and C genotypes by replica plating. The data are shown in Table 18-1. In any cotransduction experiment one must be sure that cells are not infected with more than one transducing particle. However, since cells are usually infected with about one plaque-forming P1 per 10 cells, the multiplicity of infection by transducing particles is about 10^{-4} (and for any given donor gene 10^{-6}), and by the Poisson distribution (Chapter 5) only about 1 cell in 10^7 would receive two particles. To be quite sure, one could use various ratios of phage to bacteria and show that the cotransduction frequency is independent of the multiplicity of infection; however, this is usually unnecessary. The cotransduction frequencies in Table 18-1 show that C is nearer to A than B is to A, which is consistent with either of the orders A C B or C A B. These orders can be distinguished by diagramming the recombination events needed to generate the various recombinant genotypes. This is shown in Figure 18-5. The rule to analyze the data is the usual one: the rarest recombinant class is that achieved by the maximum number of exchanges. Table 18-1 shows that the rarest class is $A^+B^+C^+$. Figure 18-5 shows that this recombinant is produced by a quadruple exchange with order 1 and a double exchange with order 2. Thus, order 1 is the correct one.

Table 18-1   Some data from a transduction experiment

Donor genotype: $A^+B^+C^-$
Recipient genotype: $A^-B^-C^+$

100 A^+ colonies selected for further analysis:

Genotypes observed	Number of colonies
$A^+B^+C^+$	5
$A^+B^+C^-$	19
$A^+B^-C^+$	49
$A^+B^-C^-$	27
	100

Cotransduction frequencies:

A^+B^+ $(5 + 19)/100 = 0.24$
A^+C^- $(19 + 27)/100 = 0.46$

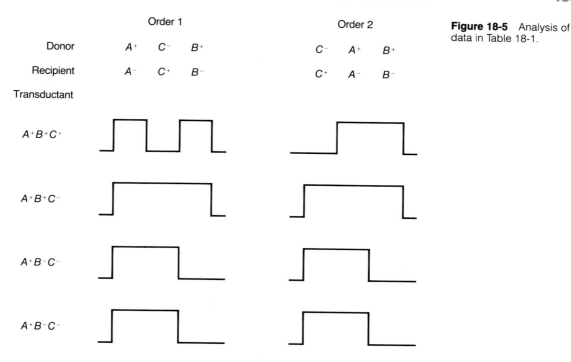

Figure 18-5 Analysis of data in Table 18-1.

The data just presented allow determination of gene order in a straightforward way but not genetic distance. The main problem in relating recombination frequencies to genetic distances is that even when three markers are near enough to each other that they can be carried by a single transducing particle, many particles will lack one of the outside markers and, clearly, infection by such a particle cannot lead to transduction of the marker that it lacks. Thus, to convert frequencies to map distances can be complex. However, the genetic map of *E. coli* is not a linkage map in the classical sense, but a temporal map, in which the map unit is one minute rather than one percent recombination. It has been possible to develop an equation that relates cotransduction frequencies to time. The Wu formula is most often used:

$$\text{Cotransduction frequency} = (1 - d/L)^3$$

in which d is the distance in minutes between a selected marker and an unselected marker in minutes, and L is the size of the transducing fragment in minutes. For P1, L is theoretically

$$\frac{\text{molecular weight of P1 DNA}}{\text{molecular weight of } E. coli \text{ DNA}} \times 100 \text{ minutes} = 2.2 \text{ minutes}$$

However, because exchange cannot occur at the ends of the transducing fragment and because the termini are probably partly degraded by nucleases, the value of 2.0 seems to be more accurate. Therefore, the distance between two markers in minutes is

$$d = 2 - 2(\text{cotransduction frequency})^{1/3}$$

Hence, for the genes A, B, and C in Table 18-1 the map is A–0.76–B–0.46–C, in which the numbers are minutes.

PROPERTIES OF SPECIALIZED TRANSDUCING PARTICLES

In the generalized transducing systems discussed so far, transducing particles form during a lytic cycle when host DNA fragments are packaged. Since fragmentation is nearly random, all possible host sequences are represented in a heterogeneous population of generalized transducing particles, if the population is large enough.

Transducing particles of a different type can also be produced during excision of an integrated prophage. These particles differ from generalized transducing particles in three ways: (1) the transducing particles contain both host and phage DNA linked in one continuous molecule; (2) only one or at most two regions of the host DNA, specifically those regions that flank the prophage, are found in these particles; and (3) a single transducing particle can serve as a template for production of a homogeneous population of identical transducing particles. These particles are called **specialized transducing particles.** They are produced only by induction of a lysogen and are produced equally frequently in $recA^-$ and $recA^+$ cells (in a $recA^-$ cell, only if a temperature-sensitive repressor can be thermally inactivated).

Formation of Specialized Transducing Particles from a λ Lysogen

The mechanism by which specialized transducing particles form is shown in Figure 18-6, which depicts the formation of the galactose- and biotin-transducing forms of phage λ—namely, λ*gal* and λ*bio*.

When λ prophage is induced, an orderly sequence of events ordinarily ensues in which the prophage DNA is precisely excised from the host DNA. In the case of phage λ this is accomplished by the combined efforts of the *int* and *xis* genes acting on the left and right prophage attachment sites. At a very low frequency—namely, in about one cell per 10^6–10^7 cells—an excision error is made; two incorrect cuts are made, one within the prophage and the other in the bacterial DNA. The pair of abnormal cuts will not always yield a length of DNA that can fit in a λ phage head—it may be too large or too small. However, if the spacing between the cuts produces a molecule between 79 percent and 106 percent of the length of a normal λ phage-DNA molecule, packaging can occur. Since the prophage is located between the

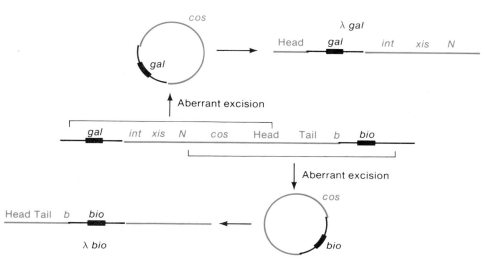

Figure 18-6 Aberrant excision leading to production of λ*gal* and λ*bio* phages.

E. coli gal and *bio* genes and because the cut in the host DNA can be either to the right or the left of the prophage, transducing particles can arise that carry the *bio* genes (cut to the right) or the *gal* genes (cut to the left).

Formation of the λ*gal*- and λ*bio*-transducing particles entails loss of λ genes. λ*gal* particles lack the tail genes, and sometimes head genes, both of which are located at the right end of the prophage; the λ*bio* particle lacks the *int, xis, . . .* genes from the left end of the prophage. The number of missing phage genes of course depends on the position of the cuts that generated the particle and thus correlates with the amount of bacterial DNA in the particle. The missing phage genes come from the prophage ends, but because of the permutation of the gene order in the prophage and the phage particle, the deleted phage genes are always from the central region of the phage DNA, as shown in the figure. The head and the tail genes are essential, so *gal*-transducing particles are unable to form plaques: they are *defective*, and this is denoted λd*gal*. The genes missing in *bio*-transducing particles are usually nonessential genes (*int, xis, red*), so λ*bio* is usually *plaque*-forming. This is denoted λp*bio*. In some *bio* phages the deletion extends into or past gene *N*, making these phage defective; these are denoted λd*bio*. *Gal*-transducing particles have been isolated from lysogens in which a large deletion has been introduced between *gal* and *att*. With these strains the *gal* locus is near enough to the prophage that *gal* can substitute for the nonessential *b2* region; these useful particles are λp*gal*. The term deletion is usually used to describe a DNA molecule in which wild-type base sequences are absent; however, in a specialized transducing particle the missing phage genes are replaced by bacterial genes. Thus, these are called **deletion-substitution particles.**

A common source of confusion about the origin of a d*gal*-transducing particle is an apparent conflict between the production of the particle and its lack of plaque-forming ability. However, notice that the transducing particle is produced by aberrant excision from a *normal* prophage. The prophage contains all of the essential genes, and hence the necessary gene products (head and tail proteins) are present in the cell. Thus, no problem exists in producing a transducing particle—only when the transducing particle infects a cell is there a deficiency of essential gene products.

As just mentioned, the size of the bacterial DNA substituted for phage DNA in a specialized transducing particle can vary, depending on the location of the aberrant exchange. The location of the exchange can be determined by a cross between a particular λ*gal* phage and several normal λ phage, each carrying a different mutation with a known location. If wild-type λ cannot arise in a cross with a particular mutant, the mutation must be in the region of the substitution in that phage (Figure 18-7(a)). Similarly, mutations can be mapped by performing crosses with λ*gal* particles having substitutions of known extent (Figure 18-7(b)). The precise physical location of the substitution can also be determined by **heteroduplex analysis.** DNA of a normal phage and a transducing particle is isolated, mixed, denatured and renatured (Chapter 2), and examined by electron microscopy. Hybrid molecules can

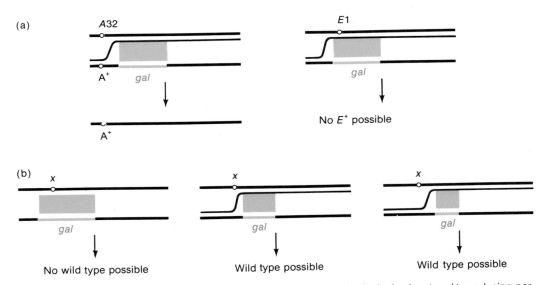

Figure 18-7   (a) Locating the left terminus of a *gal* substitution in a λ *gal* transducing particle. A cross is carried out between λ *gal* and normal λ marked either with the *A32* or *E1* mutations, whose locations are known. In the cross at the left only the relevant recombinant is shown. (b) Locating a mutation *x* by crosses with three different λ*gal* particles, selecting for *x+* phage. The shaded red area indicates the nonhomologous regions in which crossing over does not occur. Note that the right end of the *gal* substitution is always the *att* site.

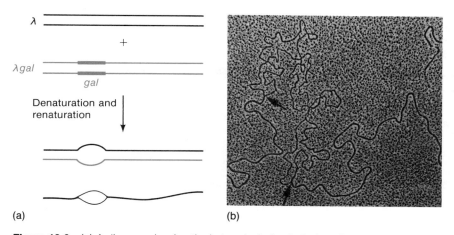

Figure 18-8 (a) A diagram showing the heteroduplexing technique for determining the size and location of a substitution. (b) An electron micrograph of λ and λ *gal* DNA. The arrows show the termini of the denaturation bubble. Single-stranded and double-stranded DNA have nearly the same width with the technique used to produce this micrograph. [Courtesy of Norman Davidson.]

then be seen in which homologous single strands have joined to form double-stranded DNA but nonhomologous single-stranded regions remain in unrenatured bubbles (Figure 18-8). Thus, the endpoints of the substitution can be localized as the branch points of the bubble.

Specialized Transduction of a Nonlysogen

There are several mechanisms by which a specialized transducing particle can transduce a mutant bacterium. The mechanisms are the same for the λ*gal* and λ*bio* particles, so we will use only λ*gal* as an example. Consider a Gal$^-$ cell that is infected with a lysate resulting from induction of a Gal$^+$ culture lysogenic for λ. This lysate will consist overwhelmingly of the normal λ phage but will contain a tiny fraction of the λd*gal*$^+$ phage. Conditions are used that lead to the establishment of immunity repression (high MOI and late log-phase cells); thus, the cell is not killed by a lytic response. If these infected cells are plated on galactose color-indicator plates (e.g., purple for Gal$^+$, white for Gal$^-$), purple colonies are found at a very low frequency (about 0.001 percent of the infected cells). These purple colonies are of two types (Figure 18-9):

Type I consists of nonlysogenic cells, all of which are Gal$^+$. If the colony is dispersed and individual bacteria are allowed to form colonies, all colonies are purple. These contain stable *gal*$^+$ bacteria and have arisen as a result of two genetic exchanges, as shown in the figure.

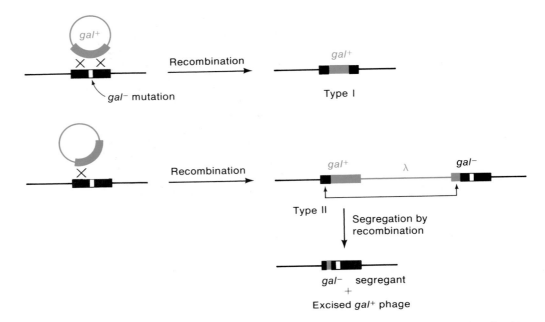

Figure 18-9 Production of type I and type II λ*gal* transductants and segregation of *gal*⁻ cells from a type II cell.

Type II cells contain a prophage. If a type II colony is dispersed and individual cells are plated, about 1 percent of the colonies are white (and hence Gal⁻) and also no longer have a prophage. Type II cells have arisen by a single exchange within the *gal* genes. These cells now contain two copies of the *gal* genes, one *gal*⁺ and one *gal*⁻. These transductants are called **heterogenotes,** or partial diploids. The type I cells, which contain only one *gal*⁺ gene, are haploids.

When there are two copies of the *gal* genes in a cell, intramolecular genetic recombination can occur with moderate efficiency. This accounts for the production of Gal⁻ nonlysogenic cells during continued growth of a type-II *gal*⁺-transductant, as shown in the figure.

Type II cells are incapable of being induced to produce progeny λd*gal* particles, because the prophage lacks the tail genes. However, transcription and synthesis of many phage-specific proteins do occur.

Heterogenotes can also be produced in another way. If cells are infected at an MOI such that many cells are infected by both a λ*gal* and a normal λ (which are present in excess in the initial lysate), a λ*gal*-λ tandem dilysogen can form. These arise by sequential insertion. First, the normal λ inserts by the usual *POP'* × *BOB'* reaction. Once the prophage is established, the λ*gal*, which has the attachment site *BOP'* can insert into the *BOP'* at the *gal* end of the prophage (this is an efficient Int-promoted exchange). Thus, the gene order in the dilysogen is gal^-(chromosome)–λgal^+–λ, with the gal^+ gene in the center of the dilysogen (the reader should work this out).

Specialized Transduction of a Lysogen

A λ lysogen can also be transduced to produce a heterogenote. This occurs at a high frequency in two ways. The first mechanism is the same as that which produces type II cells—namely, a single genetic exchange—except that the exchange can occur either in the gal operon or the prophage. Note also that conditions leading to the lysogenic response are unnecessary if the prophage has the same immunity repressor as the λdgal, because no phage development can occur. These transductants are heterogenotes both for the gal operon and the prophage. They are more unstable than the type II transductants described earlier because the probability of an intramolecular exchange leading to production of a gal^- segregant is greater in that it can occur in both the gal and prophage segments of the DNA.

The second type of transduction occurs when the prophage and the λdgal are heteroimmune (have different repressors). In this case the λdgal can express the int gene and a single genetic exchange can occur between the attachment sites of the prophage and the infecting phage (Figure 18-10).

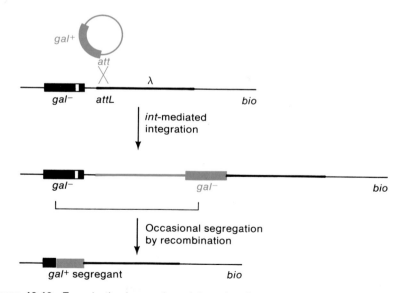

Figure 18-10 Transduction by prophage integration. Segregation by recombination in the *gal* genes occasionally occurs; this segregation can also occur at the *att* sites if some integrase is made. The infecting phage and the prophage must be heteroimmune for the insertion to occur.

This also produces an unstable heterogenote, as shown in the figure. (Note that this Int-mediated mechanism cannot occur with a λp*bio*, because the *bio* substitution replaces the *int* gene.)

High-Frequency-Transducing Lysates

The heteroimmune and homoimmune transductants just described are very useful because these dilysogens contain a λd*gal* prophage and a normal prophage that has functioning head and tail genes and, when induced, these genes can provide the structural components necessary for packaging λd*gal*. Thus, when such a dilysogen is induced, both normal and λd*gal* particles are produced by the mechanism shown in Figure 18-11. Roughly equal quantities of the two particles are produced. Since the DNA content of λd*gal* is generally different from that of normal λ DNA and the protein content is the same, the two particles can be separated by equilibrium centrifugation in a CsCl density gradient. (The densities of DNA and protein molecules are 1.7 and 1.3 g/cm^3 respectively. The density of the phage is determined by the proportion of DNA to protein.)

The dilysogens just described yield lysates half of whose phage are transducing particles. In laboratory jargon such a lysate is called a **high-frequency-transducing (HFT) lysate.** A lysate formed from a single lysogen in which aberrant excision occurs only infrequently is called a **low-frequency-transducing (LFT) lysate.**

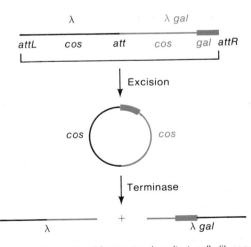

Figure 18-11   Production of λ and λ*gal* from a tandem (λ, λ*gal*) dilysogen.

SPECIALIZED TRANSDUCING PHAGE
AS A CLONING VEHICLE

Specialized transducing particles have a wide variety of uses. One of these is as a cloning vehicle. For example, a concentrated sample of λ*gal* can serve as a source of *gal* DNA for base-sequencing analysis. Now, this type of cloning is most often done by genetic engineering techniques (Chapter 20). An early experiment, now classic, used φ80*supIII*, a plaque-forming transducing phage that carries the *E. coli supIII* gene. Studies of base sequences in the DNA of this phage provided the first evidence for the generation of a suppressor tRNA by a base change in the anticodon of a normal tRNA.

Specialized transducing phage have also served as hybridization probes for identifying specific mRNA molecules. For example, early studies of the kinetics of transcription of the *lac* operon were carried out by radioactive labeling of RNA after induction of the operon and measuring the amount of *lac* mRNA by determining the amount of filter-bound radioactivity on nitrocellulose filters containing denatured φ80*lac* DNA.

PROBLEMS

1. If 10^8 phage λ infect 10^8 gal^+ bacteria and all phage adsorb, what fraction of the infected cells will yield *gal*-transducing phage?

2. Remember that a λ prophage is located between the *E. coli gal* and *bio* genes. The *lac* gene is about 10 minutes away from the *gal* gene in the *E. coli* time-of-entry map. Roughly what would be the ratio of *lac*- and *gal*-transducing particles?

3. What are the differences between a phage that produces only specialized transducing particles, one that produces only generalized transducing particles, and one that can produce both?

4. Why are λ specialized-transducing particles generated only by induction rather than by lytic infection?

5. Some temperate phages that integrate their DNA do not seem to form specialized transducing particles. Give several possible reasons for this observation.

6. Following are some facts about specialized transduction by phage P22 in *Salmonella typhimurium:* (1) Specialized transducing particles of P22 carrying the *proA* and *proB* genes can be generated. These genes are immediately adjacent to the prophage attachment site on the bacterial chromosome. The *proA* and *proB* specialized transducing particles have the following behavior: (2) Lysates of the specialized transducing particles can go through the lytic cycle, but only in mixed infections. (3) In single infections the particles transduce by substitution; however, in mixed infections, the particles transduce by lysogenization. What is the molecular basis of each of the foregoing observations?

7. Under what circumstances could a lysate of phage P1 containing transducing particles carry phage λ?

8. In a classical experiment *E. coli* cells, which are genetically unable to synthesize thymine, were fed 5-bromouracil to make their DNA "heavy" and then were infected with phage P1 and incubated in a medium containing (1) thymine, (2) 5-bromouracil, or (3) thymine plus ^{32}P until lysis occurred. Analysis of phage progeny by density-gradient centrifugation showed that transducing particles from the first two media had similar densities and that those from the third were not radioactive. From these observations, what do you conclude about the phage genetic material in transducing particles and the origin of transduced segments?

9. A bacterium with genotype $A^+B^-C^-$ was transduced by a particle carrying the linked genes $A^-B^+C^+$, and $A^+B^+C^+$ recombinants were selected. The reciprocal cross was also performed. The number of $A^+B^+C^+$ recombinants was the same in both the first cross and the reciprocal cross. What information does this give you about the order of the three genes?

10. P1 transduction analysis was carried out to determine the order of mutant sites in the *trpA* gene of *E. coli*, using a closely linked mutation, called *ant,* as an unselected marker. In each of the crosses listed below, Trp$^+$ recombinants were selected and then tested for the *ant$^+$* or *ant$^-$* allele. The donor genotype is given first, and the percent of Trp$^+$ recombinants that is *ant$^+$* is given in parentheses.

$$\text{Cross 1: } ant^+\,trpA34 \quad \times \quad ant^-\,trpA223 \quad (18\% \; ant^+)$$
$$\text{Cross 2: } ant^+\,trpA46 \quad \times \quad ant^-\,trpA223 \quad (52\% \; ant^+)$$
$$\text{Cross 3: } ant^+\,trpA223 \quad \times \quad ant^-\,trpA34 \quad (50\% \; ant^+)$$
$$\text{Cross 4: } ant^+\,trpA223 \quad \times \quad ant^-\,trpA46 \quad (20\% \; ant^+)$$

What is the order of the three *trpA* markers with respect to the *ant* gene?

11. P1 is grown on $trpC^+\,pyrF^-\,trpA^-$ cells and used to transduce $trpC^-\,pyrF^+\,trpA^+$ recipient. Trp$^+$ transductants were selected and tested for the presence of the other markers. The following data were obtained: $trpC^+\,pyrF^-\,trpA^-$, 548 colonies; $trpC^+\,pyrF^+\,trpA^-$, 579 colonies; $trpC^+\,pyrF^-\,trpA^+$, 3 colonies; $trpC^+\,pyrF^+\,trpA^+$, 90 colonies.

(a) What is the order of the three genes?
(b) What are the cotransduction frequencies for *trpC* and *pyrF, trpA* and *pyrF,* and for *trpC* and *trpA*?
(c) What is the map distance in minutes between the three markers?
(d) Why are the map distances not additive?

Chapter 19

Strain Construction

A great deal of the actual laboratory work in microbial genetics is concerned with constructing bacterial strains and phages containing particular genes or combinations of genes. A certain amount of this construction is done by mutagenesis, but as collections of mutations have grown to the tens of thousands, more often one attempts to combine existing mutations. The techniques of strain construction used by geneticists consist to some extent of a set of standard procedures but more often they are a bag of tricks, which cannot really be taught in any coherent way. In this chapter no attempt is made to present a comprehensive collection of techniques. Instead we present a potpourri in order to give the reader a flavor of the ingenuity frequently found in microbial genetics laboratories. The techniques will be given in two parts—construction of bacterial strains and of phage strains. In order to facilitate understanding these procedures an *E. coli* map showing the mutations discussed in this chapter is provided in Figure 19-1.

CONSTRUCTION OF BACTERIAL STRAINS

Novel bacterial strains are constructed for two main reasons: to provide multiply mutant bacteria for mapping purposes and to create bacteria that have special properties that vastly simplify the study of certain genetic and biochemical processes. An example of the latter is the fusion of a promoter for

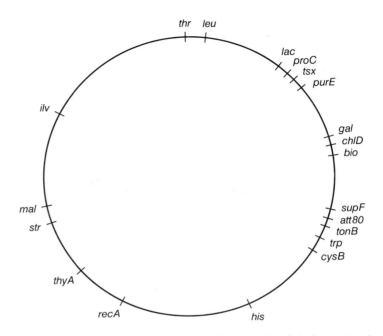

Figure 19-1 An *E. coli* map, showing genes used in examples of strain construction in this chapter.

an operon whose gene products are difficult to measure quantitatively with the easily assayed β-galactosidase gene. To accomplish these goals the geneticist has available the techniques of mutagenesis, recombination, transposition, and so on, but these must often be coupled with powerful selection techniques that make it possible to find a rare bacterium with the desired genotype. Some of these techniques have been seen earlier in this book. For example, an auxotrophic mutant might be very difficult to find were it not for the powerful mutant-enrichment procedure of penicillin selection, described in Chapter 10. Also in that chapter we saw color-indicator plates that enabled one to isolate different mutations in the *lac* operon. In this section a variety of techniques, some of which illustrate general procedures, will be described. We begin with the more straightforward methods.

Isolation of Sugar-utilization Mutants

Bacteria growing on rich (broth) plates containing the colorless dye triphenyltetrazolium chloride (usually called **tetrazolium**) reduce the dye to an intensely red compound. The reaction is inhibited at low pH. Thus, Lac$^-$ bacteria growing on lactose-tetrazolium plates yield red colonies, whereas Lac$^+$ bacteria, which produce acid by metabolizing lactose, produce white colonies. Tetrazolium plates are especially valuable in searches for sugar-utilization mutants because of the ease with which one red colony can be seen against a background of a large number of white colonies. In contrast, on EMB agar one would seek a white colony against a background of purple colonies on a purple plate, which is much harder to see. On tetrazolium plates one Lac⁻ colony can usually be detected among about 5000 colonies on one plate. Thus, by using mutagens that raise the mutation frequency to about 10^{-5}, 100 plates (not a large number) would yield 5–10 mutants.

If a multiple mutant, such as a Lac⁻Gal⁻Ara⁻ triple mutant were desired, it could be isolated by a series of selections on tetrazolium plates. First, a Lac⁻ mutant could be selected from a mutagenized wild-type culture using lactose-tetrazolium plates. A culture of this mutant would then be mutagenized and plated on galactose-tetrazolium plates to select for a Gal⁻ mutant, which would be Lac⁻Gal⁻. A third round of mutagenizing and plating on arabinose-tetrazolium plates would yield the triple mutant. The entire procedure would take less than two weeks; most of the time would be spent waiting for colonies to grow.

Temperature-sensitive *lac* mutants, which are Lac⁻ at 42°C and Lac⁺ at 30°C, can be isolated by a slight modification of the procedure. In a primary selection, a mutagenized culture is plated on lactose-tetrazolium plates, which are incubated at 42°C. Lac⁻ colonies are selected and replated on lactose-tetrazolium plates at 30°C. Colonies that are red on a 42°C plate and white on a 30°C plate have the Lac(Ts)⁻ phenotype.

Isolation of Thymine-requiring Mutants

Mutants that require a supply of thymine in the growth medium have been valuable in the study of bacterial DNA synthesis for three reasons: (1) their DNA can be made radioactive by growth in medium containing radioactive thymine (which is incorporated poorly by Thy⁺ cells), (2) a very high-density label can be introduced by adding 5-bromouracil, which substitutes for and is much denser than thymine, to the medium, and (3) by withholding thymine, DNA synthesis can be inhibited. Penicillin selection, which enriches for most auxotrophic mutants, fails with Thy⁻ mutants, because for unknown reasons Thy⁻ cells lose the ability to divide, and hence to form colonies, after about 15 minutes in a medium lacking thymine (this is called thymineless death). However, *thy*⁻ mutants can be isolated by growth in medium containing either of the substances trimethoprim or aminopterin, which interfere with biochemical reactions that use the vitamin folic acid as a cofactor. For complex biochemical reasons, Thy⁻ mutants grow more rapidly than Thy⁺ cells in a medium containing trimethoprim and thymine. After about 100 generations of growth, which requires several dilutions into fresh medium, cultures consist primarily of mutants in the *thyA* gene. The technique is unaffected by the presence of other auxotrophic mutations. For example, if a Leu⁻Thy⁻ mutant is desired, one need only start with a Leu⁻ strain and carry out the selection in medium supplemented with leucine.

Unusual Selections for Auxotrophic Deletion Mutants

In a discussion of the proof of λ prophage insertion in Chapter 17 it was pointed out that deletions in the *tonB–trp* region can be isolated by selecting a cell that is resistant to phage T1 and then seeking among them a Trp⁻ mutant. Since a double mutant would be exceedingly rare, most of these mutations turn out to be deletions. Similar techniques apply to other regions of the genome. For example, cells resistant to phage T6 are also often purine-requiring and yield deletions in this region of the *E. coli* genome.

Deletions of the histidine (*his*) genes can be isolated in a novel way. The temperate phage P2 differs from most other phages in that it can integrate at several different chromosomal sites. One of these is near the *his* operon. Frequently P2 is spontaneously excised from this site, leaving a cured cell (lacking a prophage). With fairly high frequency the excision event is an aberrant one in which a large portion of the chromosome is also removed. This process, which is called **P2 eduction**, produces a large deletion that invariably includes the *his* operon, but may include many other nearby genes. These His⁻ mutants can be enriched by penicillin selection and isolated by replica plating. However, another technique is available. If an auxotrophic mutant is plated on minimal medium containing a growth-limiting amount of the nutrient, the colony will be quite small and can be isolated on the basis of its size. This technique is not applicable to most mutations because the mutation frequency is too low; that is, for a typical auxotroph thousands of plates might be needed to find one tiny colony. This number could be reduced by vastly increasing the number of colonies per plate, but colony size is always quite variable on minimal plates containing more than a few hundred colonies. However, eduction occurs fairly frequently, so the number of His⁻ mutants in a culture is high (about 10^{-3} in an overnight culture). Thus, this simple screening technique works well for *his* deletions.

Strain Construction Using Existing Strains

New strains are most often created by transferring mutations from one strain to another. The two most common methods are bacterial conjugation and transduction with phage P1. Conjugation is most useful when exchange of large segments of the chromosome is either necessary or can be tolerated. Transduction is of value when small segments are to be transferred, and has the advantage of maintaining isogenicity. For example, suppose two variants of one strain are needed, one carrying a particular *pro⁻* mutation and another with a particular *trp⁻* mutation. Both variants could be constructed by transferring the mutations from appropriate Hfr strains. However, since there is no convenient way to control the amount of DNA that is exchanged in the recipient, the new strains could differ with respect to silent or leaky mutations present in different regions of the Hfr. However, with transduction a maximum of two percent of the chromosomal DNA can be transferred to the recipient, and an even smaller amount is usually retained after recombination has occurred; thus, it is likely that the two desired variants will remain almost isogenic, except for the purposely introduced *pro⁻* and *trp⁻* mutations.

In making strains by Hfr transfer it is convenient, though not necessary, to use an Hfr whose origin is near the desired gene. For example, if the time of entry of the desired *pro⁻* marker is less than 10 minutes, one can use a 1:1 ratio of Hfr to recipient, bypass any selective procedure, and merely collect several hundred recipient cells and test for the marker by replica plating. Of course, a counterselection would be used to prevent growth of the Hfrs. More often, markers are selected or acquired by their linkage to a selective marker. For example, if a particular *pro⁻* allele is needed, a mating can be carried out between an Hfr *pur⁺pro⁻str-s* and an *F⁻pur⁻pro⁺ str-r* strain, with plating on medium containing proline and lacking adenine to select for the Pur⁺ phenotype. Pur⁺Str-r colonies are picked and tested by replica plating for the closely linked *pro⁻* mutation (Figure 19-2).

Similar selections can be carried out with P1 transduction. For example, a Lac⁻ cell can be made Lac⁺ merely by infecting it with P1 that has been grown on Lac⁺ cells and selecting for Lac⁺ transductants on lactose-minimal medium. When using cotransduction, which is the most common technique, it is necessary to choose another marker that is within two minutes of the desired mutation in the time-of-entry map. Let us consider how we might construct a Leu⁻ strain that carries the suppressor SupF. The *leu* gene is at about 1 minute on the *E. coli* map and the *supF* gene is located at about 26 minutes in a cluster containing *tonB* (T1-resistance) and *cysB* (cysteine synthesis). Let us assume that our strain collection includes the following strains: I, *leu⁻sup0*; II, *cysB⁻*; III, *supF*. We might proceed in the following way (Figure 19-3):

1. Make a T1-resistant *(tonB⁻) cysB⁻* strain by plating T1 on strain II. Call this strain IIA.
2. Grow P1 on strain IIA and infect strain I. Select for transduction to T1-resistance by plating on a plate covered with T1. Replica-plate to medium lacking cysteine and thereby select a *leu⁻ tonB⁻ cysB⁻* strain (IA).

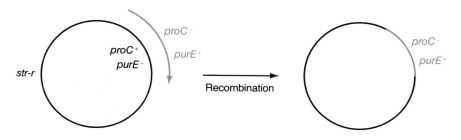

Figure 19-2 Selection of a *proC⁻* strain by mating an Hfr *purE⁺proC⁻str-s* strain with an *F⁻purE⁻ proC⁺str-r* strain. Cells are plated on medium containing proline and streptomycin and lacking a purine.

Figure 19-3 A scheme for constructing a *leu⁻ supF* cell by successive changes, starting with strains I, II, and III. The markers *cysB, tonB,* and *sup0* are within one minute of one another. The *leu* marker is actually 25 minutes away.

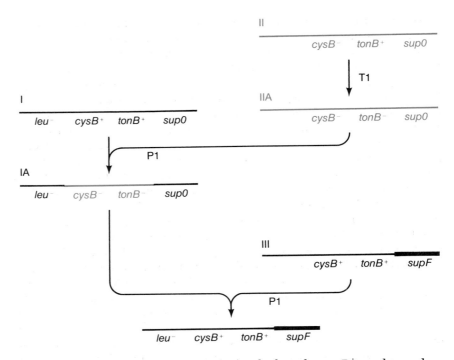

3. Grow P1 on the *supF* strain (III), which is also *cysB⁺*, and transduce the new *leu⁻ cysB⁻* strain (IA) to *leu⁻ cysB⁺*. Some of these will be *supF*, and these can be found by testing for the growth of a conditional-mutant phage that can form a plaque on *supF* but not on *sup⁻* bacteria.

A technical trick is necessary whenever P1 transduction is done. Since only about 1 particle in 10^4–10^5 carries bacterial DNA, a large number of active P1 phage are present on a transduction plate, and some means is desirable to prevent these phage from infecting and lysing or lysogenizing the transductants. The protection is provided by the requirement for the Ca^{2+} ion for adsorption by P1. Thus, initially P1 is allowed to adsorb in liquid medium to the bacteria to be transduced, using a fairly low multiplicity of infection, and the infected cells are then plated on selective plates that contain the citrate ion (which tightly binds the Ca^{2+} ion). None of the progeny phage are able to adsorb to the transductants.

The *recA* allele is necessary in a wide variety of experiments, for example, those in which recombination, SOS repair, or prophage induction are being studied. A *recA⁻* mutation cannot be selected in any way, so it is moved from one strain to another using an Hfr and obtaining the mutation as an unselected marker. What is necessary first is an Hfr that carries a *recA⁻* mutation. We begin by constructing this strain. The procedure is the following (Figure 19-4(a)):

(a) Construction of Hfr *recA⁻*. Mate Hfr *ilv⁺* with *F⁻ ilv⁻ recA⁻* , and select Ilv⁺.

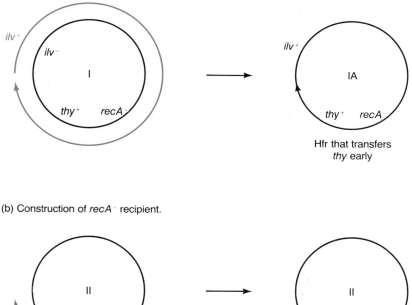

Hfr that transfers
thy early

(b) Construction of *recA⁻* recipient.

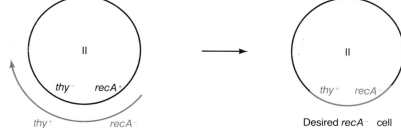

Desired *recA⁻* cell

Figure 19-4 A scheme for transferring the *recA⁻* marker from one strain (I) to another (II). (a) Strain I is converted to HfrIA, which transfers *thy* and *recA* as early markers. (b) HfrIA is mated with strain II, to yield the desired *recA⁻* strain (IIA).

1. A *recA⁻* strain is mutagenized with a chemical mutagen and a mutant that requires isoleucine *ilv⁻* is isolated. The result is a *ilv⁻ recA⁻* strain (I).

2. This strain is then mated with an Hfr that transfers the *ilv* gene as a terminal marker, and Ilv⁺ recombinants are selected. Even though the recipient is *recA⁻*, recombinants form because the *recA⁺* allele is transferred quite early by this Hfr. Since F is transferred last from an Hfr, some of these recombinants will be Hfrs. Thus, several Ilv⁺ recombinants are mixed with appropriately marked females and transfer of an early marker, such as *mal*, is sought; if transfer occurs, the strain is an Hfr (IA). These Hfr strains are then checked for sensitivity to ultraviolet radiation to ensure that they have retained the *recA⁻* mutation; recall that *recA⁻* cells are exceedingly UV-sensitive (Chapter 9).

We now proceed to transferring the *recA⁻* strain to any desired cell. The method is to select for a closely linked marker and check for the presence of the *recA⁻* allele among the recombinants. A marker commonly used for this purpose is *thy⁻*, which is only a few minutes away from *recA*. Recall that a *thy⁻* marker can be introduced into any strain by mutation and selected by growth in medium containing trimethoprim (see earlier in this chapter). The Hfr *recA⁻* is then mated with an *F⁻ thy⁻* strain (II) and Thy⁺ recombinants (IIA) are selected by growth on medium lacking thymine (Figure 19-4(b)). These recombinants are then tested for the presence of the *recA⁻* allele by the UV sensitivity of the strains. As a final test, one would use the strain in a mating with an Hfr and show that recombinants could not form.

Use of Transposons to Isolate Mutations in and Deletions of Particular Genes

Transposons are widely used to generate mutations in particular genes. The mutations are especially valuable when complete absence of gene function is required, since the gene is interrupted and the mutation is never leaky. The following procedure could be used to isolate a Gal⁻ mutation by transposon mutagenesis. Let us assume we have a λ lysogen (it could easily be made) and that the λ carries the temperature-sensitive repressor mutation *cI857*, described in Chapter 17. This cell line is infected with a λ carrying an Amp-r transposon. Because of immunity, the cell survives the infection. Growth of the cells allows time for transposition to occur, and we may assume that transposition into one of the *gal* genes in some cell occurs. The cells are plated on agar containing ampicillin. Colonies can form only if transposition has occurred from the infecting λ DNA to the chromosome, because λ DNA does not replicate in a lysogen. Thus, none of the Amp-r colonies will consist of cells containing an *amp-r* λ DNA molecule, and the *amp-r* gene must have a chromosomal location. These colonies can then be replica-plated onto galactose color-indicator plates, and the desired Gal⁻ colonies can be collected. Note that we began with a λ lysogen, so the cells could survive infection by λ. If the continued presence of this λ prophage is not desired, it can be removed by the curing protocol described in Chapter 17. That is, the culture is heated for 5 minutes at 42°C to initiate derepression and to synthesize the excision system, then grown at 30°C to reestablish repression and to allow segregation of prophage-free cells, and finally plated at 42°C. Only cells lacking the prophage will grow.

In Chapter 12 we saw that deletions are often associated with removal of transposons. These deletions, though rare, can usually be found if sufficiently sensitive selective procedures are available. Let us examine the construction of a deletion that includes a portion of a λ prophage. We begin by forming a λ lysogen in a Gal⁺ cell. Recall that a λ prophage is located between the *gal* and *bio* genes. The *tet-r* transposon Tn10 is then introduced into the cell (in any of a variety of ways) and a Tet-r Gal⁻ cell is isolated by the procedure of the preceding paragraph. This mutant will have Tn10 inserted into one of the *gal* genes. A Tet-s Gal⁻ colony is then isolated using the following trick. Penicillin and tetracycline are added to a culture of growing cells for a short period of time. Tet-r cells can grow normally in the presence of tetracycline and hence are killed by the penicillin. Tet-s cells do not grow and hence are resistant to penicillin. Furthermore, Tet-s cells are not permanently inhibited by tetracycline, so when both antibiotics are removed, the Tet-s cells can grow. Thus, the culture treated with the two antibiotics is plated, and colonies are tested by replica-plating for the Tet-s phenotype. Cells in these colonies will have lost Tn10. They are then tested on galactose color-indicator plates for the ability to utilize galactose. Excision of a transposon is quite rare, so most of these colonies are *gal*-Tn10 deletions and hence remain Gal⁻. This is, by the way, a straightforward way to obtain a *gal* deletion. The Gal⁻ cells can then be tested for the presence of various λ genes, for λ immunity, and for the Bio phenotype, to determine the extent of the deletion.

Moving the *lac* Genes

For a variety of experiments it is desirable to relocate genes. Here we see an example of a procedure for moving the *lac* genes. F′*lac* inserts into the chromosome at fairly high frequency by Rec-mediated recombination between homologous *lac* regions of the plasmid and of the chromosome. However, if the chromosome contains a deletion for the entire *lac* operon, such integration is not possible. However, integration still occurs at other sites, though at a hundredfold lower frequency, probably either by recombination between IS elements or by IS-mediated transposition. These alternative sites are scattered throughout the chromosome. Integration can be detected if an F′ is used that carries a temperature-sensitive mutation in the F replication function. Thus, one transfers F′(Ts)*lac*⁺ to a *lac*-deletion bacterium, plates the mated cells at 42°C, which does not allow replication of the F′(Ts), and selects for colonies that are Lac⁺. Since the integration frequency is low, color-indicator plates are not adequate and the selection must be done on lactose-minimal plates. The resulting strains are Hfrs, so the position of integration can be located by determining the times of entry of various genes by mating with a suitable recipient.

Often integration occurs within a gene and thereby generates a mutant. Thus, integration at particular sites can be selected for by looking for particular mutations. For example, if the original cell is sensitive to T1, a cell line in which integration occurred into the *tonB* gene can be isolated by plating the cells on plates covered with phage T1.

The low-frequency integration just described can occur in either the clockwise or counterclockwise direction. The orientation is ascertained from the time-of-entry curves.

F′(Ts)*lac* can recombine with other plasmids to generate temperature-sensitive F′ plasmids carrying genes other than *lac*. As long as a cell can be obtained in which there is a deletion covering the chromosomal gene carried on the F′, that gene can be moved to other chromosomal sites.

Gene Fusions

In studying regulation of gene expression it is usually necessary to be able to assay for products of the genes in an operon. For many genes the assays are either very tedious or inaccurate. In contrast, the assay for β-galactosidase is simple and quite precise; hence, genes are often repositioned to put the *lacZ* gene under control of the promoter of an operon of interest. Such strains are called **gene fusions.**

In Chapter 7 we saw one example of a gene fusion, namely the fusion of the *lac* and *pur* operons by introduction of an easily selected deletion between these genes. However, this fusion was not particularly valuable since it linked *pur* genes to the *lac* promoter. Nonetheless, this fusion illustrates a general method, namely, the introduction of a deletion between the promoter of one operon and a gene of another. Clearly, a deletion cannot be put between the *lac* operon and any gene; a gene that is quite distant would require a gigantic deletion that would remove a large number of essential genes. Thus, it is necessary to move the *lac* operon.

Using the preceding technique we saw how to move the *lac* genes to the *tonB* locus. This relocation puts *lac* adjacent to the attachment site for phage ϕ80. Preparation of a ϕ80 lysogen and a search for transducing particles made by inducing the lysogen has yielded a ϕ80*lac* particle. We will use this phage to isolate a fusion between the *lac* and the *trp* operons that will put the *lacZ* gene under control of the *trp* promoter. Figure 19-5 shows a portion of the genetic map surrounding the *trp* operon. Downstream from this ope-

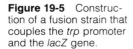

Figure 19-5 Construction of a fusion strain that couples the *trp* promoter and the *lacZ* gene.

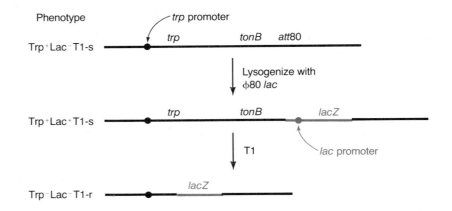

ron is the *tonB* locus and the prophage attachment site for phage φ80. Using a Lac⁻ strain, φ80*lac* is inserted into the attachment site. This yields the gene order *trp tonB lac*. In this strain the *trp* and *lac* promoters still control their respective genes. Cells are then plated on agar seeded with T1. Surviving colonies are tested for the ability to grow on minimal plates lacking tryptophan. Those that cannot grow on these plates are T1-resistant and *trp⁻* and hence are likely to have deletions that extend from the *trp* operon through the *tonB* locus. These deletions are tested further for the ability grow on lactose as a carbon source *when tryptophan is present*. Since tryptophan represses the *trp* operon, fusion strains that have removed the *lac* promoter will not be able to utilize lactose in the presence of tryptophan; if lactose can be utilized, the deletion must stop short of the *lac* promoter. Mutations in the *trp* operon can be distinguished by their ability to grow when various components of the tryptophan biosynthetic pathway are provided. By such tests the extent of the deletion into the *trp* operon can be determined.

Fusions Formed by a Mu-containing λ

The fusion procedure described in the preceding section depends on the ability to isolate deletions in particular regions. We now describe a general method for fusing a promoter of an operon and the *lacZ* gene, which can be used anywhere in the genome, as long as a gene in the operon can be detected by a plating test. This procedure, which is shown in Figure 19-6, requires a unique λ transducing phage whose origin will not be described (it was created by genetic trickery). This phage, called λp(*lac*::Mu), carries a mutant *lacP⁻* promoter, a functional *lacZ* gene, and a small portion of the transposing phage Mu. To remind the reader of the *lac* composition of the gene, we write it λp(*lacP⁻ lacZ⁺*, Mu). This phage lacks a normal λ attachment site, so it cannot insert into the bacterial attachment site.

We begin with a bacterium that is wild-type for an operon of interest. We use X to denote a gene in the operon, *pX* for the promoter for the operon, and X' to denote fragments of the operon. The technique requires that the bacterium have a complete deletion of the *lac* operon, but such a deletion is present in several Hfr cell lines as an early marker and can be transferred to the strain of interest. The procedure begins by lysogenizing the strain with a temperature-sensitive Mu-repressor mutant, Mu(Ts). An X⁻ colony is isolated; this bacterium will have Mu(Ts) inserted into the X operon. This X⁻::Mu(Ts) strain is then lysogenized with λp(*lacP⁻ lacZ⁺*, Mu). Because the λ lacks an attachment site and because the *lac* operon is deleted in the bacterium, insertion can only occur by homologous (RecA-mediated) recombination between the Mu elements in the phage and in the bacterium. The λ lysogen remains Lac⁻ because of the *lacP⁻* mutation and is temperature-sensitive because of the presence of Mu(Ts). Mu excises at very low fre-

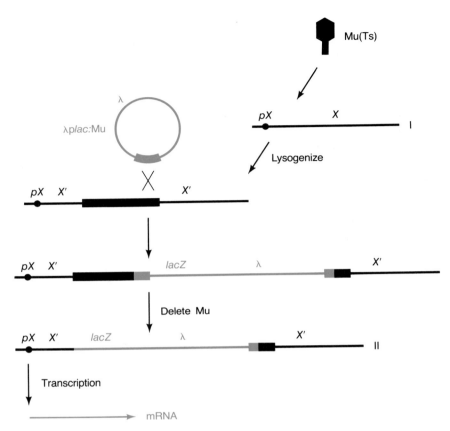

Figure 19-6 Production of gene fusions with λplac:Mu. Strain I, the starting strain, has the promoter pX for the X operon. After lysogenization with Mu(Ts) and insertion of λplac:Mu, the Mu(Ts) is deleted, yielding the strain II, in which the lacZ gene is transcribed from the promoter pX.

quency. However, since the Mu is temperature-sensitive, cells in which excision has occurred can be found by the ability to form a colony at 42°C. These 42°C survivors that are Lac⁺ when the inducer of the X operon is present contain pX fused to the lacZ gene.

Directed Mutagenesis with P1

When a culture is treated with a mutagen, mutations can arise in many genes. Of constant concern when mutants are isolated is the possible occurrence in the mutant cell of silent mutations in other genes; such mutations are important in that the strains might not be isogenic and the phenotype could result from the effects of several mutations. This problem could be avoided if the mutagen could be applied only to a small region of the bacterial chromosome. This is possible by the following trick. Let us consider introducing a *gal⁻* mutation into a wild-type cell. We begin by isolating a spontaneous mutation in a nearby gene, for example, *bio*. Phage P1 is then grown on the original *bio⁺* strain to yield a phage population that includes some particles that can transduce the *bio⁺* gene. The P1 population (not the bacteria) is then treated with a powerful chemical mutagen and used to infect the *bio⁻* strain. *Bio⁺* transductants are isolated. Since the *gal* and *bio* loci cotransduce, some of the *bio⁺* cotransductants will contain a mutagenized *gal* locus. To find these, the *bio⁺* transductants are tested on galactose color-indicator plates for the Gal⁻ phenotype. A few are usually found.

CONSTRUCTION OF PHAGE MUTANTS

As in the construction of bacterial mutants, a variety of techniques are used to construct phage mutants. Some are rather general and others make use of specific properties of certain mutations, as will be seen in the procedures that follow.

Production of Single Mutants

The production of a single mutant usually involves little other than selection following mutagenesis. In some cases one makes guesses about the properties of the mutants, as we saw in Chapters 16 and 17, when λ *red⁻*, *int⁻*, and *xis⁻* mutations were isolated. In other cases, for example, with plaque-morphology mutations, no selection is needed; that is, λ clear-plaque mutations and T4 *r* mutations are picked up just by inspection of the plates. However, often the procedure is one of brute force. For example, most of the existing λ mutations, which are chain-termination mutations, were isolated by mutagenizing a phage suspension, plating it out on *sup⁺* bacteria, and then testing tens of thousands of plaques for suppressor-sensitive mutations by stabbing into plates seeded with *sup0* cells. About 50 plaques can be tested on a single stab plate, so many plates were needed. Those plaques that did not produce clearing on restabbing were collected and classified by complementation. A similar gigantic "mutant hunt" was responsible for most of the T4 chain-termination mutations outside of the *rII* region and for most mutations in other phages.

Most phage research requires multiply marked phage. For example, these are necessary in three-factor crosses. In λ genetics one frequently needs phage that contain a clear mutation and a mutation in another gene (for example, a head gene or a replication gene), or in studying lysogeny, various combinations that include *int* or *xis* mutations. How some of these have been made will be explained in the following sections.

Production of Double Mutants of Different Classes

Perhaps the easiest type of phage mutant to make is a double mutation in which the two mutations are of different types, such as a plaque-morphology mutation and a suppressor-sensitive mutation. For example, a T4 $45^- rII$ (most of the T4 genes are designated by numbers) can be isolated by crossing rII and $45^- r^+$ and *plating on a mixture* of sup^+ and sup^- bacteria. 45^+ phage can infect both cells and produce a clear plaque, whereas 45^- mutants can only grow on the sup^+ cells and hence form a turbid plaque. The r^+ and rII mutations are identified by plaque size, so that the $45^- rII$ double mutant can be recognized as a large turbid plaque.

Production of a $\lambda cI^- O^-$ is a little more complicated; a mixed bacterial indicator does not always work because of the natural turbidity of wild-type λ. However, the following steps can be taken:

1. Cross $cI^- O^+$ with $cI^+ O^-$, and plate on sup^+ bacteria. The progeny are: $cI^+ O^-$, $cI^- O^+$ (parents); and $cI^- O^-$, $cI^+ O^+$ (recombinants).
2. Select a few hundred clear (cI^-) plaques; these are either $cI^- O^+$ or $cI^- O^-$ (recombinant).
3. Test these clear plaques for the O^- allele by stabbing into lawns of $sup0$ and sup^+ bacteria. Any stab that produces a clear spot on the sup^+ bacteria and no spot on the $sup0$ lawn is one of the desired $cI^- O^-$ double mutants.

Stabbing a few hundred plaques takes little more than about 15 minutes, so the procedure is straightforward.

Note that a λ "trick" can be used if one wants a $cI^- P^-$ phage instead. Recall that many λP^- phage will plate on GroP cells, whereas λP^+ will not. Thus, plating the progeny of the cross $cI^- P^+ \times cI^+ P^-$ on GroP cells will yield only the $cI^+ P^-$ parents (turbid plaques) and the desired $cI^- P^-$ mutant, which will produce a clear plaque.

Making λ Multiple Mutants That Include Mutations in the *int*, *xis*, or *red* Genes

An unusual interaction exists between λ and another phage called P2. Its basis is of no concern at this point. What is important is that wild-type λ will not form a plaque on a P2 lysogen, because of the presence of certain elements in the *red–gam* region of the λ map. This phenomenon serves as a selective procedure for certain mutations. λ biotin-transducing particles have a substitution starting with the *int* gene and often continuing into the region *xis-red-gam*. One λbio particle, $\lambda bio11$, with an *int–gam* substitution, is able to form a plaque on a P2 lysogen. However, recall from Chapter 16 that a $red^- gam^-$ mutant of λ (and hence *bio11*) is unable to form a plaque on

Figure 19-7 Relevant parts of the λ map for constructing λ*cI857xis⁻R⁻*.

$recA^-$ cells. This phenomenon provides the second tool for selecting mutations in the *red–gam* region. Let us consider how we might make λ*xis⁻ cI857R⁻* (a triple mutant) starting with *cI857R⁺*. Figure 19-7 shows the relevant region of the λ map. The following steps are carried out:

1. Cross *bio11c⁺* × *cI857R⁺*. Progeny are plated on a P2 lysogen (to select for *bio11*) at 42°C (a temperature at which the *cI857* mutation produces a clear plaque.
2. Select clear plaques, which will be either *bio11cI857R⁻* (single exchange between *bio11* and *cI857*) or *bio11cI857R⁺* (double crossover flanking *cI857*).
3. Identify the *R⁻* phage by stabbing plaques into agar containing *sup0* cells. Those that do not lyse the *sup0* cells are *R⁻*.
4. Perform a second cross between the *bio11cI857R⁻* just made and *c⁺xis⁻*.
5. Plate the progeny on *recA⁻sup⁺* cells at 42°C. The *recA⁻* marker selects against the *bio11* marker and selects for the region covered by *bio11*, which includes the *xis⁻* mutation. Thus, only *xis⁻* phage can grow.
6. Select clear plaques. These phage must carry the *cI857* marker. Most of these *xis⁻cI857* phage will be *R⁻* unless a double crossover occurred.
7. Identify the *R⁻* allele by stabbing into *sup0* cells, as in step 3.

The result is λ*cI857xis⁻R⁻*. Note that exactly the same procedure can be used to produce *int⁻R⁻* and *red⁻R⁻* mutants.

Construction of a Phage with Two Suppressor-sensitive Mutations

In constructing a phage with two suppressor-sensitive mutations, it is rarely possible to select for the mutations directly. For example, consider the construction of λ*D⁻Q⁻*. One cannot simply cross *D⁻* and *Q⁻* phages and select for the double mutant, because only a wild-type recombinant can be selected by growth on *sup0* cells. Also, testing progeny for inability to grow on *sup0* bacteria will not distinguish a double mutant from either of the parents, for neither parent nor double mutant will grow. The solution is to use linkage

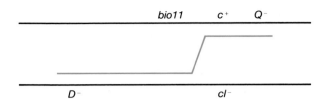

Figure 19-8 A cross that can yield $\lambda D^- Q^-$. Plating on $recA^-$ bacteria and selecting turbid (c^+) plaques demands the exchange shown in red. If only a single exchange occurs, the result is $\lambda D^- c^+ Q^-$.

procedures. For example, a cross could be made between a $cI^- D^-$ and $c^+ R^-$ *bio11* (Figure 19-8). Plating on a $recA^-$ cell and selecting turbid plaques introduces a requirement for the exchange between the *bio* and c markers shown in the figure in red. Unless there is a double crossover, the phage will be $D^- c^+ R^-$. It is of course necessary to check that the double crossover has not occurred. A simple way is to plate the phage on *sup0* cells and measure the reversion frequency. A comparison can be made to the reversion frequencies of each single mutation. Single mutants will have reversion frequencies in the range of 10^{-5} to 10^{-6}. The double mutant should yield no revertants when 10^8 phage are plated.

When making T4 double mutants the λ *bio11* trick is not available. However, note that the principle in the cross just shown is to select for an exchange between two markers. With T4 one could require an exchange between an *rII* marker and an ac^- (resistance to acridines) marker. Plating K12(λ) on agar containing acridines would accomplish this. For example, if one wanted a double mutant $a^- b^-$ and the genes have the order a *rII ac b*, the cross $a^- r^+ ac^+ \times rIIac^- b^-$ could be done (Figure 19-9). Plating on K12(λ) on acridine-containing medium would demand that all phage that form plaques be $r^+ ac^-$. Many of these plaques would be formed by $a^- r^+ ac^- b^-$ phage.

Formation of a double *rII* mutant can be accomplished by selecting for recombination between flanking markers. Assume that the recombinant *rII20rII62* is desired and that two other markers a and b have the order a

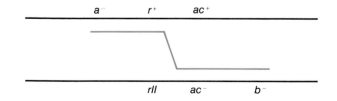

Figure 19-9 A T4 cross that can yield $a^- b^-$ phage. Plating on K12(λ) or acridine-containing agar requires that all phage forming plaques have experienced the exchange shown in red. If there are no additional exchanges, the phage will be $a^- r^+ ac^- b^-$.

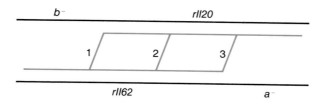

Figure 19-10 A cross that can yield an *rII* double mutant, using two flanking suppressor-sensitive markers, *a* and *b*. Plating the cross on *sup0* bacteria requires an exchange in intervals 1, 2, or 3. A cross in interval 2 yields the desired *rII* double mutant.

rII20 rII62 b. The cross $b^- rII20 \times a^- rII62$ can be carried out with selection on *sup0* bacteria (Figure 19-10). All plaques must be $a^+ b^+$, which demands an exchange between *a* and *b*. If the exchange occurs in intervals 1 or 3, a single *rII* mutation will be present; exchange in interval 2 results in a double *rII* mutant. Spot-testing recombinants for reversion on K12(λ) will distinguish a single *rII* mutant from the desired double mutant.

Isolation of λ Deletions

Ingenious techniques are often used in phage genetics to isolate mutations. Here we just present one, which exploits a physical property of λ deletions. λ phage are unstable in the absence of divalent cations, in particular Mg^{2+}. If λ phage are placed in solutions of ethylene diamine tetracetic acid (EDTA), a chelating agent that binds divalent cations and effectively removes them from solution, the phage gradually burst open, releasing their DNA. They, of course, lose the ability to form plaques. Apparently, the divalent cations serve two purposes: (1) They neutralize the charge repulsion between the negatively charged phosphates in the DNA. Without neutralization the DNA tends to expand, which increases the internal pressure in the head. (2) They strengthen the phage head. When λ is placed in EDTA, a small fraction (10^{-6}) of the phage survive and remain able to form plaques. These prove to be deletions in nonessential regions of the genome (usually in the *b2* region, see Figure 15-2). Apparently, with less DNA in the phage head, there is less internal pressure. By use of different concentrations of EDTA it is possible to isolate deletions of different lengths. The higher the concentration the larger the deletion. The maximum deletion produces a 21 percent reduction in the size of the λ DNA. Apparently smaller DNA molecules do not satisfy some requirement of the packaging system. Note that this method could not be used for any phage, like T4, that packages by the headful mechanism.

It seems that the process could be simplified by incorporating EDTA into the agar. However, *E. coli* does not grow on such agar. However, sodium pyrophosphate, a weaker chelating agent, also inactivates λ without affecting the ability of *E. coli* to form a lawn. Indeed, pyrophosphate agar (using different concentrations of pyrophosphate) can be used to select deletions.

Construction of Partial Prophage Deletions

In Chapter 17 we described studies that employed a bacterial strain containing only a part of a prophage. Several methods are available to prepare such strains. For example, consider a lysogen of λ*cI857int⁻*, which has been made by use of an *int⁺* helper phage. If this lysogen is heated to 42°C, the cells will be killed by virtue of thermal induction turning on various host-killing functions. However, about 1 cell in 10^8 will survive this treatment, and if the experiment is done on galactose color-indicator plates, some of these survivors will be Gal⁻. These cells have a deletion that removes the *gal* genes and extends into the prophage past killing functions (that is, past gene *R*). If the prophage is *R⁻*, most deletions will terminate between genes *P* and *R*.

A related method makes use of the *chlD* gene, which is located in the region *gal chlD attλ bio*. Wild-type *E. coli* cannot grow in the presence of the chlorate ion because the *chlD* gene converts it to a toxic compound. Plating a lysogen on medium containing chlorate and selecting survivors that are also Gal⁻ yields deletions that remove all or part of the *gal* locus, pass through the *chlD* gene, and generally enter the λ prophage. When a non-lysogen is exposed to chlorate, deletions that remove the entire *gal–bio* region are common. This is a useful technique to eliminate *attλ* from *E. coli*.

Isolation of a Phage Carrying a Suppressor

Lysogenic phages commonly produce specialized transducing particles. However, since the aberrant cuts made in both the prophage and the chromosome can occur at a variety of places, transducing particles may carry as little as a part of one gene or as many as several genes. In early studies of suppression an experimental approach was taken that led to the isolation of a phage that carried the *supIII* gene. This was of some importance because it became possible to determine the base sequence of the suppressor tRNA. Let us see how the particle was isolated and how it can be used.

The *supIII* gene is near the attachment site for phage φ80, so transducing particles carrying *supIII* should arise frequently. However, since specialized transducing particles are rare, a plating test is needed to enable the particle to be found. A simple color-indicator test, such as transduction of a Gal⁻ strain to Gal⁺, used to find λ*gal*, will not suffice. In the original isolation procedure a φ80 strain carrying a chain-termination mutation (in a phage gene) suppressible by *supIII* was used to lysogenize a *supIII* strain. A lysogen was isolated and then induced in order to obtain a lysate that contained some

Table 19-1 Plaque-forming ability of a φ80 mutant and its *supIII*-transducing variant on two bacterial strains

Bacteria	Plaque-forming ability		
	Phage φ80	Phage φ80(Am)	Phage φ80(Am)supIII
sup⁻	Yes	No	Yes
supIII	Yes	Yes	Yes

Note: Am refers to the presence of a chain termination mutation in a phage gene.

transducing particles. The lysate was then plated on *sup0* bacteria. Nontransducing φ80 could not form plaques because of the chain-termination mutation (Table 19-1). However, about 1 phage per 10^6 formed plaques; these phages carried a suppressor and were φ80*supIII*.

This phage was used in many bacterial strain constructions in which it was desired to introduce a suppressor into a *sup0* strain. Merely by lysogenizing with the transducing phage, the bacterium became Sup⁺.

Transducing particles carrying other suppressors have also been prepared. The usual procedure is to move genes around by techniques described above so that *attλ* or *att80* is near the suppressor of interest, and then obtain transducing particles by the techniques just described.

PROBLEMS

1. Starting with $F^- trp^- lac^-$ str-r, make a $trp^- gal^- lac^+$ str-r strain. You may make use of any of the following strains: $F^- gal^- lac^-$, $F^- gal^- lac^+$, Hfr $gal^- lac^+$, and Hfr $gal^+ lac^-$, in which the Hfrs transfer clockwise from an origin at 99 minutes.

2. You have an Hfr Δ*lac* strain (Δ represents a deletion) that transfers the deletion a few minutes after mating. Your strain collection also contains numerous Hfr lac^+ strains. Use these strains to replace a lac^- point mutation in an F^- strain by a *lac* deletion.

3. By using λ*cI857* (a λ with a temperature-sensitive *cI* repressor) as a tool, convert a gal^+ strain to one that is deleted for *attλ*.

4. The *uvrB* gene is near the *bio* locus. Starting with a prototroph, make a *uvrB* deletion. No mating or transduction need be done. Any phage may be used, as well as any tricks you can think of.

5. A highly useful lac^- mutation has been isolated in an Hfr. You would like to transfer this mutation to a particular female cell. Mating and selection for the Lac⁻ character will not work very well because the *lac* gene is transferred fairly late by this Hfr; thus, the fraction of recipients that will be Lac⁻ will be too small to detect on a color-indicator plate. Suggest a procedure that will work. Use phage T6 and any kind of plate.

6. Starting with three different T4 phages, each carrying one of the mutations *ac41*, *rII*, and *h*, make a triple mutant. Describe the bacteria and the plates that are used.

7. Prepare a *uvrA⁻ uvrB⁻ leu⁻* triple mutant, starting with *uvrA⁻* and *leu⁻* strains. The *uvrA* gene is at 91.6 minutes, and *uvrB* is in the *gal attλ bio chlD uvrB* cluster at 18–19 minutes. A useful gene will be *malB*, mutation of which makes cells resistant to λ; *malB* is within 0.1 minute of the *uvrA* gene. Be sure to include a test for the double mutant.

8. You have an *F⁻ leu⁻ str-r* strain and wish to make it *leu⁻ str-s*. Assuming you have a large collection of bacteria, how might you do it?

9. A *dna*(Ts) mutant is unable to synthesize DNA at high temperature but continues to make protein for a time equivalent to 2–3 generations. In the absence of DNA synthesis (that is, at a nonpermissive temperature) cell division is inhibited in such mutant bacteria. If the time at high temperature does not exceed a few generations, lowering the temperature allows DNA synthesis to be restored, and after a generation at a low (permissive) temperature, cell division resumes and proceeds until the proper ratio of DNA to protein is achieved. Suggest a procedure by which a *dna*(Ts) mutant might be isolated. No special plates are needed, but one or more of the following pieces of common lab equipment—centrifuge, filter, pH meter—will be helpful.

10. Isolate a Leu⁻ strain using Tn10. You may use λ*cI857* or λ carrying Tn10.

11. Select a bacterial strain in which the promoter for the arabinose operon is linked to the *lacZ* gene.

12. Prepare λ*cI857red⁻Q⁻* starting with λ*cI857Q⁻*, λ*bio11c⁺*, and λ*c⁺red⁻*. λ*Q⁻* is a suppressor-sensitive mutation.

13. Starting with λ*bio11R⁻* and λ*cI857A⁻*, isolate λ*cI857A⁻R⁻*. The gene order is *A bio11 cI R*, and the recombination frequency between *A* and *R* is about 10 percent.

14. Starting with λ*bio11c⁺A⁻* and λ*cI857R⁻*, make λ*cI857A⁻R⁻*. Note the difference between this construction and that in Problem 13. A critical point is the fairly high recombination frequency between *A* and *R* (ca. 10 percent). [Hint: this is somewhat of a brute-force selection.]

15. In this problem you will manipulate the λ *b2* gene. The relevant gene order is *b2 cI P R*, with *cI* and *P* separated by about one map unit. *P⁻* and *R⁻* are suppressor-sensitive mutations.

   (a) Isolate λ*b2R⁻*, starting with λ*b2* and λ*R⁻*.
   (b) Isolate λ*cI857b2P⁻* using λ*b2cI857* and *c⁺P⁺*.

THE NEW MICROBIAL GENETICS

Chapter 20

Genetic Engineering

In addition to advancing the understanding of natural phenomena, genetics is also important in manipulating biological systems, for scientific or economic reasons, an endeavor that has made possible the creation of organisms having new phenotypes or genotypes. The fundamental techniques for accomplishing this have been mutagenesis, gene transfer, and genetic recombination followed by selection for desired characteristics. When using such techniques, geneticists have been forced to work with the random nature of mutagenic and recombination events, which required selective procedures, often quite complex, to find an organism with the required genotype among the many types of organisms produced. Recently, a new technique has been developed by which the genotype of an organism can instead be modified in a *directed and predetermined* way. In this technique, which is alternately called **recombinant DNA technology, genetic engineering,** or **gene cloning,** purified DNA fragments are isolated and recombined by *in vitro* manipulations. Selection of a desired genotype is still necessary but the probability of success is usually many orders of magnitude greater than that with traditional procedures. The recombinant DNA technology has greatly enhanced our ability to manipulate organisms and has revolutionized the study of gene structure and regulation. The basic technique is quite simple: two DNA molecules are isolated and cut into fragments by one or more specialized enzymes and then the fragments are joined together *in a desired combination*. Finally, they are restored to a cell by the $CaCl_2$ transformation procedure for replication and reproduction.

THE JOINING OF DNA MOLECULES

The basic procedure of the recombinant DNA technique consists of two stages—(1) joining a DNA segment (which is of interest for some reason) to a DNA molecule that is able to replicate, and (2) providing a milieu that allows propagation of the joined unit (Figure 20-1). When this is done, the genes in the donor segment are said to be **cloned** and the carrier molecule is the **vector** or **cloning vehicle.** In this section we will examine this procedure further, as well as other procedures used for joining. Several types of vectors will also be described.

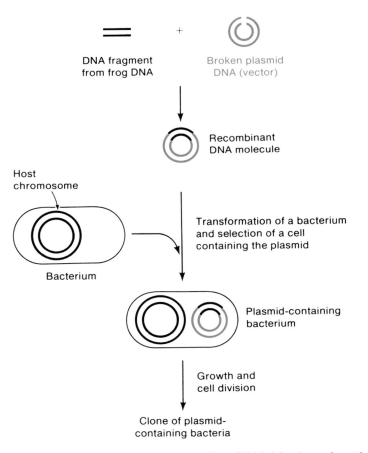

Figure 20-1   An example of cloning. A fragment of frog DNA is joined to a cleaved plasmid vector. The hybrid plasmid transforms a bacterium. Henceforth, frog DNA is carried to all progeny bacteria by replication of the plasmid.

Vectors

To be useful, a vector must have three properties:

1. It must be able to replicate.
2. There must be some way to introduce vector DNA into a cell.
3. There must be some means of detecting its presence, preferably by a plating test.

The three most common types of vectors in use are plasmids, *E. coli* phage λ, and viruses, because the DNA of each of these vectors has all three of the aforementioned properties. That is, each has a replication origin and carries genes that are identifiable by simple plating or biochemical tests, and the DNA can be made to penetrate particular cells by the $CaCl_2$ transformation technique. Only a small number of vectors are in common use; some of these vectors and their properties are listed in Table 20-1.

Restriction Enzymes

DNA restriction was discovered as an outgrowth of the study of host restriction and modification of phages, described for phage λ in Chapter 5. An attempt to account for the site specificity of the DNA degradation led to the isolation of novel nucleases called **restriction enzymes** or **restriction endonucleases.** These are enzymes that recognize a specific base sequence in a DNA molecule and make two cuts, *one in each strand*, generating 3'-OH

Table 20-1 Some cloning vehicles in use at present

Designation	Genotype or characteristics
Plasmid	
pSC101	*tet-r*
ColE1	*immE1*
pBR322	*tet-r amp-r*
pUC8/9	*amp-r lac$^+$*
pHC79	*amp-r tet-r λcos*
Phage	
λgt4·λB	Can lysogenize; is thermally inducible.
λ-Charon	*lacZ$^+$*
λΔz1	Insertion occurs in *lacZ* gene.
M13mp7	Contains *lacP, lacO,* and *lacZ*; single-stranded DNA; useful for sequencing.
Virus	
SV40	Virus infects animal cells. Maximum size of added DNA is no more than 3×10^6 molecular weight units.

and 5′-P termini. Nearly 1000 different restriction enzymes have been purified from hundreds of different microorganisms. All but a few of these enzymes recognize sequences that are nucleotide palindromes. That is, the recognition site always has a sequence whose general form is

$$
\begin{array}{c|c}
\text{A B C} & \text{C}' \text{ B}' \text{ A}' \\
\text{A}' \text{ B}' \text{ C}' & \text{C B A}
\end{array}
\quad \text{or} \quad
\begin{array}{c}
\text{A B X B}' \text{ A}' \\
\text{A}' \text{ B}' \text{ X B A}
\end{array}
\quad \text{or} \quad
\begin{array}{c|c}
\text{A B} & \text{B}' \text{ A}' \\
\text{A}' \text{ B}' & \text{B A}
\end{array}
$$

in which the capital letters represent bases, a prime indicates a complementary base, X is any base, and the dashed line is the axis of symmetry. A few sensitive sequences having more than six bases are known but none have been observed containing fewer than four bases.

There are two major types of restriction enzymes: type I enzymes, which recognize a specific sequence but make cuts elsewhere; and type II enzymes, which make cuts only *within* the recognition site. Type II enzymes are the most important class and we confine our attention to them. All restriction enzymes of this class make two single-strand breaks, one break in each strand. There are two distinct arrangements of these breaks: (1) both breaks at the center of symmetry (generating **flush** or **blunt ends**), or (2) breaks that are symmetrically placed around the line of symmetry (generating **cohesive ends**). These arrangements and their consequences are shown in Figure 20-2. Table 20-2 lists the sequences and cleavage sites for twelve useful restriction enzymes, nine of which generate cohesive sites and three of which yield flush ends.

An important point about these restriction enzymes is this: since a particular enzyme recognizes a unique sequence, *the number of cuts made in the DNA from an organism is limited*. For example, a typical bacterial DNA

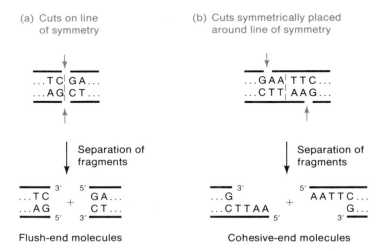

Figure 20-2 Two types of cuts made by restriction enzymes. The arrows indicate the cleavage sites. The dashed line is the center of symmetry of the sequence.

Table 20-2 Some restriction endonucleases and their cleavage sites

Microorganism	Name of enzyme	Target sequence and cleavage sites
Generates cohesive ends		
E. coli	EcoRI	G↓A A T T C C T T A A↑G
Bacillus amyloliquefaciens H	BamHI	G↓G A T C C C C T A G↑G
B. globigii	BglII	A↓G A T C T T C T A G↑A
Haemophilus aegyptius	HaeII	Pu G C G C↓Py Py↑C G C G Pu
Haemophilus influenza	HindIII	A↓A G C T T T T C G A↑A
Providencia stuartii	PstI	C T G C A↓G G↑A C G T C
Streptococcus albus G	SalI	G↓T C G A C C A G C T↑G
Thermus aquaticus	TaqI	T↓C G A A G C↑T
Generates flush ends		
Brevibacterium albidum	BalI	T G G↓C C A A C C↑G G T
Haemophilus aegyptius	HaeI	(A)G G↓C C(T) (T)C C↑G G(A)
Serratia marcescens	SmaI	C C C↓G G G G G G↑C C C

Note: The vertical dashed line indicates the axis of dyad symmetry in each sequence. Arrows indicate the sites of cutting. The enzyme TaqI yields cohesive ends consisting of two nucleotides, whereas the cohesive ends produced by the other enzymes contain four nucleotides. The enzyme HaeI recognizes the sequence GGCC whether the adjacent base pair is A·T or T·A, as long as dyad symmetry is retained. Pu and Py refer to any purine and pyrimidine, respectively.

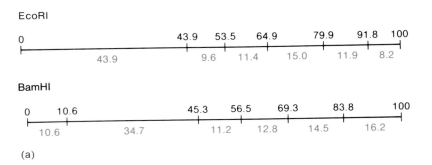

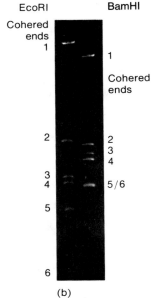

Figure 20-3   (a) Restriction maps of λ DNA for EcoRI and BamHI nucleases. The vertical bars indicate the sites of cutting. The black numbers indicate the percentage of the total length of λ DNA measured from the gene *A* end of the molecule. The red numbers are the lengths of each fragment, again expressed as percentage of total length. (b) A gel electrophoregram of EcoRI and BamHI restriction-enzyme digests of λ DNA. The bands labeled cohered ends contain molecules consisting of the two terminal fragments joined by the normal cohesive ends of λ DNA. Numbers indicate fragments in order from largest (1) to smallest (6). Bands 5 and 6 of the BamHI digest are not resolved. [Courtesy of Dennis Anderson and Lynn Enquist.]

molecule, which contains roughly 3×10^6 base pairs, is cut into a few hundred to a few thousand fragments. Smaller DNA molecules such as phage or plasmid DNA molecules may have fewer than ten sites of cutting (frequently, one or two, and often none). Because of the specificity just mentioned, *a particular restriction enzyme generates a unique family of fragments from a particular DNA molecule*. Another enzyme will generate a different family of fragments from the same DNA molecule. (Some enzymes—isoschizomers—have the same specificity and generate identical families.) Figure 20-3(a) shows the sites of cutting of *E. coli* phage λ DNA by the enzymes EcoRI and BamHI. The family of fragments generated by a single enzyme is usually detected by agarose gel electrophoresis of the digested DNA (Figure 20-3(b)). The fragments migrate at a rate that is a function of molecular weight (Chapter 3) and their molecular weights can be determined by reference to fragments of known molecular weight run concurrently.

Cloning a Restriction Fragment in a Plasmid

A particular restriction enzyme recognizes only a single base sequence. Furthermore, except for those enzymes producing blunt ends, the cuts generate at each end of a fragment single-stranded termini that are complementary. This is the basis of the joining procedure to be described—namely, the fragments generated by a particular enzyme acting on one DNA molecule have the same set of single-stranded ends as the fragments produced by the same enzyme acting on a different DNA molecule (as long as both DNA molecules have sequences recognized by the enzyme). Therefore, fragments from the DNA molecules of two different organisms (for example, a bacterium and a frog) can be joined by renaturation of complementary single-stranded termini. Furthermore, if the joint is sealed with DNA ligase (ligated) after base-pairing, the fragments are joined permanently.

The joining technique takes on special importance if one of the sources of cleaved DNA is a plasmid. Figure 20-4 shows a plasmid DNA molecule that has only one cleavage site for a particular restriction enzyme, which is also used to cleave frog DNA. If frog fragments are mixed with linearized plasmid DNA and joining is allowed to occur, a circular plasmid DNA molecule containing frog DNA can be generated. The significance of this technique is that the hybrid plasmid can be reestablished in a bacterium by $CaCl_2$ transformation. Then, the inserted DNA fragment, which replicates as part of the plasmid, is **cloned.**

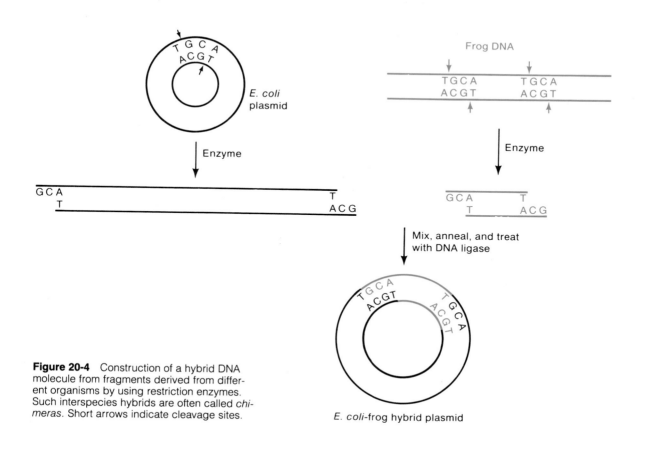

Figure 20-4   Construction of a hybrid DNA molecule from fragments derived from different organisms by using restriction enzymes. Such interspecies hybrids are often called *chimeras*. Short arrows indicate cleavage sites.

Joining of DNA Fragments by Addition of Homopolymers

The field of recombinant DNA research began in 1972, just before the properties of restriction enzymes were understood, with the development of a general method for joining any two DNA molecules. This method used the enzyme **terminal nucleotidyl transferase,** an unusual DNA polymerase obtained from animal tissue, which adds nucleotides (by means of triphosphate precursors) to the 3′-OH group of an extended single-stranded segment of a DNA chain. The reaction differs from that of ordinary polymerases in that *it does not need a template strand*. In order to generate the extended single strand with a 3′-OH terminus one need only treat the DNA molecule with a 5′-specific exonuclease to remove a few terminal nucleotides. In a reaction mixture consisting of exonuclease-treated DNA, dATP, and the enzyme, poly(dA) tails will form at both 3′-OH termini (Figure 20-5). If, instead, dTTP were provided, the DNA molecule would have poly(dT) tails. Two molecules can be joined if poly(dA) tails are put on one DNA molecule and poly(dT) tails on the second molecule and the poly(dA) is allowed to anneal to the poly(dT), as shown in the figure. Completion of the joined molecule is accomplished by gap-filling with DNA polymerase I and sealing with DNA ligase.

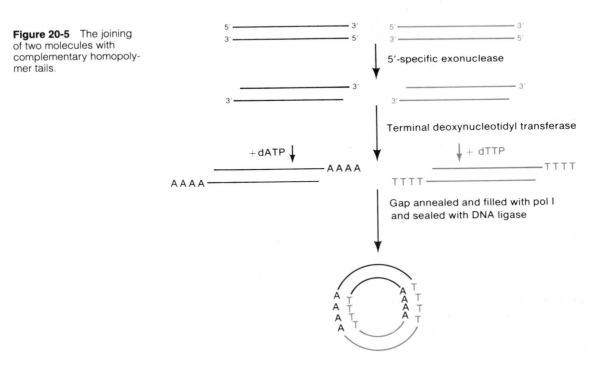

Figure 20-5 The joining of two molecules with complementary homopolymer tails.

This method, which is called **homopolymer tail joining,** is useful when DNA molecules lacking complementary ends are to be joined. Such molecules may be the result of digestion with restriction enzymes that yield blunt ends or may be prepared by mechanical breakage of large DNA molecules; c-DNA (complementary DNA), which is a DNA molecule prepared in the laboratory by copying an RNA template and which is extremely important in the recombinant DNA technology (to be described shortly), also has blunt ends.

Blunt-End Ligation

E. coli phage T4 encodes a DNA ligase, which is produced in large quantities in an infected cell. This enzyme is a typical ligase in that it seals nicks in double-stranded DNA having 3'-OH and 5'-P termini, but has the additional property of joining two DNA molecules having completely base-paired ends (Figure 20-6). How this reaction occurs is unknown. Whereas blunt-end ligation is quite useful in many situations, it has a significant disadvantage compared to homopolymer joining, namely, the two ends of any fragment can be joined to form a nonrecombinant (and generally useless) circle. Such joining does not occur with homopolymer tail joining since the two ends are identical rather than complementary. This problem can, however, be avoided by use of a high DNA concentration, as shown in the figure.

Joining with Linkers

In some cases it is useful to be able to join one molecule with blunt ends to a second molecule produced by a restriction endonuclease that generates single-stranded termini. This is possible if a short DNA segment (a **linker**) containing a restriction site is coupled to both ends of the blunt-ended mol-

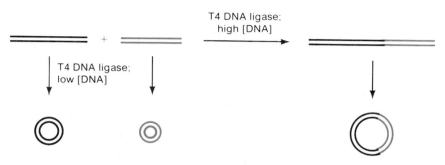

Figure 20-6 Two blunt-end joining reactions. At low concentrations of DNA intramolecular circularization is favored.

Figure 20-7 Formation of
recombinant DNA through
use of a linker. The short
arrows indicate the sites
of cutting by the EcoRI
enzyme.

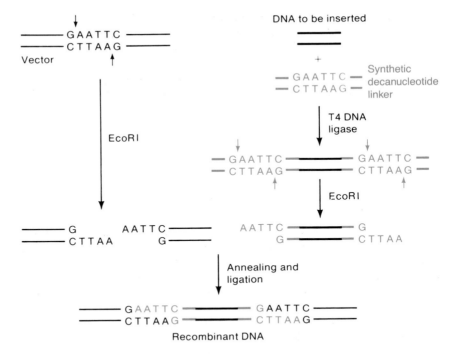

ecule. How this is done is shown in Figure 20-7, in which a blunt-ended
molecule is inserted at the EcoRI site of a vector. This procedure is useful
because at a later time the sequence contained in the blunt-ended fragment
can be recovered from the vector by treatment with a restriction enzyme
that cuts the site in the linker. Such recovery is not possible if joining is done
with homopolymer tails.

Generation of Fragments by Hydrodynamic Shear

Consider a gene G that is to be cloned in a plasmid. This is accomplished by
inserting a DNA fragment containing G into the plasmid DNA. The simplest
method is to treat both the plasmid DNA and the DNA containing G with
the same restriction enzyme and anneal the resulting fragments. However,
it may be that every known restriction enzyme that cuts the plasmid also
makes a cut within gene G and thereby inactivates the gene. Clearly some
means of cleavage other than the use of restriction enzymes is required; it
has been found that the gene inactivation caused by site-specific cleavage
can be eliminated if the donor DNA containing G is fragmented not enzymatically but mechanically by **hydrodynamic shear** forces.

  If a DNA solution is passed very rapidly through a hypodermic needle
or a small orifice or is stirred vigorously, the DNA molecules are broken into

fragments in the size range 3000–30,000 base pairs. In some DNA molecules, gene G, which might typically consist of 1000–2000 base pairs, will be cleaved but, on the average, the gene will remain intact.

The fragments resulting from shearing usually have blunt ends. In order to be joined efficiently to a plasmid broken by a restriction enzyme, an appropriate linker may be added, as was shown in Figure 20-7, though blunt-end joining can be carried out.

Scrambling of Fragments During Joining

Joining of fragments does not always produce a functional DNA sequence. For example, consider a linear DNA molecule that is cleaved into four fragments, A, B, C, and D. Reassembly could yield the original molecule, whose sequence we will assume to be ABCD, but since B and C have the same pair of cohesive ends, a molecule with the sequence ACBD could form with the same probability as ABCD. Other molecules having multiple copies of one of the fragments—for example, ABBD or ACCD—are also possible. Thus, only about one-fourth of the resulting sequences will be functional. Similarly, if there were five fragments A, B, C, D, and E, the molecules A(BCD)E, A(BDC)E, A(CBD)E, A(CDB)E, A(DBC)E, and A(DCB)E, and those with duplicated and triplicated fragments, are all equally likely. The probability of obtaining a functional sequence when reannealing fragments obtained from two different DNA molecules is of course lower. Thus, if plasmid DNA is cleaved into four fragments and frog DNA is cleaved into 1000 fragments, most assembled plasmids will be scrambled and nonfunctional. On the other hand, if a functional plasmid is selected by genetic means and if an insertion is possible, say, between B and C, the probability of having a plasmid containing any frog DNA fragment will be quite high; if a *particular* frog DNA fragment is desired, the probability will be about 1000 times lower.

INSERTION OF A PARTICULAR DNA MOLECULE INTO A VECTOR

As we have described the procedures so far, a collection of fragments obtained in various ways (for example, by restriction enzyme digestion or shearing) is allowed to anneal with a vector, yielding a large number of hybrid molecules. If a particular gene is to be cloned, the vector possessing that gene must be isolated from the class of all vectors possessing foreign DNA. Although for many genes this procedure is adequate, there are certainly many cases in which the clone of interest is either so rare or so difficult to detect that it would be preferable if, prior to hybridization, the fragment containing the gene of interest could be purified. In this section three procedures are described in which a particular DNA molecule can be cloned.

c-DNA and the Use of Reverse Transcriptase

Proof of correct insertion of a particular gene into a vector is most easily detected by observing expression of the gene. For example, the *lac* operon can be detected in a plasmid by virtue of its ability to synthesize β-galactosidase. However, if one wants to clone a eukaryotic gene (which is probably the most common goal in modern cloning research), knowing whether a particular eukaryotic gene has been inserted into a bacterial DNA vector can be a problem; the difficulty is that, in general, a eukaryotic gene cannot be transcribed and translated into a functional protein when cloned in a bacterium. By the technique described in this section the insertion of some eukaryotic genes can become a quite straightforward procedure.

Let us assume that we wish to clone the rabbit gene for β-globin (a subunit of hemoglobin) in a bacterial plasmid. (The phrase "to clone a particular gene" is common usage to mean "to form a vector containing a particular gene.") The recombinant plasmid could be formed by treating both rabbit cellular DNA and the *E. coli* plasmid DNA with the same restriction enzyme, mixing the fragments, annealing, ligating, and finally infecting *E. coli* with the DNA mixture containing the hybrid plasmid. The main problem is to identify the bacterial colony containing this plasmid. If the gene to be cloned is a bacterial *gal* gene, one can use a Gal⁻ host and select a Gal⁺ colony on an appropriate growth medium. However, it is not possible for a bacterium containing the β-globin gene to synthesize β-globin because the gene contains introns (Chapter 6), and all known bacteria lack the enzymes needed to form eukaryotic mRNA molecules from a primary transcript. Other tests might be performed to find the desired cell, such as hybridization of bacterial DNA with purified β-globin mRNA, but since only about one plasmid-containing cell in 10^5 will contain the β-globin gene, this method will be very tedious. Therefore, a more convenient procedure is desirable. One useful procedure, which is applicable to certain classes of genes, is described in the remainder of this section.

Some types of animal cells, of which one example is the reticulocyte, which produces β-globin, make only one or a very small number of proteins. In these cells the specific mRNA molecules, which are already processed, constitute a large fraction of the total mRNA synthesized in the cell and, for this reason, mRNA samples can usually be obtained that consist almost exclusively of a single mRNA species—in this case, β-globin mRNA. If genes of this type—that is, those whose gene products are the major cellular proteins—are to be cloned, the purified mRNA of each type of cell can serve as a starting point for creating a recombinant plasmid containing only the gene of interest.

An enzyme present in the coat of certain animal viruses, **reverse transcriptase,** is at the core of this technique. Reverse transcriptase can use an RNA molecule as a template and, in a multistep process, synthesize a double-stranded DNA copy (for the mechanism, see references at the end of this book). A double-stranded DNA molecule prepared in this way is called **complementary DNA** or **c-DNA.** If the introns have been removed from the RNA molecule that forms the template, before the RNA is isolated, the corresponding c-DNA will contain an uninterrupted coding sequence. The c-DNA generally has blunt ends and is usually joined by the linker method shown in Figure 20-7.

Insertion of a Particular DNA Fragment Obtained from a Large Donor Molecule by Restriction Cutting

The study of the regulation of gene expression in phages has been helped considerably by the cloning of particular phage genes in a plasmid. Cloning of a phage gene is facilitated if one can first purify the fragment containing the gene. For instance, Figure 20-3 showed an EcoRI restriction map for phage λ; suppose that one wants to clone a gene that is known to be in the 9.6 fragment. One way to accomplish this is to add the complete enzymatic digest of the DNA to a cleaved vector, allow the DNA to anneal, and seek a cell containing a plasmid that has the gene of interest. Since there are six fragments, at most one-sixth of the plasmids containing phage DNA will have the gene. In some cases, it is valuable to increase this fraction. This can be done by purifying the 9.6 fragment and using only the purified fragment in the annealing mixture. The most direct way to isolate a fragment is to extract it from a gel after electrophoretic separation. This can be done in the following way (Figure 20-8). The restriction digest is electrophoresed for a period of time, the electric field is turned off, and the DNA bands are located by

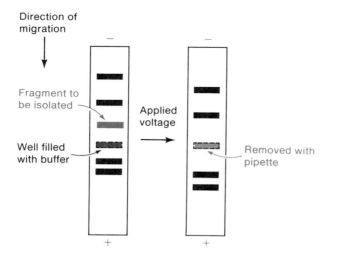

Figure 20-8 Method for isolating a particular restriction fragment.

soaking the gel in a buffer containing ethidium bromide. Recall that the fluorescence of ethidium bromide is markedly enhanced when the molecule binds to DNA, so the DNA can be located by shining near-ultraviolet light on the gel and noting the positions of intense fluorescence. A small well is cut in the agar ahead of the band of interest and filled with buffer. The electric field is applied again and the motion of the fragments is followed visually. When the fragment of interest migrates into the buffer-containing well, the electric field can be turned off again and the buffer, now containing the DNA fragment, can be removed.

DETECTION OF RECOMBINANT MOLECULES

The joining of DNA molecules is a straightforward process, as we have seen in the preceding sections. However, if a vector is cleaved by a restriction enzyme and annealed with an unfractionated collection of restriction fragments, many types of molecules result—examples are a self-annealed vector that has not acquired any fragments, a vector with one or more fragments, and a molecule consisting only of many joined fragments. Molecules of the third class cannot stably transform $CaCl_2$-treated cells because, in general, such molecules lack a replication origin and replication genes; thus, molecules of this class are not a problem. However, some means is needed to ensure that a plasmid (or a phage) detected after $CaCl_2$ transformation does possess an inserted DNA fragment. Furthermore, even if this can be demonstrated, there is no guarantee that a plasmid containing donor DNA has the DNA segment of interest. Thus, some test is required for detecting the desired plasmid. Similar problems exist for phage and viral vectors. In this section several procedures that provide solutions to these problems are described.

Methods Which Insure That a Plasmid Vector Will Contain Inserted DNA

The plasmid pBR322 is a widely used vector. It is small (4363 base pairs), and has two different antibiotic-resistance markers—namely, resistance to tetracycline (*tet-r*) and to ampicillin (*amp-r*). Thus, plasmid-containing transformants are easily detected by growth of a DNA-transformed culture on agar containing one of these antibiotics. Also, pBR322 has seven different types of restriction-enzyme cleavage sites at which foreign DNA can be inserted.

A common procedure for detecting insertion is **insertional inactivation,** which is carried out as follows (Figure 20-9). The BamHI and SalI sites are within the *tet* gene. Thus, insertion at either of these sites yields a plasmid that is *amp-r tet-s*, because the *tet* gene is inactivated. If wild-type (Amp-s Tet-s) cells are transformed with a DNA solution in which the cleaved pBR322

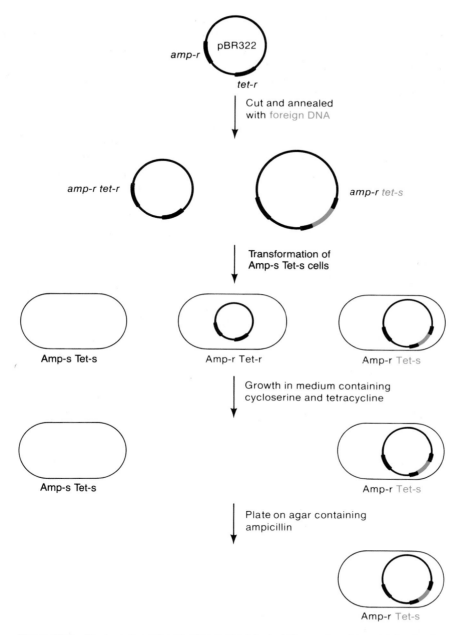

Figure 20-9 The insertional inactivation method for isolating a plasmid containing foreign DNA.

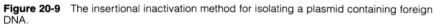

has been annealed with restriction fragments and the cells are plated on agar containing ampicillin, surviving colonies, which have to be Amp-r, must possess the plasmid. These colonies are then further tested for sensitivity to tetracycline. Because pBR322 carries the *tet-r* allele, an Amp-r colony will also be Tet-r unless the *tet-r* allele has been inactivated by insertion of foreign DNA. Thus, an Amp-r Tet-s cell must contain not only pBR322 DNA but donor DNA as well.

This procedure can be simplified by growing the transformed cells, prior to plating, in a medium containing cycloserine and tetracycline. Cycloserine kills growing cells, but Tet-s cells are merely inhibited, not killed, by tetracycline. Thus, in this growth medium Tet-r cells (which grow) are killed and Tet-s cells (which are inhibited) survive. Plating of cells treated in this way on agar containing ampicillin yields Amp-r Tet-s colonies; these all possess pBR322 containing a donor DNA fragment.

The PstI site in pBR322 is in the *amp* gene. Thus, the insertional inactivation procedure can also be used with insertion at this site, and an Amp-s Tet-r colony is selected.

Three other procedures select against reconstituted plasmids lacking foreign DNA. One is to treat a cleaved plasmid with **alkaline phosphatase,** an enzyme that removes 5′-P termini, leaving 5′-OH groups. Following this treatment a functional plasmid cannot be reconstituted unless foreign DNA is inserted, as shown in Figure 20-10. A second procedure, homopolymer tail joining, automatically selects against re-formation of the plasmid because the two 3′-OH termini of the cleaved plasmid both have homopolymers containing the same base. A third procedure utilizes a special type of plasmid called a cosmid.

Cosmids (phage λ *cos*-site-carrying plasmids) are novel vectors that combine the features of a plasmid and phage λ to increase the probability of selecting a recombinant plasmid carrying foreign DNA. A typical cosmid is a circular ColE1 plasmid (Chapter 11) carrying both the gene for resistance to the drug rifampicin (*rif-r*) and the *cos* site of phage λ (Chapter 16). The cosmid has two cleavage sites for the HindIII restriction enzyme, which separate the *rif* gene from the *cos* site and the ColE1 region (Figure 20-11).

In Chapter 16 the *cos*-site-cutting (Ter) system responsible for packaging phage λ DNA into the phage head was described. The Ter system can act on a DNA molecule only if the following two conditions are satisfied: (1) the DNA molecule must contain two *cos* sites, and (2) the *cos* sites must be separated by no less than 38,000 base pairs and no more than 54,000 base pairs. The use of cosmids as vectors is based on a technique for packaging λ DNA *in vitro* and the requirements just stated.

The cosmid shown in Figure 20-11 cannot serve as a substrate for *in vitro* packaging because the DNA has only one *cos* site. If the cosmid is treated with the HindIII restriction enzyme, the resulting linear fragments still cannot be packaged. Panel (a) of the figure shows that if the molecules are annealed with one another, among the many resulting molecules, one is

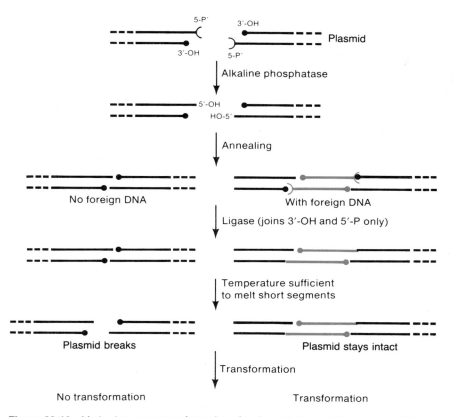

Figure 20-10 Method to prevent re-formation of a cleaved plasmid. The plasmid DNA, but not the foreign DNA, is treated with alkaline phosphatase, an enzyme that converts terminal 5'-P groups to 5'-OH groups.

a dimer (or higher multimer) containing two *cos* sites. However, the Ter system still cannot act on this molecule because the *cos* sites are separated by only a few thousand base pairs. Panel (b) shows that if a mixture of cleaved cosmid DNA and HindIII restriction fragments obtained from the DNA of another organisms (e.g., camel DNA) is allowed to reanneal, of the many molecules that arise, one is a linear dimer having two *cos* sites separated by a sufficient amount of foreign DNA that their distance is suitable for the λ packaging system. Thus, *in vitro* packaging yields a collection of transducing particles containing a linear fragment of recombinant cosmid DNA, terminated at each end by the normal cohesive ends of λ DNA. Note that packaging only occurs if the cosmid contains foreign DNA (though it would be possible to package a multimer containing many copies of the *rif* gene in tandem). Panel (c) shows the result of infecting Rif-s *E. coli* with the cosmid-carrying transducing particles. (There is no phage development because the particles

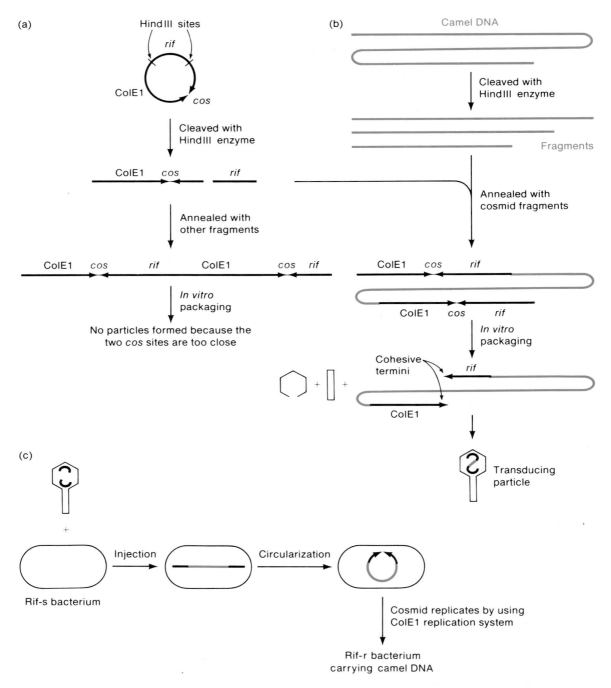

Figure 20-11 Use of a cosmid as a vector. (a) Cleavage of cosmid. (b) Formation of transducing particle. (c) Transduction.

contain no phage genes.) The cosmid DNA is injected and circularizes via the λ cohesive termini. Once the cosmid DNA has circularized and has been ligated, it can replicate, using the ColE1 replication system. Growth of the infected cells on agar containing rifampicin selects cells containing the *rif* gene, the ColE1 region, and the foreign DNA.

Methods Which Insure That a Phage Vector Will Contain Foreign DNA

There are two procedures in use for cloning genes in *E. coli* phage λ. Both are based on the fact that λ has several centrally located genes that are not needed for lytic growth (*b2, int, xis, red, gam, cIII*). These genes are in the so-called nonessential region (see Chapter 16).

The λ variant λgt4·λβ has two EcoRI restriction sites near the termini of the nonessential region (wild-type λ has five EcoRI sites but, in forming λgt4·λβ, three of these sites were removed by genetic manipulation). To use this vector the DNA is cleaved with the EcoRI enzyme and the three fragments are separated by gel electrophoresis. The terminal fragments are isolated and the central fragment is discarded (Figure 20-12). The terminal fragments can be joined via the complementary single-stranded termini but the resulting DNA molecule, whose length is 72 percent that of wild-type λ

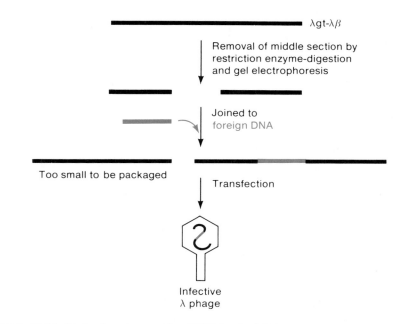

Figure 20-12 Packaging of recombinant DNA in a λ phage.

DNA, will be noninfective, because the minimum length of DNA that can be packaged in a λ phage head is 77 percent of the wild-type value. However, if foreign DNA is inserted, the resulting hybrid DNA becomes infective (that is, produces progeny phage), because the DNA can be packaged. Thus, if an *E. coli* culture is infected with DNA annealed from a mixture of the two terminal fragments and foreign DNA, any phage that is produced will contain foreign DNA.

Another λ variant contains a section of the *lac* operon that includes the *lacZ* gene, the *lacP* promoter, and the *lacO* operator. There is a restriction site in the *lac* operator, so insertion of foreign DNA into the phage at that site yields a phage with constitutive synthesis of β-galactosidase (Chapter 7). There are several substrates of β-galactosidase that are noninducing and produce colored compounds if cleaved by β-galactosidase. If such a substance is incorporated into agar, and a mixture of *lacO*$^+$ and *lacO*^{c} (constitutive) cells is plated on the agar, the *lacO*^{c} colonies will be colored because only these cells synthesize β-galactosidase. Similarly, if *lacO*$^+$ and *lacO*^{c} phages are plated on a lawn of *lacO*$^+$ cells on agar containing such a substance, *lacO*$^+$ phage will produce colorless plaques but *lacO*^{c} phage (which contain inserted foreign DNA) will yield colored plaques.

Physical Methods of Identifying a Plasmid Containing Foreign DNA

A plasmid containing a fragment of foreign DNA is larger than the original uncleaved plasmid and can be identified by this change. A particular increase in molecular weight is also a useful criterion when a specific fragment of c-DNA is to be inserted, if the c-DNA is unique, having been prepared from a purified mRNA.

A simple electrophoretic technique is useful to detect an increase in molecular weight. Lysates of single colonies of plasmid-containing cells are electrophoresed (usually sixteen different colonies at a time); there is enough plasmid DNA in a single colony to be visible as a single band moving far ahead of the chromosomal DNA. The plasmid DNA moves a distance related to its molecular weight—the larger the DNA, the smaller the distance moved in a given time interval, so the larger plasmid DNA molecules, which contain donor DNA, move more slowly than those lacking donor DNA; the colonies containing donor DNA are then easily identified.

If the plasmid is thought to contain a particular segment of foreign DNA, this can be confirmed by two different hybridization procedures, if mRNA is available. If radioactive mRNA is used and it is complementary only to the desired foreign DNA, the total DNA content from a bacterium containing the plasmid is denatured and fixed to a nitrocellulose filter, and the radioactive mRNA is added. After renaturation the filter is washed and the presence of bound radioactivity indicates that the plasmid contains integrated foreign DNA complementary to the tester mRNA (filter hybridization, Chapter 6).

When purified plasmid DNA is available, the electron-microscopic technique of **R-looping** (Chapter 6) can be used.

Mass Screening for Plasmids Containing a Particular DNA Segment

The simplest procedure for detecting a particular foreign gene is complementation. In the first such experiment reported, a particular gene in the histidine biosynthetic pathway of yeast was detected by joining a plasmid with fragments of the yeast chromosome and transforming a particular His⁻ *E. coli* mutant with a collection of recombinant plasmid DNA molecules, some of which contained the yeast gene, and then selecting for His⁺ colonies by growth on agar lacking histidine. This method works only for genes that are able to complement and is not generally successful when eukaryotic DNA is cloned in bacteria, because usually either the eukaryotic DNA is not expressed or there is no corresponding bacterial gene. The two procedures described in this section are more generally applicable.

The **colony** or ***in situ*** **hybridization assay** allows detection of the presence of a particular gene (Figure 20-13). Colonies to be tested are transferred from an agar surface to a filter paper. A portion of each colony remains on the agar, which constitutes the reference plate. The paper is treated with NaOH, which lyses the cells and releases denatured DNA, and then the paper is dried, which fixes the denatured DNA to the paper. The dry paper is flooded with ^{32}P-labeled mRNA, which is complementary to the gene being sought, and is subjected to conditions that favor renaturation. The paper is then washed to remove unbound [^{32}P]mRNA; bound radioactivity remains on the paper only if [^{32}P]mRNA has hybridized. The paper is dried and placed on autoradiographic film, and blackening of the film locates the position of the colony of interest, which can then be selected from the reference plate.

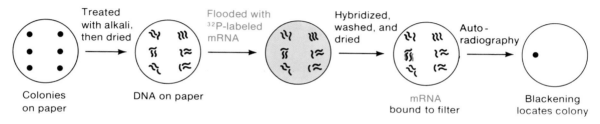

Figure 20-13 Colony hybridization. The reference plate, from which the colonies on paper were obtained, is not shown.

A similar assay is done with phage vectors. Phage are plated on a lawn of bacteria and after plaques develop, paper is placed on the agar. Some phage from each plaque stick to the paper. The paper is then treated with alkali, as above, in order to fix and denature the DNA, and the hybridization procedure just described is used to detect the plaque whose phage contain the gene of interest. The desired plaque is then removed from the reference plate.

If the protein product of the gene of interest is synthesized, two immunological techniques allow the protein-producing colony to be identified. If the protein is excreted, the **radioactive antibody test** (Figure 20-14) is used.

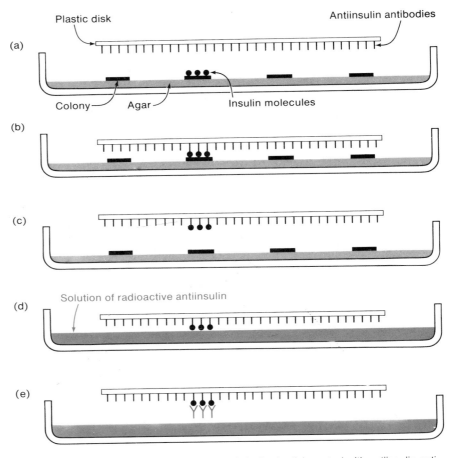

Figure 20-14 Radioactive antibody test. (a, b) A plastic disk coated with antiinsulin antibody is touched to the surface of colonies on a petri dish. (c) Insulin molecules from an insulin-producing colony bind to antibodies. (d) The disk is then dipped into a solution of radioactive antiinsulin. (e) Radioactivity adheres to the disk at positions of insulin-producing colonies.

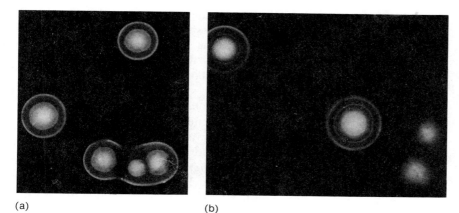

(a)   (b)

Figure 20-15 The immunoprecipitation test. Colonies making β-galactosidase are formed on agar containing antibody to the enzyme. (a) Precipitin bands form around each colony. (b) A mixture of Lac⁺ colonies (with bands) and Lac⁻ colonies (no bands).

In this test a plastic disk to which antibodies to the gene product are bound is pressed onto the colonies. If the protein is present, it will bind to the antibody. The disk is then placed in a solution containing radioactive antibody, which sticks to the bound protein on the disk. The disk is then washed and autoradiographed. The location of the radioactivity shows which colony synthesized the gene product.

An **immunoprecipitation test** can also be used to identify a protein-producing colony by adding an antibody to the protein to the agar. If the protein is excreted, an antibody-antigen precipitate called **precipitin** forms a visible white ring around the colony producing the protein (Figure 20-15). Two slight modifications of this procedure allow one to detect a protein that is not excreted. In one modification a host cell is used that is lysogenic for λ$cI857$. A copy of a reference plate containing colonies to be tested is made by transferring the colonies onto agar containing antibody. When the colonies on the antibody plate become visible, the temperature is raised to 42°C, which inactivates the $cI857$ repressor. This induces lysis of many of the cells about one hour later, thereby releasing the cell contents. The protein of interest then reacts with the antibody and forms a precipitin ring at the site of the colony. In the second modification, agar containing both antibody and the enzyme lysozyme is poured on the colonies and allowed to harden. The lysozyme lyses the cells on the surface of the colony, releasing the intracellular proteins; again, the one of interest forms a precipitin ring.

CLONING OF SINGLE-STRANDED DNA

For many purposes, for example, base-sequence analysis, it is desirable to be able to clone a particular single strand of DNA. This is possible with phage M13 as a vector. M13 is a phage containing single-stranded circular DNA. Its replication cycle consists of two steps. In the first, a double-stranded circular intermediate (called RF, for replicating form) is formed; in the second stage the RF is used as a template for synthesis of the phage DNA (the strand contained in the phage particle is called the + strand). If a foreign gene is spliced into the RF and cells are transformed (transfected) with the recombinant RF, the resulting phage will contain only one of the complementary strands (Figure 20-16).

Cloning with M13 makes use of a phenomenon called α-complementation. The active form of the *E. coli lacZ* gene product, β-galactosidase, is a tetramer consisting of four identical subunits of 1021 amino acids each. A small deletion (called ΔM15) of amino acids 11–41, which are not in the active site, prevents formation of the tetramer. The presence of certain small NH_2-terminal fragments consisting of the 92 NH_2-terminal amino acids enables the tetramer to form. Cloning with M13 is made possible by a genetically engineered form of the phage with the following features:

1. The phage carries a HindII fragment of the *lac* operon that includes the *lac* promoter and operator, and the NH_2-terminal region of the *lacZ* gene. M13 is unusual in that it does not kill its host cell. Progeny phage are instead extruded through the cell wall. However, multiplication of infected cells is slow, and very turbid plaques result from the localized slow growth of the infected cells. Thus, when the M13 mutant carrying the HindII fragment infects a Lac⁻ host cell carrying the ΔM15 deletion, cells within the plaque will be phenotypically Lac⁺. The presence of Lac⁺ cells in the plaque can be seen by adding two things to the agar: an inducer such as IPTG (Chapter 7) and a color-forming substrate that produces a blue product when cleaved by β-galactosidase. That is, the mutant M13 forms *blue plaques* on this medium.

2. The HindII fragment has been engineered to contain a short sequence with cleavage sites for the restriction enzymes EcoRI, SstI, SmaI,

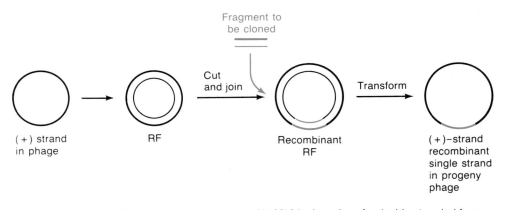

Figure 20-16   Cloning of a single strand in M13 by insertion of a double-stranded fragment into the RF.

BamHI, XbaI, SalI, PstI, and HindIII, in that order; this is called a **polycloning site** and increases the versatility of the vector by allowing the use of numerous restriction enzymes. Insertion of DNA into any of these sites destroys the α-complementing ability of the HindII fragment, and results in the M13 hybrid phage forming a *white plaque*.

The engineered M13 is used in the following way. RF obtained from cells with the phage is cleaved with two restriction enzymes active in the poly-cloning site. DNA to be cloned is also obtained by cleavage with the same enzymes. Since the restriction fragment has different single-stranded ter-mini, it can only be inserted in one orientation. The ΔM15 host is then transfected with ligase-treated DNA. White plaques are isolated, for only these contain phage with inserted DNA. Phage in one plaque are then prop-agated, and DNA is isolated from phage particles. Since the phage DNA is single-stranded, it contains only one of the complementary strands of the cloned fragment. In order to be able to isolate both individual strands, another M13 variant was engineered to contain the polycloning segment in the oppo-site orientation. Cloning of the same restriction fragment separately in both vectors yields two different recombinant phage (Figure 20-17). Isolation of DNA from each phage yields one of the complementary strands of the cloned fragment.

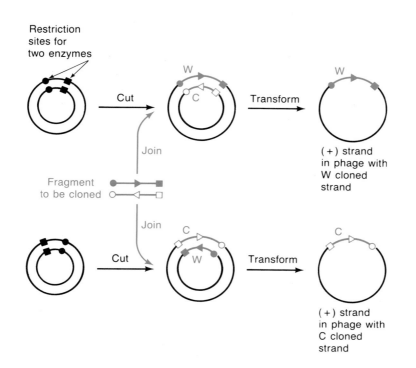

Figure 20-17 Separate cloning of complementary strands of a fragment by insertion in opposite orien-tations in M13 RF.

GENE LIBRARIES

Many laboratories utilize recombinant DNA techniques repeatedly as a means of isolating particular genes or DNA segments from a single organism. It is quite time-consuming to go through the complete cloning procedure each time a new DNA segment is needed. Thus, collections have been established of hybrid plasmid-containing bacteria that are of sufficient size that each segment of the cellular DNA is represented once (or, on occasion, twice) in the collection. Such collections are called **colony banks** or **gene libraries.** In this section we describe how a gene library is established and how to determine how many clones are needed in order that each sequence will be represented.

One of the first libraries was a collection of *E. coli* colonies containing the ColE1 plasmid in which yeast DNA had been cloned. To form the library, purified ColE1 DNA was cleaved with the EcoRI nuclease, which makes a single cut in this DNA molecule. The yeast DNA was not cleaved with a restriction enzyme; doing that would certainly have prevented many genes from being in the library as intact units inasmuch as some genes would be cleaved by EcoRI. Instead, yeast DNA was broken at random by hydrodynamic shear forces. In order to link the ColE1 and yeast DNA molecules, poly(dT) tails were added to the 3'-OH termini of the cleaved ColE1 DNA and poly(dA) tails were added to the yeast DNA fragments. The molecules were joined by homopolymer tail joining, and the bacteria were transformed. Cells containing the ColE1 factor are resistant to colicin E1 added to the agar, so plasmid-containing cells were readily isolated. The colonies were then transferred to blocks containing small wells filled with storage medium.

The following simple calculation enables one to determine how many colonies are required to make a complete library. Consider a DNA sample containing fragments of such a size that each fragment represents a fraction f of the genomic DNA. The probability P that a particular sequence is present in a collection of N colonies is $P = 1 - (1 - f)^N$ or

$$N = \ln (1 - P)/\ln (1 - f)$$

Let us assume that we want every sequence to be in the library with a probability of 0.99—that is, $P = 0.99$. If the donor DNA is haploid yeast (whose molecular weight is about 10^{10}) and the average molecular weight of the sheared DNA is 8×10^6, then $f = 8 \times 10^6/10^{10} = 0.0008$ and $N = [\ln (1 - 0.99)]/[\ln (1 - 0.0008)] = 5754$. Thus, if the library contains about 5800 colonies, there is a 99 percent probability that any yeast gene will be present in at least one colony. Furthermore, if a few colonies are selected at random for further study, it is exceedingly likely that their yeast DNA sequences do not overlap. For cells with a large DNA content (such as *Drosophila*) the required number of colonies is about 300,000; this can be reduced roughly n-fold by increasing the size of the constituent fragments by a factor of n.

If a clone containing a particular DNA segment is needed, it can in principle be found by any of the screening procedures described earlier in this chapter.

PROBLEMS

1. Restriction enzymes generate three types of termini. What are they?

2. Are the termini of a restriction fragment produced by a particular enzyme always the same or can they be different?

3. Can two different restriction enzymes act at the same site?

4. What sequence features do all restriction sites have?

5. Will the sequences 5'GGCC and 3'GGCC be cut by the same restriction enzyme?

6. Could the sequences 5'AGCGCT and 5'GGCGCC be cut by the same restriction enzyme?

7. What is the source of restriction enzymes?

8. Can a restriction enzyme cut at two sites that have different base sequences?

9. What property must two different restriction enzymes possess if they yield identical patterns of breakage?

10. What unique property is possessed by the enzyme terminal nucleotidyl transferase, with respect to its polymerizing ability?

11. What three methods can be used to join fragments with blunt ends?

12. How must a blunt-ended molecule be treated before homopolymer tail-joining can be carried out?

13. Name two methods that will ensure that a fragment will be joined to a vector in a unique direction.

14. How can one ensure that a plasmid vector does not re-form after cleavage with a restriction enzyme?

15. If a cleaved plasmid vector (with one cleavage site) and a set of restriction fragments are annealed, some circular molecules are found that are smaller than the plasmid vector. Account for these molecules.

16. What is the advantage of using a plasmid with two antibiotic-resistance genes as a cloning vehicle?

17. What two methods are used to ensure that a particular fragment has been cloned?

18. What advantage results from the use of a cosmid?

19. Restriction enzymes A and B are used to cut the DNA of phages T5 and T7, respectively. A particular T5 fragment is mixed with a particular T7 fragment and a circle forms that can be sealed by DNA ligase. What can you conclude about these enzymes?

20. Phage X82 DNA is cleaved into six fragments by the enzyme BglI. A mutant is isolated whose plaques look quite different from the wild-type plaque. DNA isolated from the mutant is cleaved into only five fragments. Account for this change. In general, discuss why often a restriction enzyme may be found that yields n fragments with wild-type phage and $n - 1$ fragments with a mutant.

21. The A^+ allele of a cell is easily selected by growth on a medium lacking substance A. However, repeated attempts to clone the A gene by digestion of cellular DNA and the vector by the enzyme EcoRI (the vector has an EcoRI site) are unsuccessful. If HaeIII is used (the vector also has an HaeIII site), clones are easily found. Explain.

22. Suppose you had a plasmid with an *amp-r* gene having an EcoRI site and a *lacZ* gene with a SalI site. If you were trying to clone a particular gene, would it be easier to use EcoRI or SalI to cleave the DNA? Think about how you would detect the recombinant plasmid.

23. A Kan-r Tet-r plasmid is treated with the BglI enzyme, which cleaves the *kan* (kanamycin) gene. The DNA is annealed with a BglI digest of *Neurospora* DNA and then used to transform *E. coli*.

(a) What antibiotic would you put in the growth medium to ensure that a colony has the plasmid?
(b) What antibiotic-resistance phenotypes will be found among the colonies?
(c) Which phenotype will have the *Neurospora* DNA?

24. A *lac⁺ tet-r* plasmid is cleaved by a restriction enzyme in the *lac* gene. The restriction site for this enzyme contains four bases and the cuts generate a two-base single-stranded end. The particular site in the *lac* gene is in the first amino acid of the chain, changes of which rarely generate a mutant. The single-stranded ends are converted to blunt ends with DNA polymerase I and then the ends are joined by blunt-end ligation. A *lac⁻ tet-s* bacterium is transformed with the DNA and Tet-r bacteria are selected. What will be the Lac phenotype of the colonies?

25. Consider a restriction enzyme that makes x cleavage sites within the *lac* operon, generating fragments containing (1) the promoter and operator, (2) the *lacZ* gene, and (3) the *lacY* gene. This collection of fragments is annealed with a vector and a *lacZ⁻ lacY⁻* cell is transformed. Among the Lac⁺ transformants some clones make more β-galactosidase than permease and others make more permease than β-galactosidase. Give two different explanations for these clones. (Think about the fact that annealing is a random process.)

26. Sometimes a cloned bacterial gene is not expressed because it has been separated from its promoter. The gene can be expressed if it can be cloned in a plasmid containing a promoter to which the gene can be linked. The gene must be oriented correctly though, because the coding sequence must be linked to the strand containing the promoter. Unfortunately, when mixing the fragments with a cleaved plasmid, joining will occur in all possible orientations. Consider a plasmid containing the *E. coli lac* operon (complete with the *lac* promoter), with a BamHI site separating the promoter and the genes. The *lac* sequence also contains restriction sites for several other enzymes. Explain how any one of these enzymes might be utilized in the cloning procedure to ensure that the gene of interest is linked to the *lac* promoter.

27. How many clones are needed to establish a gene library for monkey DNA (6×10^9 base pairs per cell) if fragments whose sizes average 2×10^4 base pairs are used and if one wishes 99 percent of the monkey genes to be in the library?

28. Explain the following phenomenon. Restriction enzyme 1 acts at all sites acted on by restriction enzyme 2, yet enzyme 2 cannot act at any of the sites for enzyme 1.

29. A restriction digest of a particular DNA produced by EcoRI always yields five fragments. In one experiment, an extra fragment, which is quite faint and migrates more slowly than all of the other bands, is seen. How might you explain this fragment?

30. A DNA molecule is cleaved by a restriction enzyme and analyzed by gel electrophoresis. Only one sharp band is seen. Explain.

31. When DNA isolated from phage J2 is treated with the enzyme SalI, eight fragments are produced, whose sizes are 1.3, 2.8, 3.6, 5.3, 7.4, 7.6, 8.1, and 11.4 kb. However, if J2 DNA is isolated from infected cells, only seven fragments are found, whose sizes are 1.3, 2.8, 7.4, 7.6, 8.1, 8.9, and 11.4. What is the likely form of the intracellular DNA?

32. If a plasmid is cleaved in five positions with a restriction enzyme that produces an extended 3' terminus, will all fragments be able to circularize?

Chapter 21

Applications of
Genetic Engineering

Current interest in genetic engineering centers on its many practical applications. Some of these are based on restriction-enzyme technology. For example, restriction enzymes can be used to generate novel maps of genomic DNA, to produce mutations at particular sites, and to identify mutant sites. However, the majority of the applications are based on cloning, in particular, the following: (1) isolation of a particular gene, part of a gene, or region of a genome; (2) production of particular RNA and protein molecules in quantities formerly thought to be unobtainable; (3) improvement in the production of biochemicals (such as enzymes and drugs) and commercially important organic chemicals; (4) production of varieties of plants having particular desirable characteristics (for example, requiring less fertilizer or resistance to disease); and (5) creation of organisms with economically important features. These and a few other applications are the subject of this final chapter.

RESTRICTION MAPPING

It is frequently of value to determine the location of restriction cuts in a particular DNA molecule. A map showing the positions is called a **restriction map.** There are several ways to obtain such a map. One method, applicable to DNA molecules with both linear and circular forms is shown in Figure 21-1. The electrophoretic patterns of enzymatic digests of the linear and circular forms of the DNA are compared. The pattern for the circles will lack the two bands corresponding to the terminal fragments and will contain a new band that is formed by the joined ends. Thus, the terminal fragments are identified as those that are missing from the circle digest.

In a typical digest one or two cuts will not have been made in every molecule; some faint bands formed by the uncut fragments will be seen. The intensity of these bands can be increased by decreasing the reaction time, generating what is called a **partial digest.** In such a digest other bands are weakened compared to the normal digest. The molecular weight of each enhanced band is invariably the sum of the molecular weights of two fragments contained in the weakened bands, which indicates that the two fragments are adjacent in the original uncleaved molecule. If the number of cuts is small (for example, six), this procedure may be sufficient for unambiguous ordering of the fragments. If the molecule does not have a circular form, the terminal fragments can be identified by labeling with ^{32}P prior to cleavage and determining which fragments (after cleavage) are radioactive. This labeling is done by the polynucleotide kinase reaction, an enzymatic reaction in which only the 5' terminus of a DNA strand is labeled.

The position of some fragments can also be identified by examining genetic deletion mutants. If the deletion does not eliminate a restriction site,

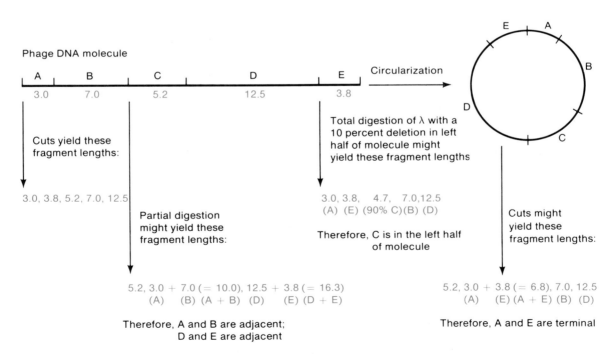

Figure 21-1 A possible scheme for determining the positions of restriction cuts in a phage λ DNA molecule, which has a linear and a circular form.

a new fragment (some of whose DNA is deleted) will appear in the digest of the deletion mutant and will be smaller, by the size of the deleted DNA, than the corresponding fragment from the wild-type undeleted molecule. If the deletion instead spans a restriction site, two bands will be absent and a new band will appear; the molecular weight of the new band equals the sum of the molecular weights of the missing fragments minus that of the deletion.

Although the methods just described usually work, the most common way to obtain a restriction map is the **double-digest procedure,** which uses two different restriction enzymes. The procedure is to take three samples of a particular DNA species, treat each of two of these with a separate enzyme and one sample with both enzymes (a **double digest**), and then compare the three sets of fragments. An example of the sets of fragments that might be generated is shown in Figure 21-2. The value of this procedure is that it yields restriction maps for both enzymes simultaneously. Let us assume, as would usually be the case, that the terminal fragments in each digest are identified either by the circularization procedure or by labeling the 5′ termini with ^{32}P. With this information one can then identify the fragments that are adjacent to the terminal fragments. Let us first determine which fragment in digest I is adjacent to fragment A. In the double digest the cut forming fragment F (of digest II) must divide the fragment of digest I adjacent to A, thereby generating a small fragment whose length is $2.3 - 1 = 1.3$ and another fragment of length x; thus, $1.3 + x$ must equal the length of fragment B, C, or D. Excluding the terminal fragments in the double digest, namely, those of length 1 and 1.7, the only possible value of x is 0.7; thus,

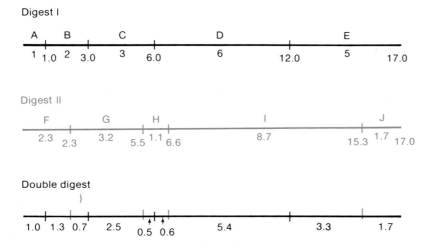

Figure 21-2 Three restriction digests useful in determining restriction maps. Enzyme I yields digest I, enzyme II yields digest II, and enzymes I and II together yield the double digest. To simplify the discussion, the fragments are given in order, though the order would be unknown until the data were analyzed.

the fragment adjacent to A must have a length of $1.3 + 0.7 = 2$, which is the length of B. Hence, B is adjacent to A. Now we identify the fragment adjacent to F in digest II. The right end of B must divide the fragment adjacent to F. Since $A + B = 3$ and $F = 2.3$, then $3 - 2.3$ + a fragment of length p must equal the length of a nonterminal fragment in digest II. The only possible value of p is 2.5, so the fragment adjacent to F has a length of 3.2 and must be G. We now turn our attention to the right end of the molecule. E divides the fragment adjacent to J. Thus, for some fragment of length q in the double digest, $5 - 1.7 + q$ must equal the length of a fragment in digest II. The only possible value of q is 5.4 and the length of the fragment adjacent to J is 8.7—that is, fragment I. This process can be continued to generate the complete restriction maps for both enzymes, as the reader should verify. The process is sometimes more complicated if different fragments having the same size are present.

SITE-SPECIFIC MUTAGENESIS

Mutations are introduced into organisms in the laboratory for a variety of reasons. When a mutation is needed solely as a genetic marker, methods that introduce mutations at random sites are generally sufficient. However, in modern molecular genetics to achieve a deep understanding of the molecular mechanisms underlying such phenomena as gene expression and gene regulation often requires mutations in particular base sequences (for example, binding sites). This need has led to the development of a variety of procedures for introducing mutations at specific sites. A few of these procedures will be outlined in this section. In each procedure the gene to be altered is carried on a plasmid. The plasmid is isolated, manipulated, and then returned to a bacterium for replication and gene expression. (If the gene is not naturally carried on a plasmid, ordinary genetic methods or recombinant DNA techniques can be used to produce such a plasmid.)

1. *Production of deletions.* Several procedures are available for generating deletions at a particular position. The simplest method consists of treating a plasmid with a restriction endonuclease to make two cuts surrounding a site to be deleted. The fragment is isolated and discarded, and the ends of the larger piece are rejoined. The gene of interest now contains a deletion. With detailed restriction maps for many restriction enzymes it is also possible to delete a small amount of bases in a particular region. The disadvantage of this technique is that deletions are usually quite large, being determined only by the spacing between restriction cuts. In another procedure, the plasmid is cut by an enzyme that makes a single cut in the gene of interest. A nuclease, S1, which acts only on single-stranded DNA, removes the short cohesive termini generated by the restriction enzyme, leaving blunt ends. The plasmid can then be resealed by blunt-end ligation with T4 DNA ligase.

This method introduces deletions of 2–4 base pairs. Larger deletions can also be made with other nucleases having special properties.

2. *Point mutations at random sites in a particular region of DNA*. As in (1), a small fragment is cleaved out and purified. This fragment is then exposed to a chemical mutagen that induces base changes of the desired type. It is then rejoined to the larger portion of the plasmid, which has not been treated with the mutagen. Cells are transformed by the $CaCl_2$ technique, and after two rounds of DNA replication, the gene contains one or more point mutations in the desired region. In a related method cleaved DNA is digested by a nuclease that removes nucleotides from the double-stranded region adjacent to the overhanging single strand produced by the restriction enzyme. The DNA is then exposed to sodium bisulfite, a mutagen that converts cytosine to uracil in single-stranded DNA. The single-stranded ends, which still have short complementary termini, are then reannealed, and the gap is filled in with a DNA polymerase. An adenine is incorporated opposite the mutated cytosine (which is now a uracil). Transformation and replication *in vivo* yields an AT pair at the site of a former GC pair. The method is nonspecific in that all cytosines in the single-stranded region may be altered.

3. *Point mutations at a particular base pair* (Figure 21-3). The plasmid is nicked with an endonuclease, using conditions that yield on the average one single-strand break per plasmid. The DNA is then denatured and intact circles are isolated. Short single-stranded oligonucleotides with defined sequences are easily synthesized, and one such oligonucleotide is made that has a mutant base at the desired site. This oligonucleotide is renatured with the intact single-stranded circle and a fragment of DNA polymerase I that lacks $5' \rightarrow 3'$ exonuclease activity (this is called the Klenow fragment) is added with the four nucleoside triphosphates. The oligonucleotide primes the circles and the circles are copied, yielding a double-stranded circle. (The circles that are not complementary to the oligonucleotide are not primed and hence are not replicated.) The replicated circles are sealed with DNA ligase and covalent circles are isolated. Bacteria are then transformed by these circles, and DNA replication occurs. Each circle contains a base-pair mismatch at the mutant site. Mismatch repair occurs at random, generating some wildtype sequences and some with a mutation at a particular site. Genetic tests are used to detect the cells with the mutant sequence.

PRODUCTION OF PROTEINS FROM CLONED GENES

One of the goals of genetic engineering is the production of a large quantity of a particular protein that is otherwise difficult to obtain (for example, proteins for which there are normally only a few molecules per cell). In principle

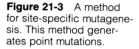

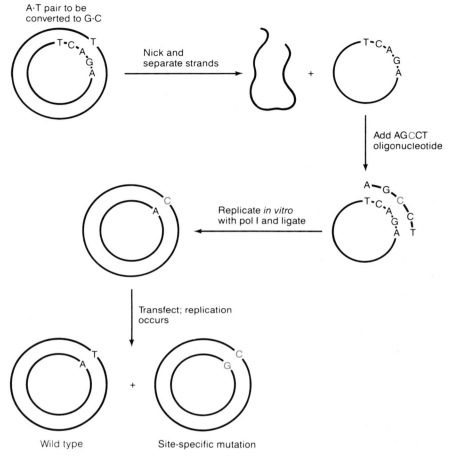

Figure 21-3 A method for site-specific mutagenesis. This method generates point mutations.

the method is straightforward: the gene is inserted adjacent to a promoter and tested to determine whether it is oriented such that the coding strand will be transcribed. With a high-copy-number plasmid or an actively replicating phage, synthesis of a gene product may reach a concentration of about 1 to 5 percent of the cellular protein.

In practice, the goal of producing large quantities of a desired protein is not always met, because of a variety of theoretical and technical problems. Some of these problems and how one seeks solutions are explained in the following sections. Some successful procedures will also be described.

Expression of a Cloned Gene

Expression of a cloned gene requires that the gene be both transcribed and translated. Let us see how one designs a system to be sure that this occurs.

Often a restriction-enzyme cleavage site separates the coding portion of a gene from its promoter, and unless another promoter is provided, the gene cannot be transcribed after it is inserted into a vector. In that case vectors engineered to contain a ribosome binding site and a promoter near the insertion site are used. Usually the easily regulated promoter-operator region of the *E. coli lac* operon (Chapter 7) or the *cI* repressor-*oL*-*pL* region of *E. coli* phage λ (Chapter 16) are used. For instance, in the first case, transcription can be initiated by the addition of an inducer of the *lac* operon. With the λ system, the temperature-sensitive *cI857* repressor is used.

Coupling a restriction fragment containing a gene of interest to a functioning promoter and a ribosomal binding site does not guarantee that the gene will be correctly expressed (Figure 21-4). A problem is that both ends of a restriction fragment have the same single-stranded termini, so insertion into a vector can occur in two possible orientations. Since only one DNA strand of any gene is a coding strand, only one orientation will yield useful mRNA. Insertion in both orientations occurs with equal probability; therefore, one must select a recombinant vector that has the gene in the correct orientation. If synthesis of the desired gene product occurs, the correct orientation is detected easily. However, if there is no synthesis, methods described in earlier sections must be used.

The problem of gene expression is more severe when eukaryotic genes are to be cloned. Particular problems are the following:

1. The eukaryotic promoter may not be recognized by a bacterial RNA polymerase.
2. The mRNA transcribed from eukaryotic genes lacks a specific nucleotide sequence needed for binding to bacterial ribosomes.
3. The mRNA may contain introns that must be excised (RNA processing).
4. The protein product of eukaryotic genes often must be altered (protein processing). For example, proinsulin requires particular proteolytic cleavages before it becomes active insulin.

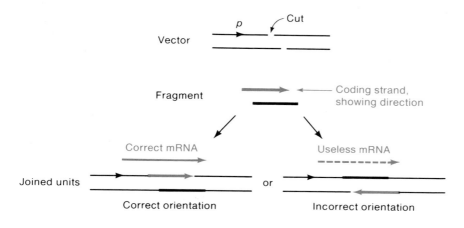

Figure 21-4 Insertion of a cloned gene in two orientations with respect to a promoter *p* in the vector.

5. Eukaryotic proteins are often recognized by bacterial proteases as foreign and are cleaved.

Problems 1 and 2 can usually be eliminated by coupling the gene to the *lac* or λ promoter, as just described. The intron problem is avoidable if c-DNA can be prepared. However, for most eukaryotic genes the necessary intron-free mRNA cannot be isolated in pure form. The fact that many finished proteins are modified forms of the initial translation product limits the usefulness of bacteria for the production of eukaryotic proteins, though in some cases (soon to be described) the processing can be carried out *in vitro* after the protein is isolated. Bacterial proteases that destroy eukaryotic proteins can presumably be eliminated (partially, at least) by mutation. Some examples of animal proteins synthesized in *E. coli* will be described shortly.

Cloning in yeast (*Saccharomyces cerevisiae*) seems to solve most of these problems. Yeast RNA polymerase is able to recognize many animal promoters and processing systems in yeast can often remove introns from animal genes and process some animal proteins. Furthermore, there is much less degradation of foreign eukaryotic proteins. Some cloning in yeast has been carried out with a yeast plasmid. However, with the advanced technology of genetic engineering in *E. coli* it has proved more valuable to engineer a plasmid capable of multiplying in both *E. coli* and *Saccharomyces*. Such a plasmid, called a **shuttle vector,** is described in the next section.

Shuttle Vectors

A **shuttle vector** is a vector that can replicate in different organisms. The first shuttle vector, which contained *E. coli* and yeast components, was used to clone yeast genes. If a yeast gene were cloned into an *E. coli* plasmid and then *E. coli* cells were transformed by the recombinant plasmid, in general the yeast gene would not be expressed, for the usual reasons—lack of recognition of yeast promoters in *E. coli*, incorrect processing, and several other problems. When cloning of yeast genes was first attempted, these difficulties limited the techniques for detecting colonies containing yeast DNA to *in situ* hybridization. Shuttle vectors, which contain sequences from an *E. coli* plasmid and a particular region of the yeast genome, were designed to avoid this problem (Figure 21-5). Essential features of this shuttle vector are: two replication origins (one active in yeast and one in *E. coli*), two selective markers (*trp*, detectable in yeast, and *amp-r*, detectable in *E. coli*), and restriction sites next to a yeast promoter. Such a vector can be cleaved and and a yeast DNA fragment can be inserted, but, most important, the hybrid vector will transform *trp*$^-$ yeast cells, producing Trp$^+$ cells. The gene of interest carried on the fragment can also be detected by virtue of its expression in yeast.

A problem with these vectors is that they are not particularly stable in yeast, because they lack a centromere (the portion of a chromosome by which the chromosome is attached to the mitotic spindle); thus, plasmid replicas

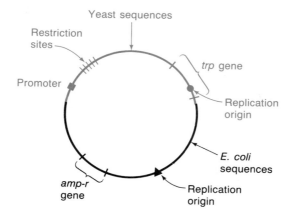

Figure 21-5   A yeast-*E. coli* shuttle vector.

are not efficiently segregated into daughter cells. To avoid this problem, the recombinant vector is isolated from a culture obtained from a colony of successfully transformed yeast and then used to transform *E. coli*, in which its presence can be detected by an antibiotic-resistance marker (Figure 21-5). In *E. coli* the recombinant vector can be maintained indefinitely. The sequence of steps in the use of this shuttle vector is listed below:

1. Insert a eukaryotic gene in the cleaved restriction site in the yeast segment.
2. Transform Trp⁻ yeast and plate on a medium lacking tryptophan.
3. Select Trp⁺ colonies.
4. Test Trp⁺ colonies for expression of the eukaryotic gene.
5. Isolate a colony with the expressed gene.
6. Isolate the plasmid.
7. Transform Amp-s *E. coli*.
8. Select an Amp-r colony.

The Amp-r colony will contain the shuttle vector with the inserted eukaryotic gene.

Synthesis of Somatostatin Using a Synthetic Gene

In this section a successful attempt to synthesize a eukaryotic protein is described. The student should note how detailed knowledge of gene expression gained from fundamental research on bacterial regulation and protein synthesis contributed to the success.

Somatostatin is a 14-residue polypeptide hormone synthesized in the hypothalamus. It has been made in *E. coli* by a procedure that is applicable to most short polypeptides. Through chemical techniques, a double-stranded

DNA molecule was synthesized that contained 51 base pairs; the base sequence of the corresponding mRNA is

AUG-(42 bases encoding somatostatin)-UGAUAG.

This mRNA molecule has a methionine codon (which will not be used for initiation), a coding sequence for somatostatin, and two stop codons. The vector was a plasmid containing the *lac* promoter and the NH$_2$-terminal segment of the *lac* gene. The vector was cleaved in the *lacZ* segment (Figure 21-6) by a restriction enzyme that leaves blunt ends; the synthetic DNA molecule was then blunt-end-ligated to the cleaved plasmid. Addition of a *lac* inducer yielded a protein consisting of the NH$_2$-terminal segment of β-galactosidase *coupled by methionine* to somatostatin. This protein was purified and treated with cyanogen bromide, a reagent that cleaves proteins only at the carboxyl side of methionine. Thus, the methionine linker remained attached to the β-galactosidase fragment and somatostatin was released. Fol-

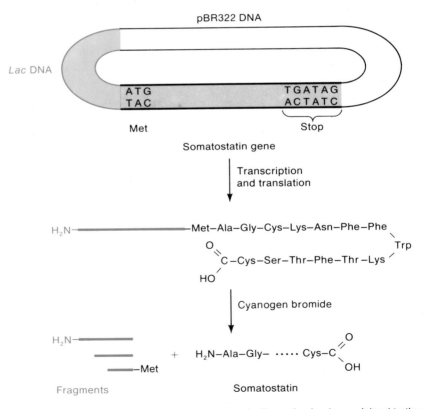

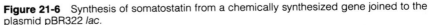

Figure 21-6 Synthesis of somatostatin from a chemically synthesized gene joined to the plasmid pBR322 *lac*.

lowing the methionine-coupling by cyanogen bromide cleavage is a useful trick for separating any polypeptide from a bacterial protein to which it is fused, as long as the polypeptide itself does not contain methionine. Another more generally useful linker is $(Asp)_4$-Lys. In a sequence $(Asp)_4$-Lys-X, in which X is another amino acid, the protease enterokinase cleaves between lysine and X. Since such a sequence is not widespread, this linker is more valuable than methionine.

APPLICATIONS OF GENETIC ENGINEERING

In the preceding sections, we have discussed the utility of the recombinant DNA technology mostly in terms of the production of useful proteins, and this is certainly of great importance. Cloned genes have also been used used as probes in hybridization experiments to detect RNA species, just as $\phi80lac$ was used to study the kinetics of synthesis of *lac* mRNA. This technique is particularly valuable in the study of regulation in eukaryotes, in which transducing phage are not available. Genetic engineering also has other uses that we document in this section.

     1. *Studies of regulation by subcloning*. A variant of the cloning technique, **subcloning,** is used to identify regulatory sequences in both bacteria and eukaryotes. The technique begins with cloning of a functional genetic unit, which typically contains the coding sequence, regulatory sequences, and often irrelevant adjacent DNA. The cloned sequence is then reisolated and screened for cutting sites by several restriction enzymes. The isolated sequence is trimmed to successively smaller sizes by one or more enzymes and then recloned in a suitable plasmid. The plasmid DNA is used in a transformation experiment, and the phenotype of the recipient cell is observed. A change in phenotype occurs when regulatory regions or structural genes are cut. For example, if the *E. coli lac* operon were trimmed in this way, one would identify an element adjacent to the structural genes (the *lacI* gene) as a regulatory element, because when removed, *lac* expression would become constitutive. Other trimming could yield a Lac$^-$ phenotype and loss of ability to make *lac* mRNA and thereby identify the promoter.

     2. *Construction of industrially important bacteria*. Bacteria with novel phenotypes can be produced by genetic engineering, sometimes by combining the features of several other bacteria. For example, several genes from different bacteria have been inserted into a single plasmid that has then been placed in a marine bacterium, yielding an organism capable of metabolizing petroleum; this organism has been used to clean up oil spills in the oceans. Furthermore, many biotechnology companies are at work designing bacteria that can synthesize industrially important chemicals. Bacteria have been designed that are able to compost waste more efficiently and to fix nitrogen

(to improve the fertility of soil), and an enormous effort is currently being expended to create organisms that can convert biological waste to alcohol. An interesting bacterium is a strain of *Pseudomonas fluorescens*, which lives in association with maize and soybean roots. A lethal gene from *Bacillus thuriengis*, a bacterium pathogenic to the black cutworm, has been engineered into this bacterium. The black cutworm causes extensive crop damage and is usually combatted with noxious insecticides. In preliminary studies inoculation of soil with the engineered *Ps. fluorescens* resulted in death of the cutworm. This type of genetic engineering will surely have a great impact on world economy and environmental quality.

3. *Genetic engineering of plants*. Altering the genotypes of plants is an important application of recombinant DNA technology. In Chapter 11 we discussed the bacterium *Agrobacterium tumefaciens* and its plasmid Ti, which produces crown gall tumors in plants. These tumors result from integration of the plasmid DNA into the plant chromosome. It is possible by genetic engineering to introduce genes from one plant into this plasmid and then, by infecting a second plant with the bacterium, transfer the genes of the first plant to the second plant. (Actually genes are first cloned in an *E. coli* plasmid and then recloned in Ti.) Attempts are being made to perform plant breeding in this way. An example is the attempted alteration of the surface structure of the roots of grains such as wheat, by introducing certain genes from legumes (peas, beans), in order to give grains the ability of the legumes to establish root nodules of nitrogen-fixing bacteria. If successful, this would eliminate the need for the addition of nitrogenous fertilizers to grain-growing soils.

The first engineered recombinant plant of commercial value was developed in 1985. An economically important herbicide (weed killer) is glyphosate, which inhibits a particular essential enzyme in many plants. However, most herbicides cannot be applied to fields growing crops because both the crop and the weeds would be killed. (The chlorinated acids that selectively kill dicotyledonous plants but not the grasses, such as maize and the cereals, are out of favor because of their persistence in soil, toxicity to animals, and possible carcinogenicity in humans.) The target gene of glyphosate is also present in the bacterium *S. typhimurium*. A resistant form of the gene was obtained by mutagenesis and growth of *Salmonella* in the presence of glyphosate; the gene was cloned in *E. coli* and then recloned in *Agrobacterium*. Infection of plants with purified Ti containing the glyphosate-resistance gene has yielded varieties of maize, cotton, and tobacco that are resistant to glyphosate. Thus, fields of these crops can be sprayed with glyphosate at any stage of growth of the crop.

4. *Production of drugs*. Genetic engineering is having an impact on clinical medicine. The initial focus was on developing organisms that would overproduce antibiotics, thereby reducing production costs, and this has been

accomplished for several antibiotics. Of greater significance is the production of biologically active compounds that hitherto could only be isolated in minuscule amounts from enormous amounts of tissue. We have already seen the example of somatostatin. Another example is the antiviral agent α-interferon, which could not even be tested clinically before genetic engineering provided a source of pure material. This substance reduces the duration of viral infection, is effective against herpesvirus infection of the eye, reduces the incidence of attacks of multiple sclerosis, suppresses atherosclerosis in rats on high-cholesterol diets, and is being examined for its antitumor activity. In 1985, tumor necrosis factor, a powerful anticancer substance in rats was cloned, and animal studies began shortly afterward. Interleukin II, a substance that stimulates multiplication of certain cells in the immune system, has also been cloned; it is being tested on patients with AIDS and other viral diseases and shows promise in shrinking certain cancers in humans.

     5. *Synthetic vaccines*. A major breakthrough in disease prevention has been the development of synthetic vaccines. Production of certain vaccines such as anti-hepatitis B, has been difficult because of the extreme hazards of working with large quantities of the virus. The danger would be avoided if the viral antigen could be cloned and purified in *E. coli* or yeast, because the pure antigen could be given as a vaccine. Several viral antigens have been cloned, but because of either thermal instability or poor antigenicity when pure, attempts to make vaccines in this way have been unsuccessful. However, the use of vaccinia virus (the anti-smallpox agent) as a carrier has been fruitful. The procedure makes use of the fact that viral antigens are on the surface of virus particles and that some of these antigens can be engineered into the coat of vaccinia. The use of vaccinia virus is not straightforward because the virus is huge, containing a single DNA molecule with a molecular weight of 160×10^6, far too large for controlled cutting by restriction enzymes and subsequent splicing. Therefore, the viral antigen genes are first cloned in an *E. coli* plasmid. Animal cell lines that support the growth of vaccinia are then infected with both normal vaccinia DNA and the plasmid DNA containing the viral-antigen gene. Genetic recombination occurs in the infected cells, and some progeny vaccinia particles are formed that contain the cloned gene and hence the foreign viral antigen in the vaccinia coat. Various procedures are used to isolate these vaccinia hybrids. By 1985 vaccinia hybrids with surface antigens of hepatitis B, influenza virus, and vesicular stomatitis virus (which kills cattle, horses, and pigs) had been prepared and shown to be useful vaccines in animal tests. A surface antigen of *Plasmodium falciparum,* the parasite that causes malaria, has also been placed in the vaccinia coat; this may lead to an antimalaria vaccine.

     Since the early 1980s an increasingly greater fraction of researchers in biology have turned their attention toward genetic engineering. This is exemplified by the formation of more than 200 biotechnology companies since that time and the creation of scientific journals and newspapers devoted exclusively to genetic engineering. With the legal consequences of developing new products, patent lawyers are even being required to take courses in molecular biology and microbial genetics. It is likely that this growing billion-dollar industry will someday employ many readers of this book.

PROBLEMS

In the problems below, the abbreviation kb refers to kilobase pairs.

1. A bacterial gene is cloned in a vector next to a promoter. Hybridization with a radioactive mRNA probe indicates that the gene is present in the cell; However, the gene product is not made. What is the most probable reason for the lack of synthesis of the gene product?

2. Give several reasons why a cloned prokaryotic gene might not be expressed in a prokaryote. Assume that the vector does not provide any particular means for expression of cloned genes.

3. What method may be used to avoid the difficulties in Problem 2?

4. Give two reasons why a simply cloned eukaryotic gene will not usually yield functional mRNA in a bacterial host.

5. Assuming the difficulties in Problem 4 are solved, give three reasons why a desired protein may not be produced from a eukaryotic gene cloned in a bacterium.

6. What element is usually lacking in a yeast-*E. coli* shuttle vector that requires that the plasmid be maintained in *E. coli*?

7. When a yeast-*E. coli* shuttle vector is used, is the initial selection by transformation carried out in yeast or in *E. coli,* and why?

8. The DNA of temperate phage P4 is linear, double-stranded, 11.5-kb long, and has cohesive ends. Digestion with the BamHI restriction endonuclease yields fragments 6.4, 4.1 and 1.0 kb in length. Partial digestion with the BamHI enzyme yields fragments 10.5, 7.4, 6.4, 4.1, and 1.0 kb in length. Circular P4 DNA made with DNA ligase can be digested with the BamHI enzyme to yield fragments 6.4 and 5.1 kb in length.

   (a) What is the order of fragments in the DNA?
   (b) DNA is extracted from cells lysogenic for P4 and digested completely with the BamHI enzyme. Each of the fragments is tested for hybridization with radioactive denatured P4 phage DNA. Fragments whose lengths are 15.0, 12.5, and 6.4 kb are labeled with radioactivity. What conclusions can you draw from this experiment about prophage structure?

9. Wild-type phage X4 DNA is treated with a restriction enzyme and five fragments—I, II, III, IV, and V—are separated and purified by gel electrophoresis. It has been decided to map these fragments by comparison to the genetic map of X4. This is done by performing twenty-five transfection experiments using the five mutations, a^-, b^-, c^-, d^-, and e^-, which are equally spaced from along the

DNA. The experiment consists of transformation of a Rec$^+$ cell with intact DNA from a single mutant and a purified fragment. The data are shown below:

	I	II	III	IV	V
a^-	21	14	21,842	3	8
b^-	6	0	4	4856	2
c^-	32,681	16	8	11	13
d^-	1	3	0	1	1825
e^-	22	17,813	27	18	30

The values indicate the number of plaques per microgram of intact DNA molecules. What is the order of the fragments starting at the left end of the map?

10. A DNA fragment has been isolated that contains the coding sequences of two genes, A and B, of E. coli. In an effort to make large quantities of the gene products, the fragment is joined to a standard plasmid and the DNA is used to transform an A^-B^- bacterium. Colonies containing the plasmid are isolated easily, but none are A^+B^+. However, A^+B^- and A^-B^+ colonies are found. Explain this observation.

11. The restriction enzyme EcoRI cleaves a particular phage DNA molecule into five fragments of sizes 2.9, 4.5, 6.2, 7.4, and 8.0 kb. If the DNA has previously been radioactively labeled at the 5′ ends, the 6.2-kb and 8.0-kb fragments are found to be labeled. BamHI cleaves the same molecule into three fragments, whose sizes are 6.0, 10.1, and 12.9 kb, with the 6.0-kb and 10.1-kb fragments labeled with ^{32}P. When both enzymes are used together, the fragments have sizes 1.0, 2.0, 2.9, 3.5, 6.0, 6.2, and 7.4 kb. What are the restriction maps for the two enzymes?

12. In analyzing DNA by treatment with a restriction endonuclease followed by gel electrophoresis, in a particular experiment it is observed that in addition to the expected bands, DNA is present in all regions of the gel starting from the position of the most slowly-moving fragment and past that of the most rapidly-moving fragment. If the DNA is analyzed before treatment with the restriction enzyme, no such material is seen. Suggest an explanation.

13. Can the c-DNA technique be used with any eukaryotic gene? This is a practical and not a theoretical question.

14. Plasmid pBR607 DNA is circular and double-stranded and has a molecular weight of 2.6×10^6. This plasmid carries two genes whose protein products confer resistance to tetracycline (Tet-r) and ampicillin (Amp-r) in host bacteria. The DNA has a single site for each of the following restriction enzymes: EcoRI, BamHI, HindIII, PstI, and SalI. Cloning DNA into the EcoRI site does not affect resistance to either drug. Cloning DNA into the BamHI, HindIII and SalI sites abolishes tetracycline-resistance. Cloning into the PstI site abolishes ampicillin-resistance. Digestion with the following mixtures of restriction enzymes yields

fragments with the sizes (in millions of molecular weight units) indicated in the column at the right:

EcoRI, PstI	0.46, 2.14
EcoRI, BamHI	0.2, 2.4
EcoRI, HindIII	0.05, 2.55
EcoRI, SalI	0.55, 2.05
EcoRI, BamHI, PstI	0.2, 0.46, 1.94

Position the PstI, BamHI, HindIII and SalI cleavage sites on a restriction map, relative to the EcoRI cleavage site.

15. The DNA of the plasmid pHUB1 is circular, double-stranded, and contains 5.7 kilobase pairs (5.7 kb). This plasmid carries a gene whose protein product confers resistance to tetracycline (Tet-r) on the host bacterium. The DNA has one site each for the following restriction enzymes: EcoRI, HpaI, BamHI, PstI, SalI, and BglII. Cloning into the BamHI and SalI sites abolishes tetracycline-resistance; cloning into the other sites does not. Digestion with various restriction enzymes and combinations of enzymes yields DNA fragments whose sizes in kilobase pairs are shown below.

EcoRI	5.7
EcoRI, BamHI	0.4, 5.3
EcoRI, HpaI	0.5, 5.2
EcoRI, SalI	0.7, 5.0
EcoRI, BglII	1.1, 4.6
EcoRI, PstI	2.4, 3.3
PstI, BglII	1.3, 4.4
BglII, HpaI	0.6, 5.1

Position the cleavage sites for all these enzymes on a map of the pHUB1 DNA. Draw the map as a circle with 5.7 kb marked off on the circumference.

References

Chapter 1

Fincham, J. R. S. 1966. *Genetic Complementation*. Benjamin-Cummings.

Hayes, W. 1968. *The Genetics of Bacteria and Their Viruses*. Wiley.

Snyder, L. A., D. Freifelder, and D. L. Hartl. 1985. *General Genetics*. Jones and Bartlett.

Srb, A. M., R. D. Owen, and R. S. Edgar. 1965. *General Genetics*. W. H. Freeman.

Chapter 2

Abdel-Monem, M., and H. Hoffmann-Berling. 1980. "DNA unwinding enzymes." *Trends Biochem. Sci.*, 5, 128.

Bauer, W. R. 1978. "Structure and reactions of closed duplex DNA." *Ann. Rev. Biophys. Bioeng.*, 7, 287.

Bauer, W. R., et al. 1980. "Supercoiled DNA." *Scient. Amer.*, July, p. 118.

Cooper, T. G. 1977. *The Tools of Biochemistry*. Wiley.

Cozzarelli, N. 1980. "DNA gyrase and supercoiling of DNA." *Science*, 207, 953.

Crick, F. H. C. 1974. "The double helix: a personal view." *Nature*, 248, 766.

Crick, F. H. C. 1976. "Linking numbers and nucleosomes." *Proc. Nat. Acad. Sci.*, 73, 2639.

Davidson, J. N. 1977. *The Biochemistry of the Nucleic Acids*. Academic Press.

Freifelder, D. (ed.). 1978. *The DNA Molecule: Structure and Properties*. W. H. Freeman.

Freifelder, D. 1987. *Molecular Biology*. 2nd ed. Chapter 4. Jones and Bartlett.

Freifelder, D. 1982. *Physical Biochemistry*. W. H. Freeman.

Kolata, G. 1983. "Z-DNA moves toward real biology." *Science,* 222, 496.

Linn, S. M., and R. J. Roberts (eds.). 1982. *Nucleases*. Cold Spring Harbor.

Maxam, A. M., and W. Gilbert. 1977. "A new method for sequencing DNA." *Proc. Nat. Acad. Sci.,* 74, 560.

Meselson, M., F. W. Stahl, and J. Vinograd. 1957. "Equilibrium sedimentation of macromolecules in density gradients." *Proc. Nat. Acad. Sci.,* 43, 581.

Nordheim, A., et al. 1981. "Antibodies to left-handed Z-DNA bind to interband regions of *Drosophila* polytene chromosomes." *Nature,* 294, 417.

Rickwood, D., and B. D. Hames (eds.). 1982. *Gel Electrophoresis of Nucleic Acids*. IRL Press.

Sanger, F., S. Nicklen, and A. R. Coulson. 1977. "DNA sequencing with chain-terminating inhibitors." *Proc. Nat. Acad. Sci.,* 74, 5463.

Ts'o, P. O. P. 1974. *Basic Principles in Nucleic Acid Chemistry*. Academic Press.

Vinograd, J., R. Radloff, and W. Bauer. 1967. "A dye-buoyant density method for the detection and isolation of closed circular duplex DNA." *Proc. Nat. Acad. Sci.,* 57, 1514.

Wang, J. 1982. "DNA topoisomerases." *Scient. Amer.,* July, p. 94.

Watson, J. D. 1968. *The Double Helix*. Atheneum.

Watson, J. D., and F. H. C. Crick. 1953. "Molecular structure of nucleic acid. A structure for deoxyribose nucleic acid." *Nature,* 171, 737.

Younghusband, H. B., and R. B. Inman. 1974. "The electron microscopy of DNA." *Ann. Rev. Biochem.,* 43, 605.

Chapter 3

Anfinsen, C. G. 1973. "Principles that govern the folding of protein molecules." *Science,* 181, 223.

Baldwin, R. L. 1978. "The pathway of protein folding." *Trends Biochem. Sci.,* 3, 66.

Bernhard, S. 1968. *The Structure and Function of Enzymes*. Benjamin/Cummings.

Blake, C. C. F., and L. N. Johnson. 1984. "Protein structure." *Trends Biochem. Sci.,* 9, 147.

Clothia, C. 1984. "Principles that determine the structure of proteins." *Ann. Rev. Biochem.,* 53, 537.

Creighton, T. E. 1983. *Proteins: Structure and Molecular Properties*. W. H. Freeman.

Dickerson, R. E., and I. Geis. 1980. *The Structure and Action of Proteins*. Harper & Row.

Freifelder, D. 1982. *Physical Biochemistry*. W. H. Freeman.

Kendrew, J. C. 1961. "The three-dimensional structure of a protein molecule." *Scient. Amer.,* December, p. 96.

Kim, P. S., and R. L. Baldwin. 1982. "Specific intermediates in the folding reactions of small proteins and the mechanism of protein folding." *Ann. Rev. Biochem.,* 51, 459.

Koshland, D. E. 1973. "Protein shape and biological control." *Scient. Amer.*, October, p. 52.

Lehninger, A. L. 1982. *Principles of Biochemistry*. Worth.

Lesk, A. M. 1984. "Themes and contrasts in protein structure." *Trends Biochem. Sci.*, 9, 5.

Monod, J., J. P. Changeux, and F. Jacob. 1963. "Allosteric proteins and cellular control systems." *J. Molec. Biol.*, 6, 306.

Pauling, L., R. B. Corey, and H. R. Branson. 1951. "The structure of proteins: two hydrogen-bonded helical configurations of the polypeptide chain." *Proc. Nat. Acad. Sci.*, 37, 205.

Perutz, M. 1964. "The hemoglobin molecule." *Scient. Amer.*, November, p. 64.

Perutz, M. 1978. "Hemoglobin structure and respiratory control." *Scient. Amer.*, December, p. 92.

Phillips, D. C. 1966. "The three-dimensional structure of an enzyme molecule." *Scient. Amer.*, November, p. 78.

Schachman, H. K. 1974. "Anatomy and physiology of a regulatory enzyme." *Harvey Lectures*, 68, 67.

Schultz, G. E., and R. H. Schirmer. 1979. *Principles of Protein Structure*. Springer-Verlag.

Srere, P. A. 1984. "Why are enzymes so big?" *Trends Biochem. Sci.*, 9, 387.

Stryer, L. 1981. *Biochemistry*. Chapter 2. W. H. Freeman.

Wold, F. 1971. *Macromolecules: Structure and Function*. Prentice-Hall.

Chapter 4

Begg, K. J., and W. D. Donachie. 1977. "Growth of *Escherichia coli* cell surface." *J. Bacter.*, 129, 1524.

Cooper, S., and C. E. Helmstetter. 1968. "Chromosome replication and the division cycle of *E. coli* B/r." *J. Molec. Biol.*, 31, 519.

Davis, B. D., et al. 1980. *Microbiology*. Harper & Row.

Demerec, M., et al. 1966. "A proposal for a uniform nomenclature in bacterial genetics." *Genetics*, 54, 61.

Donachie, W. D. 1979. "The cell cycle of *E. coli*." In *Developmental Biology of Prokaryotes* (J. H. Parish, ed.). Univ. of California Press.

Glass, R. E. 1982. *Gene Function: E. coli and its Heritable Elements*. Univ. of California Press.

Helmstetter, C. E., and D. J. Cummings. 1963. "Bacterial synchronization by selection of cells at division." *Proc. Nat. Acad. Sci.*, 56, 707.

Helmstetter, C. E., and S. Cooper. 1968. "DNA synthesis during the division cycle of rapidly growing *E. coli* B/r." *J. Molec. Biol.*, 31, 507.

Henning, U. 1975. "Determination of cell shape in bacteria." *Ann. Rev. Microbiol.*, 29, 45.

Ingraham, J. L., O. Maaløe, and F. C. Neidhardt. 1983. *Growth of the Bacterial Cell*. Sinauer.

Novick, A., and L. Szilard. 1950. "Experiments with the chemostat on spontaneous mutations in bacteria." *Proc. Nat. Acad. Sci.*, 36, 708.

Pettijohn, D. E. 1976. "Prokaryotic DNA in nucleoid structure." *CRC Crit. Rev. Biochem.*, 4, 175.

Powell, E. O. 1956. "Growth rate and generation time of bacteria, with special reference to continuous culture." *J. Gen. Microbiol.*, 15, 492.

Shehata, T. E., and A. G. Marr. 1970. "Synchronous growth of enteric bacteria." *J. Bacter.*, 103, 789.

Sinden, R. R., and D. E. Pettijohn. 1981. "Chromosomes in living *E. coli* cells are segregated into domains of supercoiling." *Proc. Nat. Acad. Sci.*, 78, 224.

Stanier, R. Y., E. A. Adelberg, and J. Ingraham. 1976. *The Microbial World*. Prentice-Hall.

Steck, T., and K. Drlica. 1984. "Bacterial chromosome segregation: evidence for DNA gyrase involvement in decatenation." *Cell,* 36, 1081.

Stent, G. S., and R. Calendar. 1978. *Molecular Genetics*. Chapter 2. W. H. Freeman.

Worcel, A., and E. Burgi. 1972. "On the structure of the folded chromosome of *E. coli.*" *J. Mol. Biol.*, 71, 127.

Chapter 5

Adams, M. 1959. *Bacteriophages*. Interscience.

Arber, W. 1974. "DNA modification and restriction." *Prog. Nucleic Acid Res.*, 14, 1.

Arber, W., and S. Linn. 1967. "DNA modification and restriction." *Ann. Rev. Biochem.*, 38, 467.

Arber, W. 1971. "Host-controlled variation." In *The Bacteriophage Lambda* (A. D. Hershey, ed.). Cold Spring Harbor.

Bertani, G., and J. J. Weigle. 1953. "Host-controlled variation in bacterial viruses." *J. Bacter.*, 65, 113.

Campbell, A. 1969, *Episomes*. Harper & Row.

Casjens, S., and J. King. 1975. "Virus assembly." *Ann. Rev. Biochem.*, 44, 555.

Hershey, A. D. 1946. "Spontaneous mutations in bacterial viruses." *Cold Spring Harb. Symp. Quant. Biol.*, 11, 67.

Hershey, A. D., and M. Chase. 1952. "Independent function of viral protein and nucleic acid in growth of bacteriophage." *J. Gen. Physiol.*, 36, 39.

Horne, R. W. 1963. "The structure of viruses." *Scient. Amer.*, January, p. 48.

Kellenberger, E., and J. Sechaud. 1957. "Electron microscopic studies of phage replication. II. Production of phage-related structures during multipliction of phages T2 and T4." *Virol.*, 3, 256.

Luria, S. E. 1953. "Host-induced modifications of bacterial viruses." *Cold Spring Harb. Symp. Quant. Biol.*, 18, 237.

Luria, S. E. 1970. "The recognition of DNA in bacteria." *Scient. Amer.*, January, p. 88.

Luria, S. E., et al. 1978. *General Virology*. Wiley.

Matthews, C. K. 1971. *Bacteriophage Biochemistry.* Van Nostrand Reinhold.

Simon, L. D., and T. F. Anderson. 1967. "The infection of *E. coli* by T2 and T4 bacteriophages as seen in the electron microscope. I. Attachment and penetration." *Virol.,* 32, 279. "II. Structure and function of the baseplate." *Virol.,* 32, 298.

Sinsheimer, R. L. 1959. "A single-stranded deoxyribonucleic acid from bacteriophage φX174." *J. Molec. Biol.,* 1, 43.

Stent, G. S. 1963. *Molecular Biology of Bacterial Viruses.* W. H. Freeman.

Williams, R. C., and H. W. Fisher. 1974. *An Electron Microscopic Atlas of Viruses.* C. C. Thomas.

Chapter 6

Adhya, S., and M. Gottesman. 1978. "Control of transcription termination." *Ann. Rev. Biochem.,* 47, 967.

Altman, S. (ed.). 1978. *Transfer RNA.* MIT Press.

Barrell, B. G., G. M. Aire, and C. A. Hutchison III. 1976. "Overlapping genes in bacteriophage φX174." *Nature,* 264, 34.

Benzer, S., and S. P. Champe. 1962. "A change from nonsense to sense in the genetic code." *Proc. Nat. Acad. Sci.,* 48, 1114.

Bermek, E. 1978. "Mechanisms in polypeptide chain elongation on ribosomes." *Prog. Nucl. Acid Res. Mol. Biol.,* 2, 63.

Bremer, H., et al. 1965. "Direction of chain growth in enzymic RNA synthesis." *J. Molec. Biol.,* 13, 540.

Brenner, S., F. Jacob, and M. Meselson. 1961. "An unstable intermediate carrying information from genes to ribosomes." *Nature,* 190, 576.

Brimacombe, R., et al. 1976. "The ribosome of *E. coli.*" *Prog. Nucl. Acid Res. Mol. Biol.,* 18, 1.

Bujard, H. 1980. "The interaction of *E. coli* RNA polymerase with promoters." *Trends Biochem. Sci.,* 5, 274.

Burgess, R. R., et al. 1969. "Factor-stimulating transcription by RNA polymerase." *Nature,* 221, 43.

Capecchi, M., and G. N. Gussin. 1965. "Suppression *in vitro*: identification of a serine tRNA as a nonsense suppressor." *Science,* 149, 417.

Caskey, C. T. 1980. "Peptide chain termination." *Trends Biochem. Sci.,* 5, 234.

Celis, J. E., and J. D. Smith. 1979. *Nonsense Mutations and tRNA Suppression.* Academic Press.

Chambliss, G. (ed.). 1980. *Ribosomes: Structure, Function, and Genetics.* University Park Press.

Chambon, P. 1981. "Split genes." *Scient. Amer.,* May, p. 60.

Chappeville, F., et al. 1962. "On the role of soluble ribonucleic acid in coding for amino acids." *Proc. Nat. Acad. Sci.,* 48, 1086.

Clark, B. 1980. "The elongation step of protein biosynthesis." *Trends Biochem. Sci.,* 5, 207.

Clark, B. F. C. 1977. "Correlation of biological activities with structural features of transfer RNA." *Prog. Nucl. Acid Res. Mol. Biol.*, 20, 1.

Cold Spring Harbor Laboratory. 1966. *The Genetic Code*. Vol. 31, Symposium on Quantitative Biology.

Crick, F. H. C., et al. 1961. "General nature of the genetic code for proteins." *Nature*, 192, 1227.

Freifelder, D. 1987. *Molecular Biology*. 2nd ed. Chapters 17 and 18. Jones and Bartlett.

Garen, A. 1968. "Sense and nonsense in the genetic code." *Science*, 160, 149.

Glass, R. E. 1982. *Gene Function: E. coli and its Heritable Elements*. Univ. of California Press.

Gold, L., et al. 1981. "Translational initiation in prokaryotes." *Ann. Rev. Microbiol.*, 35, 365.

Goodman, H. M., et al. 1968. "Amber suppression: a nucleotide change in the anticodon of tyrosine transfer RNA." *Nature*, 217, 1019.

Grunberg-Manago, M., and F. Gros. 1977. "Initiation mechanisms of protein synthesis." *Prog. Nucl. Acid Res. Mol. Biol.*, 20, 209.

Hanna, M. M., and C. F. Meares. 1983. "Topography of transcription: path of the leading end of nascent RNA through the *E. coli* transcription complex." *Proc. Nat. Acad. Sci.*, 80, 4238.

Hunt, T. 1980. "The initiation of protein synthesis." *Trends Biochem. Sci.*, 5, 178.

Holley, R., et al. 1965. "Structure of a ribonucleic acid." *Science*, 147, 1462.

Jukes, T. 1978. "The amino acid code." *Adv. Enzym.*, 47, 375.

Jukes, T. 1983. "Evolution of the amino acid code: inferences from mitochondrial codes." *J. Molec. Evol.*, 19, 219.

Khorana, H. G. 1968. "Nucleic acid synthesis in the study of the genetic code." In *Nobel Lectures: Physiology or Medicine*. Vol. 4. American Elsevier.

Kim, S. 1978. "Three-dimensional structure of transfer RNA and its functional implications." *Adv. Enzymol.*, 46, 279.

Krayefsky, A. A., and M. K. Kukhenova. 1979. "The peptidyl transferase center of ribosomes." *Prog. Nucl. Acid Res. Mol. Biol.*, 23, 2.

Kurland, C. G. 1977. "Structure and function of the bacterial ribosome." *Ann. Rev. Biochem.*, 46, 173.

Lake, J. 1980. "Ribosome structure and tRNA binding sites." In *Ribosomes: Structure, Function, and Genetics* (G. Chambliss, ed.). University Park Press.

Lake, J. 1981. "The Ribosome." *Scient. Amer.*, August, p. 84.

Lake, J. A. 1983. "Evolving ribosome structure: domains in archaebacteria, eubacteria, and eukaryotes," *Cell*, 33, 318.

Lathe, R. 1978. "RNA polymerase of *E. coli*." *Curr. Topics Microbiol. Immunol.*, 83, 37.

Lewin, B. 1985. *Genes II*. Chapter 20. Wiley.

Losick, R., and M. Chamberlin 1976. *RNA Polymerase*. Cold Spring Harbor.

McClure, W. R. 1985. "Mechanism and control of transcription initiation in prokaryotes." *Ann. Rev. Biochem.*, 54, 171.

Moses, P. B., and P. Model. 1984. "A Rho-dependent transcription termination signal in bacteriophage f1." *J. Molec. Biol.*, 172, 1.

Nierhaus, K. N. 1982. "Structure, assembly, and function of ribosomes." *Curr. Topics Micro. Immunol.*, 97, 81.

Nirenberg, M. 1963. "The genetic code." *Scient. Amer.*, March, p. 80.

Nirenberg, M., and P. Leder. 1964. "RNA codewords and protein synthesis." *Science*, 145, 1399.

Nirenberg, M., and J. H. Matthei. 1961. "The dependence of cell-free protein synthesis in *E. coli* upon naturally occurring or synthetic polyribonucleotides." *Proc. Nat. Acad. Sci.*, 47, 1588.

Noller, H. F. 1984. "Structure of ribosomal RNA." *Ann. Rev. Biochem.*, 53, 119.

Nomura, M. 1973. "Assembly of bacterial ribosomes." *Science*, 179, 864.

Nomura, M., A. Tissières, and P. Lengyel (eds.). 1974. *Ribosomes*. Cold Spring Harbor.

Pribnow, D. 1975. "Nucleotide sequence of an RNA polymerase binding site at an early T7 promoter." *Proc. Nat. Acad. Sci.*, 72, 784.

Prince, J. B., R. B. Gutell, and R. A. Garrett. 1983. "A consensus model of the *E. coli* ribosome." *Trends Bioch. Sci.*, 8, 359.

Rich, A., and S. Kim. 1978. "The three-dimensional structure of transfer RNA." *Scient. Amer.*, January, p. 52.

Roberts, J. 1969. "Termination factor for RNA polymerase." *Nature*, 224, 1168.

Rosenberg, M., and D. Court. 1979. "Regulatory sequences involved in the promotion and termination of RNA transcription." *Ann. Rev. Genet.*, 13, 319.

Schimmel, P. R., S. Putney, and R. Starzyk. 1983. "RNA and DNA sequence recognition and structure-function of aminoacyl tRNA synthetases." In *DNA Makes RNA Makes Protein* (T. Hunt et al., eds.). Elsevier Biomedical.

Schimmel, P. R., and D. Soll. 1979. "Aminoacyl-tRNA synthetases: general features and recognition of transfer RNAs." *Ann. Rev. Biochem.*, 48, 601.

Shine, J., and L. Dalgarno. 1974, "The 3' terminal sequence of *E. coli* 16S ribosomal RNA: complementarity to nonsense triplets and ribosomal binding sites." *Proc. Nat. Acad. Sci.*, 71, 1342.

Soll, D., J. Abelson, and P. Schimmel (eds.). 1980. *Transfer RNA*. Volumes 1 and 2. Cold Spring Harbor.

Spirin, A. S. 1985. "Ribosomal translocation: factors and models." *Prog. Nucl. Acid Res. Mol. Biol.*, 32, 75.

Steitz, J. A., and K. Jakes. 1975. "How ribosomes select initiation regions in mRNA: base pair formation between the 3' terminus of 16S rRNA and the mRNA during initiation of protein synthesis in *E. coli*." *Proc. Nat. Acad. Sci.*, 72, 4734.

Travers, A. A., and R. P. Burgess. 1969. "Cyclic reuse of RNA polymerase sigma factor." *Nature*, 222, 537.

Weissbach, H., and S. Pestka (eds.). 1977. *Molecular Mechanisms in Protein Synthesis*. Academic Press.

Chapter 7

Beckwith, J., and P. Rossow. 1974. "Analysis of genetic regulatory mechanisms." *Ann. Rev. Genet.*, 8, 1.

Bertrand, K., C. Squires, and C. Yanofsky. 1976. "Transcription termination *in vitro* in the leader region of the tryptophan operon of *E. coli.*" *J. Molec. Biol.*, 103, 319.

Casadaban, M. J., and S. N. Cohen. 1980 "Analysis of gene control signals by DNA fusion and cloning in *E. coli.*" *J. Molec. Biol.*, 138, 179.

Clark, B. F. C., et al. (eds.). 1978. *Gene Expression.* Pergamon.

Crawford, I. P., and G. V. Stauffer. 1980. "Regulation of tryptophan biosynthesis." *Ann. Rev. Biochem.*, 49, 163.

Freifelder, D. 1987. *Molecular Biology.* Chapter 15. Jones and Bartlett.

Gilbert, W., and A. Maxam. 1973. "The nucleotide sequence of the *lac* operator." *Proc. Nat. Acad. Sci.*, 70, 3581.

Gilbert, W., and B. Muller-Hill. 1966. "Isolation of the *lac* repressor." *Proc. Nat. Acad. Sci.*, 58, 2415.

Glass, R. E. 1982. *Gene Function: E. coli and its Heritable Elements.* Univ. of California Press.

Gottesman, S. 1984. "Bacterial regulation: global regulatory networks." *Ann. Rev. Genet.*, 18, 415.

Jacob, F., and J. Monod. 1961. "Genetic regulatory mechanisms in the synthesis of proteins." *J. Molec. Biol.*, 3, 318.

Jobe, A., and S. Bourgeois. 1972. "*Lac* repressor-operator interaction. VI. The natural inducer of the *lac* operon." *J. Molec. Biol.*, 69, 397.

Keller, E. B., and J. M. Calve. 1979. "Alternative secondary structure of leader RNAs and the regulation of the *trp, phe, his, thr,* and *leu* operons." *Proc. Nat. Acad. Sci.*, 76, 6186.

Little, J., et al. 1980. "Cleavage of the *E. coli lexA* protein by the *recA* protease." *Proc. Nat. Acad. Sci.*, 77, 3225.

Majors, J. 1975. "Specific binding of CAP factor to *lac* promoter DNA." *Proc. Nat. Acad. Sci.*, 256, 672.

Maniatis, T., and M. Ptashne. 1976. "A DNA operator." *Scient. Amer.*, January, p. 64.

Miller, J. H. 1980. "Genetic analysis of the *lac* repressor." *Curr. Topics Microbiol. Immunol.*, 90, 1.

Miller, J. H., and W. S. Reznikoff (eds.). 1978. *The Operon.* Cold Spring Harbor.

Nierlich, D., W. Rutter, and C. F. Fox (eds.). 1976. *Molecular Mechanisms in the Control of Gene Expression.* Academic Press.

Oxender, D. L., G. Zurawski, and C. Yanofsky. 1979. "Attenuation in the *E. coli* tryptophan operon: role of RNA secondary structure involving the tryptophan coding region." *Proc. Nat. Acad. Sci.*, 76, 5524.

Pardee, A. B., F. Jacob, and J. Monod. 1959. "The genetic control and cytoplasmic expression of inducibility in the synthesis of β-galactosidase by *E. coli.*" *J. Molec. Biol.*, 1, 165.

Platt, T. 1978. "Regulation of gene expression in the tryptophan operon in *E. coli.*" In *The Operon* (J. H. Miller and W. S. Reznikoff, eds.). Cold Spring Harbor.

Ptashne, M., and W. Gilbert. 1970. "Genetic repression." *Scient. Amer.*, June, p. 36.

Reznikoff, W., R. Winter, and B. C. Hurley. 1974. "The location of the repressor-binding site in the *lac* operon." *Proc. Acad. Sci.*, 71, 2314.

Stark, G. R., and G. M. Wahl. 1984. "Gene amplification." *Ann. Rev. Biochem.*, 53, 447.

Ullman, A., and A. Danchin. 1980. "Role of cyclic AMP in regulatory mechanisms of bacteria." *Trends Biochem. Sci.*, 5, 95.

Watson, M. D. 1983. "Attenuation: translational control of transcription termination." In *DNA Makes RNA Makes Protein* (T. Hunt et al., eds.). Elsevier Biomedical.

Yanofsky, C. 1981. "Attenuation in the control of expression of bacterial operons." *Nature*, 289, 751.

Yanofsky, C., R. L. Kelley, and V. Horn. 1984. "Repression is relieved before attenuation in the *trp* operon of *E. coli* as tryptophan starvation becomes increasingly severe." *J. Bacter.*, 158, 1018.

Chapter 8

Abdel-Monem, M., and H. Hoffman-Berling. 1980. "DNA unwinding enzymes." *Trends Biochem. Sci.*, 5, 128.

Alberts, B. M., and R. Sternglantz. 1977. "Recent excitement in the DNA replication problem." *Nature*, 269, 655.

Cairns, J. 1963. "The chromosome of *E. coli.*" *Cold Spring Harb. Symp. Quant. Biol.*, 28, 43.

Cairns, J. 1966. "The bacterial chromosome." *Scient. Amer.*, January, p. 36.

DeLucia, P., and J. Cairns. 1969. "Isolation of an *E. coli* strain with a mutation affecting DNA polymerase." *Nature*, 224, 1164.

Geider, K., and H. Hoffman-Berling. 1981. "Proteins controlling the helical structure of DNA." *Ann. Rev. Biochem.*, 50, 233.

Gellert, M. 1981. "DNA topoisomerases." *Ann. Rev. Biochem.*, 50, 879.

Gilbert, W., and D. Dressler. 1968. "DNA replication: the rolling circle model." *Cold Spring Harb. Symp. Quant. Biol.*, 33, 473.

Konrad, E. B., and I. R. Lehman. 1974. "A conditional lethal mutant of *E. coli* defective in the $5' \rightarrow 3'$ exonuclease associated with DNA polymerase I." *Proc. Nat. Acad. Sci.*, 71, 2048.

Kornberg, A. 1960. "Biological synthesis of deoxyribonucleic acid." *Science*, 131, 1503.

Kornberg, A. 1980. *DNA Replication;* also, *1982 Supplement.* W. H. Freeman.

Lehman, I. R. 1974. "DNA ligase: structure, mechanism, and function." *Science*, 186, 790.

Lehman, I. R., and D. Uyemura. 1976. "DNA polymerase I: essential replication enzyme." *Science*, 193, 963.

McHenry, C., and A. Kornberg. 1977. "DNA polymerase III holoenzyme of *E. coli*: purification and resolution into subunits." *J. Biol. Chem.*, 252, 6478.

Meselson, M., and F. W. Stahl. 1957. "The replication of DNA in *E. coli.*" *Proc. Nat. Acad. Sci.,* 44, 671.

Morrison, A., and N. R. Cozzarelli. 1981. "Contact between DNA gyrase and its binding site on DNA: features of symmetry and symmetry revealed by protection from nuclease." *Proc. Nat. Acad. Sci.,* 78, 1416.

Nossal, N. 1983. "Prokaryotic DNA replication systems." *Ann. Rev. Biochem.,* 52, 581.

Ogawa, T., and T. Okazaki. 1980. "Discontinuous DNA replication." *Ann. Rev. Biochem.,* 49, 421.

Okazaki, R. T., et al. 1968. "Mechanism of DNA chain growth. I. Possible discontinuity and unusual secondary structure of newly synthesized chains." *Proc. Nat. Acad. Sci.,* 59, 598.

Prescott, D. M., and P. L. Kuempel. 1972. "Bidirectional replication of the chromosome in *E. coli.*" *Proc. Nat. Acad. Sci.,* 69, 2842.

Riley, D., and H. Weintraub. 1979. "Conservative segregation of parental histones during replication in the presence of cycloheximide." *Proc. Nat. Acad. Sci.,* 76, 328.

Rowen, L., and A. Kornberg. 1978. "Primase, the *dnaG* protein of *E. coli*: an enzyme which starts DNA chains." *J. Biol. Chem.,* 253, 758.

Scheuermann, R. H., and H. Echols. 1984. "A separate editing exonuclease for DNA replication: the ε subunit of *E. coli* DNA polymerase III holoenzyme." *Proc. Nat. Acad. Sci.,* 81, 7747.

Schnös, M., and R. Inman. 1970. "Position of branch points in replication of lambda DNA." *J. Molec. Biol.,* 51, 61.

Tomizawa, J. 1978. "Replication of colicin E1 plasmid DNA *in vitro.*" In *DNA Synthesis: Present and Future* (I. Molineux and K. Kohiyama, eds.). Plenum.

Tomizawa, J., and G. Selzer. 1979. "Initiation of DNA synthesis in *E. coli.*" *Ann. Rev. Biochem.,* 48, 999.

Tye, B. K., et al. 1977. "Transient accumulation of Okazaki fragments as a result of uracil incorporation into nascent DNA." *Proc. Nat. Acad. Sci.,* 74, 154.

Valenzuela, M., et al. 1976. "Lack of a unique termination site in lambda DNA replication." *J. Molec. Biol.,* 102, 569.

Wang, J. 1985. "DNA topoisomerases." *Ann. Rev. Biochem.,* 54, 665.

Watson, J. D., and F. H. C. Crick. 1953. "Genetic implications of the structure of desoxyribonucleic acid." *Nature,* 171, 964.

Wickner, S. H. 1978. "DNA replication proteins of *E. coli.*" *Ann. Rev. Biochem.,* 49, 421.

Chapter 9

Boyce, R. P., and P. Howard-Flanders. 1964. "The release of UV-induced thymine dimers from DNA in *E. coli* K12." *Proc. Nat. Acad. Sci.,* 51, 293.

Freifelder, D. 1966 "Lethal changes in bacteriophage DNA produced by x rays." *Radiat. Res. Suppl.,* 6, 80.

Friedberg, E. C. 1985. *DNA Repair.* W. H. Freeman.

Grossman, L., et al. 1975. "Enzymatic repair of DNA." *Ann. Rev. Biochem.*, 44, 19.

Hanawalt, P. C., et al. 1979. "DNA repair in bacteria and mammalian cells." *Ann. Rev. Biochem.*, 48, 783.

Hanawalt, P. C., E. C. Friedberg, and C. F. Fox. 1978. *DNA Repair Mechanisms.* Academic Press.

Hanawalt, P. C., and R. B. Setlow. 1975. *Molecular Mechanisms for Repair of DNA.* Plenum.

Haseltine, W. A. 1983. "Ultraviolet light repair and mutagenesis revisited." *Cell*, 33, 13.

Little, J. W., et al. 1980. "Cleavage of the *E. coli lexA* protein by the *recA* protease." *Proc. Nat. Acad. Sci.*, 77, 3225.

Howard-Flanders, P. 1981. "Inducible repair of DNA. " *Scient. Amer.*, March. p. 72.

Little, J. W., and D. W. Mount. 1982. "The SOS regulatory system of *E. coli*." *Cell*, 29, 11.

Livneh, Z., and I. R. Lehman. 1982. "Recombinational bypass of pyrimidine dimers by the RecA protein of *E. coli*." *Proc. Nat. Acad. Sci.*, 79, 3171.

McPartland, A., L. Green, and H. Echols. 1980. "Control of *recA* gene RNA in *E. coli*: regulatory and signal genes." *Cell*, 20, 731.

Roberts, J. J. 1978. "The repair of DNA modified by cytotoxic, mutagenic, and carcinogenic chemicals." *Adv. Radiat. Biology*, 6, 212.

Setlow, R. B. 1966. "Cyclobutane-type pyrimidine dimers in polynucleotides." *Science*, 153, 379.

Setlow, R. B., and W. L. Carrier. 1964. "The disappearance of thymine dimers from DNA. An error correcting mechanism." *Proc. Nat. Acad. Sci.*, 51, 226.

Sutherland, J. 1978. "Mechanism of action of the photoreactivation enzyme from *E. coli*: recent results." In *DNA Repair Mechanisms* (P. C. Hanawalt et al., eds.). p. 137. Academic Press.

Walker, G. C. 1985. "Inducible DNA repair systems." *Ann. Rev. Biochem.*, 54, 425.

Ward, J. 1975. "Molecular mechanisms of radiation-induced damage to nucleic acids." *Adv. Radiat. Biol.*, 5, 182.

West, S. C., et al. 1982. "Postreplication repair in *E. coli*: strand-exchange reactions of gapped DNA by RecA protein." *Molec. Gen.*, 187, 209.

Willets, N., and A. J. Clark. 1969. "Characteristics of some multiply recombinant-deficient strains." *J. Bacter.*, 10, 231.

Chapter 10

Ames, B. W. 1979. "Identifying environmental chemicals causing mutations and cancer." *Science*, 204, 587.

Ames, B. W., et al. 1973. "Carcinogens are mutagens: a simple test system combining liver homogenates for activation and bacteria for detection." *Proc. Nat. Acad. Sci.*, 70, 2381.

Coulondre, R., et al. 1978. "Molecular basis of base-substitution hotspots in *E. coli*." *Nature*, 274, 775.

Cox, E. C. 1976. "Bacterial mutator genes and the control of spontaneous mutation." *Ann. Rev. Genet.*, 10, 135.

Davis, B. D. 1948. "Isolation of biochemically deficient mutants of bacteria by penicillin." *J. Amer. Chem. Soc.*, 70, 4267.

Drake, J. W. 1969. "Comparative rates of spontaneous mutation." *Nature*, 221, 1132.

Drake, J. W. 1970. *The Molecular Basis of Mutation.* Holden-Day.

Drake, J. W., and R. H. Baltz. 1976. "The biochemistry of mutagenesis." *Ann. Rev. Biochem.*, 45, 11.

Freese, E. 1959. "The specific mutagenic effect of base analogues on phage T4." *J. Molec. Biol.*, 1, 87.

Gorini, L., and H. Kaufman. 1960. "Selecting bacterial mutants by the penicillin method." *Science*, 131, 604.

Heidelberger, C. 1975. "Chemical carcinogens." *Ann. Rev. Biochem.*, 44, 79.

Hong, J. S., and B. N. Ames. 1971. "Localized mutagenesis of any small region of the bacterial chromosome." *Proc. Nat. Acad. Sci.*, 68, 3158.

Kohno, T., and J. R. Roth. 1974. "Proflavin mutagenesis of bacteria." *J. Molec. Biol.*, 89, 17.

Kunz, B. A. 1982. "Genetic effects of deoxyribonucleotide pool imbalances." *Environ. Mutag.*, 4, 695.

Lederberg, J., and E. Lederberg. 1952. "Replica plating and indirect selection of bacterial mutants." *J. Bacter.*, 63, 399.

Luria, S. E., and M. Delbrück. 1943. "Mutations of bacteria from virus sensitivity to virus resistance." *Genetics*, 28, 491.

Novick, A., and L. Szilard. 1951. "Experiments with the chemostat on spontaneous mutations in bacteria." *Proc. Nat. Acad. Sci.*, 36, 708.

Orgel, L. E. 1965. "The chemical basis of mutation." *Adv. Enzymol.*, 27, 289.

Rossi, J. J., and C. M. Berg. 1971. "Differential recovery of auxotrophs after penicillin enrichment in *E. coli*." *J. Bacter.*, 106, 297.

Roth, J. R. 1974. "Frameshift mutations." *Ann. Rev. Genet.*, 8, 319.

Schuster, H. "The reaction of nitrous acid with deoxyribonucleic acid." *Biochem. Biophys. Res. Commun.*, 2, 320.

Singer, B., and J. T. Kusmierek. 1982. "Chemical mutagenesis." *Ann. Rev. Biochem.*, 51, 655.

Smith, M. 1982. "Site-directed mutagenesis." *Trends Biochem. Sci.*, 7, 440.

Streisinger, G., et al. 1966. "Frameshift mutations and the genetic code." *Cold Spring Harb. Symp. Quant. Biol.*, 31, 77.

Sugimura, T., S. Kondo, and H. Takebe (eds.). 1982. *Environmental Mutagens and Carcinogens.*" Alan Liss.

Chapter 11

Achtman, M. 1973. "Genetics of the F sex factor in Enterobacteriaceae." *Curr. Top. Microbiol. and Immunol.*, 60, 79.

Achtman, M., N. Willets, and A. J. Clark. 1972. "Conjugal complementation analysis of transfer-deficient mutants of F'*lac* in *E. coli.*" *J. Bacter.*, 110, 831.

Bradley, D. E. 1980. "Determination of pili by conjugative bacterial drug resistance plasmids of incompatibility groups B, C, H, J, K, M, V, and X." *J. Bacter.*, 141, 828.

Broda, P. 1979. *Plasmids*. W. H. Freeman.

Bukhari, A. I., J. A. Shapiro, and S. L. Adhya. 1978. *DNA Insertion Elements and Plasmids*. Cold Spring Harbor.

Chakrabarty, A. M. 1976. "Plasmids in *Pseudomonas*." *Ann. Rev. Genet.*, 10, 7.

Clark, A. J., and G. J. Warren. 1979. "Conjugal transmission of plasmids." *Ann. Rev. Genetics*, 13, 99.

Clowes, R. C. 1972. "Molecular structure of bacterial plasmids." *Bact. Rev.*, 36, 361.

Clowes, R. C. 1973. "The molecule of infectious drug resistance." *Scient. Amer.*, April, p. 18.

Everett, R., and N. Willets. 1980. "Characterization of an *in vivo* system for nicking at the origin of conjugal DNA transfer of the sex factor F." *J. Molec. Biol.*, 136, 129.

Freifelder, D. 1967. "Role of the female in bacterial conjugation in *E. coli.*" *J. Bacter.*, 94, 370.

Freifelder, D. 1968. "Studies on *E. coli* sex factors. III. Covalently closed F'*lac* DNA molecules." *J. Molec. Biol.*, 34, 31.

Glass, R. E. 1982. *Gene Function: E. coli and its Heritable Elements*. University of California Press.

Gurney, D. G., and D. Helinski. 1975. "Relaxation complexes of plasmid DNA and protein. III. Association of protein with the 5' terminus of the broken DNA strand in the relaxed complex of plasmid ColE1." *J. Biol. Chem.*, 250, 8796.

Hardy, K. G. 1975. "Colicinogeny and related phenomena." *Bact. Rev.*, 39, 464.

Hardy, K. 1981. *Bacterial Plasmids*. Nelson.

Holloway, B. W., and V. Krishnapillai. 1975. "Bacteriophages and bacteriocins." In *Genetics and Biochemistry of Pseudomonas* (Clarke, P. H., and M. H. Richmond, eds.). Wiley.

Holloway, B. W. 1979. "Plasmids that mobilize bacterial chromosomes." *Plasmid*, 2, 1.

Hooykaas, P. J. J., and R. A. Schilperoort. 1985. "The Ti plasmid of *Agrobacterium tumefaciens*: a natural genetic engineer." *Trends Biochem. Sci.*, 10, 307.

Kingsman, A., and N. Willets. 1978. "The requirement for conjugal DNA synthesis in the donor strand during F'lac transfer." *J. Molec. Biol.*, 122, 287.

Lewin, B. 1977. *Gene Expression. 3. Plasmids and Phage*. Wiley-Interscience.

Low, K. B. 1972. "*E. coli* K-12 F' factors, old and new." *Bacteriol. Rev.*, 36, 587.

Meynell, E. 1972. *Bacterial Plasmids*. Macmillan.

Mitsuhashi, S. (ed.). 1980. *R Factor: Drug Resistance Plasmid*. University Park Press.

Novick, R. P. 1980. "Plasmids." *Scient. Amer.*, December, p. 102.

Novick, R. P., et al. 1976. "Uniform nomenclature for bacterial plasmids: a proposal." *Bact. Rev.*, 40, 168.

Rownd, R. H., et al. 1978. "Dissociation, amplification, and reassociation of composite R-plasmid DNA." In *Microbiology 1978* (D. Schlessinger, ed.). p. 33. American Society for Microbiology.

Scaife, J., D. Leach, and A. Galizzi. 1985. *Genetics of Bacteria*. Academic Press.

Schlessinger, D. (ed.). 1974. *Microbiology 1974*. American Society for Microbiology. Entire volume is devoted to plasmids.

Smith, H. W. 1978. "Antibiotic resistance in bacteria and associated problems in farm animals before and after the 1969 Swann report." In *Antibiotics and Antibiosis in Agriculture* (M. Woodbine, ed.). p. 345. Butterworth.

Smith, M. D., et al. 1980. "Transfer of plasmids by conjugation in *Streptococcus pneumoniae*." *Plasmid*, 3, 70.

Summers, O, and S. Silver. 1978. "Microbial transformation of metals." *Ann. Rev. Microbiol.*, 32, 637.

Willets, N. 1972. "The genetics of transmissible plasmids." *Ann. Rev. Genetics*, 6, 257.

Willets, N. 1978. "Control of conjugation in F-like plasmids." In *Microbiology 1978* (D. Schlessinger, ed.). p. 211. American Society for Microbiology.

Willets, N., and R. Skurray. 1980. "The conjugation system of F-like plasmids." *Ann. Rev. Genetics*, 14, 41.

Chapter 12

Bukhari, A. I. 1981. "Models of DNA transposition." *Trends Biochem. Sci.*, 6, 56.

Bukhari, A. I., J. A. Shapiro, and S. L. Adhya. 1978. *DNA Insertion Elements and Plasmids*. Cold Spring Harbor.

Calos, M. P., and J. H. Miller. 1980. "Transposable elements." *Cell*, 20, 579.

Campbell, A. 1981. "Evolutionary significance of accessory DNA elements in bacteria." *Ann. Rev. Microbiol.*, 35, 55.

Campbell, A., et al. 1979. "Viruses and inserting elements in chromosomal evolution." In *Concepts of the Structure and Function of DNA, Chromatin, and Chromosomes* (A. S. Dion, ed.) Symposia Specialists.

Chao, L., et al. "Transposable elements as mutator genes in evolution." *Nature*, 303, 633.

Chow, L. T., and A. I. Bukhari. 1977. "Bacteriophage Mu genome: structural studies in Mu DNA and Mu mutants carrying insertions." In *DNA Insertion Elements, Plasmids, and Episomes* (Bukhari, A. I., et al., eds.). p. 295. Cold Spring Harbor.

Cohen, S. N., and J. A. Shapiro. 1980. "Transposable genetic elements." *Scient. Amer.*, February, p. 40.

Grindley, N. D. F., and R. R. Reed. 1985. "Transpositional recombination in prokaryotes." *Ann. Rev. Biochem.*, 54, 863.

Kleckner, N. 1977. "Translocatable elements in procaryotes." *Cell*, 11, 11.

Kleckner, N. 1981. "Transposable genetic elements." *Ann. Rev. Genetics*, 15, 341.

Nevers, P., and H. Saedler, 1977. "Transposable genetic elements as agents of gene instability and chromosomal rearrangements." *Nature*, 268, 109.

Rogers, J. 1985. "Origins of repeated DNA." *Nature*, 317, 765.

Shapiro, J. A. 1979. "Molecular model for the transposition and replication of bacteriophage Mu and other transposable elements." *Proc. Nat. Acad. Sci.*, 76, 1933.

Shapiro, J. A. 1983. *Mobile Genetic Elements*. Academic Press.

Starlinger, P. 1977. "DNA rearrangements in procaryotes." *Ann. Rev. Genetics,* 11, 103.

Starlinger, P., and H. Saedler. 1976. "IS elements in microorganisms." *Curr. Topics Microbiol. Immunol.*, 75, 111.

Chapter 13

Archer, L. J. 1973. *Bacterial Transformation*. Academic Press.

Avery, O. T., C. M. MacLeod, and M. McCarty. 1944. "Studies on the chemical nature of the substance inducing transformation of pneumococcal types." *J. Exp. Med.*, 79, 137.

Cosloy, S. D., and M. Oishi. 1973. "Genetic transformation in *E. coli* K-12." *Proc. Nat. Acad. Sci.*, 71, 84.

Dubnau, D. 1976. "Genetic transformation of *Bacillus subtilis*: review with emphasis on the recombination mechanism." In *Microbiology 1976* (D. Schlessinger, ed.). American Society for Microbiology.

Henner, D. J., and J. A. Hoch. 1980. "The *B. subtilis* chromosome." *Microbiol. Reviews*, 44, 57.

Hotchkiss, R. D., and J. Marmur. 1954. "Double marker transformations as evidence of linked factors in desoxyribonucleate transforming agents." *Proc. Nat. Acad. Sci.*, 40, 55.

Lacks, S. 1962. "Molecular fate of DNA in genetic transformation in *Pneumococcus*." *J. Mol Biol.*, 5, 119.

Mandel, M., and A. Higa. 1970. "Calcium-dependent bacteriophage DNA infection." *J. Molec. Biol.*, 53, 159.

McCarty, M. 1985. *The Transforming Principle: Discovering that Genes Are Made of DNA*. Norton.

Notani, N. K., and J. K. Setlow. 1974. "Mechanism of bacterial transformation and transfection." *Prog. Nucleic Acid Res. Mol. Biol.*, 14, 39.

Portoles, A., R. Lopez, and M. Espinosa (eds.). 1977. *Modern Trends in Bacterial Transformation*. North Holland.

Smith, H. O., et al. 1981. "Genetic transformation." *Ann. Rev. Biochem.*, 50, 41.

Chapter 14

Bachman, B. J., and K. B. Low. 1980. "Linkage map of *E. coli* K-12, edition 6." *Microbiol. Reviews*, 44, 1.

Bianchi, M. E., and C. M. Radding. 1983. "Insertions, deletions, and mismatches in heteroduplex DNA made by RecA protein." *Cell*, 35, 511.

Birge, E. A. 1981. *Bacterial and Bacteriophage Genetics*. Springer-Verlag.

Clark, A. J., and A. D. Margulies. "Isolation and characterization of recombination-deficient mutants of *E. coli* K-12." *Proc. Nat. Acad. Sci.*, 53, 451.

Cullum, J., and P. Broda. 1979. "Chromosome transfer and Hfr formation by F in *rec*⁺ and *recA* strains of *E. coli* K-12." *Plasmid*, 2, 358.

Cunningham, R. P., et al. 1979. "Single strands induce RecA protein to unwind duplex DNA for homologous pairing." *Nature*, 281, 191.

Cunningham, R. P., et al. 1980. "Homologous pairing in genetic recombination: RecA protein makes joint molecules of gapped circular DNA and closed circular DNA." *Cell*, 20, 223.

Deonier, R. C., and N. G. Davidson. 1976. "The sequence organization of the integrated F plasmid in two different Hfr strains of *E. coli*." *J. Molec. Biol.*, 107, 207.

Eisenstark, A. 1977. "Genetic recombination in bacteria." *Ann. Rev. Genetics*, 11, 369.

Freifelder, D. 1966. "Replication of DNA during F'*lac* transfer." *Biochem. Biophys. Res. Commun.*, 23, 576.

Glass, R. E. 1982. *Gene Function: E. coli and its Heritable Elements*. University of California Press.

Goldmark, P. J., and S. Linn. 1972. "Purification and properties of the recBC DNase of *E. coli* K-12." *J. Biol. Chem.*, 247, 1849.

Hayes, W. 1968. *The Genetics of Bacteria and Their Viruses*. Blackwell.

Holloway, B. W., V. Krishnapillai, and A. F. Morgan. 1979. "Chromosomal genetics of *Pseudomonas*." *Microbiol. Reviews*, 43, 73.

Jacob, F., and E. L. Wollman. 1961. *Sexuality and the Genetics of Bacteria*. Academic Press.

Low, K. B., and R. D. Porter. 1978. "Modes of gene transfer and recombination in bacteria." *Ann. Rev. Genet.*, 12, 249.

McEntee, K., et al. 1979. "Initiation of genetic recombination catalyzed *in vitro* by the RecA protein of *E. coli*." *Proc. Nat. Acad. Sci.*, 76, 2615.

Norkin, L. C. 1970. "Marker specific effects in recombination." *J. Molec. Biol.*, 57, 633.

Potter, H., and D. Dressler. 1976. "On the mechanism of genetic recombination: electron-microscopic observation of recombination intermediates." *Proc. Nat. Acad. Sci.*, 73, 3000.

Radding, C. M. 1978. "Genetic recombination: strand transfer and mismatch repair." *Ann. Rev. Biochem.*, 47, 847.

Radding, C. M. 1982. "Homologous pairing and strand exchange in genetic recombination." *Ann. Rev. Genet.*, 16, 405.

Radding, C. M., et al. 1983. "Three phases in homologous pairing: polymerization of *recA* protein on single-stranded DNA, synapsis, and polar strand exchange." *Cold Spring Harb. Symp. Quant. Biol.*, 47, 821.

Riley, M., and A. Anilionis. 1978. "Evolution of the bacterial genome." *Ann. Rev. Microbiol.*, 32, 519.

Sanderson, K. E., and P. E. Hartman. 1978. Linkage map of *Salmonella typhimurium*, edition V." *Microbiol. Rev.*, 42, 471.

Sarathy, P. V., and O. Siddiqi. 1973. "DNA synthesis during bacterial mating. II. Is DNA replication in the Hfr obligatory for chromosome transfer?" *J. Molec. Biol.*, 78, 443.

Scaife, J., D. Leach, and A. Galizzi. 1985. *Genetics of Bacteria*. Academic Press.

Sigal, N., and B. Alberts. 1972. "Genetic recombination: the nature of a crossed-strand exchange between two homologous DNA molecules." *J. Molec. Biol.*, 71, 789.

Stahl, F. W. 1979. "Specific sites in generalized recombination." *Ann. Rev. Genetics*, 13, 7.

Valenzuela, M., and R. B. Inman. 1975. "Visualization of a novel junction in bacteriophage lambda DNA." *Proc. Nat. Acad. Sci.*, 72, 3024.

West, S. C., et al. 1981. "Homologous pairing can occur before DNA strand separation in generalized genetic recombination." *Nature*, 290, 29.

Wood, T. H. 1968. "Effects of temperature, agitation, and donor strain on chromosome transfer in *E. coli* K-12." *J. Bacter.*, 96, 2077.

Wu, A. M., et al. 1983. "Unwinding associated with synapsis of DNA molecules by *recA* protein." *Proc. Nat. Acad. Sci.*, 80, 1256.

Chapter 15

Benzer, S. 1961. "On the topography of the genetic fine structure of T4." *Proc. Nat. Acad. Sci.*, 47, 403.

Casjens, S. 1985. *Virus Assembly*. Jones and Bartlett.

Caspar, D. L. D., and A. Klug. 1962. "Physical principles in the construction of regular viruses." *Cold Spring Harb. Symp. Quant. Biol.*, 27, 1.

Davison, P. F. et al., 1961. "The structural unity of the DNA of T2 bacteriophage." *Proc. Nat. Acad. Sci.*, 47, 1123.

Doermann, A. H. 1952. "The intracellular growth of bacteriophages. I. Liberation of intracellular bacteriophage by premature lysis with another phage." *J. Gen. Physiol.*, 35, 645.

Doermann, A. H. 1983. "Introduction to the early years of bacteriophage T4." In *Bacteriophage T4* (C. K. Matthews et al., eds.). p. 1. American Society for Microbiology.

Earnshaw, W. C., and S. R. Casjens. 1980. "DNA packaging by the double-stranded DNA bacteriophages." *Cell*, 21, 319.

Eiserling, F. A. 1979. "Bacteriophage structure." *Comprehensive Virol.*, 13, 69.

Ellis, E. L., and M. Delbrück. 1939. "The growth of bacteriophage." *J. Gen. Physiol.*, 22, 365.

Hershey, A. D. 1946. "Spontaneous mutations in bacterial viruses." *Cold Spring Harb. Symp. Quant. Biol.*, 11, 67.

Hershey, A. D., and R. Rotman. 1949. "Genetic recombination between host-range and plaque-type mutants of bacteriophage in single bacterial cells." *Genetics*, 34, 44.

Huberman, J. A. 1968. "Visualization of replicating mammalian and T4 bacteriophage DNA." *Cold Spring Harb. Symp. Quant. Biol.*, 33, 509.

Kikuchi, Y., and J. King. 1975. "Assembly of the tail of bacteriophage T4." *J. Supramol. Struct.*, 3, 24.

Levinthal, C. 1954. "Recombination in phage T2; its relation to heterozygosis and growth." *Genetics*, 39, 169.

Lewin, B. 1979. *Gene Expression. III. Plasmids and Phages.* Wiley-Interscience.

MacHattie, L. et al., 1967. "Terminal repetition in permuted bacteriophage DNA molecules." *J. Molec. Biol.*, 23, 355.

Matthews, C. K. 1977. "Reproduction of large virulent bacteriophages." *Comprehensive Virol.*, 7, 179.

Matthews, C. K., et al. 1983. *Bacteriophage T4.* American Society for Microbiology.

Meselson, M., and J. Weigle. 1961. "Chromosome breakage accompanying genetic recombination in bacteriophage." *Proc. Nat. Acad. Sci.*, 47, 857.

Rabussay, D., and E. P. Geiduschek. 1977. "Regulation of gene action in the development of lytic bacteriophage." *Comprehensive Virol.*, 8, 1.

Richards, K. E., R. C. Williams, and R. Calendar. 1973. "Mode of DNA packaging within bacteriophage heads." *J. Molec. Biol.*, 78, 255.

Sechaud, J., et al. 1965. "Chromosome structure in phage T4. II. Terminal redundancy and heterozygosis." *Proc. Nat. Acad. Sci.*, 54, 1333.

Streisinger, G., R. S. Edgar, and D. Denhardt. 1964. "The chromosome structure in phage T4. I. The circularity of the linkage map." *Proc. Nat. Acad. Sci.*, 51, 775.

Streisinger, G., J. Emrich, and M. M. Stahl. 1967. "Chromosome structure in phage T4. III. Terminal redundancy and length determination." *Proc. Nat. Acad. Sci.*, 57, 292.

Thomas, C. A. 1964. "Circular T2 DNA molecules." *Proc. Nat. Acad. Sci.*, 52, 1297.

Williams, R. C., and H. W. Fisher. 1974. *An Electron Micrographic Atlas of Viruses.* C. C. Thomas.

Wood, W. B., and J. King. 1979. "Genetic control of complex bacteriophage assembly." *Comprehensive Virol.*, 13, 581.

Zinder, N. (ed.). 1975. *RNA Phages.* Cold Spring Harbor.

Chapter 16

Adhya, S., et al. 1981. "Regulatory circuits of bacteriophage lambda." *Prog. Nucl. Acid Res. Mol. Biol.*, 26, 103.

Arber, W. 1983. "A beginner's guide to lambda biology." In *Lambda II* (R. W. Hendrix et al., eds.). p. 381. Cold Spring Harbor.

Campbell, A. 1983. "Bacteriophage λ." In *Mobile Genetic Elements* (J. A. Shapiro et al, eds.). p. 65. Academic Press.

Dove, W. F. 1968. "The genetics of lambdoid phages." *Ann. Rev. Genet.*, 2, 305.

Echols, H. 1980. "Bacteriophage development." In *Molecular Genetics of Development* (T. Leighton and W. F. Loomis, eds.). Academic Press.

Echols, H., and H. Murialdo. 1978. "Genetic map of bacteriophage lambda." *Microbiol. Reviews*, 42, 577.

Epp, C., and M. L. Pearson. 1976. "Association of bacteriophage λ *N* gene protein with *E. coli* RNA polymerase." In *RNA Polymerase* (R. Losick and M. Chamberlin, eds.). Cold Spring Harbor.

Freifelder, D., L. Chud, and E. Levine. 1974. "Requirement for maturation of *E. coli* phage λ." *J. Molec. Biol.*, 83, 563.

Freifelder, D., and I. Kirschner. 1971. "A phage lambda endonuclease controlled by genes *O* and *P*." *Virol.*, 44, 233.

Friedman, D., and M. Gottesman. 1983. "Lytic mode of lambda development." In *Lambda II* (R. W. Hendrix et al., eds.). p. 21. Cold Spring Harbor.

Furth, M., and S. Wickner. 1983. "Lambda DNA replication." In *Lambda II* (R. W. Hendrix et al., eds.). p. 145. Cold Spring Harbor.

Hendrix, R. W., et al. (eds.). 1983. *Lambda II*. Cold Spring Harbor.

Hershey, A. D. (ed.). 1971. *The Bacteriophage Lambda*. Cold Spring Harbor.

Hershey, A. D., and E. Burgi. 1965. "Complementary structure of interacting sites at the ends of lambda DNA molecules." *Proc. Nat. Acad. Sci.*, 53, 325.

Hohn, T., and I. Katsura. 1977. "Structure and assembly of bacteriophage λ." *Curr. Topics Microbiol. Immunol.*, 78, 69.

Hu, S.-L., and W. Szybalski. 1979. "Control of rightward transcription in coliphage lambda by the regulatory functions of phage genes *N* and *cro*." *Virol.*, 98, 424

Kaiser, A. D., M. Syvanen, and T. Masuda. 1975. "DNA packaging steps in bacteriophage lambda head assembly." *J. Molec. Biol.*, 91, 175.

Kellenberger, G., M. L. Zichichi, and J. J. Weigle. 1961. "A mutation affecting the DNA content of bacteriophage lambda and its lysogenizing properties." *J. Molec. Biol.*, 3, 399.

Lewin, B. 1977. *Gene Expression. III. Plasmids and Phages*. Wiley-Interscience.

Luzzati, D. 1970. "Regulation of lambda exonuclease synthesis: role of the *N* gene product and λ repressor." *J. Molec. Biol.*, 49, 515.

Pero, J. 1971. "Deletion mapping of the site of action of the *tof* gene product." In *The Bacteriophage Lambda* (A. D. Hershey, ed.). p. 599. Cold Spring Harbor.

Ross, D., and D. Freifelder. 1976. "Maturation of a single λ phage particle from a circular dimer of DNA." *Virol.*, 74, 414.

Signer, E., et al. 1968. "The general recombination system of bacteriophage lambda." *Cold Spring Harb. Symp. Quant. Biol.*, 33, 711.

Szybalski, W. 1977. "Initiation and regulation of transcription and DNA replication in coliphage lambda." In *Regulatory Biology* (J. C. Copeland and G. A. Marzluff, eds.). Ohio State University Press.

Zissler, J., et al. 1971. "The role of recombination in growth of bacteriophage λ." In *The Bacteriophage Lambda* (A. D. Hershey, ed.). pp. 455, 469. Cold Spring Harbor.

Chapter 17

Campbell, A. 1976. "How viruses insert their DNA into the DNA of a host cell." *Scient. Amer.*, December, p. 102.

Craig, N., and H. A. Nash. 1983. "The mechanism of phage λ site-specific recombination: site-specific breakage of DNA by Int topoisomerase." *Cell*, 35, 795.

Devoret, R. 1979. "Bacterial test for potential carcinogens." *Scient. Amer.*, August, p. 40.

Echols, H., and L. Green. 1979. "Some properties of site-specific and general recombination inferred from *int*-initiated exchanges by bacteriophage λ." *Genetics*, 93, 297.

Echols, H., and G. Guarneros. 1983. "Control of integration and excision. In *Lambda II* (R. W. Hendrix et al., eds.). p. 75. Cold Spring Harbor.

Franklin, N. C., W. F. Dove, and C. Yanofsky. 1965. "A linear insertion of a prophage into the chromosome of *E. coli* shown by deletion mapping." *Biochem. Biophys. Res. Commun.*, 18, 910.

Folkmanis, A., and D. Freifelder. 1972. "Studies on lysogeny with *E. coli* phage λ. I. Physical observation of the insertion process." *J. Molec. Biol.*, 65, 63.

Freifelder, D., and I. Kirschner. 1971. "The formation of homoimmune double lysogens of phage λ and the segregation of single lysogens from them." *Virol.*, 44, 633.

Freifelder, D., and M. Meselson. 1970. "Topological relationship of prophage λ to the bacterial chromosome in lysogenic cells." *Proc. Nat. Acad. Sci.*, 65, 200.

Gottesman, M., and R. Weisberg. 1971. "Prophage insertion and excision." In *The Bacteriophage Lambda* (A. D. Hershey, ed.). p. 113. Cold Spring Harbor.

Hendrix, R. W., et al. 1983. *Lambda II*. Cold Spring Harbor.

Hershey, A. D. (ed.). 1971. *The Bacteriophage Lambda*. Cold Spring Harbor.

Herskowitz, I., and D. Hagen. 1980. "The lysis-lysogeny decision of phage λ: explicit programming and responsiveness." *Ann. Rev. Genet.*, 14, 399.

Hochschild, A., N. Irwin, and M. Ptashne. 1983. "Repressor structure and the mechanism of positive control." *Cell*, 32, 319.

Hopkins, N., and M. Ptashne. 1971. In *The Bacteriophage Lambda* (A. D. Hershey, ed.). p. 571. Cold Spring Harbor.

Hsu, P.-L., W. Ross, and A. Landy. 1980. "The lambda phage *att* site: functional limits and interaction with *int* protein." *Nature*, 285, 85.

Jacob, F., and E. L. Wollman. 1953. "Induction of phage development in lysogenic bacteria." *Cold Spring Harb. Symp. Quant. Biol.*, 18, 101.

Johnson, A. D., B. J. Meyer, and M. Ptashne. 1979. "Interactions between DNA-bound repressors govern regulation by the λ phage repressor." *Proc. Nat. Acad. Sci.*, 76, 5061.

Kaiser, A. D. 1957. "Mutations in a temperate bacteriophage affecting its ability to lysogenize *E. coli*." *Virol.*, 3, 42.

Landy, A., and W. Ross. 1977. "Viral integration and excision: structure of the lambda *att* site." *Science*, 197, 1147.

Lederberg, E. 1951. "Lysogenicity in *E. coli* K-12." *Genetics*, 36, 560.

Lieb, M. 1953. "Establishment of lysogenicity in *E. coli*." *J. Bacter.*, 65, 642.

Miller, H. I., et al. 1979. "Site-specific recombination of bacteriophage λ: the role of host gene products." *Cold Spring Harb. Symp. Quant. Biol.*, 43, 1121.

Miller, H. I., et al. 1980. "Regulation of the integration-excision reaction by bacteriophage λ." *Cold Spring Harb. Symp. Quant. Biol.*, 45, 439.

Miller, H., and D. I. Friedman. 1977. "Isolation of *E. coli* mutants unable to support lambda integrative recombination." In *DNA Insertion Elements, Plasmids, and Episomes* (A. I. Bukhari et al., eds.). p. 349. Cold Spring Harbor.

Miller, H. I., and H. A. Nash. 1981. "Direct role of the *himA* gene product in phage λ." *Nature*, 290, 523.

Mizuuchi, M., and K. Mizuuchi. 1980. "Integrative recombination of bacteriophage λ: extent of the DNA sequence involved in the attachment site function." *Proc. Nat. Acad. Sci.*, 77, 3220.

Nash, H. A. 1978. "Integration and excision of bacteriophage λ." *Curr. Topics Microbiol. Immun.*, 78, 171.

Nash, H. A., et al. 1980. "Strand exchange in λ integrative recombination: genetics, biochemistry, and models." *Cold Spring Harb. Symp. Quant. Biol.*, 45, 417.

Oppenheim, A., and A. B. Oppenheim. 1978. "Regulation of the *int* gene of bacteriophage λ: activation by the *cII* and *cIII* gene products and the roles of the *pI* and *pL* promoters." *Molec. Gen. Genet.*, 165, 39.

Ptashne, M. 1967. "Isolation of the λ repressor." *Proc. Nat. Acad. Sci.*, 57, 306.

Ptashne, M., et al. 1980. "How the λ repressor and Cro work." *Cell*, 19, 1.

Reichard, L., and A. D. Kaiser. 1971. "Control of lambda repressor synthesis." *Proc. Nat. Acad. Sci.*, 68, 2185.

Roberts, J. W., C. W. Roberts, and D. W. Mount. 1977. "Inactivation and proteolytic cleavage of phage lambda repressor *in vitro* in an ATP-dependent reaction." *Proc. Nat. Acad. Sci.*, 74, 2283.

Roberts, J., and R. Devoret. 1983. Lysogenic induction." In *Lambda II* (R. W. Hendrix et al., eds.). p. 123. Cold Spring Harbor.

Ross, W., and A. Landy. 1983. "Patterns of λ Int recognition in the regions of strand exchange." *Cell*, 33, 261.

Rothman, J. L. 1965. "Transductional studies on the relation between prophage and host chromosome." *J. Molec. Biol.*, 12, 892.

Schulman, M., and M. Gottesman. 1971. "Lambda *att*^{2}: a transducing phage capable of intramolecular *int-xis* promoted recombination." In *The Bacteriophage Lambda* (A. D. Hershey, ed.). p. 477. Cold Spring Harbor.

Signer, E. R., and J. Weil. 1968. "Site-specific recombination in bacteriophage λ." *Cold Spring Harb. Symp. Quant. Biol.*, 33, 715.

Weisberg, R. A., et al. 1977. "Bacteriophage λ: the lysogenic pathway." *Comprehensive Virol.*, 8, 197.

Weisberg, R. A., and M. Gottesman. 1971. "The stability of Int and Xis functions." In *The Bacteriophage Lambda* (A. D. Hershey, ed.). p. 489. Cold Spring Harbor.

Weisberg, R. A., and A. Landy. 1983. "Site-specific recombination in phage lambda." In *Lambda II* (R. W. Hendrix et al., eds.). p. 211. Cold Spring Harbor.

Chapter 18

Campbell, A. 1977. "Defective bacteriophages and incomplete prophages." *Comprehensive Virol.*, 8, 259.

Ebel-Tsipis, J., D. Botstein, and M. S. Fox. 1972. "Generalized transduction by phage P22 in *S. typhimurium*. I. Molecular origin of transducing DNA." *J. Molec. Biol.*, 71, 433.

Ebel-Tsipis, J., M. S. Fox, and D. Botstein. 1972. "Generalized transduction by bacteriophage P22 in *S. typhimurium*. II. Mechanism of integration of transducing DNA." *J. Molec. Biol.*, 71, 449.

Echols, H., and D. Court. 1971. "The role of helper phage in *gal* transduction." In *The Bacteriophage Lambda* (A. D. Hershey, ed.). p. 701. Cold Spring Harbor.

Gottesman, S., and J. R. Beckwith. 1969. "Directed transposition of the arabinose operon: a technique for the isolation of specialized transducing bacteriophages for any *E. coli* gene." *J. Molec. Biol.*, 44, 117.

Morse, M. L., E. M. Lederberg, and J. Lederberg. 1956. "Transduction in *Escherichia coli* K12." *Genetics*, 41, 142.

Schmieger, H. "Phage P22 with increased or decreased transduction abilities." *Mol. Gen. Genet.*, 119, 75.

Shimada, K., R. A. Weisberg, and M. E. Gottesman. 1972. "Prophage λ at unusual chromosomal locations." *J. Molec. Biol.*, 63, 483.

Wall, J. D., and P. D. Harriman. 1974. "Phage P1 mutants with altered transducing abilities for *E. coli*." *Virol.*, 59, 532.

Zinder, N. D. 1953. "Infective heredity in bacteria." *Cold Spring Harb. Symp. Quant. Biol.*, 18, 261.

Zinder, N. D., and J. Lederberg. 1952. "Genetic exchange in *Salmonella*." *J. Bacter.*, 64, 679.

Chapter 19

Bonhoeffer, F., and H. Schaller. 1965. "A method for selective enrichment of mutants based on the high UV sensitivity of DNA containing 5-bromouracil." *Biochem. Biophys. Res. Commun.*, 20, 93.

Casadaban, M. J., et al. 1977. "Construction and use of gene fusions directed by bacteriophage Mu insertions." In *DNA Insertion Elements, Plasmids, and Episomes* (A. I. Bukhari et al., eds.). p. 531. Cold Spring Harbor.

Davis, R. W., D. Botstein, and J. R. Roth. 1980. *A Manual for Genetic Engineering. Advanced Bacterial Genetics*. Cold Spring Harbor.

Faelen, M., et al. 1977. "In vivo genetic engineering: the Mu-mediated transposition of chromosomal DNA segments onto transmissible plasmids." In *DNA Insertion Elements, Plasmids, and Episomes* (A. I. Bukhari et al, eds.). p. 521. Cold Spring Harbor.

Franklin, N. 1978. "Genetic fusions for operon analysis." *Ann. Rev. Genet.*, 12, 193.

Gottesman, S., and J. R. Beckwith. 1969. "Directed transposition of the arabinose operon: a technique for the isolation of specialized transducing bacteriophages for any *E. coli* gene." *J. Molec. Biol.*, 44, 117.

Miller, J. H. 1972. *Experiments in Molecular Genetics*. Cold Spring Harbor.

Zissler, J., et al. 1971. "The role of recombination in the growth of bacteriophage λ. II. Inhibition of growth by prophage P2." In *The Bacteriophage Lambda* (A. D. Hershey, ed.). p. 469. Cold Spring Harbor.

Chapter 20

Abelson, J., and E. Butz. 1980. "Recombinant DNA." *Science*, 209, 1317.

Beers, R. F., and E. G. Bassett (eds.). 1977. *Recombinant Molecules: Impact on Science and Society*. Raven.

Berg, P. 1981. "Dissections and reconstructions of genes and chromosomes." *Science*, 213, 296.

Brown, D. D. 1973. "The isolation of genes." *Scient. Amer.*, August, p. 20.

Chilton, M.-D. 1983. "A vector for introducing new genes into plants." *Scient. Amer.*, June. p. 50.

Cohen, S. N., et al. 1973. "Construction of biologically functional bacterial plasmids *in vitro*." *Proc. Nat. Acad. Sci.*, 70, 3240.

Curtiss, R. 1976. "Genetic manipulation of microorganisms: potential benefits and hazards." *Ann. Rev. Microbiol.*, 30, 507.

Davis, R. W., D. Botstein, and J. R. Roth. 1980. *A Manual for Genetic Engineering. Advanced Bacterial Genetics*. Cold Spring Harbor.

Dillon, J. R., A. Nasim, and E. R. Nestmann. 1985. *Recombinant DNA Methodology*. Wiley.

Drlica, Karl. 1984. *Understanding Gene Cloning*. Wiley.

Garland, P. B., and R. Williamson (eds.). 1979. *Biochemistry of Genetic Engineering*. Biochemical Society, London.

Glover, D. M. 1984. *Gene Cloning: The Mechanics of DNA Manipulation*. Chapman and Hall.

Grunstein, M., and D. S. Hogness. 1975. "Colony hybridization: a method for the isolation of cloned DNAs that contain a specific gene." *Proc. Nat. Acad. Sci.*, 72, 3961.

Hacket, P. B., J. A. Fuchs, and J. W. Messing. 1984. *An Introduction to Recombinant DNA Techniques*. Benjamin/Cummings.

Helinski, D. R. 1978. "Plasmids as vehicles for gene cloning: impact on basic and applied research." *Trends Biochem. Sci.*, 3, 10.

Lobban, P., and A. D. Kaiser. 1973. "Enzymatic end-to-end joining of DNA molecules." *J. Mol Biol.*, 78, 453.

Maniatis, T., E. F. Fritsch, and J. Sambrook. 1982. *Molecular Cloning*. Cold Spring Harbor.

Maniatis, T., et al. 1978. "The isolation of structural genes from libraries of eucaryotic DNA." *Cell*, 15, 687.

Mertz, J., and R. Davis. 1972. "Cleavage of DNA: RI restriction enzyme generates cohesive ends." *Proc. Nat. Acad. Sci.*, 69, 3370.

Murray, N. 1983. "Phage lambda and molecular cloning." In *Lambda II* (R. W. Hendrix et al., eds.). p. 395. Cold Spring Harbor.

Murray, N. 1983. "Lambda vectors." In *Lambda II* (R. W. Hendrix et al., eds.). p. 677. Cold Spring Harbor.

Nathans, D. and H. O. Smith. 1975. "Restriction endonucleases in the analysis and restructuring of DNA molecules." *Ann. Rev. Biochem.*, 44, 273.

Roberts, R. J. 1982. "Restriction and modification enzymes and their recognition sequences." *Nucl. Acid Res.*, 10, 117.

Rodriguez, R. L., and R. C. Tait. 1983. *Recombinant DNA Techniques*. Addison-Wesley.

Schell, J., and M. van Montague. 1983. "The Ti plasmids as natural and as practical gene vectors for plants." *Biotechnol.*, April, 175.

Scott, W. A., and R. Werner. 1977. *Molecular Cloning of Recombinant DNA*. Academic Press.

Smith, D. H. 1979. "Nucleotide sequence specificity of restriction enzymes." *Science*, 205, 455.

Verma, I. M. 1977. "The reverse transcriptase." *Biochim. Biophys. Acta*, 473, 1.

Wu, R. (ed.). 1979. *Recombinant DNA. Methods Enzymol.*, Vol. 68. Academic Press.

Chapter 21

Anderson, W. F., and E. G. Diacumakos. 1981. "Genetic engineering in mammalian cells." *Scient. Amer.*, July, p. 106.

Anderson, W. F. 1984. "Prospects for human gene therapy." *Science*, 226, 401.

Casadaban, M. J., and S. N. Cohen. 1980. "Analysis of gene control signals by DNA fusion and cloning in *E. coli.*" *J. Molec. Biol.*, 138, 179.

Caskey, C. T., and R. L. White. 1983. *Recombinant DNA Applications to Human Disease*. Cold Spring Harbor.

Curtiss, R., III. 1976. "Genetic manipulation of microorganisms: potential benefits and biohazards." *Ann. Rev. Microbiol.*, 30, 507.

Gilbert, W., and L. Villa-Komaroff. 1980. "Useful proteins from recombinant bacteria." *Scient. Amer.*, April, p. 74.

Gillam, S., M. Zoller, and M. S. Smith. 1985. "Mutant construction by *in vitro* mutagenesis." In *Recombinant DNA Methodology* (J. R. Dillon et al., eds.). p. 157. Wiley.

Klingmuller, W. 1979. "Genetic engineering for practical application." *Naturwiss.*, 66, 182.

Seeberg, P. H., et al. 1978. "Synthesis of growth hormone by bacteria." *Nature*, 276, 795.

Shortle, D., and D. Nathans. 1978. "Local mutagenesis: a method for generating viral mutants with base substitutions in preselected regions of the viral genome." *Proc. Nat. Acad. Sci.*, 72, 2170.

Smith, H. O. and M. L. Bernstiel. 1976. "A simple method for DNA restriction site mapping." *Nucl. Acid Res.*, 3, 2387.

Zoller, M. J., and M. Smith. 1982. "Oligonucleotide-directed mutagenesis using M13-derived vectors: an efficient and general procedure for the production of point mutations in any fragment of DNA." *Nucl. Acid Res.*, 10, 6487.

Answers

Chapter 1

1. Genotype.

2. Pen-r.

3. No.

4. Pro$^-$. Note that it is capitalized and roman.

5. An absolute defective mutant has a mutant phenotype in all conditions; a conditional mutant has a mutant phenotype only in certain conditions.

6. A Pen-r mutant contains a *pen-r* mutation.

7. The phenotype is Leu$^+$, because the + allele is dominant. It is not temperature-sensitive.

8. $(2 \times 10^{-6})(8 \times 10^{-5}) = 1.6 \times 10^{-10}$.

9. The reversion frequency for the point mutation is the highest; the frequency for the deletion is, for all practical purposes, zero.

10. No. If it was, Vase would be temperature-sensitive.

11. R is a precursor to Z, perhaps occurring immediately before Z in the biosynthetic pathway. The mutation inactivates an enzyme that causes conversion of R to the next step in the pathway.

12. x^+y^+ (wild-type) and x^-y^- (double mutant).

13. *a d b*. If *a* and *d* were on opposite sides of *b*, *a–d* would be 4%. Thus, the order must be *a d b* or *d a b*. For the first order, *a–d* would be 1.2%; for the second order, *b–d* would have to be larger than *a–b*, which is not the case.

14. No, unless a third gene has been given as a reference.

15. Remember that the central marker will be exchanged. Thus, the genotypes are *ABd* and *abD*.

16. In general, at least four.

17. The mutations complement. At least two genes are required.

18. The enzyme is probably not the product of the gene.

19. That the *G*-gene and *H*-gene products probably join together to form an active macromolecular complex. The regions in which mutations occur (which can be compensated or are compensatory) are probably in the binding sites.

20. Two more: for example, one to locate *c* to the right or left of *a* and another to locate *d* to the right or left of *c*.

21. (a) The gene order is *a c b d*. (b) $0.15 \times 0.04 = 0.006$.

22. The three classes of frequencies represent single, double, and triple exchanges. If *ABCd* and *abcD* are the products of a single exchange, the order must be *A B C D*.

23. (a) In a cross *AbDe* $\times$ *aBdE*, *AE* can result from a exchange anywhere between *A* and *E*. Since the map distance from *A* to *E* is $0.01 + 0.02 + 0.03 = 0.06$, *AE* will occur at a frequency of 0.06. (b) To get *ABDE*, one needs exchanges in the intervals *A–B*, *B–D*, and *D–E*; the probability of getting all three exchanges is $0.01 \times 0.02 \times 0.03 = 0.000006$. Therefore, the fraction of *AE* that will be *ABDE* is $0.000006/0.06 = 0.0001$.

24. *e g f* is the gene order. The spacing between *e* and *g* is three times that between *f* and *g*.

25. This can be worked out most easily by writing down the cross with the three possible gene orders: (I) *A B D*, (II) *A D B*, and (III) *B A D*. Recalling that single crossovers are much more probable than double crossovers, we can predict that for I *AbD* occurs much less frequently than *Abd*, for II *AbD* occurs with comparable frequency with *Abd*, and that for III *AbD* occurs much more frequently than *Abd*. Hence, order III is the only one that agrees with the data.

26. Two. One is Jase, and the other is the product of the gene in which the j^-(Ts) mutation resides.

27. (a) *kyuQ*. (b) A regulatory mutant that prevents synthesis of each gene product.

28. Four genes. The groups are: (1,7), (2,4), (5), and (3,6).

Chapter 2

1. A nucleotide is a nucleoside phosphate.

2. Purines: adenine, guanine. Pyrimidines: cytosine, thymine, uracil.

3. Thymine, DNA; uracil, RNA.

4. Ribose has an OH group on the 2′ carbon, and 2′-deoxyribose has a H atom at the same position.

5. An *N*-glycosidic bond.

6. The 3′ and 5′ carbon atoms.

7. One 3′-OH and one 5′-P group.

8. One.

9. AT, 2; GC, 3.

10. A 5′-phosphate and a 3′-hydroxyl group.

11. 45/10 = 4.5.

12. The linear density of DNA is roughly 2×10^6 molecular weight units per micrometer, and the average weight of a base pair is 660. Thus, the number of base pairs is $4 \times 10^6/660 = 6060$. One could also calculate it from the fact that the base pairs are separated by 3.4 Å.

13. The first is the sum of the molar concentrations of the purines and the second is the sum for the pyrimidines; thus, the two sums are equal.

14. If a viewer watched an object move along a right-handed helix, the object would follow a clockwise path and move away from the viewer; a counterclockwise path would cause the object to move away in a left-handed helix. The DNA helix is right-handed.

15. The sugars have opposite orientations in the two strands; thus, at one end of a double-stranded DNA molecule one strand terminates with a 3′-OH group and the other strand with a 5′-P group.

16. It is an inverted repeat or a palindrome.

17. A nuclease is an enzyme capable of breaking a phosphodiester bond. An endonuclease can break any phosphodiester bond, whereas an exonuclease can only remove a terminal nucleotide or a short terminal oligonucleotide.

18. It could not because there would be no free ends, which an exonuclease requires. The supercoiling is irrelevant; only the fact that it is covalently closed is important.

19. Only twisting followed by joining.

20. If $[A]/[C] = \frac{1}{3}$, then $([A] + [T])/([G] + [C]) = \frac{1}{3}$, and the DNA is 25% A+T, or 12.5% A.

21. If 18% is A, 18% is also T, for a total of 36% for A+T. Therefore, G+C will be 64%, and C will be half that, or 32%.

22. Yes. The base-pairing rule would be satisfied perfectly and the two components of the double-stranded portion would be antiparallel.

23. The mass/length of double-stranded DNA is approximately $2 \times 10^6/\mu m$; hence, the molecular weight is 32.8×10^6.

24. Molecule 1, because of the long tract of GC pairs in molecule 2.

25. Pairing will occur randomly. Thus, there are three bands, in the ratios 1 ^{15}N^{15}N : 2 ^{14}N^{15}N : 1 ^{14}N^{14}N.

26. The mixture is 45/50 HH and 5/50 LL before denaturation. After renaturation there will be $(0.9)^2 = 0.81$ HH, $(0.1)^2 = 0.01$ LL, and $2(0.9)(0.1) = 0.18$ HL, or 40.5, 9.0 and 0.5 μg of HH, HL, and LL, respectively.

27. (a) $4 \times 10 = 40$ base pairs. (b) Four nodes.

28. Ten base pairs are contained in one turn of a DNA helix. Thus, the molecule will be a supercoil having one node, that is, a figure-8.

29. Naturally occurring DNA is negatively supercoiled. Ethidium bromide adds positive supercoiling. Thus, the compact supercoil progressively loses its supercoiling until a slower moving nontwisted circle has formed. As more ethidium bromide is bound, the DNA becomes progressively positively supercoiled and hence more compact, so the value of s increases.

30. The charge per nucleotide is the same for molecules of all sizes, so the net force on all molecules is in a single direction. The pore size of the gel determines the rate at which a molecule of a particular size can move through the gel. Larger molecules cannot penetrate the pores as easily as smaller molecules and hence move more slowly through the gel.

31. The 5′-terminal bases are obtained from the 3′ termini of the complementary strand. Thus, the two strands are 5′-TGATACGACGAAGTACTGG and 3′-ACTATGCTGCTTCATGACC.

Chapter 3

1. The group, other than the amino or carboxyl group, that is covalently linked to the α carbon.

2. Glycine.

3. An amino group at one end and a carboxyl group at the other.

4. The side chain of cysteine, which has a terminal SH group.

5. A random coil has no intramolecular interactions. Collisions with solvent molecules causes continual fluctuations in the shape of the molecule.

6. Two molecules that are poorly soluble, that is, interact poorly with water, can reduce contact with water by clustering, because clustering minimizes the ratio of surface to volume. More precisely, clustering reduces the number of molecules that become ordered by the presence of a a poorly soluble molecule.

7. A hydrophobic interaction.

8. Cysteine: disulfide bonds; lysine: ionic bonds and hydrogen bonds; isoleucine: hydrophobic bonds; glutamic acid: ionic bonds and hydrogen bonds. All could participate in van der Waals bonds.

9. A carbon and a nitrogen.

10. Alanine, which is nonpolar, will tend to be internal; serine is polar and will tend to be on the surface.

11. The peptide bond.

12. Two cysteines. Methionine, which also contains an SH group, does not form disulfide bonds.

13. Set 3, in a hydrophobic cluster.

14. All of the peptide bonds.

15. Only hydrogen bonds.

16. Structural proteins.

17. The hydrophobic interaction. Van der Waals bonds can also be important.

18. In organic solvents the hydrophobic interaction, which is very important in protein folding, breaks down almost completely.

19. As the size of an object increases, the ratio of its surface area to its volume decreases. Since polar amino acids are more often located on the surface than within a folded polypeptide chain and the reverse is true of nonpolar amino acids, the ratio of polar to nonpolar amino acids also decreases.

20. The side chain of glycine (a hydrogen atom) is so small that there is no steric interference with tight folding.

21. Both will tend to be near an amino acid of opposite charge.

Chapter 4

1. Glucose.

2. Log phase.

3. 2 hours = 6 doubling times. Thus, $2^6 = 64$.

4. Just one, leucine.

5. The total dilution is $(100)(100) = 10^4$. Thus, the cell density is $72/[(0.1)](10^4) = 7.2 \times 10^6$ per ml.

6. The microscopic count shows that the cell density is $(453)(1000)/0.001 = 4.53 \times 10^8$ per ml. Since plating yields a much smaller number, it is clear that many of the cells are dead.

7. Draw a curve relating absorbance and cell density; it will yield a straight line. You will find that 0.7 corresponds to 7×10^8 cells/ml.

8. I shows that Leu is required. II implies that Arg is needed. III indicates that Trp is not needed. Thus, the genotype is $arg^- leu^-$.

9. The most likely explanation is that the required substance exists in an insufficient concentration in tryptone for the maximum growth rate and more is present in nutrient broth. This is not uncommon, since acid hydrolysis can destroy some molecules (for example, tryptophan).

10. $75/0.3 = 250$ minutes.

11. The mutant cannot use the maltose. It metabolizes the lactose in preference to the amino acids in the medium, so the colonies are purple.

12. Purple, since all cells can use glucose.

13. The most common reason for such an observation is that the concentration of the required nutrient is not high enough.

14. B is either a precursor to A, or at least B can be converted to A.

15. A single DNA molecule and numerous protein molecules.

16. Probably none.

17. One single-strand break causes loss of supercoiling of one loop rather than of the entire DNA molecule.

18. From the total molecular weight you would know the total length of the DNA in a nucleus. Even allowing for up to 100 distinct DNA molecules per cell, the length of each molecule would still be very great compared to the diameter of a nucleus. You would be forced to conclude that each DNA molecule must fold back on itself repeatedly.

Chapter 5

1. One.

2. Tailed icosahedra, nontailed icosahedra, filaments.

3. In biological experiments one is concerned with the number of viable phage, so they are counted by plating and counting plaques. However, for physical experiments one may need to know the total number of particles, viable and nonviable. This is not a trivial measurement, though a rough idea (no more than a 50 percent error) can be obtained by electron microscopy. The best method is to measure the DNA concentration of a solution and divide this value by the molecular weight of the phage DNA molecule, if it is known.

4. Just one.

5. Lysis.

6. One. If the bacterium has not lysed, all progeny will be in the same position.

7. Use the Poisson equation. With a 1:1 ratio of phage to bacteria, the fraction uninfected is e^{-1}.

8. Adsorption, entry of nucleic acid, transcription, production of phage proteins, production of phage nucleic acid, assembly of particles (including encapsidation of nucleic acid), release of progeny particles.

9. Phage-host specificity refers to the fact that a particular phage can grow on only one or a small number of bacterial species or strains. The most frequent cause is the inability of a phage to adsorb, except to certain bacteria.

10. In general, the bacterium will lack an adsorption site.

11. No. Phage can develop only in a metabolizing bacterium.

12. A phage can multiply only in a growing bacterium, and the nutrients in the growth medium become depleted.

13. Host restriction.

14. 50 to a few hundred.

15. Clearly, the receptors never evolved in the first place in order that the bacterium could be infected. Thus, the receptors must be important to the bacterium. For example, the receptor for phage T1 is a transport system for the Fe^{2+} ion. Receptor sites might gradually change in time to reduce the accessibility of certain phages but the change will be limited by the number of amino acids that can be altered without loss of essential activity.

16. The bacteria had stopped growing when the phage were added, so phage growth is not possible.

17. T4 will be turbid, because half of the bacteria will never be infected; that is, B/4 can continue to grow in the plaque. T4h will be clear, because all bacteria will be infected.

18. The cell culture lyses visibly at 60 minutes. This time is 20 minutes after the MOI becomes greater than 1 phage per cell. The calculation is the following: After 20 min there are $200 \times 5000 = 10^6$ phage/ml. At $t = 30$ min the bacteria have doubled to ca. 4×10^7 per ml. At $t = 40$ the number of phage per ml is $200 \times 10^6 = 2 \times 10^8$, which is greater than the number of bacteria. Thus, lysis will occur one phage life cycle, or 20 minutes later, at $t = 60$.

19. After adsorption, divide the mixture in half. Add antiserum to one half to kill unadsorbed phage. Measure the number of infected cells by their ability to produce phage; this is the number of infective centers. To the other half, add chloroform to kill infected cells and measure unadsorbed phage. The MOI equals

    (Phage added − Unadsorbed phage)/Infected cells

20. From the Poisson distribution, $e^{-2} = 0.135$ of the bacteria are uninfected. Therefore the number of plaques is $200(1 - 0.135) = 173$.

21. (a) These are bacteria to which no phage adsorb. Since the total MOI = 5, the fraction of cells which are uninfected is $e^{-5} = 0.007$. The number of uninfected cells is 7×10^5. (b) This number includes bacteria to which P4, but not P2, adsorb, as well as the infected cells. The MOI of P2 is 3; the fraction of cells getting no P2 is $e^{-3} = 0.05$. Since 0.007 get neither P2 nor P4, then $0.05 - 0.007 = 0.043$ get one or more P4 but no P2. Hence, $0.043 \times 10^8 = 4.3 \times 10^6$ bacteria get P4 but no P2. (c) These are the bacteria to which P2 adsorb but no P4 adsorb. Using the above reasoning, $e^{-2} = 0.135$ get no P4, and 0.007 get neither; hence $0.135 - 0.007 = 0.128$, and 1.28×10^7 cells adsorb P2 but no P4. (d) These are the bacteria infected by at least one P2 and one P4. The number of cells getting at least one P4 and one P2 is $10^8 - (1.28 \times 10^7) - (4.3 \times 10^6) - (7 \times 10^5) = 8.22 \times 10^7$.

22. A temperate phage may choose between a lytic and lysogenic cycle; a virulent phage can engage only in a lytic cycle.

23. High.

24. Stationary phase.

25. The plaque has a turbid center surrounded by a clear area.

26. Bacterial, *BOB'*; phage, *POP'*.

Chapter 6

1. A 5′-triphosphate and a 3′-OH.

2. A stem-and-loop and a tract of U's.

3. Not at all.

4. From the 5′ end to the 3′ end.

5. Antiparallel.

6. In a given DNA segment both DNA strands cannot serve as a template for transcription. Actually there are a few known exceptions to this rule in which transcribed regions overlap slightly.

7. A coding strand is a segment of a DNA strand that is copied by RNA polymerase. An antisense strand is a DNA strand that is complementary to a coding strand.

8. The Pribnow box and the -35 sequence.

9. PPP-5′-AGCUGCAAUG-3′ and PPP-5′-CAUUGCAGCU-3′.

10. Leaders, spacers, and the unnamed region following the last stop codon of an mRNA are untranslated regions.

11. A ribosome is a particle that binds mRNA and two tRNA molecules such that amino acids are aligned in the order that the codons are aligned in the mRNA.

12. $4^4 = 256$.

13. UUG, leucine; AAU, asparagine.

14. With G at the 3′ terminus the poly(U) will terminate with UUG and the polypeptide will end with leucine; if it is at the 5′ terminus, the poly(U) will start with GUU, so the polypeptide will initiate with glycine.

15. UAA, UAG, UGA.

16. AUG; methionine.

17. A, U, C.

18. R′Q′P′, because both codon and anticodon are by convention written with the 5′ end at the left.

19. 1.

20. UAG.

21. The minimum is 3; the actual number would be larger since at least one of the proteins would contain at least one methionine.

22. $(3 \times 124) + 3$ (termination) $+ 3$ (initiation) $= 378$.

23. There is no way of knowing because both codons could be altered to yield the stop codons UAA and UAG.

24. (1), (3), (4).

25. Two, in each case.

26. No. 70S ribosomes dissociate when translation is completed and do not re-form until the 30S subunit has bound mRNA, fMet-tRNA, and initiation factors.

27. The 5′ end is read first.

28. The amino end is made first.

29. The 70S ribosome forms when a 50S subunit joins the 30S preinitiation complex.

30. The A site.

31. RNA polymerase to transcribe mRNA and tRNA, a few enzymes for processing tRNA, about 20 tRNA synthetases, several enzymes for regenerating GTP, one or a few for forming a peptide bond, one or a few for releasing a finished polypeptide and a few for protein processing—perhaps 40 to 50 enzymes in all. This represents about 2 percent of the *E. coli* enzymes.

32. NH_2-Arg-Leu-COOH, when reading starts exactly at the 5′ terminus. Reading can begin at other positions also yielding NH_2-Gly-Tyr-Arg-Lys-COOH and NH_2-Val-Ile-Gly-Lys-COOH and fragments thereof.

33. Synthesis would have to await the completion of a molecule of mRNA. In the existing system protein synthesis can occur while the mRNA is being copied from the DNA. Thus, protein synthesis can start earlier than would be possible with reverse polarity, and the mRNA is relatively resistant to nuclease attack.

34. The 5′ end, since ribosomes move along the mRNA from the 5′ end to the 3′ end.

Chapter 7

1. Turning on and off a set of genes as a unit, simultaneously or by a single signal.

2. The ratio of the amount of each components is maintained constant without having each one separately regulated.

3. Negative regulation.

4. Not always, especially if needed in large quantities. If it were regulated, it would usually be autoregulated.

5. It binds to a repressor.

6. The promoter is inaccessible to RNA polymerase.

7. β-Galactosidase (*lacZ* product), lactose permease (*lacY* product), and the transacetylase (*lacA* product).

8. It is constitutive.

9. Inducible, because the *lacI* product is diffusible.

10. Constitutive.

11. The major reason is that all promoters do not have the same strength, so the rate of initiation of transcription varies from one gene to the next.

12. Constitutively.

13. No, because the repressor is a diffusible protein.

14. Yes, since binding of the repressor to the operator interferes with binding of RNA polymerase to the promoter.

15. Low.

16. No. In the first case, it could not make cAMP. In the second case, it could not make CRP.

17. No.

18. No. There is often a polar effect in that the relative amounts of each protein decrease toward the 3′ end of the mRNA.

19. With a repressor the molecule is usually an early (generally the first) reactant in the pathway. The repressor is inactivated by the binding. With an aporepressor the molecule is usually the product of the pathway. An aporepressor is activated by the binding.

20. No. Nothing is specifically bound to an attenuator.

21. In attenuation the structure of the termination sequence is altered such that RNA polymerase either does or does not see it. In antitermination RNA polymerase is altered in order that it can ignore certain termination sites.

22. At very low levels all inducible operons are transcribed—for example, a repressor might come off the operator for an instant and RNA polymerase will get on. Hence each cell will contain a few β-galactosidase and permease molecules. The permease will bring in a few lactose molecules, so derepression can occur.

23. (a) Yes, I, yes; (b) yes, I, yes; (c) yes, C, yes; (d) yes, I , yes; (e) yes, C, yes; (f) yes, I, yes; (g) no, neither, no; (h) yes, I , yes; (i) no, neither, no; (j) yes, constitutive, no.

24. If neither glucose nor lactose is present, two proteins (the repressor and CRP-cAMP) are bound to the DNA. If glucose alone is present, only one protein (the repressor) is bound to the DNA.

25. The cell cannot be derepressed. The simplest explanation is that the repressor cannot bind the inducer. The partial diploid given at the end of the problem

would not make enzyme, because the constitutive component makes no enzyme and the other component cannot be induced.

26. (a) In general, when a substance X is lacking from the medium, an enzyme that degrades it has little value to the cell. Cells in which 20 percent of their protein is useless must grow slightly more slowly. If the culture medium does not limit growth for other reasons, growth will be limited by how fast the necessary quantities of the required proteins can be made. The repressorless constitutive strain should grow 20 percent slower; that 20 percent disadvantage will be compounded in each generation. When the inducible strain has undergone 30 doublings, the constitutive one will have undergone only $0.8 \times 30 = 24$ doublings. At that time, for every 2^{30} cells of the inducible strain, there are only 2^{24} cells of the constitutive strains, that is, 2^6 times as many cells of the former as of the latter. Therefore, the ratio of wild-type (inducible) cells to repressorless (constitutive) cells is $2^6 = 64$. (b) A noninducible mutant, owing to a repressor that is defective in binding the gratuitous inducer. The reasoning is the following. The gratuitous inducer is not a substrate of the enzyme; that is, it is not a carbon source. Ability to remain uninduced would confer a growth advantage on a cell in these special conditions, so if such a mutation arose spontaneously, it would be selected for in these circumstances.

27. (a) The strain is constitutive, that is, $lacI^-$ or $lacO^c$. (b) The second mutant must be $lacZ^-$, since no β-galactosidase is made; also, it must have a normal repressor-operator system, since permease synthesis responds to lactose. That is, its genotype is $lacI^+ lacO^+ lacZ^- lacY^+$. Since the partial diploid is regulated, the operator in the operon with a functional $lacZ$ gene must be wild-type. Hence, the first mutant must be $lacI^-$.

28. Glucose is irrelevant to the functioning of the *trp* operon. When tryptophan is present (1, 3), the repressor is active and is bound to the DNA.

29. It clearly lacks a general system that regulates sugar metabolism. The only thing we know about is the cAMP-dependent system, namely, that several mutations prevent activity of this system. For example, the mutant could make (1) either a defective CRP protein, and either fail to bind to the promoter or be unresponsive to cAMP, or (2) an inactive adenyl cyclase, making no cAMP.

30. The operator is probably next to gene E. Deletions that remove the operator make transcription constitutive. The promoter cannot be between the operator and gene E, because in that case all deletions that remove the operator would also remove the promoter, and no constitutive deletions would be found. The order is E o p, which is consistent with the fact that some deletions are not constitutive; these would be larger deletions that include both the operator and the promoter. Thus, the gene order (transcription from left to right) is p o E D C B A.

31. Since addition of Q causes Qase to be made, the system is regulated. Deletion of *kyu* prevents induction, suggesting that *kyu* encodes a positive regulatory element. A partial diploid with both wild-type *kyu* and the deletion is regulated, indicating that the Kyu product is present and that it is a positive regulator activated by an inducer. Examination of mutants confirms this view. The *kyu1* mutant never makes Qase but a partial diploid with wild-type is inducible, indi-

cating that the *kyuI* mutation is recessive. It seems likely that the Kyu product binds Q and is activated. The *kyu2* mutation is constitutive and dominant, suggesting that it is able to turn on transcription of Qase without Q. The Kyu1 product probably fails to bind Q and the active positive regulator is a Q-Kyu complex. The Kyu2 product is probably able to bind to the promoter without first binding Q.

32. (a) Three kinds of repressor mutations might occur: (1) a repressor that cannot bind X—the operon is on; (2) a repressor that cannot bind to operator even when X is present—the operon is on; (3) a repressor that binds to the operator without X—the operon is off. (b) Phenotypes of partial diploid with wild-type and each of the mutants are: first mutant, inducible; second mutant, inducible; third mutant, unable to synthesize X under all conditions.

Chapter 8

1. Template.

2. All, ½, ¼; respectively.

3. θ replication.

4. Rolling circle replication.

5. No. After one round of replication it has hybrid density but an ^{15}N strand always remains in the circle.

6. In rolling circle replication only one strand of the parental circular molecules serves as a template; the second strand of the branch is copied from a progeny strand. In θ replication both parental strands are templates.

7. Positive supercoiling.

8. Negative supercoiling.

9. Yes.

10. DNA gyrase.

11. Polymerizing activity, 5′-3′ exonuclease, 3′-5′ exonuclease.

12. From the 3′ end to the 5′ end.

13. Deoxynucleoside triphosphates.

14. To a 3′-OH group.

15. The $5' \rightarrow 3'$ activity removes ribonucleotides from the 5′ termini of precursor fragments, and the $3' \rightarrow 5'$ exonuclease removes a base that has been incorrectly added to the growing end of a DNA strand.

16. Pol III requires a helicase to unwind the DNA; Pol I can unwind a helix without any accessory protein.

17. Pol III is responsible for the addition of all nucleotides in the growing fork; Pol I removes RNA from the 5′ end of precursor fragments and replaces it with deoxynucleotides.

18. A helicase unwinds the parental helix in the replication fork.

19. It prevents reformation of double-stranded DNA after the parental DNA is unwound by a helicase.

20. The $3' \rightarrow 5'$ exonuclease activity.

21. DNA polymerases join a 5'-*tri*phosphate to a 3'-OH group and in so doing remove two phosphates, so that the phosphodiester bond contains one phosphate; a ligase joins a 5'-*mono*phosphate to a 3'-OH group.

22. One strand is copied from the 3' end to the 5' end, and the other strand is copied in the direction opposite to that of the movement of the replication fork by synthesis in short pieces.

23. The lagging strand.

24. The lagging strand.

25. Remember that the group at the growing end of the leading strand is always a 3'-OH group because DNA polymerases can only add nucleotides to such a group. Because DNA is antiparallel, the opposite end of the leading strand is a 5'-P group. Hence: a, 3'-OH; b, 3'-OH; c, 5'-P.

26. It remains completely hydrogen-bonded to the DNA template.

27. The RNA primer must be removed from the precursor fragment that was made first, because DNA ligase cannot join DNA to RNA.

28. The $5' \rightarrow 3'$ exonuclease.

29. Rolling circle replication must be initiated by a single-strand break. θ replication does not need such a break.

30. (a) One. (b) There is no maximum. (c) One of the parental single strands.

31. After one round of replication, all will be $^{15}N^{14}N$; after the second round, $\frac{2}{3}$ will be $^{15}N^{14}N$; after the third round, 50 percent $^{15}N^{14}N$, 50 percent $^{14}N^{14}N$.

32. (a) In rolling circle replication one parental strand remains in the circular portion and the other is at the terminus of the branch. Therefore, only half of the parental radioactivity can appear in progeny. (b) If there is no recombination between progeny DNA molecules, one progeny phage will be radioactive. If recombination occurs once, two particles will be radioactive. If recombination is frequent and occurs at random, the radioactivity will be distributed among the progeny.

33. The molecular weight of a nucleotide pair is about 660, so the DNA molecule contains 38,000 nucleotide pairs. The replication rate of *E. coli* DNA is 50,000 nucleotide pairs per minute per replication fork. Thus, the time to replicate the phage DNA is $(1/2)(38,000/50,000) = 0.38$ min, or 22 sec.

34. Assume bidirectional replication and that both replication forks move at the same rate. Then the order is determined by the distance of each gene from the origin. Thus, the order is *AGBFCDE*.

Chapter 9

1. In the same strand.

2. Increased.

3. It will be unaffected since phage particles do not contain any repair enzymes.

4. Photoreactivation.

5. The Uvr endonuclease, polymerase I, and DNA ligase.

6. Incision refers to the production of one single-strand break, usually near a nucleotide that is to be removed. Excision refers to the second break and subsequent removal of a nucleotide or an oligonucleotide.

7. It is inducible and it is error-prone.

8. Like all light-absorbing phenomena, there is a photoreceptor that absorbs in a particular range of wavelengths. Thus, we should expect a wavelength-response curve.

9. Statement 2.

10. (a) Either the damage to T4 DNA is, for some reason, barely repairable by the Uvr system, or T4 possesses its own repair system. (b) The mutant probably lacks a T4-encoded repair system.

11. (a) X is not inducible, because the enzymes were present before protein synthesis was blocked by chloramphenicol. (b) X would probably be considered inducible. The residual 5 percent could be due to either a second noninducible system, which can excise thymine dimers, or to a small amount of synthesis of X proteins when chloramphenicol is present.

12. The phage apparently makes its own incision enzyme.

13. (a) The larger the DNA molecule, the greater is the probability that the DNA will receive a lethal alteration. (b) The phage may possess its own repair system.

14. A simple nonrepair explanation is that in a log-phase culture many cells are paired and both cells have to be killed to prevent colony formation. An alternative is that the repair systems are more active in rapidly multiplying cells.

15. The dose that yields, on the average, one lethal hit per organism.

16. The molecular weights are inversely proportional to the dose yielding a survival of $1/e$, if the inactivation mechanisms are the same, as is usually the case with x irradiation. From phages I, II, and IV the rate of production of lethal hits is $1/1/e$-survival/(molecular weight), or 5.35×10^{-7} lethal hits per rad per million molecular weight units. Thus, the molecular weight of the DNA of phage III is probably $10^{6}/(5.35 \times 10^{-7})(4.14 \times 10^{4}) = 45 \times 10^{6}$.

Chapter 10

1. Yes, because, strictly speaking, that is the definition of a mutation.

2. No. The change does not always cause an amino acid replacement, and an amino acid change does not always alter protein function.

3. Transition, 2; transversion, 1, 3.

4. Transitions.

5. A particular base pair, such as AT, can undergo two transversions, namely, AT → TA and AT → CG, but only one transition, AT → GC. This pattern is true for all base pairs. Thus, the theoretical ratio of transversions to transitions is 2:1. Actually, transversions are quite rare because all base changes are not equally frequent; in fact, Pu → Py and Pu → Py changes are fairly rare.

6. A missense mutation results in an amino acid substitution. A nonsense mutation results in premature termination of a protein.

7. An amino acid substitution in the protein product causes weakening of the structure such that a shape change occurs in the protein above or below a critical temperature.

8. A tautomer can cause mutations if it has two forms in equilibrium with another and each form has different base-pairing properties. A tautomeric molecule can be incorporated in an aberrant form and then switch to a form that cannot pair with the base in the DNA. Alternatively, when in the template strand, it can temporarily switch its base-pairing properties and allow incorporation of a base that would not have been added to the end of the growing chain if the switch had not occurred.

9. It tautomerizes, which allows base substitutions, and it affects nucleotide pools, which increases incorporation errors.

10. Yes, because the entire frame of reading of the bases is changed, which results in alterations of most amino acids downstream from the mutation. Only two types of base addition or deletion would not cause a phenotypic change: one that occurs in the space between genes and one that occurs very near the end of a gene.

11. Three. Otherwise, there would have been a frameshift and all downstream bases would have been changed.

12. A frameshift or a deletion.

13. Chain termination mutations, since the reversion event would have to eliminate the chain termination site. Of course, since a chain termination codon contains three bases, it is possible that a second-site reversion could occur in one of the two positions of the chain termination codon other than that which had originally been altered. However, this would certainly produce an amino acid change at the mutated site, so reversion would be unlikely.

14. Many substances are not mutagenic by themselves but are converted to mutagens by enzymes in the liver.

15. This mutant could not be isolated because there is no temperature at which it could grow.

16. The ratio of polymerizing activity to exonuclease function will decrease in the antimutator since errors will be removed more often.

17. They clearly interact. Since a change in the sign of the charge yields a mutant and a second sign-change in another amino acid yields a revertant, amino acids 28 and 76 are probably held together by an ionic bond.

18. $2000 \times (1 \times 10^{-5}) = 0.02$ mutation per cell per generation.

19. $\mu = -\ln (5/12)/(5 \times 10^8) = 1.8 \times 10^{-9}$.

20. A: $-\ln (22/40)/(5.6 \times 10^8) = 1.1 \times 10^{-9}$. B: $-\ln (15/37)/(5 \times 10^8) = 1.8 \times 10^{-9}$.

21. Ease is of course often a matter of opinion, but in this case there is little doubt that the reverse mutation (pro^- to pro^+) is the more easily detected. Simply by placing bacteria or other microorganisms on two types of solid medium, one with proline and the other lacking proline, rare pro^+ cells can be detected by their ability to grow on both media.

22. Proline determines the way the polypeptide backbone folds, so any change from proline (change 1) usually yields a mutant phenotype. Arginine to lysine (change 2) is probably ineffective because both have the same charge and nearly the same size. Substitution of threonine, which is weakly charged, by isoleucine, which is quite hydrophobic (change 3), would very likely cause mutation. Valine to isoleucine (change 4), both of which are nonpolar and have nearly the same size, would probably have little or no effect. Glycine to alanine (change 5), both of which are nonpolar, would have little effect unless there was insufficient space for the slightly larger alanine side chain. Histidine to tyrosine (change 6) probably would cause a phenotypic change because of the differences in charge and lack of flexibility, though in many proteins the change might be ineffective. Glycine to phenylalanine (change 7) would probably have an effect because phenylalanine is so bulky that it is likely to cause a change in shape of the protein. For any of the changes mentioned, it is always possible that the shape change is in a part of the molecule that is fairly tolerant of changes.

23. Amino acid substitutions at many positions in a protein have little or no effect on its activity; only certain substitutions occurring at critical positions have a detectable effect.

24. It will be incorporated into DNA opposite a T. In a later round of replication it will pair mostly with T but occasionally with C, leading to a GC pair, at the site of an AT pair. Therefore, the change is an AT-to-GC transition.

25. The site of the removed guanine serves as a template for an A; thus, the change is GC to TA, a transversion.

26. Yes, because nitrous acid causes both AT-to-GC and GC-to-AT transitions.

27. (a) A double point mutation. (b) No. Frequencies give no information about the type of base change.

28. Although a second-site mutation can restore activity, the new interaction that replaces the original intramolecular interaction destroyed by the mutation may not be as strong as the wild-type interaction and hence may be more easily disrupted by thermal motion.

29. The mutation frequency in the Lac$^-$ culture is 9×10^{-9}. The probability of production of a suppressor is independent of the presence of a suppressor-sensitive mutation (which the Lac$^-$ mutation is), so the mutation rate in the Lac$^+$ cell is also 9×10^{-9}.

Chapter 11

1. Not really. The largest plasmids have a low copy number but the small plasmids can exist in any number.

2. A plasmid to which plasmid-specific proteins, one of which has nuclease activity, are bound.

3. Nicking of a relaxation complex occurs in one strand and at a particular site.

4. θ replication.

5. Yes, during transfer.

6. In both. Transfer replication occurs in the donor and conversion of the transferred single strand to double-stranded DNA occurs in the recipient.

7. A nick at the transfer origin to generate a 3'-OH group.

8. The donor.

9. The genes reside in the plasmid. Thus, F pili are made by pili genes in F, and R pili are made by genes in an R factor.

10. (1) They are involved in recognition of donor and recipient; and (2) the donor pilus attaches to the recipient cell, retracts and brings the two cells in close contact.

11. The genes required for transfer—for example, pili genes, and replication genes.

12. A conjugative plasmid is able to form a donor-recipient pair. A mobilizable plasmid carries genes needed for DNA transfer. A self-transmissible plasmid is both conjugative and mobilizable.

13. All three terms describe F. ColE1 is only conjugative.

14. It cannot mediate effective contact (formation of donor-recipient pairs).

15. The origin from the high-copy-number plasmid. Hence, the copy number will be high.

16. They must have the same, or at least similar, inhibitors of replication initiation.

17. F contains DNA sequences that are homologous to sequences in the chromosome. Recombination between these sequences gives rise to insertion.

18. No. ColE1 is nonconjugative, so the cell could not transfer its DNA unless F were also present.

19. RTF, which carries the replication and transfer genes, and the r determinant, which carries the genes for antibiotic resistance.

20. Yes. The simplest plasmid would be a tiny DNA fragment carrying no information other than a copy of the replication origin of the host cell.

21. The plasmid has a temperature-sensitive mutation in some replication function.

22. Since an RNA primer is not needed to initiate rolling circle replication, a reasonable guess would be that transcription of an inducible gene is needed for transfer.

23. Synthesis in the donor provides the single strand that is copied. Synthesis in the recipient converts the transferred strand to double-stranded DNA.

24. (a) Lac$^+$. (b) Yes. (c) F'lac has integrated into the chromosome. The chromosome is now replicating from the origin of F and hence does not need the DnaA protein. (d) No.

25. (a) Lac$^-$. (b) F'(Ts)lac^+ has integrated into the bacterial chromosome. (c) Integration has occurred within a gal gene.

26. 5.

27. Early in the development of the colony a cell has divided to yield a daughter cell lacking F'lac.

28. No. The donor cannot make the enzyme because of the streptomycin. The females also cannot, because they are lac^-. Cells that have received the plasmid also are unable to make the enzyme, because without functional primase, the transferred circle (transferred by looped rolling circle replication) will not be converted to double-stranded DNA. Hence, the lac gene of the transferred plasmid will not be transcribed, and enzyme will not be made.

Chapter 12

1. Terminal inverted repeats, and a gene encoding a transposase.

2. A unique sequence flanked by a sequence in direct or inverted repeat (the termini of the transposable element), the entire three-component unit flanked by a short sequence in direct repeat (the target sequence).

3. No. Each target sequence will be different.

4. The number of nucleotide pairs.

5. 5 to 10 nucleotide pairs.

6. A transposable element consisting of a unique sequence flanked by two copies of a simple transposon. The simple transposons can be in an inverted- or direct-repeat array.

7. Yes. Although the termini of simple transposons can be in direct repeat, these transposons have inverted-repeat termini, so there will always be inverted terminal sequences.

8. No.

9. Many of the IS elements contain transcription termination sites. Hence if they are downstream from a promoter, whether within a gene or between genes of a polycistronic region, they will cause premature termination and hence loss of expression of downstream genes.

10. When transposition occurs, a copy of the transposable element appears at a new position without loss of the original element.

11. The transposon itself and the target sequence.

12. Yes, since within a given DNA segment, for example, the *E. coli lac* operon, certain sites are occupied more often than others. It is likely that transposition can occur to any target sequence, yet certain sequences are more likely than others.

13. The fused plasmid is a cointegrate; it contains two copies of the transposon.

14. The transposase.

15. The enzyme is a resolvase. There are two copies of the transposable element, as is always the case with transposition in bacteria.

16. Replicon fusion, deletion, inversion, transposition, and mutation.

17. The total amount of DNA always increases when transposition occurs, because of the duplication of the target sequence. With most transposons the transposon itself is also duplicated.

18. An effect of sequence on the probability of the first breakage event seems most likely.

19. (1) If it produces within a gene an insertion that is not a multiple of three base pairs, it can eliminate a natural stop codon. (2) If it is quite long and inserts between two cistrons in a polycistronic mRNA, the probability of initiation of translation of downstream cistrons can be reduced. (3) It can contain an in-phase stop codon quite far from the start codon of a subsequent cistron.

20. Heat denaturation followed by quick cooling and electron microscopic viewing might show a stem-and-loop structure. This would be suggestive but by no means would it be proof. Only base sequencing can answer the question, because if a transposon is present in the sequence, a target sequence must be duplicated. If a short duplication is seen, this is again only suggestive. If a duplication is not contained in the sequence, one can say with certainty that no transposon (of the types known) is present.

21. (a) Integration of the plasmid into the chromosome by homologous recombination; reversion to temperature-insensitivity; transposition of the *amp-r* gene from the plasmid to the chromosome. (b) Integration (if homology-dependent).

22. You could start by forming heteroduplexes with a λ DNA molecule lacking the insertion. Each insertion will produce a single-stranded loop. If the loops are not the same size, the transposons must be different. If the loops are the same size, the two transposons are probably the same. A better test would be to move one of the transposons (A) to a nonhomologous DNA molecule. Then the DNA containing transposon A can be renatured with DNA containing transposon B. Only if the transposons are the same will heteroduplexes form. Radioactive RNA could also be obtained by *in vitro* transcription of the DNA containing A, and this RNA could be used in a hybridization experiment with the DNA containing B. Successful hybridization would indicate that the transposons are the same.

23. (a) The IS elements are inserted in two different orientations, and there is a transcription stop sequence in only one orientation. (b) Heteroduplexes between the DNA of the two phages would indicate that the transposons are the same or different.

24. (a) A high frequency of adjacent mutations suggests that a deletion including these genes has occurred. The deletion frequency is sufficiently high that one might reasonably suspect that a transposon is adjacent to the genes and that the phenomenon is an example of transposon-mediated deletion. (b) Since many mutations result from SOS repair, which is absent in $recA^-$ cells, and since the transposition is independent of RecA function, study of the phenomenon in a $recA^-$ cell would be useful. If the deletion frequency were the same in Rec$^+$ and Rec$^-$ cells, a transposon would be implicated. (c) Most mutagens would increase the frequency of chk^- mutations with little or no effect on deletion frequency, if transposition were involved. There might be an effect if the site represented a mutational hot spot and the phenotype was a result of a pair of point mutations.

25. (1) Precise excision of the transposon; (2) a deletion that either just removes part or all of the *tet* gene or the transposon plus adjacent DNA; (3) a mutation in the *tet* gene.

26. No. The phage DNA sequences are inserted but the terminal bacterial DNA sequences are not.

27. No, because transposition occurs throughout the life cycle.

28. Certainly some should exist, but at exceedingly low frequency because the bacterial sequences at the termini are different. For example, suppose the termini, which include a few thousand base pairs, never contain overlapping bacterial sequences; then, the number of different termini would be about 100. Hence, only 1 molecule in $(100)^2$, or 1 in 10^4, would have unfrayed ends.

29. No. First of all, most bacterial genes are normally turned off, and most bacterial gene products are not needed by Mu. In addition, if a gene product were needed by Mu, it is unlikely that it would have to be induced by the infection. Hence, the gene product would exist in the cell at the time of infection.

Chapter 13

1. Progeny of a transformed cell all had the new character; in genetic terms, the cells bred true.

2. The ability of a cell extract to transform was not reduced by treatment of the extract with enzymes that depolymerize proteins or RNA.

3. (a) The solution containing 10^{-7} mg/ml DNA would contain 0.02 percent of 10^{-7} mg $= 2 \times 10^{-14}$ g protein. The molecular weight of a protein containing 300 amino acids is 300 times the average molecular weight of an amino acid, or approximately $300 \times 100 = 3 \times 10^4$. Thus, the number of protein molecules is $(6 \times 10^{23})(2 \times 10^{-14})/(3 \times 10^4) = 4 \times 10^5$. (b) The numbers do not exclude the possibility that the transformation is protein-mediated, since the maximum

estimate of the number of protein molecules (4×10^5) exceeds the number of transformants 400-fold.

4. Isolate DNA and centrifuge it in a sucrose gradient. Fractionate the centrifuge tube and test each sample for transformability using a very low DNA concentration. Try to find two genetic markers that co-transform, that is, two for which the recipient bacterium can acquire both markers even when the DNA is very dilute. You will observe that the genetic element that carries linked markers always sediments more rapidly than the bulk of the elements that carry only one marker. If this can be shown for several pairs of markers, you will have shown that transformability sediments with DNA and that co-transformability requires DNA of higher molecular weight. This should be pretty convincing. You could then also break the DNA molecules and show that as the average molecular weight of the DNA decreases, there is no loss of transformability of individual markers but that linkage is lost.

5. Show that transformability as a function of pH follows curves for DNA denaturation rather than protein denaturation. Show that transformation occurs at an ionic strength (e.g., 0.3 M) at which DNA-protein interactions are very weak. Separate protein and DNA by gel electrophoresis or gel chromatography, and test purified fractions for transformability.

6. Isolate the DNA and centrifuge to equilibrium in buoyant CsCl. Since DNA and protein have different densities, they will come to equilibrium at different positions in the gradient. Fractionate the centrifuge tube. The fractions containing DNA only or protein only can be tested separately for transforming activity.

7. White (nontransformant, or a cell in which the lac^+ DNA was integrated and in which there was a conversion from + to −), red (one lac^+ strand integrated and a conversion from − to +), and sectored (one lac^+ strand integrated and no conversion).

8. Mix the bacteria in the presence of a large amount of a DNase. If recombination resulted from transformation, all transformation would be eliminated, because the DNase would destroy all of the free DNA in the culture. Conjugation would be completely resistant to the enzyme.

9. The first selection says that B is nearer to A than C is to A. Thus, the order is either $C\ A\ B$ or $A\ B\ C$. The second selection says that B and C can cotransform. If $C\ A\ B$ were the correct order, then cotransformation of B and C would mean that A and C should also cotransform. The latter is not the case, so the order must be $A\ B\ C$.

10. A tenfold dilution of the DNA decreased the number of A^+ transformants by a factor of 10, as expected. However, the cotransformants decreased considerably more. Thus, A and B are not linked.

11. In the $lacZ1/lac^+$ heterozygous region either no mismatch repair occurs or correction tends toward the + genotype. In contrast, mismatch repair usually corrects a $lacZ2/lac^+$ mismatch to the $lacZ2$ sequence.

12. Since one strand of every DNA fragment is digested by nucleotides, only half of the DNA (5000 cpm), will be associated with the cells.

13. From the first part of the experiment one can obtain the following cotransformation frequencies: A^+B^-, $(12 + 3)/250 = 0.06$; A^+B^+, $(12 + 100)/250 = 0.45$. The second part of the experiment yields the frequencies B^-C^+, $(17 + 13)/250 = 0.12$; A^+C^+, $(17 + 85)/250 = 0.41$. Therefore, A and C are near one another, and B is nearer to C than B is to A. Furthermore, B cannot be between A and C, because A is nearer to C than B is to C. Thus, the gene order is $A\ C\ B$.

Chapter 14

1. A donor.

2. In an Hfr cell F is integrated into the chromosome.

3. Many. Each integration site produces a particular Hfr cell line.

4. It prevents the Hfr from producing a colony.

5. An Hfr cell can transfer more DNA, but usually only a fragment of the chromosome is transferred. An F^+ or F′ cell generally transfers the intact plasmid.

6. The values are extrapolated back to the time axis; the time corresponding to a frequency of zero is the time of entry.

7. Hfr and F^- cells are broken apart before chromosomal transfer is complete; thus, a gene transferred later will be present in the recipient at a lower frequency than a gene transferred earlier.

8. The order is $p\ f\ q$ and the map distances are origin–7–p–4–f–8–q.

9. Aberrant excision of F from an Hfr cell, such that adjacent bacterial genes are contained in the circular plasmid.

10. (a) *Met*⁻ is the counterselective marker, and prevents growth of the Hfr cells. (b) *His, leu, trp*, in that order. (c) An interrupted mating, that is, a time-of-entry experiment.

11. (a) True, because *a* enters before *c*. (b) False, because *b* enters before *c*. (c) True, because some exchanges will be between *a* and *b*. (d) False; see (c). (e) True, because *b* is between *a* and *c*. (f) True, because the *a*⁻ and *b*⁻ markers are near, and on, the same DNA molecule. (g) True, because *b* enters after *a*.

12. The larger the fraction the nearer the particular *gal* mutation is to the *bio* locus. Thus, the gene order is *galA galB galC bio*.

13. Between *c* and *d*, because *x* is closely linked to both genes.

14. *a*, 10; *b*, 15; *c*, 20; *d*, 30. Gene *d* is probably very near the *str* locus.

15. The *arg*⁻ mutations have different map locations and must be in different genes; one is transferred before *met*; the other is transferred after *met*.

16. *Pro* is a terminal marker. F is closely linked to *pro* (that is, all *pro*⁺ recombinants are donors) and hence F^+.

17. DNA-binding, proteolysis.

18. Presynaptic binding and conjunction. Drifting could occur but homologous alignment would never be achieved.

19. No. Base-pairing occurs in the next stage.

20. DNA-binding is necessary for conjunction. Helicase activity is necessary for strand assimilation.

21. No.

22. $F'f$, $F'ef$, . . . , by the technique of looking for early transfer of a later marker. $F'g$ could be obtained if a $recA^-$ recipient were used.

23. Genes h and tet must be very near one another, so selection for the h marker of the Hfr tends to select also for the tet marker of the Hfr.

24. All receive the marker, but the positions of the exchange determine whether the marker is present in the recombinant.

25. V13 contains an XP1 prophage between e and f. When the prophage is transferred to the female, no repressor is present and phage induction occurs. Thus, recombinants can form only if mating is interrupted before gene f enters the female. Strain S132 probably also contains an XP1 prophage, so the repressor contained in this female strain prevents induction.

Chapter 15

1. In a simultaneous infection the T4 nuclease destroys the T7 DNA.

2. To prevent incorporation of cytosine into T4 DNA.

3. The T4-induced nucleases will destroy all newly synthesized phage DNA molecules since, in cases 1 and 2, glucosylation cannot occur.

4. No. The DNA is terminally redundant, so the injection sequence differs for each particle.

5. Modify $E. coli$ RNA polymerase (1) to prevent recognition of promoters that have been used and are no longer needed and (2) to permit recognition of the next promoter that is needed.

6. (1), (3), and (4).

7. Yes. If the headful rule is followed, the deletion will produce terminal redundancy.

8. The early transcripts always contain genes for DNA replication and genes that (in varied ways) regulate transcription. Late mRNA generally encodes the structural protein genes and the lysis system. T7 is an exception in that the replication and structural protein genes are initially on a single transcript.

9. The r^+ phage will lyse both cells and produce a clear plaque. The rII phage will only lyse the B cells and hence will produce a turbid plaque.

10. Recall that only r^+ recombinants plate on K12(λ). Thus, since recombination is reciprocal in mass lysates, the number of recombinants is twice the number of r^+ phage, or 52. Thus, the recombination frequency is 52/482, or 10.8 percent,

11. (a) Cross 4 shows that *A* and *D* overlap, and 6 says that *B* and *D* overlap. Cross 1 indicates that *A* and *B* do not overlap. Thus, these three deletions have the order *A D B*. Cross 5 demonstrates that *B* and *C* overlap, and 3 shows that *C* and *D* do not. Thus, *B*, *C*, and *D* have the order *D B C*. The complete order is *A D B C*, with overlap between each pair of adjacent deletions. (b) Between the termini of *A* and *B*.

12. The structural proteins from the late transcripts are needed in stoichiometric amounts. The early proteins are mostly enzymes needed in tiny amounts. Since both early and late mRNA molecules have nearly the same half-lives, more copies of late mRNA are needed than early mRNA.

13. *B A D*.

14. Between *B* and *D* and covered by *A*.

15. It is either a double mutant or a deletion.

16. Since there are no plaques in the background, the *rIIA* mutation is a deletion. Mutant 1 complements and hence is in the *rIIB* cistron. Mutant 2 recombines and hence contains an *rIIA* mutation that is not covered by the tester *rIIA* deletion.

17. The deletion spans the junction of the *rIIA* and *rIIB* cistrons, removing a portion of each cistron.

18. The mutation is a deletion of $1.1 - 0.6 = 0.5$ μm, which is $0.5/57 = 0.88$ percent of the genome.

19. In the wild-type DNA the terminal redundancy is two genes long. Two genes are deleted, so the terminal redundancy should increase to four genes in the deletion—that is, set 2.

20. Use the Poisson distribution. The fraction of cells infected by two or more phage $= 1 - P(0) - P(1) = 0.8$. Therefore 80 percent of 10^8 bacteria (8×10^7 bacteria) are infected by two or more phage.

21. There are several approaches, each of which involve isolating phage mutants that eliminate the activity of X. It is also necessary to have an *in vitro* assay for X; this might be enzymatic or immunological, or might require isolation of the protein. First, consider a heat-sensitive mutant. With this, it can be shown that at high temperature, enzymatic activity or perhaps immunological reactivity (not as good a test) is substantially reduced. Second, consider a chain-termination mutation. Here it can be shown by sedimentation or by gel electrophoresis that the protein has a lower molecular weight than the wild-type protein. Note the greater power of the second approach. With the heat-sensitive mutant, one merely correlates a property *in vitro* with that *in vivo*, but with a chain-termination mutant one observes not only the absence of the wild-type protein but the appearance of a new and smaller molecule.

22. The best way is to use a temperature-sensitive mutant and raise the temperature of the infected cell at various times. If the gene product is needed only in the interval between T_1 and T_2, then raising the temperature later than T_2 will not affect phage development, and a period of heating at any time between zero time and $T = T_1$, followed by cooling, will usually not affect phage development.

Chapter 16

1. The linear molecule spontaneously circularizes via the single-stranded termini, and the single-strand breaks are sealed by DNA ligase.

2. θ and rolling circle replication.

3. T4, cut to fill a head; λ, cut at *cos* sites.

4. DNA ligase, DNA gyrase.

5. L1, R1, R2, and R4.

6. No late mRNA is made. Therefore, there are no structural proteins, DNA concatemers are not cleaved, and the cells do not lyse.

7. (a) Messenger RNA terminates at the first leftward terminator *tL1* of λ, and N protein is needed to prevent termination. This conclusion is correct as long as the *trp* operon lacks a promoter. (b) Yes. If part of gene *N* is deleted, *tL1* must also be deleted and the mRNA from the main leftward promoter will extend through the *trp* operon.

8. For the dimer, after the first DNA molecule is cut out and packaged, the remaining DNA has no *cos* site and hence cannot be packaged. Thus, only one particle can be formed from a dimeric circle. By the same reasoning, two particles are packaged from a trimeric circle.

9. Recombination produces a circular dimer in some of the infected cells.

10. The E^- lysate contains active tails. These tails can attach to live *E. coli*. Filled heads from the J^- lysate attach to the preadsorbed tails. Following attachment, the λ J^- DNA can inject and the infection can proceed, producing J^- progeny. The genotype is J^-, because this is the genotype of the DNA contained in the tailless heads.

11. No, because T4 can be transcribed and the T4-encoded nucleases will destroy the λ DNA.

12. Probably a mutation in the termination sequence *tR2*. This mutation allows Q protein to be made without N protein. The plaques are small because in the absence of an active N protein, the *O* and *P* genes are not transcribed at a normal level.

13. Since the density is lower, the mutant has either less DNA or more protein. Since λ packages by cutting *cos* sites and since the mutation is known to be a deletion, the mutant clearly has less DNA. To calculate the size of the deletion, we let x be the fraction (in the deletion phage) of the normal amount of DNA. The density of the phage, that is, $1.492 = (1.708x + 1.300)/(1 + x)$. Solving for x yields $x = 0.889$. Thus, the deletion represents 11.1 percent of the genome.

14. λ cannot be plated on the *dnaA* bacteria at a nonpermissive temperature because the gene is essential for bacterial multiplication; hence, at high temperature there will be no lawn. The method is to adsorb λ to the *dnaA* mutant at 42°C at an MOI < 1 and plate the infected cells on a wild-type indicator at 42°C. If DnaA protein is not needed, each infected cell will produce a plaque. If it is needed, no plaques will result. In fact, DnaA protein is not needed.

15. The mutant probably has a very leaky mutation in gene N in the region of the N protein that interacts with RNA polymerase.

16. The mutant has a chain-termination mutation. The amino acid put in by either *supI* or *supII* cells does not create a functional protein, but the amino acid inserted at the mutant site by *supIII* cells does yield an active protein.

17. The branch grows in length until the Ter system becomes active. At this point, λ units are cleaved from the branch at roughly the same rate as the growth of the branch. This accounts for the fairly constant length of the branch. Cleavage cannot occur unless a head is assembled. Since E protein is the major head protein, the Ter system is inactive with E^- mutants. Thus, the branch will continue to increase in length.

18. The $λR^-$ phage can acquire the R^+ allele from the cryptic strain by recombination and produce R^+ progeny that can lyse both bacteria, yielding a clear spot. The $λred^-$ cannot pick up the R^+ allele and hence can grow only on the cryptic strain; therefore, it produces a turbid spot.

19. In the absence of N protein, the bacteria-killing functions (such as lysozyme production) are not activated. Thus, the cells survive. However, the fact that progeny cells receive a λ means that λ DNA must be able to replicate *somewhat* in the absence of N. We know that N stimulates transcription of the *OP* region by antiterminating at *tR1*. If replication occurs without N, the terminator *tR1* must be a weak one, allowing a small amount of transcription of the *O* and *P* genes. A small amount of transcription may lead to synthesis of just enough O and P protein for one round of DNA replication per cell generation, which would enable the λ plasmid to be maintained from one cell generation to the next. One would expect fairly frequent segregation of λ-free cells, which is indeed the case.

Chapter 17

1. λ*imm434* and λ*imm21*.

2. λ*immλ*, λ*imm21*.

3. High.

4. Generally their repressors and repressor-binding sites (operators) must be the same, though it is sufficient if the repressor of each phage can recognize the operator of the other.

5. A phage that can plate on a homoimmune lysogen.

6. Both make clear plaques.

7. No difference.

8. At the normal bacterial attachment site for λ, since the hybrid phage has λ *att*.

9. Bacterial, *BOB'*; phage, *POP'*; *gal*, *BOP'*; *bio*, *POB'*.

10. *D E F A B C*.

11. Int and cI.

12. Integration—integrase and the bacterial integration host factor; excision—the phage Int and Xis proteins.

13. *Pre* is activated by the cII protein; *prm* is activated by the cI protein.

14. Mutations in the *cI, cII, cIII,* and *int* genes would prevent lysogenization. Since an active N product is needed to produce the cII and cIII products in appreciable quantities, *N* mutants will lysogenize quite poorly. Mutations in *att* will also prevent lysogenization. *Cro* mutants also have a greatly reduced lysogenization frequency, but the reasons are fairly complex.

15. No, because an operator mutation is *cis*-dominant.

16. The left. If it inserted in the right, the order would be *gal BOP' R A POP' R A gal BOB' bio*.

17. (a) Repression or establishment of lysogeny. (b) Complementation between two proteins was needed to establish repression.

18. Only N^-, because it fails to make the cII protein.

19. No, because the *bio* substitution eliminates the *int* gene. Int-promoted recombination is the only possible recombination mechanism that could carry out the exchange if the other recombination systems were absent.

20. Phage production is in each case a result of prophage induction. Induction by ultraviolet light cannot occur in a *recA*$^-$ cell, because the RecA product is responsible for cleavage of the cI repressor. Induction occurs with a temperature-sensitive repressor in both mutant and wild-type since it is the high temperature that is inactivating the repressor.

21. If enough λ DNA molecules enter a bacterium, the intracellular repressor will be titrated.

22. (a) No, because in the absence of active repressor the *prm* promoter is inactive. (b) Yes, because the renatured repressor activates *prm*. (c) After 10 minutes enough DNA replication has occurred without repressor synthesis that there is insufficient repressor to bind all of the operators. After 5 minutes repressor is still sufficient.

23. The genotype will be $cI857P^+A^+$, because the only mode of phage production is Ter-mediated excision at the two *cos* sites.

24. The phage has a chain-termination mutation in one of the genes *cI, cII,* or *cIII,* and strain B (but not A) has a nonsense suppressor.

25. (a) The frequency with which a lysogen for a wild-type phage is spontaneously induced is relatively low, and the phage released form turbid plaques. However, if a mutation has occurred in the gene making the cI repressor, a lysogen containing this mutant (which will form a clear plaque) will always produce phage. (b) The gene *cI* makes repressor. Genes *cII* and *cIII* are needed in the wild-type state for the establishment of lysogeny, but not for its maintenance. Thus, a *cII*$^-$ or *cIII*$^-$ mutation in a prophage does not result in escape from repression.

26. Using the integrase of λ*b2* (λ*bio* has no *int* gene), an Int-promoted recombinational event has occurred yielding a dimeric circle containing Δ*OB'* and *POP'*, namely,

$$\Delta OP' \times POB' \longrightarrow \Delta OB' + POP'$$

and the dimer has integrated using the newly generated *POP'* site.

27. (a) Prophage DNA can replicate, even though excision does not occur. (b) The replication forks can leave the prophage and enter the bacterial DNA region (c) Excision must be fairly slow, because often the replication forks leave the prophage before excision occurs.

28. Integration into the right attachment site would require an exchange between *BOP'* and *POB'*, which is the reaction that requires the Xis protein. In an infection, the Xis protein is not made to any great extent.

29. A phage-resistant mutant is usually formed by plating about 10^8 phage on a lawn of about the same number of bacteria and picking surviving colonies. This will not work with a λ lysogen because all cells will survive. However, if λ*imm434* were used, only adsorption-defective bacteria would form colonies.

Chapter 18

1. None, because specialized transducing particles are produced only by lysogens.

2. There will be no *lac*-transducing particles, because the *lac* locus is too far away to be included in a λ phage head.

3. Specialized only: integrates into the chromosome and either does not cause host DNA fragmentation during phage development or cannot package host DNA fragments. Generalized only: fragments host DNA, can package host DNA fragments, and probably does not integrate. Both: integrates, fragments host chromosome, and can package host fragments.

4. They are only formed by aberrant excision of a prophage. In a lytic infection λ DNA is not inserted in the *E. coli* chromosome.

5. (1) Their excision system is perfect (unlikely). (2) When aberrant excision occurs the fragment size is too large or too small to be packaged. (3) No known host genes are near enough to the prophage to be picked up.

6. At each end of the prophage map essential genes are present that are absent in both types of transducing particles. Thus, lytic growth is only possible when a helper phage providing the missing functions is present. In single infection, transduction occurs by substitution because neither the right nor left prophage attachment sites can recombine with the bacterial attachment site. In mixed infection, either the transducing DNA and the helper phage DNA recombine or the attachment sites of the transducing DNA can recombine with a newly generated prophage attachment site.

7. If obtained by infecting a λ lysogen; the phage DNA would then be no different from any other bacterial DNA.

8. The DNA in the transducing particles was replicated before infection. Also, either there is no replication of bacterial DNA after infection, or replicated DNA never gets into transducing particles.

9. Three orders are possible in such a cross: *A B C* (I), *B A C* (II), and *A C B* (III). Drawing the two crosses for each of the orders shows that order II can be eliminated because a quadruple exchange would be needed in the first cross and a double exchange in the reciprocal cross, which would not yield the observed equal frequencies. Double exchanges yield wild-type transductants with both orders I and III in both crosses. Thus, orders I and III cannot be distinguished by the data. Selection of another recombinant class would yield the order.

10. Crosses 1 and 3 are reciprocal crosses. With order *ant trpA34 trpA213* cross 3 would have more cotransductants of Ant$^+$ with Trp$^+$ than would the order *ant trpA223 trpA34*. The data then yield the order *ant trpA34 trpA233*. Crosses 2 and 4 are also reciprocal crosses. The same reasoning yields the order *ant trpA223 trpA46*. Thus, the overall order is *ant trpA34 trpA223 trpA46*.

11. (a) Drawing out the three orders shows that only the order *pyrF trpA trpC* yields the rarest recombinant, *trpC$^+$ pyrF$^-$ trpA$^+$*, as the product of a quadruple exchange. (b) The cotransduction frequencies are: *pyrF–trpC*, $(548 + 3)/1220 = 0.452$; *pyrF–trpA*, $548/1220 = 0.443$; *trpA–trpC*, $(548 + 579)/1220 = 0.92$. (c) Use the Wu formula. The values are: *pyrF–trpA*, 0.466 minutes; *trpA–trpC*, 0.055 minutes; *pyrF–trpC*, 0.476 minutes. (d) The map appears to be *pyrF*–0.466–*trpA*–0.055–*trpC*, with *pyrF–trpC* also being 0.476 minutes. Note that $0.466 + 0.055$ does not equal 0.476. The value of 0.055 is accurate since it is determined from a large number of colonies. However, the difference between 0.466 and 0.476 represents only 3 colonies, which is not statistically significant.

Chapter 19

1. Mate Hfr *gal$^-$ lac$^+$* × F$^-$ *trp$^-$ lac$^-$ str-r* for 25 minutes (longer may allow the *trp$^+$* allele to be transferred. Select Lac$^+$ Str-r colonies on lactose-EMB agar and test these for the Gal$^-$ phenotype by replica-plating onto galactose-EMB (or tetrazolium) plates. When completed, be sure to check that the strain is *trp$^-$* by plating on minimal plates without tryptophan.

2. First, mate an Hfr *lac$^+$ str-s* strain with the *lac$^-$* (point)*str-s* strain and select a Lac$^+$ Str-r strain on lactose-EMB plates containing streptomycin. Then, mate the Hfr Δ*lac* with this new female. Since the deletion is transferred early, no special selection is needed. Plating the cells on lactose-EMB plates with streptomycin will be sufficient in that between 1 and 10 percent of the colonies will be Lac$^-$ and each will carry the deletion.

3. Lysogenize the *gal$^+$* strain with λ*cI857*. Grow the lysogens at 42°C on galactose-tetrazolium plates. Pick survivors (which must have lost at least the repressor gene) that are Gal$^-$. These will be deletions that extend from the *gal* locus through the prophage at least past the *cI* gene. Test the Gal$^-$ survivors for the *bio* gene by plating on medium lacking biotin. Gal$^-$ Bio$^-$ colonies will have a deletion that extends from the *gal* locus to the *bio* locus and includes *att*λ.

4. As in Problem 3, lysogenize the *gal*⁺ strain with λ*cI857*. Grow the lysogens at 42°C on galactose-tetrazolium plates. Pick survivors (which must have lost at least the repressor gene) that are Gal⁻. These will be deletions that extend from the *gal* locus through the prophage at least past the *cI* gene. Test the Gal⁻ survivors for UV sensitivity. These will be deletions that at least enter the *uvrB* gene. An alternative procedure, which does not use λ, is the following. Plate the original strain on plates containing potassium chlorate. Colonies that grow will be chlorate-resistant. There are several chlorate-resistance genes, but selection for Gal⁻ on galactose-tetrazolium plates ensures that one of the *chl* genes near *uvrB* will be deleted. Test these for UV sensitivity. This second procedure is the fastest, because two days are saved, namely, the days required to lysogenize and test the lysogen. On the other hand, the chlorate test is inconvenient, as chlorate resistance is only manifested in the strict absence of molecular O_2, a condition that is often difficult to achieve in a typical laboratory.

5. Plate the recipient with an excess of T6 and pick Tsx-r colonies. Recall that some of these contain deletions that extend into *pur*. Thus, select Pur⁻ colonies by replica plating. Mate the Hfr *lac*⁻ strain with the new *pur*⁻ female and plate on minimal plates lacking a purine. The *pur* and *lac* loci are sufficient near that the Pur⁺ recombinants can be tested for the linked *lac*⁻ mutation by replica-plating onto lactose-EMB plates.

6. First, cross T4*ac41* and T4*rII*, plate on strain B on plates containing an acridine, and select large plaques. These are *ac41 rII* recombinants. Test by plating on K12(λ). Then, prepare *E. coli* B/4 (T4-resistant cells) by plating B cells on a plate spread with T4, and picking survivors. Then, cross T4*ac41 rII* and T4*h* and plate on B/4 on plates containing an acridine. Pick large plaques, which should be *ac41 rII h* recombinants. It is worthwhile to make a final check for the *rII* mutation by plating on K12(λ) to ensure that the combinations *ac41 h* or *ac r*⁺ *h* do not produce a large plaque.

7. Grow the *leu*⁻ strain on medium containing chlorate and select Chl-r cells. Test these on galactose-tetrazolium plates to isolate Gal⁻Chl-r double mutants. Most of these will be *gal–chlD* deletions. Check these for UV sensitivity to find a deletion that extends from *gal* to *uvrB*. Plate this strain on a plate spread with 10^8 λ*cI*⁻ mutants (*cI*⁻ to avoid lysogenization) and select survivors, which are λ-resistant. Test these on maltose-tetrazolium plates to be sure that they are *malB*⁻. Grow P1 on the *uvrA*⁻ mutant, which is also *mal*⁺, and transduce the *leu*⁻ *uvrB*(deletion)*malB*⁻ strain to Mal⁺ by selection on maltose-minimal plates containing leucine. Check Mal⁺ transductants for UV sensitivity. It is not obvious that the double mutant will be more sensitive than the single mutant. However, it is worth comparing the UV sensitivity of the Mal⁺ transductants to that of either *uvrA*⁻ or *uvrB*⁻ cells. If the UV sensitivity were greater than either of the single mutants, you could feel confident that you have the double mutant. The best test would be to cross a *uvrB*⁺ allele into several Mal⁺ transductants (either by transduction or Hfr crosses, selecting for Gal⁺ or Bio⁺) and see whether the recombinants retain the sensitivity of a *uvrA*⁻ cell. If so, the Mal⁺ transductant is also *uvrA*⁻.

8. First, make it *thy⁻ str-r* using the trimethoprim selection. Then, make it Tsx-r by plating with an excess of phage T6 (*tsx* is on the opposite side of the map from both *thyA* and *str*, which are quite near one another). Take an Hfr that transfers *thy* fairly early in either of the orders *thy–str* or *str–thy* and mate it with the F⁻ cells. If the *thy–str* Hfr is used, keep the mating shorter than the time of entry of *leu*. Select against the Hfr with T6 and for the Thy⁺ trait by plating on leucine-minimal plates lacking thymine. Test Thy⁺ recombinants for streptomycin-sensitivity.

9. Continued protein synthesis without cell division will cause the cells to elongate. In fact, one expects the cells to become twice as long in the first generation time, three times as long in the second generation time, and so forth. This does occur, and such filamentous bacteria are often called "snakes." In a culture containing a few *dna⁻*(Ts) mutants only the mutants will form snakes at elevated temperature; the wild-type cells will grow and divide normally. The culture is then passed through a nitrocellulose filter with pores about 2 μ in diameter. Normal cells will pass through the filter, but mutants will be retained on the filter. The filter is then washed and the cells removed from it are plated at 30°C. Colonies are replica-plated onto fresh plates and tested for the ability to grow at 42°C. Some of the colonies are *dna*(Ts) mutants and some are Ts cell-division mutants. These are distinguished by measuring the ability of a liquid culture to incorporate radioactive thymidine at 42°C. Those that cannot are *dna*(Ts) mutants. This was the screening technique used in the 1960s in the original isolation of *dna*(Ts) mutants.

10. Prepare a Leu⁻ lysogen with λ*cI857*. Superinfect a culture of the lysogen with λ::Tn10 and plate on medium containing ampicillin. Since λ::Tn10 cannot multiply in the lysogen, these cells will contain Tn10 in the chromosome. Test these for Leu⁻ by replica-plating to a pair of minimal plates, one of which contains leucine. Select a Leu⁻ colony, prepare a liquid culture, and cure it of the prophage by growth at 42° for 5 minutes, and several hours at 30°C. Plate at 42°C and pick survivors. These are Leu⁻ nonlysogens.

11. Infect an Ara⁺ strain containing a *lac* deletion with Mu(Ts) and plate at 30°C. Cross-streak colonies against Mu to select for Mu(Ts) lysogens. Replica-plate Mu(Ts) lysogens to arabinose-tetrazolium plates to select for Ara⁻ cells (red colonies). These cells presumably contain a Mu(Ts) prophage in an *ara* gene. Lysogenize these cells with λp(*lacP⁻ lacZ⁺,Mu*) and test for λ lysogeny by the λ cross-streak test. Plate a λ lysogen at 42°C on lactose-EMB plates containing arabinose (an inducer of the arabinose operon) and select purple (*Lac⁺*) colonies. Replica-plate them to lactose-EMB plates lacking arabinose; all should be Lac⁻. Colonies that are Lac⁺ on (lactose-arabinose)-EMB plates and Lac⁻ on lactose-EMB plates have the *ara* promoter fused to the *lacZ* gene.

12. Cross *bio11c⁺* with *cI857Q⁻* and plate on a P2 lysogen at 42°C. Pick clear plaques and test for *Q⁻* by stabbing into *sup⁺* and *sup0* bacteria (*Q⁻* phage grow on *sup⁺* but not *sup0*). You now have *bio11cI857Q⁻*. Cross with *red⁻ c⁺* and plate on *recA⁻ sup⁺* cells at 42°C. Select clear plaques (λ*cI857red⁻*) and test for the *Q⁻* marker by stabbing into *sup⁺* and *sup0* bacteria.

13. Cross the two phages and plate on $recA^- sup^+$ cells at 42°C. Select clear plaques. Test for $cI857A^+R^+$ double exchanges by stabbing into $sup0$ and sup^+ bacteria. Discard those that grow on $sup0$ cells. A single exchange between A and $bio11$ will give $cI857A^-R^-$. However, A and R are so far apart that multiple exchanges may occur. Plate A^- and R^- phage on $sup0$ bacteria to determine the reversion frequency. Do the same with the clear mutant that failed to grow on $sup0$ cells. Those that produce no revertants (i.e., frequency $< 10^{-10}$) are the double mutants.

14. Cross the phage and plate on $recA^- sup^+$ cells at 42°C, and pick clear plaques. The parent $\lambda cI857R^-$ and the recombinant $\lambda cI857A^-R^-$ will both grow but the $bio11c^+A^-$ parent will not. There will also be $\lambda cI857A^+R^+$ recombinants. Test all clear plaques by stabbing them into sup^+ and $sup0$ bacteria and discard those that grow on $sup0$ cells (the A^+R^+ recombinants). Since the recombination frequency is 10 percent, about 10 percent of the clear-plaque phage that fail to grow on $sup0$ cells will be A^-R^-. These can be identified by measuring the reversion frequency; if about 20 of the plaques are tested, a nonreverting double mutant should be found.

15. (a) Cross the two phage. Progeny are $b2R^+$, b^+R^-, $b2R^-$, and b^+R^+. Treat the lysate with EDTA, which kills b^+R^- and b^+R^+ phage. Plate on sup^+ bacteria and stab about 50 plaques into sup^+ and $sup0$ cells. Those that fail to grow on the $sup0$ cells are R^-. Isolate and recheck resistance to EDTA. (b) Cross the two phage and treat the lysate with EDTA to destroy all b^+ phage. The brute-force method of stabbing 50 plaques into $sup0$ and sup^+ bacteria will not be a good idea; an exchange is needed between the $cI857$ and P^- markers, and the markers are fairly close. However, plating on GroP cells provides the selection for the P^- marker. Thus, after EDTA treatment the lysate can be plated on $groPsup^+$ cells at 42°C. Selection of clear plaques yields $\lambda cI857b2P^-$ phage.

Chapter 20

1. Blunt ends, cohesive ends with 3′ extensions, and cohesive ends with 5′ extensions.

2. Always the same.

3. Yes. Several restriction enzymes from different organisms are known that have the same site specificity.

4. They are palindromes.

5. No.

6. Yes, note that the sequences have the form of Pu GCGC Py, which is the HaeII site.

7. They are found in most, if not all, bacterial species.

8. No, they are sequence-specific enzymes.

9. They must both recognize the same base sequence.

10. It can add nucleotides to an extended single-stranded 3′ terminus of a DNA molecule without the need of a template.

11. Blunt-end ligation, addition of linkers, homopolymer tail-joining.

12. Terminal nucleotidyl transferase will not add the tail to the blunt end. First, the DNA must be treated with a 5′-P-specific nuclease to remove a few nucleotides from the 5′ end, leaving a 3′ overhang.

13. Prepare fragments by cleavage with two different restriction enzymes such that their termini are different; put different linkers on a blunt-ended molecule.

14. Treat the cleaved plasmid with alkaline phosphatase. After joining and treatment with DNA ligase, heat the molecules; treated vectors will melt open.

15. These are circularized fragments.

16. One gene can be used to detect the plasmid in a transformation experiment; and if there is a restriction site in the other gene, lack of resistance to that antibiotic can be used to show that insertion has occurred.

17. A particular fragment is isolated from a gel after electrophoretic separation, or a particular c-DNA is used.

18. The transducing phage produced by the joining techniques must contain inserted DNA.

19. Both enzymes recognize the same base sequence. They might even be the same enzyme.

20. A mutation may alter one base in a restriction site and thereby cause two potential fragments to remain uncleaved.

21. The *A* gene probably contains an EcoRI site, so the gene is destroyed. It does not contain an HaeIII site.

22. In principle, either could be used. If you used EcoRI, the plasmid could be detected by growth on medium using lactose as the sole carbon source. Colonies that grow could then be tested for antibiotic-sensitivity to find those contained DNA inserted in the *amp-r* gene. Note that this process would take two steps, and, in fact, two days of growth. If you used SalI, transformants with inserted DNA could be isolated in a single step by plating transformed bacteria on a lactose color-indicator medium containing ampicillin.

23. (a) The *tet-r* gene has not been cleaved, so addition of tetracycline to the medium will require that the colonies be Tet-r and hence carry the plasmid. (b) Tet-r Kan-r and Tet-r Kan-s. (c) Tet-r Kan-s, because insertion will occur in the cleaved *kan* gene.

24. A frameshift of two bases is generated, so all colonies will be Lac⁻.

25. Since annealing occurs at random, some clones will contain more than one copy of the *lacZ* gene and one copy of *lacY*, and other will contain more than one copy of *lacY* and one copy of *lacZ*. Furthermore, some will have the gene order, *lacP lacZ lacY* and others the gene order *lacP lacY lacZ*. Because of the polarity of synthesis of the gene products of a polycistronic mRNA, the former will make more *lacZ* and the latter will make more *lacY*.

26. Select an enzyme whose restriction site is further from the *lac* promoter than the BamHI site. Let us assume it is HaeI. If the gene of interest does not contain

a HaeI site, do the following. Cleave the plasmid with both BamHI and HaeI and retain the fragment (there will be two) that contains the replication origin and other essential genes. One end of the fragment will have a BamHI terminus, and the other end will have a HaeI terminus. Cleave the donor DNA with both enzymes and join the fragments to the isolated plasmid fragment. Since each fragment has one BamHI terminus and one HaeI terminus, joining can only occur in a particular orientation.

27. $f = (2 \times 10^4)/(3 \times 10^9) = 6.67 \times 10^{-6}$. Therefore, $N = \ln (0.01)/\ln (1 - 6.67 \times 10^{-6}) = 6.91 \times 10^6$.

28. The target sequence of enzyme 1 is part of the target sequence for enzyme 2. For example, enzyme 2 might cut in the sequence GATATC and enzyme 1 might act at ATAT, which is part of the target sequence for enzyme 2.

29. Most likely, the enzymatic digestion was not carried to completion and the extra fragment is a result of incomplete cutting at a particular site. The fragment moves slowly because its size is equal to the sum of two adjacent fragments.

30. There are three possibilities, of which the first two are quite unlikely: (1) A single cut is made precisely in the center of the molecule; (2) several cuts are made in positions such that all fragments have sizes that are indistinguishable by gel electrophoresis; (3) the molecule is a circle and a single cut is made.

31. Note that the 3.6-kb and 5.3-kb fragments are now joined to form the 8.9-kb fragment. This suggests that they are terminal fragments of a linear molecule and that the intracellular DNA is circular.

32. The plasmid is circular (that is, no free end), so all termini generated by the enzyme will have the same cohesive end.

Chapter 21

1. It has been inserted in the incorrect orientation.

2. The gene may have been separated from its promoter or from its ribosome binding site.

3. A standard procedure is to design a vector so that it contains a strong promoter and a ribosome binding site (commonly that for the *E. coli lac* operon), with an adjacent downstream restriction site.

4. (1) The bacterium does not recognize a eukaryotic promoter. (2) If a primary transcript is actually made, it will typically have introns and these cannot be removed in a bacterium.

5. (1) The mRNA lacks a ribosome binding site. (2) The protein must be processed. (3) The bacterium destroys the protein, having recognized it as foreign protein.

6. A centromere, without which progeny molecules would not always find their way into daughter cells.

7. In yeast. These shuttle vectors are used for cloning eukaryotic genes, which will usually not be expressed in *E. coli,* but may be expressed in yeast. Selection must be carried out in a cell in which the gene is expressed, that is, yeast.

8. (a) The order is 1.0−6.4−4.1. The reasoning is the following. First, in the linear DNA the 6.4-kb fragment can be linked to either of the other fragments, when digestion with BamHI is partial; therefore, the 6.4-kb fragment must be in the middle. Second, in the circular DNA, where the cohesive ends are ligated together, the 1.0-kb and 4.1-kb fragments are joined, showing that they are at the ends. (b) The P4 phage attachment site in the phage DNA lies in either the 1.0-kb or 4.1-kb BamHI fragment and not in the 6.4-kb fragment, which is preserved after integration of prophage. The BamHI sites in the host DNA that are nearest the prophage are placed to give host-phage fragments of 15.0 and 12.5 kb.

9. III-IV-I-V-II.

10. The fragments have eliminated either the promoters or the ribosome binding sites of both genes. The plasmid contains both of these, but they are on the same strand of the plasmid DNA. If genes *A* and *B* are transcribed in opposite directions, only one can be expressed from the plasmid.

11. The labeled fragments indicate the termini, which are 6.2 and 8.0 for EcoRI, and 6.0 and 10.1 for BamHI. Therefore, the BamHI map is 6.0, 12.9, 10.1, in which we arbitrarily place the 6.0 fragment at the left. If the 6.2-kb terminus of the EcoRI map were at the left, a 0.2-kb fragment would be in the double digest, but no such fragment is present. If the 8.0-kb terminus were at the left, a 2.0-kb fragment would be present, as it is. Thus, the 8.0-kb fragment is at the left, and the 6.2-kb fragment is at the right. Now, consider the 4.5-kb fragment. If it were next to the 6.2-kb fragment, the double digest would have a 0.6-kb fragment, which is not present. Also, if it were next to the 8.0-kb fragment, the double digest would have a 6.5-kb fragment, which it does not. Therefore, the 4.5-kb fragment cannot be next to either the 6.2-kb or the 8.0-kb fragment and must be in the center of the molecule. Now, consider the 7.4-kb fragment. If it were next to the 8.0-kb fragment, there would be a 2.5-kb fragment, which there is. If it were instead next to the 6.2-kb fragment, there would be a 3.5-kb fragment, which is not present. Thus, the 7.4-kb and 6.2-kb fragments are adjacent. Analysis of the position of the 2.9-kb fragment shows that it is next to the 8.0-kb fragment, which agrees with the position of the 7.4-kb fragment. Therefore, if the BamHI map has the order 6.0−12.9−10.1, the EcoRI map has the order 8.0−7.4−4.5−2.9−6.2, with the same orientation.

12. The restriction enzyme is impure and contains at least one nuclease that is not site-specific.

13. No, because a source of fairly pure mRNA is needed. This is really only practical with particular mRNA molecules made in very large quantities by particular cells, such as the cells making hemoglobin. If the mRNA is a minor species, there might be no way to isolate or identify it. Normally, one would do this by hybridization to a DNA molecule. However, often there is no way to obtain the DNA molecule except by cloning, and the c-DNA method may be the only way to do the cloning.

14.

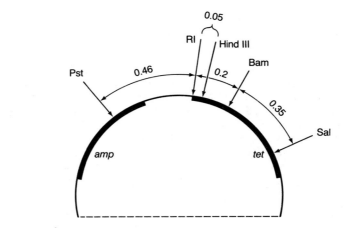

15.

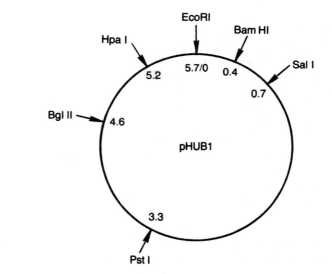

Index